Conservation Agricultural Practices for Improving Crop Production and Quality

Guest Editors

Mariola Staniak
Ewa Szpunar-Krok
Małgorzata Szostek

Basel • Beijing • Wuhan • Barcelona • Belgrade • Novi Sad • Cluj • Manchester

Guest Editors

Mariola Staniak
Department of Crops and
Yield Quality
Institute of Soil Science and
Plant Cultivation
Puławy
Poland

Ewa Szpunar-Krok
Department of Crop
Production
University of Rzeszów
Rzeszów
Poland

Małgorzata Szostek
Faculty of Technological and
Natural Sciences
University of Rzeszów
Rzeszów
Poland

Editorial Office
MDPI AG
Grosspeteranlage 5
4052 Basel, Switzerland

This is a reprint of the Special Issue, published open access by the journal *Agronomy* (ISSN 2073-4395), freely accessible at: www.mdpi.com/journal/agronomy/special_issues/H4FE1QROA2.

For citation purposes, cite each article independently as indicated on the article page online and using the guide below:

Lastname, A.A.; Lastname, B.B. Article Title. *Journal Name* **Year**, *Volume Number*, Page Range.

ISBN 978-3-7258-3646-8 (Hbk)
ISBN 978-3-7258-3645-1 (PDF)
https://doi.org/10.3390/books978-3-7258-3645-1

Contents

About the Editors

Mariola Staniak

Prof. Dr. Mariola Staniak has been a researcher at the Institute of Soil Science and Plant Cultivation in Puławy since 2003. Her research interests include the biological basis of the productivity of various crop species used for the production of food and animal feed. She is involved in improving agricultural technology, introducing innovations, and adapting treatments to the needs related to climate change. She conducts extensive research on the physiological state of crops depending on abiotic stress factors and agrotechnical treatments. She is a recognized specialist in soybean cultivation in Poland. She is also interested in issues related to the impact of agriculture on the biodiversity of flora on arable land. During the years 2016-2024, she was the head of the Department of Fodder Crop Production. She is the author and co-author of over 300 publications, including approx. 150 scientific papers; is the supervisor of three doctoral dissertations and three diploma theses; and participates in numerous scientific projects.

Ewa Szpunar-Krok

Prof. Dr. Ewa Szpunar-Krok is head of the Department of Plant Production at the University of Rzeszów (Poland). She conducts research on the optimization of field crop production (cereals, legumes, oilseeds). Currently, she is conducting research on the possibility of the agricultural use of waste such as ash from biomass combustion and municipal sewage sludge for fertilizing crops, as well as their effect on plant growth and development and the physical and chemical properties of soil. She also deals with issues related to biological, physiological, and agrotechnical aspects of plant response to abiotic and biotic stresses and the possibility of mitigating the effects of these stresses through the use of biopreparations. She is the author or co-author of 122 scientific publications; is the supervisor of four doctoral theses; and participates in numerous scientific projects.

Małgorzata Szostek

Dr. Małgorzata Szostek, Associate Professor, is a research and teaching staff member at the Institute of Agricultural Sciences, Environmental Protection, and Land Management, Faculty of Technological and Natural Sciences, University of Rzeszów. Her scientific activity primarily focuses on issues related to soil properties. Her main research interests include the impact of waste on changes in soil environmental properties, with a particular emphasis on the properties of soil organic matter. She is the author or co-author of 47 articles indexed in the JCR. Additionally, she has co-authored nine chapters in scientific monographs, as well as three patent applications and one patent. She participates in numerous scientific projects, mainly focused on research and implementation. To date, she has supervised 8 master's theses and 10 engineering theses and has served as an assistant supervisor in 2 doctoral dissertations.

agronomy

Editorial

Conservation Agricultural Practices for Improving Crop Production and Quality

Mariola Staniak [1,*,†] and Ewa Szpunar-Krok [2,*,†]

[1] Department of Crops and Yield Quality, Institute of Soil Science and Plant Cultivation-State Research Institute, Czartoryskich 8, 24-100 Puławy, Poland
[2] Department of Crop Production, University of Rzeszów, Zelwerowicza 4, 35-601 Rzeszów, Poland
* Correspondence: staniakm@iung.pulawy.pl (M.S.); eszpunar@ur.edu.pl (E.S.-K.)
† These authors contributed equally to this work.

1. Introduction

Modern agriculture faces many challenges, the most important of which are the effects of climate change, soil degradation and fertility decline, pressure on water resources, and food insecurity. In this context, it is crucial to increase crop production while reducing negative environmental impacts. One approach that favors the sustainable intensification of production is the use of conservation agriculture (CA) practices, which include various techniques that reduce the depth and intensity of mechanical tillage (strip-till, no-tillage, reduced tillage) and associated practices (cover crops, mulching, crop rotation, cultivar selection) [1,2]. CA aims to maintain the natural soil structure, reduce erosion and improve the water balance and organic matter content of soil to improve crop yield and quality [3]. The primary purpose of this Special Issue was to assemble studies that demonstrate the research progress on agricultural practices that combine the high production of quality raw materials with the provision of environmental services.

2. The Influence of Conservation Agricultural Practices on Soil Properties

With the worsening of the effects of climate change and increasing pressure on water resources, there is a growing need for adaptive management strategies, such as CA, which could be beneficial for soil quality and water use efficiency. However, the effectiveness of CA can vary considerably depending on the practice implemented and soil and climatic conditions [4], with the short-term effects of this farming system differing significantly from those observed in long-term studies [5].

In a study conducted in Mediterranean climatic conditions, Dominguez-Bohorquez et al. showed that CA practices, such as no-tillage (NT), cover crops (CC) and crop rotation, applied under irrigated and non-irrigated conditions from the first year, clearly affected soil properties, water flow and crop yields (maize, sorghum and soybean). Conservation agricultural practices resulted in increased bulk density and soil resistance penetration, leading to decreased quasi-steady ponded infiltration in the surface horizon. This phenomenon was observed especially under subsurface-drip-irrigation and no-irrigation conditions. Crop residues in this cropping system reduced soil water evaporation, especially under sprinkler irrigation, but this benefit decreased as crop residues decomposed. In a study by Alhammad et al., NT accompanied by crop residues significantly improved soil chemical and biological properties by increasing soil organic carbon content, improving nutrient availability and increasing microbial populations, such as *Azotobacter*, *Pseudomonas* and

check for
updates

Received: 28 February 2025
Accepted: 6 March 2025
Published: 10 March 2025

Citation: Staniak, M.; Szpunar-Krok, E. Conservation Agricultural Practices for Improving Crop Production and Quality. *Agronomy* **2025**, *15*, 673. https://doi.org/10.3390/agronomy15030673

Bacillus. The results of their research indicated that using the transplanting method for rice, followed by NT for wheat and green gram, increases yields, profitability and soil health.

In saline alluvial meadow soils of an arid region, Nurbekov et al. showed that the application of NT combined with legume-based short crop rotation resulted in higher retention of crop residues (by 20.9%) and root residues (by 25%) than in the conventional system, with beneficial effects on soil structure and soil carbon (C), nitrogen (N) and phosphorus (P) abundance. In this cropping system, increased retention of crop residues had a beneficial effect on soil porosity, structural stability and water retention, while soil salinity in the 0–25 cm and 75–100 cm soil profiles and soil temperature were reduced. According to the authors, the application of the NT system can counteract ongoing land degradation in dryland areas affected by salinity.

Wilczewski et al., in a review article, highlight the positive effects of CA, particularly the presence of CC, on the physical, chemical and biological properties of soil. The authors indicate an increase in soil microbial abundance and an improvement in soil enzyme activity with reduced tillage (RT). In addition, CC biomass increases the capacity of the sorption complex, which has a beneficial effect on soil moisture. CC cultivation is perceived by farmers as an effective way to improve soil structure and increase soil organic matter content and is becoming a regular part of farming in Finland, as indicated by a survey conducted by Peltonen-Sainio et al.

The importance of straw return as an effective management practice to improve the physical and chemical properties of saline–sodic soils in Northeast China was highlighted by Xie et al. They showed that in rice cultivation, carrying out straw return in the autumn is more efficient than doing so in the spring, as there is a longer and slower decomposition period. Akwakwa and Xiaoyan describe the positive effect of rice–wheat straw incorporation, combined with different nitrogen rates, on soil physico-chemical properties in the Jianghan plain of the Yangtze River basin (China). The authors showed that the incorporation of straw into the soil, combined with N fertilization, leads to varying levels of soil N, P, nitrate (NO_3^-), ammonium (NH_4^+), potassium (K), moisture, and pH, as well as wheat grain yield. In addition, the introduction of straw reduced soil bulk density and increased water retention.

3. The Impact of Conservation Agriculture Practices on the Quantity and Quality of Crop Yields

Improving the physical, chemical and biological properties of soil through CA practices can have a positive impact on yield quantity and quality, although this is also dependent on climatic conditions, soil type and crop species. Adequate management of crop residues and CC is also crucial to the effectiveness of CA. The literature reports that during the initial period of implementing CA practice, crop yields may be lower compared to conventional plow cultivation (CON), which may be due to a slower rate of N mineralization and accumulation of organic matter mainly in the surface soil layer, resulting in lower plant nutrient availability [6]. This is confirmed by the study of Nieman et al., who obtained a lower or similar yield of sorghum (depending on weather conditions) when implementing NT compared to cultivation with a chisel plow. The authors attributed the decrease in yield in dry years to a shortage of water during germination and plant emergence, while in years with more rainfall, the yield did not depend on the cropping system.

The literature reports that, in the long term, CA stabilizes organic matter content and improves the relationship between soil structure and water, resulting in increased yields and improved crop quality [7,8]. This is confirmed by the results presented by Dong et al., who showed an increase in the weight of 1000 grains and the number of grains per spike for winter wheat grown through NT compared to CON, which the authors attribute to higher soil moisture at a depth of 20 cm at the grain-filling stage. Although the number of wheat ears when implementing NT was 10.8% lower, the grain yield was 2.7% higher than that when implementing CON. The beneficial effect of higher soil moisture using CA on plant yield was demonstrated by De Santis et al. in a year with limited rainfall; growing durum wheat through NT increased its yield by 15%, improved its grain quality and led to a higher economic return compared to that grown through CON. Also, in the study by Nurbekov et al., compared to crops grown through CON, the NT cultivation of winter wheat, millet, chickpea and maize in rotation in saline alluvial meadow soils of the arid region in Uzbekistan resulted in a significant, sustained increase in the yield of all species grown in two 3-year cycles. In addition, the authors demonstrated the beneficial effects of NT on the soil environment, and concluded that the NT system plays an important role in sustainable land management in arid areas.

4. Integration of Conservation Agricultural Practices with Breeding of Resistant Varieties

CA benefits the soil environment by reducing degradation, improving fertility and increasing water retention, but its effectivity can be limited by various abiotic stresses, such as drought or salinity. The selection of cultivars with increased stress tolerance is therefore an important component of CA implementation [9–11].

Modern plant breeding adapted to cultivation in conservation systems focuses on sowing cultivars with a deep, strong and well-developed root system so that plants can effectively utilize water stored in deeper soil layers and their roots can easily penetrate compacted soil layers [12]. A study by Choi et al. showed that the late-flowering sorghum cultivar Greenstar had a stronger root system that made better use of available soil resources even at lower N rates compared to the earlier-flowering cultivar Honeyew. Better root development and higher rates of root survival are features that can be used in NT.

Soil salinity is a major agricultural problem, especially in areas with limited water availability. In NT, an increase in salinity is observed, especially in the topsoil, so breeding cultivars resistant to this stress factor is crucial to increase the stability of crop yields. Gerakari et al. analyzed the potential of using wild relatives of tomato to improve yield size and quality under high-soil-salinity conditions. Of the nine genotypes, the line IL6-6 showed the highest stress tolerance, maintaining stable photosynthetic parameters, yield levels and fruit quality even at the highest salinity levels. These results indicate that this genotype could be valuable for breeding salinity-tolerant cultivars. Also noteworthy is a study by El-Ata et al., in which the authors analyzed genotype–environment interaction. The study showed that the yield level of twelve rice cultivars was mainly determined by environmental conditions (sowing date, weather conditions), with two cultivars showing greater yield stability under different conditions compared to other cultivars; these can also be used in breeding programs that take into account the adaptation of cultivars to changing conditions.

5. Conclusions

CA comprises agricultural practices that benefit soil's fertility and properties and increase the efficiency of crop production in a sustainable manner. The results of the articles in this Special Issue demonstrate that the use of CA practices, such as NT, RT, mulching, CC and diversified crop rotations, significantly affect the physical, chemical and biological properties of soil, improve soil water retention and increase the macronutrient content available to plants. In addition, CA affects yield quantity and quality, producing high and stable yields in the long term. The use of CA has also been found to be particularly valid in environments exposed to abiotic stresses, such as drought or salinity, and the combination of CA with the breeding of cultivars with higher stress tolerance makes it possible to increase the efficiency and stability of crop production in a sustainable manner.

Author Contributions: M.S. and E.S.-K. contributed equally to this article. All authors have read and agreed to the published version of the manuscript.

Acknowledgments: The Guest Editors wish to thank all the Authors for their contribution to this Special Issue. We also want to thank the Reviewers, Editorial Managers and Editors who assisted in developing this Special Issue.

Conflicts of Interest: The authors declare no conflicts of interest.

List of Contributions

1. Dominguez-Bohorquez, J.D.; Wittling, C.; Cheviron, B.; Bouarfa, S.; Urruty, N.; Lopez, J.-M.; Dejean, C. Early-stage impacts of irrigated conservation agriculture on soil physical properties and crop performance in a French Mediterranean system. *Agronomy* **2025**, *15*, 299. https://doi.org/10.3390/agronomy15020299.

2. Dong, Z.; Yang, S.; Li, S.; Fan, P.; Wu, J.; Liu, Y.; Wang, X.; Zhang, J.; Zhai, C. Effects of no-tillage on field microclimate and yield of winter wheat. *Agronomy* **2024**, *14*, 3075. https://doi.org/10.3390/agronomy14123075.

3. Gerakari, M.; Kyriakoudi, A.; Nokas, D.; Mourtzinos, I.; Chronopoulou, E.G.; Tani, E.; Avdikos, I. Evaluation of the potential use of wild relatives of tomato (*Solanum pennellii*) to improve yield and fruit quality under low-input and high-salinity cultivation conditions. *Agronomy* **2024**, *14*, 3042. https://doi.org/10.3390/agronomy14123042.

4. De Santis, M.A.; Giuzio, L.; Tozzi, D.; Soccio, M.; Flagella, Z. Impact of no tillage and low emission N fertilization on durum wheat sustainability, profitability and quality. *Agronomy* **2024**, *14*, 2794. https://doi.org/10.3390/agronomy14122794.

5. Xie, Y.; Zhang, X.; Gao, Y.; Li, J.; Geng, Y.; Guo, L.; Shao, X.; Ran, C. Effects of different straw returning periods and nitrogen fertilizer combinations on rice roots and yield in saline–sodic soil. *Agronomy* **2024**, *14*, 2463. https://doi.org/10.3390/agronomy14112463.

6. El-Aty, M.S.A.; Abo-Youssef, M.I.; Sorour, F.A.; Salem, M.; Gomma, M.A.; Ibrahim, O.M.; Yaghoubi Khanghahi, M.; Al-Qahtani, W.H.; Abdel-Maksoud, M.A.; El-Tahan, A.M. Performance and stability for grain yield and its components of some rice cultivars under various environments. *Agronomy* **2024**, *14*, 2137. https://doi.org/10.3390/agronomy14092137.

7. Choi, N.; Choi, M.; Lee, S.; Jo, C.; Kim, G.; Jeong, Y.; Lee, J.; Na, C. Effects of ecotypes and reduced N fertilization on root growth and aboveground development of ratooning Sorghum × Sudangrass hybrids. *Agronomy* **2024**, *14*, 2073. https://doi.org/10.3390/agronomy14092073.

8. Nieman, C.C.; Franco, J.G.; Raper, R.L. Inconsistent yield response of forage Sorghum to tillage and row arrangement. *Agronomy* **2024**, *14*, 1510. https://doi.org/10.3390/agronomy14071510.

9. Nurbekov, A.; Kosimov, M.; Shaumarov, M.; Khaitov, B.; Qodirova, D.; Mardonov, H.; Yulda-sheva, Z. Short crop rotation under no-till improves crop productivity and soil quality in salt affected areas. *Agronomy* **2023**, *13*, 2974. https://doi.org/10.3390/agronomy13122974.

10. Akwakwa, G.H.; Xiaoyan, W. Impact of rice–wheat straw incorporation and varying nitrogen fertilizer rates on soil physicochemical properties and wheat grain yield. *Agronomy* **2023**, *13*, 2363. https://doi.org/10.3390/agronomy13092363.

11. Peltonen-Sainio, P.; Jauhiainen, L.; Känkänen, H. Finnish farmers feel they have succeeded in adopting cover crops but need down-to-earth support from research. *Agronomy* **2023**, *13*, 2326. https://doi.org/10.3390/agronomy13092326.

12. Alhammad, B.A.; Roy, D.K.; Ranjan, S.; Padhan, S.R.; Sow, S.; Nath, D.; Seleiman, M.F.; Gitari, H. Conservation tillage and weed management influencing weed dynamics, crop performance, soil properties, and profitability in a rice–wheat–green gram system in the Eastern Indo-Gangetic plain. *Agronomy* **2023**, *13*, 1953. https://doi.org/10.3390/agronomy13071953.

13. Wilczewski, E.; Jug, I.; Szpunar-Krok, E.; Staniak, M.; Jug, D. Shaping soil properties and yield of cereals using cover crops under conservation soil tillage. *Agronomy* **2024**, *14*, 2104. https://doi.org/10.3390/agronomy14092104.

References

1. Hobbs, P.R.; Sayre, K.; Gupta, R. The role of conservation agriculture in sustainable agriculture. *Phil. Trans. R. Soc. B.* **2008**, *363*, 43–555. [CrossRef] [PubMed]
2. Lal, R. Restoring soil quality to mitigate soil degradation. *Sustainability* **2015**, *7*, 5875–5895. [CrossRef]
3. Derpsch, R.; Friedrich, T.; Kassam, A.; Hongwen, L. Current status of adoption of no-till farming in the world and some of its main benefits. *Int. J. Agric. Biol. Eng.* **2010**, *7*, 1–25.
4. Salem, H.M.; Valero, C.; Muñoz, M.Á.; Rodríguez, M.G.; Silva, L.L. Short-term effects of four tillage practices on soil physical properties, soil water potential, and maize yield. *Geoderma* **2015**, *237–238*, 60–70. [CrossRef]
5. Giambalvo, D.; Amato, G.; Badagliacca, G.; Ingraffia, R.; Di Miceli, G.; Frenda, A.S.; Plaia, A.; Venezia, G.; Ruisi, P. Switching from conventional tillage to no-tillage: Soil N availability, N uptake, ^{15}N fertilizer recovery, and grain yield of durum wheat. *Field Crops Res.* **2018**, *218*, 171–181. [CrossRef]
6. Pittelkow, C.M.; Liang, X.; Linquist, B.A.; Van Groenigen, K.J.; Lee, J.; Lundy, M.E.; van Gestel, N.; Six, J.; Venterea, R.T.; van Kessel, C. Productivity limits and potentials of the principles of conservation agriculture. *Nature* **2015**, *517*, 365–368. [CrossRef] [PubMed]
7. Franzluebbers, A.J. Soil organic matter stratification ratio as an indicator of soil quality. *Soil Till. Res.* **2002**, *66*, 95–106. [CrossRef]
8. Peigné, J.; Ball, B.C.; Roger-Estrade, J.; David, C. Is conservation tillage suitable for organic farming? A review. *Soil Use Manag.* **2007**, *23*, 129–144. [CrossRef]
9. Langridge, P.; Reynolds, M.P. Genomic tools to assist breeding for drought tolerance. *Curr. Opin. Biotechnol.* **2015**, *32*, 130–135. [CrossRef] [PubMed]
10. Brummer, E.C.; Barber, W.T.; Collier, S.M.; Cox, T.S.; Johnson, R.; Murray, S.C.; Olsen, R.T.; Pratt, R.C.; Thro, A.M. Plant breeding for harmony between agriculture and the environment. *Front. Ecol. Environ.* **2011**, *9*, 561–568. [CrossRef]
11. Cooper, M.; Messina, C.D.; Podlich, D.; Totir, L.R.; Baumgarten, A.; Hausmann, N.J.; Wright, D.; Graham, G. Predicting the future of plant breeding: Complementing empirical evaluation with genetic prediction. *Crop Past. Sci.* **2014**, *65*, 311–336. [CrossRef]
12. Comas, L.H.; Becker, S.R.; Cruz, V.M.V.; Byrne, P.F.; Dierig, D.A. Root traits contributing to plant productivity under drought. *Front. Plant Sci.* **2013**, *4*, 442. [CrossRef] [PubMed]

Article

Impact of Rice–Wheat Straw Incorporation and Varying Nitrogen Fertilizer Rates on Soil Physicochemical Properties and Wheat Grain Yield

Gabriel Hopla Akwakwa [1,2] **and Wang Xiaoyan** [1,*]

1 Department of Crop Science, College of Agriculture, Yangtze University, Jingzhou 434025, China; gabrielakwakwa21@yangtzeu.edu.cn
2 Engineering Research Center of Ecology and Agricultural Use of Wetland, Ministry of Education, Jingzhou 434025, China
* Correspondence: wamail_wang@163.com; Tel.: +86-189-8666-1561

Abstract: Straw return (SR) is crucial for the comprehensive and efficient utilization of resources within agroecosystems; however, its impact on soils and wheat grain yield in the Jianghan Plain of the Yangtze River Basin, Hubei Province of China, is not fully known. Therefore, the present study was undertaken to assess the impact of returning rice–wheat straw, along with different nitrogen (N) fertilizer applications, on soil physicochemical properties and wheat grain yield. The Yangmai 23 wheat variety was cultivated in the Experimental Farms of Yangtze University in the Yangtze River Basin, with three rates of rice SR (0, 50 and 100%) and four N fertilizer rates (0, 33.3, 70 and 100%) with 180 kg/ha urea. The integrated use of SR- and N-fertilizer rates significantly altered soil nitrogen, nitrate, ammonium, phosphorus, potassium, pH and moisture within the 20 cm depth before the seeding, jointing and maturation stages of the wheat. The grain yields of $6408 \pm 110 - 8290.00 \pm 298$ and $4726 \pm 62 - 6758.00 \pm 196$ kg/ha were obtained in the 2021–2022 and 2022–2023 seasons, respectively. The studied soil physicochemical properties either before seeding, or at the jointing and maturation stages had a significant effect on final grain yield. These results underscore the combined effect of SR- and N-fertilizer application to improve wheat productivity in the Yangtze River Basin. However, further studies are ongoing to assess the impact of these treatments on the soil microbial community, as well as on wheat grain quality.

Keywords: inorganic fertilizers; soil fertility management; soil conservation; straw management; sustainable wheat production

---

Citation: Akwakwa, G.H.; Xiaoyan, W. Impact of Rice–Wheat Straw Incorporation and Varying Nitrogen Fertilizer Rates on Soil Physicochemical Properties and Wheat Grain Yield. *Agronomy* **2023**, *13*, 2363. https://doi.org/10.3390/agronomy13092363

Academic Editors: Mariola Staniak, Ewa Szpunar-Krok and Małgorzata Szostek

Received: 28 July 2023
Revised: 30 August 2023
Accepted: 4 September 2023
Published: 12 September 2023

1. Introduction

Wheat (*Triticum aestivum* L.) is a highly cultivated crop on a global scale, with China serving as the leading producer, accounting for 17% of the world's total wheat production [1]. According to several studies [2–4], winter wheat constitutes approximately 95% of the overall wheat production in China, encompassing both winter and spring varieties. The combined global production of wheat and the aggregate volume of wheat traded amount to 770 and 481 million tons, respectively. This substantial quantity of wheat serves as a primary food source for approximately 35% of the global population, contributing to over 20% of the total caloric intake for humans and supplying approximately 20% of the protein consumed worldwide [2,5,6]. According to the most recent data by FAO [1], enhanced wheat production prospects have contributed to a rise in overall global grain production. However, despite this improvement, the projected grain production for the 2021/22 period failed to meet the anticipated consumer demand, leading to a decrease in global stocks. China, being a nation with a long-standing history in agriculture, annually generates approximately 1.04 billion metric tons of crop straw [7].

In recent years, there has been a notable rise in the production of rice (*Oryza sativa* L.) and wheat (*T. aestivum* L.) straws, which can be attributed to consistent growth in rice and wheat cultivation. As a consequence of economic progress and enhanced quality of life, crop straw has undergone a transition from its previous role as a source of energy and animal feed to being classified as agricultural waste [8,9]. In the course of agricultural production, farmers engage in the practice of burning crop straw to maintain the pace of their farming operations, minimize labor and material resource usage, and alleviate unnecessary exertion. However, this practice has adverse consequences on the environment and has a detrimental impact on both human productivity and quality of life. Furthermore, it leads to the squandering of valuable natural resources and exerts considerable strain on the soil ecosystem [10–12]. Hence, the utilization of burnt rice and wheat straw resources has emerged as a significant environmental concern.

Crop straw is abundant in organic matter and essential nutrient elements that are crucial for promoting plant growth. The act of reintroducing straw back into the field has the potential to introduce a substantial quantity of organic matter into the soil [13–15]. This process triggers a sequence of intricate biochemical reactions within the soil ecosystem, including nitrogen mineralization and the priming effect resulting from organic matter decomposition. These reactions have the capacity to markedly enhance soil quality [16–18]. Crop straw is a carbonaceous energy resource that possesses significant quantities of nitrogen, phosphorus, potassium, and other essential nutrients required for crop development [6,19]. Its utilization plays a crucial role in mitigating imbalances in the nitrogen, phosphorus, and potassium ratios within agricultural soil, as well as compensating for deficiencies in phosphorus and potash mineralization [6]. Straw returning is a highly efficient method for addressing the issue of excessive straw treatment, while also mitigating the environmental pollution associated with straw burning. Simultaneously, the process of straw decomposition yields nutrients that can be utilized to enhance soil fertility, create favorable conditions for microbial proliferation in the soil, stimulate crop growth, and ultimately augment crop yield [12,20,21]. This practice is considered crucial for the comprehensive and efficient utilization of resources within agroecosystems, as it significantly contributes to the maintenance of a robust soil fertility cycle and the promotion of sustainable agricultural development.

Soil organic carbon encompasses carbon present in humus, as well as animal and plant residues, and microorganisms that are generated through microbial processes within the soil [18,22]. It serves as the primary source of carbon nutrients essential for the sustenance of plant and biological life within the soil. Furthermore, soil organic carbon significantly influences the physicochemical properties of the soil. Its concentration within the soil is heavily influenced by the composition and abundance of soil microorganisms. The presence of soil organic carbon plays a crucial role in governing its physicochemical and biological properties/reactions, thereby enhancing soil fertility. The accumulation and conversion of soil organic carbon can have direct or indirect impacts on soil water dynamics, nutrient availability, gas exchange, thermal regulation, and biochemical transformation [23–26]. Additionally, the levels of soil organic carbon are intricately linked to both soil quality and agricultural productivity [23–26].

Hence, it is imperative to study the fluctuations in the reservoir of soil organic carbon to uphold the sustainability of the global agroecosystem. The impact of straw return on soil has been extensively studied in terms of various environmental factors, such as soil water potential [15], temperature [3,15,27], enzyme activities [16,20], fractions of soil organic matter [7,9,22], soil quality and crop productivity [16–18], emissions of greenhouse gases from soil [20,28,29], soil physicochemical properties [19,30], and soil microbial communities [7,18,31,32]. Several studies have suggested that the practice of straw return has the potential to enhance the ecological environment of soils [4,11,17,29,33,34]. This is achieved through the promotion of antagonistic microbes, disruption of pathogen growth, and enhancement of plant resistance to pathogens; consequently contributing to the effective control of soil-borne plant diseases [26,29,35].

A study by Jaćimović et al. [4] demonstrated that the practice of straw return has the potential to alter the composition of soil microbial communities. The application of maize (*Zea mays* L.) straw at a rate of 9000 kg ha^{-1} resulted in an increase in total phospholipid fatty acid content when compared to the treatment without straw return. Additionally, this application of maize straw caused alterations in the structure of the microbial community [4,31]. The rapid decomposition of straw has a significant influence on soil microbes as it alters the microenvironment of the soil [14,15]. The combination of inorganic fertilizers and the incorporation of rice straw into the soil has been found to have a substantial impact on bacterial abundance, leading to changes in the composition of the bacterial community [7,33,36]. The application of chemical fertilizer leads to alterations in the microbial community structure, particularly in the presence of high rates of straw [37]. The addition of nitrogen (N) fertilizer to wheat and maize straw that was returned to the soil was found to enhance soil fertility and enzyme activity compared to the sole return of straw. This, in turn, leads to an improved composition of the bacterial community [7,37]. Prior study has demonstrated that the utilization of fertilizers and straw has the potential to enhance both the abundance and functionality of microorganisms, including fungi and bacteria, within soil. Furthermore, it has been observed that the composition and arrangement of the microbial community vary depending on the specific treatment applied [7,33,36,38].

The presence of straw residue has an impact on the tiller count per plant during the wheat seedling stage. Previous studies have provided evidence regarding the impact of maize straw return on soil physicochemical properties and microorganisms in various regions [4,26,28,39]. For instance, Arul et al. [39] recently reported varied impacts of maize straw on bacterial community and carbon stability at different soil depths. However, limited information exists regarding the influence of combining different topdressing nitrogen fertilizers and rice straw return on soil physicochemical properties and wheat productivity in the middle–lower Yangtze River Basin, specifically the Jianghan Plain. Consequently, an experimental study was undertaken to evaluate the interconnected impacts of wheat–rice straw incorporation and varying nitrogen fertilizer levels on wheat productivity in the Jianghan Plain. The practice of rice–wheat crop rotation dominates in Jianghan Plain [40]. However, the current study reports only the effect of rice–wheat straw and nitrogen fertilization on soil physicochemical properties and the grain yield of winter wheat. The primary aims of this study were (i) to assess the impact of combined SR and varied N fertilizer application with N fertilizer topdressing on soil quality and fertility through changes in physicochemical soil properties; and (ii) to assess the impact of returning rice–wheat straw on wheat yield.

2. Materials and Methods

2.1. Experimental Site Description and Planting Material

This study was conducted in the Jianghan Plain of Jinzhou City, Hubei Province, People's Republic of China (PRC), at the Experimental Farm Station (30036′ N, 112008′ E) of Yangtze University from 2021 to 2023, thus 2021–2022 and 2022–2023 winter wheat-cropping seasons, according to the local cropping calendar. The field is used for wheat and rice rotation. The Jianghan Plain is one of major grain production bases in Hubei Province, and forms a key part of the wheat-growing areas in the Yangtze River Basin [41]. The Experimental Farm Station is characterized by a typical subtropical monsoon climate in the middle–lower Yangtze River Basin of China. The area experiences relatively the same amount of rainfall and mean daily temperature during the two winter wheat growing seasons (Figure 1A,B). The soil type is calcareous alluvial with sandy loam texture [42]. For this study, the variety Yangmai 23 (YM23) was cultivated. Thus, YM23 is one of the main wheat varieties cultivated in both the middle and lower reaches of the Yangtze River in China [43]. Pure and healthy seeds of the YM23 variety (pedigree Yangmai 16 × Yangfu 93–11) were obtained from Jiangsu Golden Land Seed Industry Co., Ltd. (Jiangsu, China).

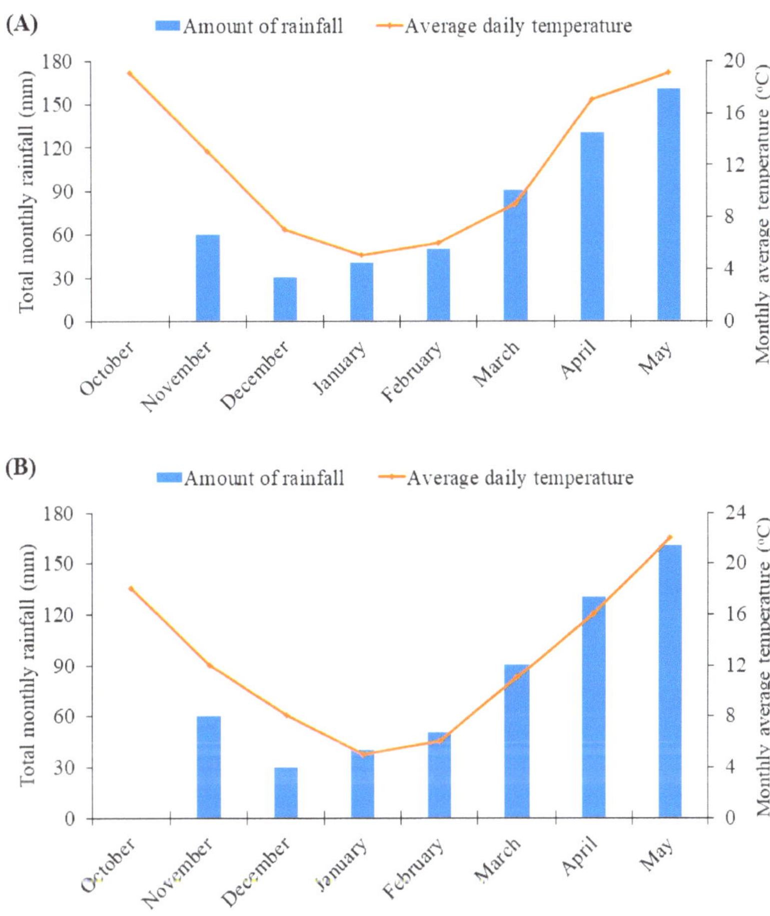

Figure 1. (**A**) Monthly total rainfall and (**B**) monthly mean temperatures recorded on-site in 2021–2022 and 2022–2023 winter wheat-growing seasons in the study area.

2.2. Experimental Design and Treatments

The experimental treatments were arranged in a randomized complete block design with three replications. In this study, 12 treatment combinations were used, including nitrogen fertilization (N) and straw return (SR) rates. Briefly, the N rates consisted of 0, 33.3, 70, and 100% N from urea (46% N), applied at different rates, each totaling 180 kg N/ha (Table 1), while the SR rates comprised 0, 50 and 100%. The experimental field was previously used for rice cultivation and the rice grains were harvested. Briefly, at 0 SR, all rice straw (both below and above ground) was removed and the field was cleaned manually (Figure 2A). In the case of 50% SR, all the straw was collected from the field, weighed, and 50% of it was returned to the field (Figure 2A). In the 100% SR treatment, no straw was collected from the field (Figure 2B). A tractor was employed to plow and harrow the demarcated plots of 0, 50 and 100% to a depth of 15 cm (Figure 2C). In all, the 12 treatments were coded T1–T12 (for details see Table 1). The total experimental area was 78 m in length and 29 m in width. Planting and other cultural practices were carried out according to the conventional practices in the study area. In brief, plots were supplied with phosphorus (P) fertilizer at the rate of 105 kg P_2O_4/ha (with calcium superphosphate) and potassium (K), at the rate of 105 kg K_2O/ha with K sulfate fertilizer, during the sowing time. Other operations such as herbicide, pesticide and fungicide application were carried out according to conventional practices, to prevent yield losses.

Table 1. Treatment combination structure of N (180 kg N/ha) and SR rates in this study.

N Rates (%)	Stage and N Amount Applied (kg N/ha)			Rice Straw Return Rate (%)		
	Sowing	Wintering	Jointing	0	50	100
0.00	0	0	0	T1	T2	T3
33.30	60	60	60	T4	T5	T6
70.00	126	54	0	T7	T8	T9
100.00	180	0	0	T10	T11	T12

Figure 2. Experimental field preparation. (**A**). Zero rice-paddy straw return (SR) plot (left) and 50% SR plot (right). (**B**). 100% SR plot. (**C**). Ploughed and harrowed plot of 100% SR.

2.3. Data Collection

2.3.1. Soil Sampling and Physiochemical Analyses

Soil samples were collected at three stages, i.e., before seeding (BF), and at the jointing stage (JT) and maturation stage (MT), at the depth of 20 cm, following recommended procedures [31]. In brief, five soil samples from each plot were obtained and mixed as composite sample. Firstly, the soil water content was quantified by drying the soil samples in an oven at 105 °C until reaching a constant weight [20,44], and the soil moisture content was obtained by the formula below:

$$Soil\ moisture\ (\%) = \frac{Fw - Dw}{Dw} \times 100\%$$

where Fw and Dw represent the fresh and dried weight of soil sample.

The soil pH was determined in water (1:2.5 w/v) using a pH meter (PHS-3C, INESA Scientific Instrument Co. Ltd., Shanghai, China). The total nitrogen, nitrate (NO_3^-) and ammonium (NH_4^+) contents were determined using semi-microkelvin, the ultraviolet spectrophotometry–colorimetric method, and the indophenol blue colorimetric method, respectively, with minor modifications [44]. The total available phosphorus content was determined using the sodium hydroxide melting–molybdenum antimony colorimetric method, with slight modifications [21,44]. Available potassium (K) in soil was quantified by 1 M NH_4OAc (pH 7.0) with a liquid-to-solid ratio of 10:1, and the K level was finally determined by flame photometer [45]. All soil analyses were carried out in the soil chemistry laboratory of College of Agriculture, Yangtze University. All analyses were carried out for the three independently replicated plots.

2.3.2. Grain Weight per Spike, 1000-Grain Weight and Grain Yield

Grain yield and its related traits were recorded from an average of two 2 m^2 harvest areas in the middle of each plot. All spikes were manually harvested, threshed and weighed. The average grain weight per spike was estimated from 30 spikes. The average weight of grains obtained from the two 2 m^2 harvest areas was converted to kg/ha. One thousand (1000) random grains from each harvested grain were weighed and recorded as 1000-grain weight after the grain moisture was monitored and measured using a grain analyzer (Infratec^{TM}, Foss, Denmark). The 1000-grain weight and grain yield were adjusted to 13% moisture. The same procedures were repeated for grains from each of the three replicates.

2.4. Statistical Analyses

The data recorded in this study were collated, cleaned and formatted in Microsoft Excel, 2018 version (Microsoft Corporation, 2018; Washington, DC, USA). At the beginning, all datasets were tested according to the basic assumptions of analysis of variance (ANOVA) at a 95% confidence interval. In addition, the normality distribution between the grain yield and soil properties was carried out using the Shapiro–Wilk test. Treatment combinations (T1–T12) were finally subjected to a one-way ANOVA, and post hoc mean separation was performed using the Tukey method with 95% confidence interval in R (version 4.2.2), with the *agricolae* package [46]. Bar graphs with means $\pm$ standard error of means were generated in GraphPad Prism, version 8.0.1 (GraphPad Software, San Diego, CA, USA). In addition, the effect of soil physicochemical properties at BF, JT and MT on grain yield was analyzed using the *lm* package in R [47,48].

3. Results

3.1. Changes in Soil Physiochemical Properties Due to Straw Return and N Fertilizer

3.1.1. Nitrogen Content in the Soil before Seeding, and at Both Jointing and Maturation Stages of Wheat

Soil N differed significantly ($p \leq 0.05$) among the 12 treatments in the two seasons (2021–2022 and 2022–2023), with the exception of JT in 2021–2022 (Table 2). The soil N at BF ranged from $0.30 \pm 0.05 - 0.56 \pm 0.11$ g N/kg in 2021–2022 and 0.24 ± 0.19 g N/kg. In addition, soil N at JT was $0.49 \pm 0.04 - 0.60 \pm 0.07$ g N/kg and $0.25 \pm 0.03 - 0.97 \pm 0.41$ g N/kg in the 2021–2022 and 2022–2023 seasons, respectively (Table 2). At MT, soil N recorded in the 2021–2022 cropping season was least for T6 (0.31 ± 0.07 g N/kg) and maximum for T8 (0.87 ± 0.03 g N/kg), while in the 2022–2023 season, the least soil N (0.35 ± 0.09 g N/kg) was recorded for T1 and T9 and the maximum (0.59 ± 0.15 g N/kg) for T2 (Table 2). The level of significant variation suggests that the inclusion of rice–wheat straw as crop residue, together N fertilization, altered the N content in the soil. For example, T8 (70% N rate combined with 50% SR) and T11 (100% N rate combined with 50% SR) had the highest and statistically similar soil N content, of 0.87 ± 0.03 and 0.86 ± 0.06 g N/kg, respectively (Table 2). Surprisingly, in the 2022–2023 cropping season, T2 plots which had only 50% SR retained 0.59 ± 0.15 g N/kg in the soil at MT, which was statistically different from the only-T1 plots (thus no N and SR) and the T9 plots (70% N and 100% SR) (Table 2).

Table 2. Influence of combined used of nitrogen fertilizer and rice-straw return rates on soil nitrogen content (g/kg) under winter wheat cultivation in 2021–2022 and 2022–2023 seasons.

Treatments	2021–2022			2022–2023		
	BF	JT	MT	BF	JT	MT
T1	0.45 ± 0.00 [a–c]	0.56 ± 0.02 [a]	0.45 ± 0.03 [c–e]	0.34 ± 0.09 [b–d]	0.41 ± 0.14 [cd]	0.35 ± 0.09 [b]
T2	0.56 ± 0.11 [a]	0.48 ± 0.05 [a]	0.43 ± 0.06 [c–e]	0.44 ± 0.01 [a–d]	0.48 ± 0.08 [cd]	0.59 ± 0.15 [a]
T3	0.50 ± 0.05 [ab]	0.59 ± 0.06 [a]	0.45 ± 0.05 [c–e]	0.41 ± 0.03 [a–d]	0.42 ± 0.13 [cd]	0.44 ± 0.00 [ab]
T4	0.53 ± 0.08 [a]	0.50 ± 0.04 [a]	0.41 ± 0.03 [d–f]	0.61 ± 0.18 [a]	0.72 ± 0.17 [b]	0.42 ± 0.03 [ab]
T5	0.35 ± 0.10 [bc]	0.49 ± 0.04 [a]	0.72 ± 0.11 [b]	0.38 ± 0.05 [a–d]	0.51 ± 0.05 [cd]	0.49 ± 0.04 [ab]
T6	0.53 ± 0.08 [a]	0.49 ± 0.05 [a]	0.31 ± 0.07 [f]	0.51 ± 0.07 [a–c]	0.35 ± 0.02 [de]	0.44 ± 0.01 [ab]
T7	0.41 ± 0.04 [a–c]	0.49 ± 0.04 [a]	0.52 ± 0.07 [c]	0.30 ± 0.13 [cd]	0.97 ± 0.41 [a]	0.51 ± 0.07 [ab]
T8	0.30 ± 0.05 [c]	0.68 ± 0.14 [a]	0.87 ± 0.03 [a]	0.41 ± 0.02 [a–d]	0.69 ± 0.13 [b]	0.45 ± 0.01 [ab]
T9	0.56 ± 0.11 [a]	0.50 ± 0.03 [a]	0.42 ± 0.02 [c–e]	0.56 ± 0.13 [ab]	0.25 ± 0.03 [e]	0.35 ± 0.09 [b]
T10	0.39 ± 0.06 [a–c]	0.60 ± 0.07 [a]	0.36 ± 0.05 [ef]	0.24 ± 0.19 [d]	0.96 ± 0.04 [a]	0.37 ± 0.07 [ab]
T11	0.35 ± 0.10 [bc]	0.53 ± 0.00 [a]	0.86 ± 0.06 [a]	0.38 ± 0.06 [a–d]	0.51 ± 0.04 [c]	0.54 ± 0.10 [ab]
T12	0.49 ± 0.04 [ab]	0.51 ± 0.03 [a]	0.49 ± 0.09 [cd]	0.60 ± 0.17 [a]	0.38 ± 0.02 [c–e]	0.37 ± 0.07 [ab]

Three stages of soil sampling comprised before seeding (BF), jointing stage (JT) and maturation stage (MT). Treatments comprised 0 kg N /ha with 0% SR (T1), 50% SR (T2) and T3 (100% SR); 33.3% N applied with 0% SR (T4), 50% SR (T5) and T6 (100% SR); 70% N applied with 0% SR (T7), 50% SR (T8) and T9 (100% SR); and 100% N applied with 0% SR (T1) 0, 50% SR (T11) and T12 (100% SR). The nitrogen rate (180 kg N/ha of urea with 46%N) was applied in proportion at sowing, wintering, and jointing stages. Means from three replications and independent plots and their standard errors of mean (SEM) at each soil sampling stage are shown in a column. Mean $\pm$ SEM in a column with the same column with a common letter indicate no significant difference ($p > 0.05$), while those with no common alphabet indicate significant difference ($p \leq 0.05$) using Tukey's post hoc mean separation.

3.1.2. Soil Nitrate Level before Seeding, and at Both Jointing and Maturation Stages of Winter Wheat under N Fertilization and Straw Return

The soil NO_3^- levels varied significantly ($p \leq 0.05$) in the two winter wheat-cropping seasons at BF, JT and MT in both seasons (Figure 3A,B). In year one (2021–2022), T7 (70% N and no SR) had the highest amount (23.95 ± 1.61 g NO_3^-/kg), and this was different from only T8 (20.96 ± 1.38 g NO_3^-/kg) and T9 (20.60 ± 1.74 g NO_3^-/kg) (Figure 3A) at BF; however the NO_3^- recorded on the T9 plot differed significantly from only T6 and T11 (Figure 3A). At JT, soil NO_3^- was highest (22.90 ± 2.45 g NO_3^-/kg) on the T5 plot (33.30% N and 50% SR) but was not different from T4 and T10, indicating that integrated use of N fertilization and crop residue could make N in the form of NO_3 available to plants in almost the way as N synthetic fertilizer. Likewise, at MT, the maximum NO_3 (13.96 ± 0.93 g NO_3^-/kg) was obtained from the plots treated with T8, and this was not statistically different from other treatments, except T3 (Figure 3A).

On the other hand, the NO_3 levels in the second year of cropping (2022–2023) were mostly lowest on the plots treated with no N or no SR (Figure 3B). For instance, T1, T4, T7 and T10 had $14.00 \pm 5.40 - 14.64 \pm 4.76$ g NO_3^-/kg, and these were lower and different from those recorded on plots that received both N fertilization and SR ($19.38 \pm 0.02 - 23.66 \pm 4.26$ g NO_3^-/kg) at BF. These indicate that the inclusion of rice–wheat straw combined with N fertilizer allows the retention of a significant amount of soil NO_3^- reserves, as was the case at both JT and MT (Figure 3B).

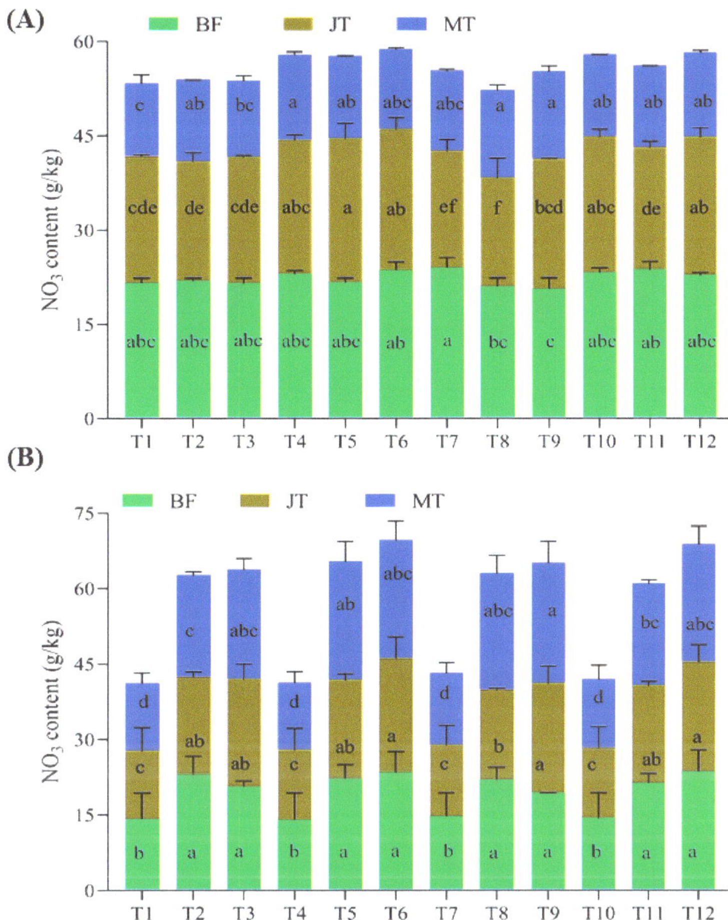

Figure 3. Soil nitrate (NO_3) levels as affected by combined application of straw return (SR) and nitrogen fertilizer rates of winter wheat grown in 2021–2022 season (**A**), and 2022–2023 season (**B**) before seeding (BF, green portion) and at jointing stage (JT, brown portion) and maturation stage (MT, blue portion). Treatments comprised 0 kg N /ha with 0% SR (T1), 50% SR (T2) and T3 (100% SR); 33.3% N applied with 0% SR (T4), 50% SR (T5) and T6 (100% SR); 70% N applied with 0% SR (T7), 50% SR (T8) and T9 (100% SR); and 100% N applied with 0% SR (T1) 0, 50% SR (T11) and T12 (100% SR). The nitrogen rate (180 kg N/ha of urea with 46% N) was applied in proportion at sowing, wintering, and jointing stages. Columns represent means from three replications and independent plots. The error bars represent standard error of means. Bars with the same color with a common letter indicate no significant difference ($p > 0.05$), while those with no common letter indicate significant difference ($p \leq 0.05$), using Tukey's post hoc mean separation [4,16,19,32,49].

3.1.3. Content of Soil Ammonium in Rice–Wheat Rotation Fields with Different N Fertilization and Straw Return

The level of soil NH_4^+ differed significantly in the 2021–2022 winter wheat-cropping season at BF, JT and MT (Figure 4A). The highest content of soil NH_4^+ was recorded in plots treated with T10 (4.99 ± 0.17 g NH_4^+/kg), which was not different from those recorded for T4, T5, T6, and T9 at BF (Figure 4A). Similarly, at JT, T5 had 1.02 ± 0.13 g NH_4^+/kg of soil, which was statistically similar to plots treated with only 50% and 100% rice–wheat straw returned (Figure 4A). Plots treated with T5 retained soil NH_4^+ at the level of

4.99 ± 0.67 g NH_4^+/kg soil, and this was not different from T6 (3.84 ± 0.53 g NH_4^+/kg soil), T11 (3.90 ± 0.19 g NH_4^+/kg soil) and T12 (4.54 ± 0.13 g NH_4^+/kg soil) at MT.

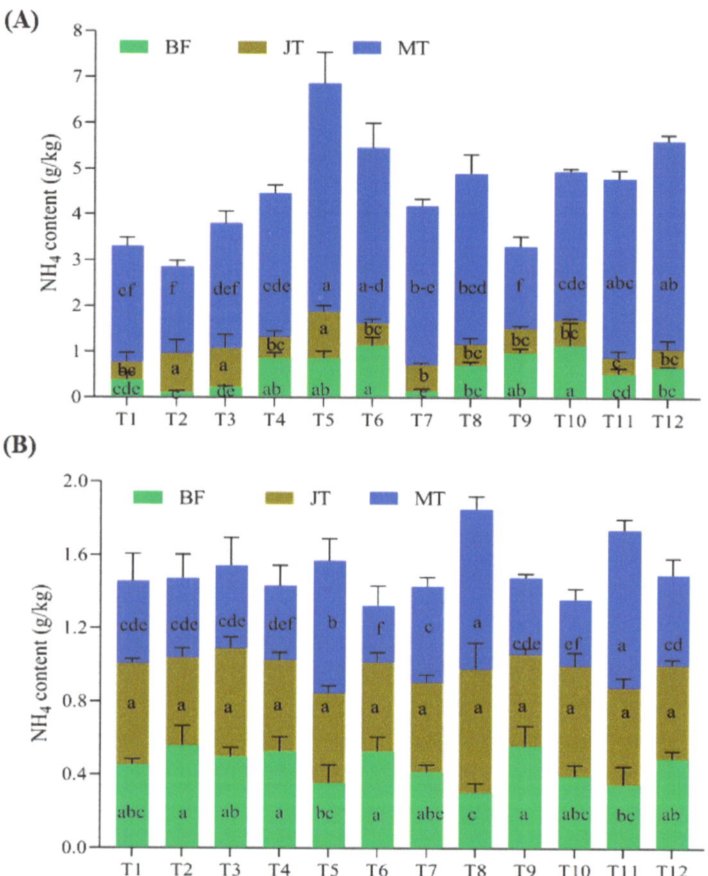

Figure 4. Soil ammonium (NH_4) levels as affected by combined application of straw return (SR) and nitrogen fertilization levels on winter wheat grown in 2021–2022 season (**A**), and 2022–2023 season (**B**) before seeding (BF, green portion), and at jointing stage (JT, brown portion) and maturation stage (MT, blue portion). Treatments comprised 0 kg N/ha with 0% SR (T1), 50% SR (T2) and T3 (100% SR); 33.3% N applied with 0% SR (T4), 50% SR (T5) and T6 (100% SR); 70% N applied with 0% SR (T7), 50% SR (T8) and T9 (100% SR); and 100% N applied with 0% SR (T1) 0, 50% SR (T11) and T12 (100% SR). The nitrogen rate (180 kg N/ha of urea with 46% N) was applied in proportion at sowing, wintering, and jointing stages. Columns represent means from three replications and independent plots. The error bars represent standard error of means. Bars with the same color with a common letter indicate no significant difference ($p > 0.05$), while those with no common letter indicate significant difference ($p \leq 0.05$), using Tukey's post hoc mean separation.

In the 2022–2023 winter wheat-cropping season, the soil NH_4^+ differed significantly ($p \leq 0.05$) at BF and MT (Figure 3B). At BF, the maximum level of soil NH_4^+ (0.56 ± 0.11 g NH_4^+/kg soil) was recorded for plots treated with T2 and T9, which were not different from those recorded by T3, T4, T6, T9 and T10 (Figure 4B), while at MT, the highest soil NH_4^+ (0.87 ± 0.07 and 0.86 ± 0.06 g NH_4^+/kg soil) was recorded for the T8 and T11 plots, respectively; these were followed by T5 (0.72 ± 0.12 g NH_4^+/kg soil).

Overall, the above results on soil N, NO_3^- and NH_4^+ largely suggest that the incorporation of crop residues improves the availability and accessibility of soil N, as well as its usable forms (NO_3^- and NH_4^+), to plant roots.

3.1.4. Variation in Soil Phosphorus in Wheat Fields before Seeding and at Both Jointing and Maturation Stages as Affected by Rice Straw Return and N Fertilization

Straw return is reported to provide more available phosphorus (P) than mineral fertilization [19]. This, according to Kubar et al. [6], may enhance plant P uptake because of the straw effect on the activity of P with low solubility. With this in mind, we quantified the P content in the soil at three stages (ST, JT and MT) in both the 2021–2022 and 2022–2023 winter wheat-cropping seasons (Table 3). The maximum soil P (71.70 ± 14.26 mg P/kg) was recorded for plots treated with T9 at BF, while 70.50 ± 9.28 mg P/kg was obtained from plots treated with T5 at JT and 72.13 ± 15.53 mg P/kg at MT, all in the 2021–2022 season (Table 1). In the same way, the T8 and T5 plots had the maximum P levels at BF (62.49 ± 2.50 mg P/kg) and JT (70.17 ± 14.90 mg P/kg) (Table 3). Surprisingly, the highest P content was recorded for plots treated with T10, which was not different from plots treated with the combined use of N and SR (Table 3). This highlights the fact that incorporation of crop residues, together with N fertilization, could enhance the availability of P in the soil matrix [37].

Table 3. Impact of nitrogen fertilizer and rice straw return rates on soil phosphorus content (mg/kg) under winter wheat cultivation in 2021–2022 and 2022–2023 seasons.

Treatments	2021–2022			2022–2023		
	BF	JT	MT	BF	JT	MT
T1	61.83 ± 4.39 [ab]	60.50 ± 0.02 [bc]	49.23 ± 7.37 [ef]	47.43 ± 7.40 [bc]	43.36 ± 12.00 [d]	57.70 ± 5.60 [cd]
T2	44.93 ± 12.51 [d]	58.63 ± 2.59 [bc]	72.13 ± 15.53 [a]	61.71 ± 1.70 [a]	56.048 ± 0.70 [bc]	59.72 ± 3.60 [a–d]
T3	60.20 ± 2.76 [bc]	60.40 ± 0.82 [bc]	54.90 ± 1.70 [de]	56.50 ± 6.50 [ab]	43.62 ± 11.70 [d]	59.71 ± 3.60 [a–d]
T4	57.57 ± 0.12 [bc]	60.93 ± 0.29 [bc]	49.00 ± 7.60 [ef]	52.65 ± 2.60 [abc]	51.92 ± 3.40 [bcd]	63.73 ± 0.50 [a–d]
T5	56.67 ± 0.78 [bc]	70.50 ± 9.28 [a]	43.87 ± 12.74 [f]	52.22 ± 2.20 [abc]	70.17 ± 14.90 [a]	64.18 ± 0.90 [a–d]
T6	56.93 ± 0.51 [bc]	55.97 ± 5.25 [c]	45.60 ± 11.00 [f]	56.89 ± 6.90 [ab]	54.84 ± 0.50 [bcd]	57.31 ± 6.00 [cd]
T7	59.50 ± 2.06 [bc]	65.53 ± 4.3 [ab]	58.60 ± 2.00 [cd]	48.79 ± 8.80 [abc]	55.45 ± 0.11 [bcd]	72.05 ± 8.80 [ab]
T8	49.57 ± 7.88 [cd]	64.53 ± 4.3 [ab]	62.60 ± 6.00 [bc]	62.49 ± 2.50 [a]	60.97 ± 5.70 [abc]	67.58 ± 4.30 [abc]
T9	71.70 ± 14.26 [a]	58.77 ± 2.45 [bc]	65.63 ± 9.03 [b]	53.42 ± 3.40 [abc]	49.36 ± 5.90 [cd]	72.80 ± 9.50 [ab]
T10	59.50 ± 2.06 [bc]	59.87 ± 1.35 [bc]	62.30 ± 5.70 [bc]	41.85 ± 1.90 [c]	54.20 ± 1.10 [bcd]	73.18 ± 9.90 [a]
T11	58.83 ± 1.39 [bc]	59.97 ± 1.25 [bc]	57.97 ± 1.36 [cd]	54.02 ± 5.40 [abc]	62.40 ± 7.10 [ab]	59.34 ± 3.90 [bcd]
T12	55.17 ± 2.28 [bcd]	59.03 ± 2.19 [bc]	57.40 ± 0.80 [cd]	60.25 ± 6.30 [ab]	61.36 ± 6.10 [abc]	51.98 ± 11.30 [d]

Three stages of soil sampling comprised before seeding (BF), jointing (JT) and maturation stages (MT). Treatments comprised 0 kg N/ha with 0% SR (T1), 50% SR (T2) and T3 (100% SR); 33.3% N applied with 0% SR (T4), 50% SR (T5) and T6 (100% SR); 70% N applied with 0% SR (T7), 50% SR (T8) and T9 (100% SR); and 100% N applied with 0% SR (T1)0, 50% SR (T11) and T12 (100% SR). The nitrogen rate (180 kg N/ha of urea with 46% N) was applied in proportion at sowing, wintering, and jointing stages. Means from three replicated independent plots and their standard errors of mean (SEM) at each soil sampling stage are shown in the columns. Mean $\pm$ SEM in a column with a common letter indicate no significant difference ($p > 0.05$), while those with no common letter indicate significant difference ($p \leq 0.05$), using Tukey's post hoc mean separation.

3.1.5. Changes in K Content in Winter Wheat Fields before Seeding and at Both Jointing and Maturation Stages as Influenced by Rice–Wheat Straw Return and N Fertilization

The available K in the soil differed significantly ($p \leq 0.05$) across BF, JT and MT due to the integrated use of rice–wheat straw return and N fertilizer in the 2021–2022 and 2022–2023 seasons (Figure 5A,B). In the 2021–2022 season, the maximum amount of soil K was (203.50 ± 29.40 mg/kg) at BF, and this statistically differed from only T1, T2 and T11 (Figure 5A). At JT, plots treated with T6 had the maximum soil K (191.80 ± 10.99 mg/kg), and this was not different from T2, T5, T8, T10 and T11 ($186.20 \pm 4.99 - 191.40 \pm 10.19$ mg/kg), whereas at MT, T1 and T2 had statistically similar but higher soil K levels of 157.40 ± 24.50 and 149.50 ± 27.00 mg/kg, respectively, and these were different from the other treatments (Figure 5A).

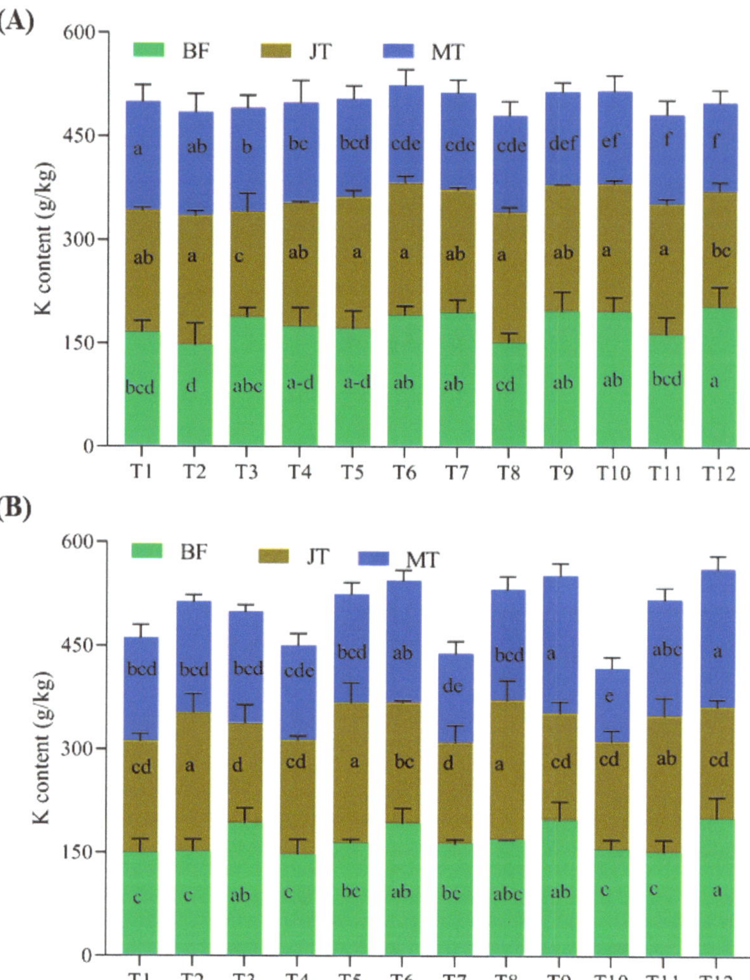

Figure 5. Soil potassium (K) levels as affected by combined application of straw return (SR) and nitrogen fertilization on winter wheat grown in 2021–2022 season (**A**), and 2022–2023 season (**B**) before seeding (BF, green portion), and at jointing stage (JT, brown portion) and maturation stage (MT, blue portion). Treatments comprised 0 kg N/ha with 0% SR (T1), 50% SR (T2) and T3 (100% SR); 33.3% N applied with 0% SR (T4), 50% SR (T5) and T6 (100% SR); 70% N applied with 0% SR (T7), 50% SR (T8) and T9 (100% SR); and 100% N applied with 0% SR (T1) 0, 50% SR (T11) and T12 (100% SR). The nitrogen rate (180 kg N/ha of urea with 46% N) was applied in proportion at sowing, wintering, and jointing stages. Columns represent means from three replicated independent plots. The error bars represent standard error of means. Bars with the same color with a common letter indicate no significant difference ($p > 0.05$), while those with no common letter indicate significant difference ($p \leq 0.05$), using Tukey's post hoc mean separation.

Interestingly, in 2022–2023, the plots treated with both N fertilization and rice–wheat straws or only rice–wheat straws had maximum amount of soil K (Figure 5B). Specifically, at BF, T12 plots had highest level of soil K (of 200.20 ± 31.10 mg/kg), while T5 and T12 had 202.70 ± 29.90 and 199.40 ± 19.40 mg/kg, respectively, which were the highest in JT and

MT. All these results suggest that inclusion of rice–wheat straw with N fertilization could enhance K supply and balance in the soil for sustainable wheat production.

3.1.6. Impact of Combined Use of Rice–Wheat Straw Return and N Fertilization on Soil pH and Moisture Content in Winter Wheat-Growing Seasons

The soil pH varied significantly ($p \leq 0.05$) due to the combined effect of SR and N fertilization at BF, JT and MT in only the 2021–2022 winter wheat-cropping season (Table 4). The pH content of soils at BF ranged from $7.47 \pm 0.10 - 7.87 \pm 0.09$, with $7.43 \pm 0.09 - 7.70 \pm 0.06$ and $7.13 \pm 0.17 - 7.53 \pm 0.23$ at JT and MT (Table 4). These results indicate that soil pH (alkalinity) reduced marginally due to the integrated use of crop residues and N fertilization.

Table 4. Effect of nitrogen fertilization and rice straw return on soil pH in winter wheat cultivation during 2021–2022 and 2022–2023 seasons.

Treatments	2021–2022			2022–2023		
	BF	JT	MT	BF	JT	MT
T1	7.67 ± 0.02 [a–c]	7.70 ± 0.06 [a]	7.33 ± 0.03 [a–c]	7.47 ± 0.04 [a]	7.53 ± 0.02 [a]	7.43 ± 0.04 [a]
T2	7.70 ± 0.01 [ab]	7.55 ± 0.08 [b–d]	7.13 ± 0.17 [c]	7.50 ± 0.01 [a]	7.67 ± 0.11 [a]	7.57 ± 0.09 [a]
T3	7.60 ± 0.02 [bc]	7.60 ± 0.20 [a–c]	7.53 ± 0.23 [a]	7.53 ± 0.02 [a]	7.60 ± 0.04 [a]	7.53 ± 0.06 [a]
T4	7.47 ± 0.10 [c]	7.53 ± 0.08 [b–d]	7.20 ± 0.11 [bc]	7.33 ± 0.18 [a]	7.63 ± 0.08 [a]	7.37 ± 0.11 [a]
T5	7.67 ± 0.09 [a–c]	7.50 ± 0.05 [cd]	7.20 ± 0.11 [bc]	7.53 ± 0.02 [a]	7.57 ± 0.01 [a]	7.43 ± 0.04 [a]
T6	7.83 ± 0.05 [a]	7.43 ± 0.04 [d]	7.30 ± 0.01 [a–c]	7.63 ± 0.12 [a]	7.47 ± 0.09 [a]	7.40 ± 0.07 [a]
T7	7.57 ± 0.15 [ab]	7.63 ± 0.11 [ab]	7.33 ± 0.03 [a–c]	7.60 ± 0.09 [a]	7.60 ± 0.04 [a]	7.50 ± 0.03 [a]
T8	7.73 ± 0.05 [ab]	7.43 ± 0.09 [d]	7.40 ± 0.09 [ab]	7.63 ± 0.12 [a]	7.43 ± 0.12 [a]	7.40 ± 0.07 [a]
T9	7.70 ± 0.16 [ab]	7.60 ± 0.07 [abc]	7.33 ± 0.03 [a–c]	7.50 ± 0.01 [a]	7.50 ± 0.06 [a]	7.50 ± 0.03 [a]
T10	7.57 ± 0.02 [bc]	7.67 ± 0.06 [a]	7.37 ± 0.06 [a–c]	7.33 ± 0.18 [a]	7.60 ± 0.04 [a]	7.43 ± 0.04 [a]
T11	7.87 ± 0.09 [a]	7.43 ± 0.11 [d]	7.13 ± 0.17 [c]	7.53 ± 0.02 [a]	7.53 ± 0.02 [a]	7.57 ± 0.09 [a]
T12	7.73 ± 0.06 [ab]	7.67 ± 0.05 [a]	7.40 ± 0.09 [ab]	7.53 ± 0.02 [a]	7.53 ± 0.02 [a]	7.53 ± 0.06 [a]

Three stages of soil sampling comprised before seeding (BF), jointing stage (JT) and maturation stage (MT). Treatments comprised 0 kg N /ha with 0% SR (T1), 50% SR (T2) and T3 (100% SR); 33.3% N applied with 0% SR (T4), 50% SR (T5) and T6 (100% SR); 70% N applied with 0% SR (T7), 50% SR (T8) and T9 (100% SR); and 100% N applied with 0% SR (T1) 0, 50% SR (T11) and T12 (100% SR). The nitrogen rate (180 kg N/ha of urea with 46% N) was applied in proportion at sowing, wintering, and jointing stages. Means from three replicated independent plots and their standard errors of mean (SEM) at each soil sampling stage are shown in the columns. Mean $\pm$ SEM in a column with a common letter indicate no significant difference ($p > 0.05$), while those with no common letter indicate significant difference ($p \leq 0.05$), using Tukey's post hoc mean separation.

The combined effect of SR and N fertilizer rates was significant ($p \leq 0.05$) on soil moisture recorded at only BF and JT in both the 2021–2022 and 2022–2023 seasons (Figure 6A,B). The highest range of soil moisture was recorded for T1 and T2 ($30.03 \pm 2.07 - 34.40 \pm 2.15\%$) at BF in 2021–2022, while $26.43 \pm 0.38 - 30.52 \pm 3.54\%$ were recorded for nine treatments, with the exception of T2, T3 and T11 at JT (Figure 6A). At BF in 2022–2023, soils treated with N and rice–wheat largely had high moisture contents (Figure 6B). However in the same season, there were some discrepancies in soil moisture recorded at JT (Figure 5B) [18,26].

3.2. Effect of Combined Use of Rice Straw and N Fertilization on Wheat Grain Yield

Wheat grain yield is influenced by a number of components, including grain weight per spike and 1000-grain weight [5,6]. Grain weight per spike, 1000-grain weight and grain yield differed significantly ($p \leq 0.05$) among the treatments in both the 2021–2022 and 2022–2023 seasons (Figure 7A–C). The highest grain weight of 63.00 ± 10.83 g per spike was obtained from T2, which was not different from T3 (62.33 ± 10.17 g), T10 (58.33 ± 6.17 g) and T12 (57.33 ± 5.17 g) in 2021–2022 (Figure 6A), while in 2022–2023, the highest grain weight of 49.71 ± 12.60 g per spike was produced by plots treated with either T7 or T9 (Figure 7A).

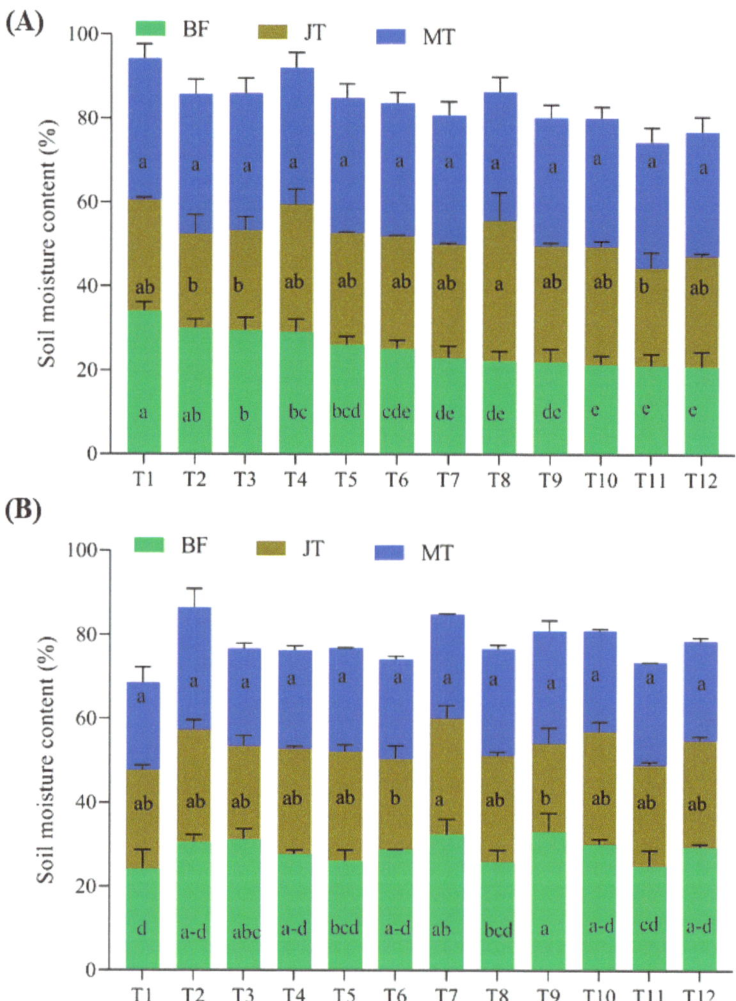

Figure 6. Soil moisture content (%) as influenced by combined application of straw return (SR) and nitrogen fertilization on winter wheat grown in 2021–2022 season (**A**), and 2022–2023 season (**B**) before seeding (BF, green portion), and at jointing stage (JT, brown portion) and maturation stage (MT, blue portion). Treatments comprised 0 kg N/ha with 0% SR (T1), 50% SR (T2) and T3 (100% SR); 33.3% N applied with 0% SR (T4), 50% SR (T5) and T6 (100% SR); 70% N applied with 0% SR (T7), 50% SR (T8) and T9 (100% SR); and 100% N applied with 0% SR (T1) 0, 50% SR (T11) and T12 (100% SR). The nitrogen rate (180 kg N/ha of urea with 46% N) was applied in proportion at sowing, wintering, and jointing stages. Columns represent means from three replicated independent plots. The error bars represent standard errors of means. Bars with the same color and a common letter indicate no significant difference ($p > 0.05$), while those with no common letter indicate significant difference ($p \leq 0.05$), using Tukey's post hoc mean separation.

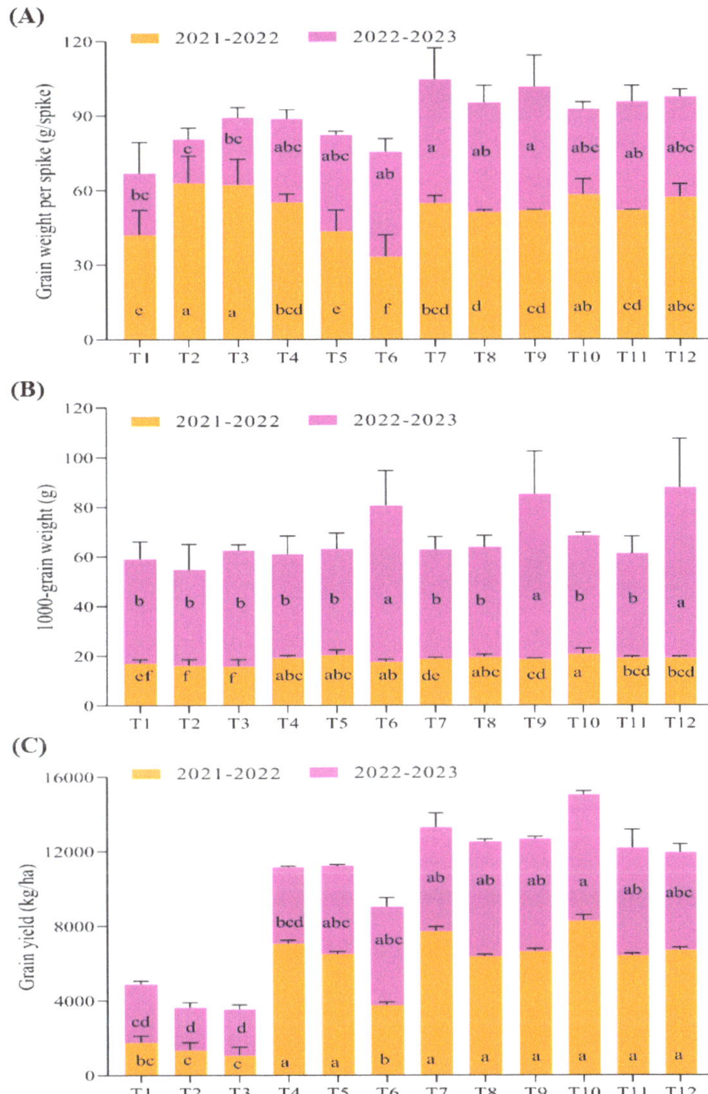

Figure 7. Influence of integrated use of nitrogen fertilizer and rice straw return (SR) on grain weight per spike (**A**), 1000-grain weight (**B**) and grain yield (**C**) of winter wheat grown in 2021–2022 and 2022–2023 seasons. Treatments comprised 0 kg N/ha with 0% SR (T1), 50% SR (T2) and T3 (100% SR); 33.3% N applied with 0% SR (T4), 50% SR (T5) and T6 (100% SR); 70% N applied with 0% SR (T7), 50% SR (T8) and T9 (100% SR); and 100% N applied with 0% SR (T1) 0, 50% SR (T11) and T12 (100% SR). The nitrogen rate (180 kg N/ha of urea with 46% N) was applied in proportion at sowing, wintering, and jointing stages. Columns represent means from three replicated independent plots. The error bars represent standard errors of means. Bars with the same color and a common letter indicate no significant difference ($p > 0.05$), while those with no common letter indicate significant difference ($p \leq 0.05$), using Tukey's post hoc mean separation.

The highest 1000-grain weight of 20.77 ± 2.18 g was obtained from plots treated with T10 followed by T5 (20.27 ± 1.88 g), T8 (19.57 ± 0.98 g) and T4 (19.37 ± 0.78 g); of these, none was statistically different from the other in the 2021–2022 season (Figure 7B), while in the 2022–2023 season, T6, T9 and T12 treatments resulted in a statistically similar 1000-grain weight, ranging from 63.00 ± 14.03 to 68.67 ± 19.69 g, and these treatments were different from the other treatments (Figure 7B).

With regard to grain yield, T4, T5 and T7–T12 plots produced 6337 ± 133 − 8299 ± 298 kg/ha, of which neither was statistically different from T1, T2, T3 and T6 (1079 ± 422 − 3760 ± 154 kg/ha) (Figure 7C). Also, in the 2022–2023 season, T5–T12 treatments produced 4726.00 ± 62 − 6758 ± 196 kg of grains per ha (Figure 6C). These results suggest that the combined use of straw return and N fertilization results in improved grain yield and its components (grain weight per ear and 1000-grain weight) (Figure 7A–C).

We also conducted regression analyses to assess the impact of the soil physicochemical properties at BF, JT and MT on grain yield of winter wheat at three stages (Figures 8 and 9). The studied soil properties had diverse effects on grain yield ($p \leq 0.05$). In the 2021–2022 cropping season for instance, the soil N at BF had a 37% reduction in grain yield, while the soil moisture at MT also reduced the grain yield by 34% (Figure 8). On the contrary, soil NH_4 at BF and MT had a 49 and 44% increase in grain yield, respectively (Figure 8). Also, soil NO_3 and P contents resulted in a 57 and 66% rise in grain yield. In the 2022–2023 winter season, soil NH_4 and P levels at JT led to a rise of 34 and 35% in grain yield, respectively (Figure 9). In addition, soil P level resulted in a 47% rise in grain yield at MT (Figure 9). Strikingly, soil pH at JT resulted in a 40% decline in grain yield (Figure 9). Altogether, these results add to the extensive literature on the effect of soil nutrient abundance on wheat grain yield.

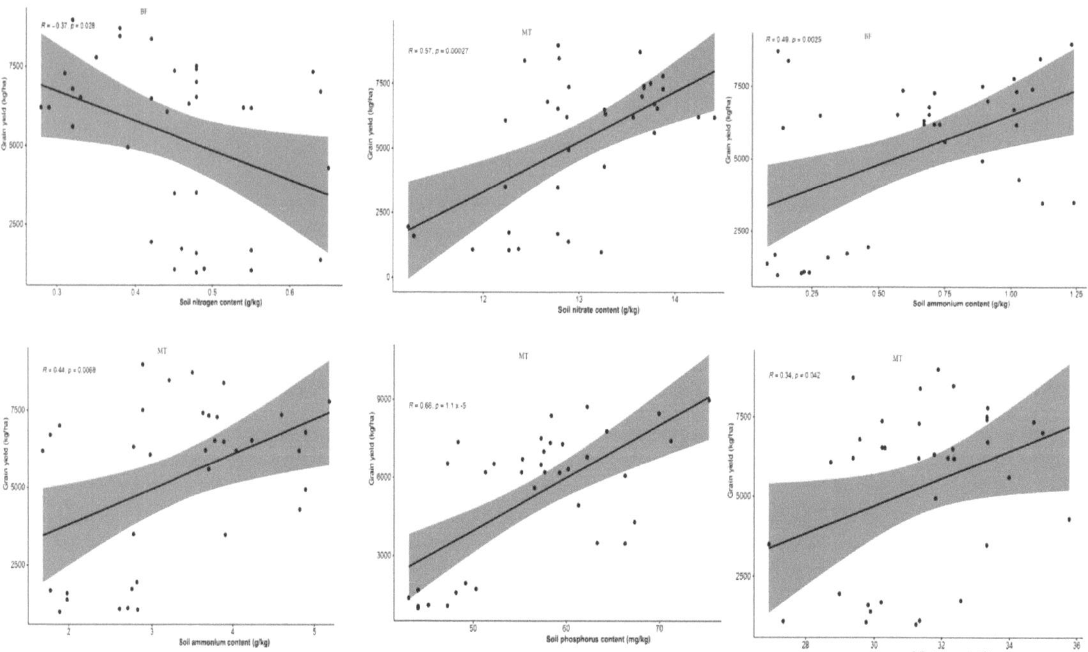

Figure 8. Soil physicochemical properties with significant ($p \leq 0.05$) effect on grain yield of winter wheat cultivated in 2021–2022 season. The three stages of soil samples used for analyses are shown at the top of each figure: before seeding (BF), jointing stage (JT) and maturation stage (MT). The R-value on each figure represents variation in grain yield explained by each of the soil physicochemical properties.

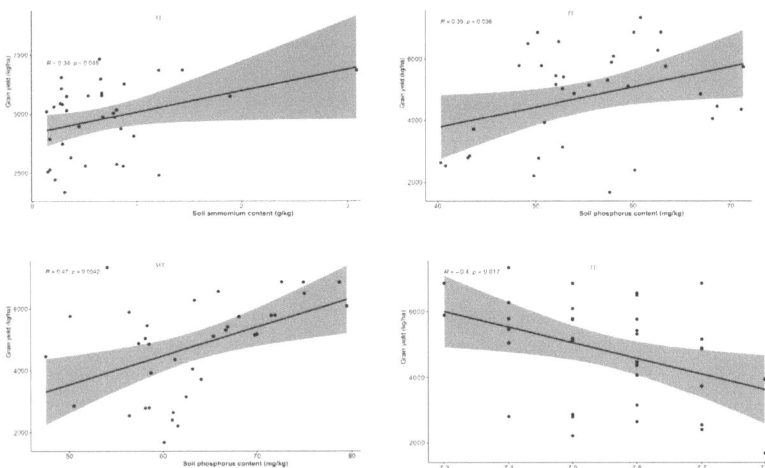

Figure 9. Soil physicochemical properties with significant ($p \leq 0.05$) effect on grain yield of winter wheat cultivated in 2022–2023 season. The three stages of soil samples used for analyses are shown at the top of each figure: jointing stage (JT) and maturation stage (MT). The R-value on each figure represents variation in grain yield explained by each of the soil physicochemical properties.

4. Discussion

Globally, the predominant approach for managing crop straw is through the practice of straw returning (SR). Nevertheless, the prevalent practice of SR in the contemporary rice–wheat rotation system has been found to have detrimental impacts on soil fertility and crop yield [44,45]. This study presents findings that support the notion that incorporating rice straw at a rate of 50 and 100% straw return kg/ha with 33.3, 70 and 100% N fertilization and topdressing application rates in the rice–wheat rotation system has a positive impact on the soil physicochemical properties and wheat productivity, ultimately leading to enhanced wheat grain yields. The traditional practice of incorporating straw into the field has the potential to rapidly enhance soil health and fertility. The utilization of concentrated ditch-buried straw return for an extended period presents benefits compared to alternative methods of SR in terms of enhancing the buildup of essential nutrients and soil organic matter. The synergistic effect of minimal- or zero-tillage practices, alongside SR, contributes to soil health and overall soil fertility [26,42].

4.1. Effect of Rice Straw and N Fertilization on Soil Physicochemical Properties

The burning of straw is prohibited in China, due to its detrimental impact on the environment, including pollution and a significant increase in ground temperature. Increased temperatures have the potential to directly impair the viability of beneficial microorganisms present in soil, thereby compromising their beneficial functions [31,40]. Furthermore, these high temperatures can also have a detrimental impact on the optimal uptake of essential soil nutrients by crops, thereby potentially hindering their overall nutrient absorption capacity [4,5,18,37]. Prior meta-analysis [30] has demonstrated that nitrogen uptake is necessary to support soil microbial activities [15,19,50]. China, being a country with a long-standing history of agriculture, annually generates approximately 1.04 billion metric tonnes of crop straw [18,29,30]. Straw returning is an efficacious method which is being promoted for addressing the excessive straw generation and for mitigating the environmental pollution associated with straw burning. Previous studies have demonstrated that the practice of returning straw to the field significantly impacts on the physicochemical properties of the soil, as well as on the microbial community [3,15,41]. The findings of our study indicate that the varied rates of nitrogen application and rice straw return differentially impacted soil physicochemical properties and wheat growth at the different growth stages, and thus influenced wheat grain yield (Tables 1–3). Furthermore, it was observed that the utilization of nitrogen fertilizer with topdressing had a more favorable impact on SR in the rice–wheat crop rotation system (Figures 3–5). These findings may be ascribed to the combined effect of SR and N fertilizer application (N rate). The variation in soil nitrogen indicates that the combination of crop residue (rice straw) and synthetic nitrogen fertilizer influences the abundance of soil nitrogen, even in its usable form (nitrates) [6,13,25,26,42,51].

Crop straw is rich in essential nutrients that play a crucial role in the growth and development of plants [18,19,35]. Nevertheless, SR possesses the capacity to replenish the essential nutrients that plants may deplete as a result of intensive cultivation, imbalanced application of fertilizers, and inadequate water supply [20,50]. Our study revealed that SR application boosted accessibility of nitrogen (N), phosphorous (P), nitrates (NO_3^-), and ammonium (NH_4^+) within 20 cm soil depth (Figures 3–5). This is consistent with earlier reports by Zhang et al. [50] and Wang et al. [27]. The study revealed a significant decrease in soil phosphorus (P) availability in the SR treatment as compared to the CK treatment, specifically at depths ranging from 20 to 30 cm. This can be attributed to the incorporation of straw into the uppermost 20 cm of soil, which effectively impedes leaching of P. Xie et al. [19] reported comparable findings in a rice–wheat experiment of extended duration [52]. The process of increasing soil organic carbon (SOC) requires a comparable duration of time [9,18,22]. Nevertheless, the agronomic practice of strip tillage exhibited a significant SOC level over the duration of the two-year study. The increase in SOC can be attributed to the occurrence of more humid weather conditions, the rice–wheat crop rotation system, and the increased application of nitrogen (N) fertilizers. The acceleration of crop residue decomposition can likely be attributed to these factors [9,18].

Puddling in a rice–wheat crop rotation system, in the absence of straw incorporation, releases essential nutrients for optimal plant growth [15,16]. A substantial increase in soil moisture content was observed. The highest soil moisture was recorded for the plots treated with 50 or 100 SR at JT and MT; this is consistent with earlier reports that straw returned to the soil minimizes the evaporation rate of soil water, and enhances soil water-holding capacity and water use efficiency [18,26]. The soil exhibited a sandy loam texture, characterized by a high bulk density and a decrease in the percentage of water-stable aggregates [43]. Nevertheless, there was no significant difference observed in the soil moisture content between the two seasons. The presence of organic matter has been found to have several effects on soil properties. Firstly, it has been observed to decrease bulk density and soil compactibility [52]. Additionally, organic matter has been shown to increase porosity and infiltration rate [13,23].

Moreover, it has been found to raise soil moisture content at field capacity [12]. These effects can be attributed to the low bulk density of organic matter and its ability to enhance soil aggregate stability. In this study, it was observed that reintroducing rice straw into the field resulted in a significant reduction in soil bulk density. The study by Song et al. [52] supports the observation that a greater number of water-stable aggregates with a size larger than 0.25 mm were formed under the soil tillage regime (STR) compared to the conventional tillage regime (CK). The rapid fluctuations in soil physical characteristics can be partially ascribed to the location-specific climate and agronomic practices. The observed variations in soil pH, which were found to be statistically significant ($p \leq 0.05$), can be attributed to the combined influence of SR and nitrogen (N) fertilization during the different growth stages of wheat: BF, JT, and MT. The results suggest that the soil pH (alkalinity) experienced a slight but significant decrease when comparing wheat fields treated with crop residues and N fertilization in the BF and MT conditions.

4.2. Effect of Rice Straw and N Fertilization with Topdressing on Wheat Grain Yield

Contradictory results have been reported by previous studies about the influence of SR on crop productivity in different climatic and soil conditions [8,9,20]. Our study found that returning crop straw to the field differentially improved wheat grain yield for the two seasons (Figure 7). This finding aligns with prior studies [4,4,35,50]. The improved wheat grain yield can be attributed to various factors. Primarily, SR replenished a substantial amount of the nutrients that were depleted by the aerial components of the wheat plants. Additionally, the presence of organic matter in crop straw contributes to improved physicochemical and biological properties of the soil, resulting in increased wheat grain productivity (Figures 6 and 7). However, earlier studies have indicated that SR had no significant effect [16,25,27] or resulted in a decline in crop productivity [15,16]. This was largely ascribed to SR-induced nitrogen immobilization, resulting in decreased crop yields [16,16,27].

The study finds that the wet climate and rice–wheat crop rotation system contribute to the decomposition of straw and the release of nutrients, which help to alleviate the disadvantages of SR. Moreover, the varied nitrogen (N) fertilization rates in this study serve to prevent the occurrence of inorganic N deficiency resulting from N immobilization [32]. The results also showed a varied wheat yield over time, for both seasons. The decrease in yield observed during the experiment can be attributed to potassium deficiency. This is supported by the low initial soil potassium content and the absence of potassium fertilizer application [45,50]. Wheat is more susceptible to potassium deficiency than rice. This is because rice can uptake potassium from irrigation water and access soil P more easily when the soil is soaked [12,16].

5. Conclusions

The outcomes of SR are contingent upon various factors, including soil type, weather conditions, climate patterns, field management practices, and cropping methods. Therefore, the present study explored the incorporation of straw into soil and assessed its impact on both soil fertility indices and crop yield within a rice–wheat rotation system over a two-year period. Our study demonstrated that the complete incorporation of straw culminated in varied levels of soil nitrogen (N), phosphorus (P), nitrate (NO_3^-), ammonium (NH_4^+), potassium, moisture, pH and wheat grain yield in comparison to the control group. The combined SR and N fertilization resulted in varied improvements in the physicochemical properties of the soil, including a reduction in bulk density and an increase in water retention capacity. Thus, a combined SR and N fertilization with topdressing may be an effective approach for enhancing soil fertility and increasing wheat grain productivity in the middle–lower Yangtze River Basin.

Author Contributions: Conceptualization, W.X.; methodology, W.X.; formal analysis, G.H.A.; investigation, G.H.A.; resources, W.X.; writing—original draft preparation, G.H.A.; writing—review and editing, G.H.A. and W.X.; supervision, W.X.; project administration, W.X.; funding acquisition, W.X. All authors have read and agreed to the published version of the manuscript.

Funding: This study was funded by the National Natural Science Foundation of China (grant number 31871578). This work was supported by the Engineering Research Center of Ecology and Agricultural Use of Wetland, Ministry of Education.

Data Availability Statement: All data generated and analyzed are in the Results section.

Acknowledgments: This paper benefited from generous support provided by individuals, groups, and institutions. The authors are grateful to the Yangtze University (China), the National Key Research and Development Program of China (China) and the National Natural Science Foundation of China. The authors also wish to record sincere appreciation for the lab colleagues, teachers, and local keepers at the greenhouse, and we thank "anonymous" reviewers for their insights.

Conflicts of Interest: The authors declare no conflict of interest.

References

1. FAO. *FAO Publications Catalogue 2020*; FAO: Rome, Italy, 2019.
2. Lu, C.; Fan, L. Winter Wheat Yield Potentials and Yield Gaps in the North China Plain. *Field Crops Res.* **2013**, *143*, 98–105. [CrossRef]
3. Wu, W.; Li, C.; Ma, B.; Shah, F.; Liu, Y.; Liao, Y. Genetic Progress in Wheat Yield and Associated Traits in China since 1945 and Future Prospects. *Euphytica* **2014**, *196*, 155–168. [CrossRef]

4. Jaćimović, G.; Aćin, V.; Mirosavljević, M.; Brbaklić, L.; Vujić, S.; Dunđerski, D.; Šeremešić, S. Effects of Combined Long-Term Straw Return and Nitrogen Fertilization on Wheat Productivity and Soil Properties in the Wheat-Maize-Soybean Rotation System in the Pannonian Plain. *Agronomy* **2023**, *13*, 1529. [CrossRef]

5. Irfan Ullah, M.; Mahpara, S.; Bibi, R.; Ullah Shah, R.; Ullah, R.; Abbas, S.; Ihsan Ullah, M.; Hassan, A.M.; El-Shehawi, A.M.; Brestic, M.; et al. Grain Yield and Correlated Traits of Bread Wheat Lines: Implications for Yield Improvement. *Saudi J. Biol. Sci.* **2021**, *28*, 5714–5719. [CrossRef]

6. Saleem Kubar, M.; Feng, M.; Sayed, S.; Hussain Shar, A.; Ali Rind, N.; Ullah, H.; Ali Kalhoro, S.; Xie, Y.; Yang, C.; Yang, W.; et al. Agronomical Traits Associated with Yield and Yield Components of Winter Wheat as Affected by Nitrogen Managements. *Saudi J. Biol. Sci.* **2021**, *28*, 4852–4858. [CrossRef]

7. Chen, Y.; Xin, L.; Liu, J.; Yuan, M.; Liu, S.; Jiang, W.; Chen, J. Changes in Bacterial Community of Soil Induced by Long-Term Straw Returning. *Sci. Agric.* **2017**, *74*, 349–356. [CrossRef]

8. Gummert, M.; Van Hung, N.; Chivenge, P.; Douthwaite, B. (Eds.) *Sustainable Rice Straw Management*, 1st ed.; Springer: Cham, Switzerland, 2020.

9. Karlsson, H.; Ahlgren, S.; Sandgren, M.; Passoth, V.; Wallberg, O.; Hansson, P.A. Greenhouse Gas Performance of Biochemical Biodiesel Production from Straw: Soil Organic Carbon Changes and Time-Dependent Climate Impact. *Biotechnol. Biofuels* **2017**, *10*, 217. [CrossRef]

10. Bellamy, P.H.; Loveland, P.J.; Bradley, R.I.; Lark, R.M.; Kirk, G.J.D. Carbon Losses from All Soils across England and Wales 1978–2003. *Nature* **2005**, *437*, 245–248. [CrossRef]

11. Lal, R. Carbon Sequestration in Dryland Ecosystems of West Asia and North Africa. *Land Degrad. Dev.* **2002**, *13*, 45–59. [CrossRef]

12. Suriyagoda, L.; De Costa, W.A.J.M.; Lambers, H. Growth and Phosphorus Nutrition of Rice When Inorganic Fertiliser Application is Partly Replaced by Straw under Varying Moisture Availability in Sandy and Clay Soils. *Plant Soil* **2014**, *384*, 53–68. [CrossRef]

13. Cusack, D.F.; Silver, W.L.; Torn, M.S.; Burton, S.D.; Firestone, M.K. Changes in Microbial Community Characteristics and Soil Organic Matter with Nitrogen Additions in Two Tropical Forests. *Ecology* **2011**, *92*, 621–632. [CrossRef] [PubMed]

14. Miura, T.; Owada, K.; Nishina, K.; Utomo, M.; Niswati, A.; Kaneko, N.; Fujie, K. The Effects of Nitrogen Fertilizer on Soil Microbial Communities under Conventional and Conservation Agricultural Managements in a Tropical Clay-Rich Ultisol. *Soil Sci.* **2016**, *181*, 68–74. [CrossRef]

15. Yang, H.; Feng, J.; Zhai, S.; Dai, Y.; Xu, M.; Wu, J.; Shen, M.; Bian, X.; Koide, R.T.; Liu, J. Long-Term Ditch-Buried Straw Return Alters Soil Water Potential, Temperature, and Microbial Communities in a Rice-Wheat Rotation System. *Soil Tillage Res.* **2016**, *163*, 21–31. [CrossRef]

16. Zhao, X.; Yuan, G.; Wang, H.; Lu, D.; Chen, X.; Zhou, J. Effects of Full Straw Incorporation on Soil Fertility and Crop Yield in Rice-Wheat Rotation for Silty Clay Loamy Cropland. *Agronomy* **2019**, *9*, 133. [CrossRef]

17. Zhang, P.; Wei, T.; Jia, Z.; Han, Q.; Ren, X. Soil Aggregate and Crop Yield Changes with Different Rates of Straw Incorporation in Semiarid Areas of Northwest China. *Geoderma* **2014**, *230–231*, 41–49. [CrossRef]

18. Liu, N.; Li, Y.; Cong, P.; Wang, J.; Guo, W.; Pang, H.; Zhang, L. Depth of Straw Incorporation Significantly Alters Crop Yield, Soil Organic Carbon and Total Nitrogen in the North China Plain. *Soil Tillage Res.* **2021**, *205*, 104772. [CrossRef]

19. Xie, J.; Evgenia, B.; Zhang, Y.; Wan, Y.; Hu, Q.-J.; Zhang, C.-M.; Wang, J.; Zhang, Y.-Q.; Shi, X.-J. Substituting Nitrogen and Phosphorus Fertilizer with Optimal Amount of Crop Straw Improves Rice Grain Yield, Nutrient Use Efficiency and Soil Carbon Sequestration. *J. Integr. Agric.* **2022**, *21*, 3345–3355. [CrossRef]

20. Ji, B.; Hu, H.; Zhao, Y.; Mu, X.; Liu, K.; Li, C. Effects of Deep Tillage and Straw Returning on Soil Microorganism and Enzyme Activities. *Sci. World J.* **2014**, *2014*, 451493. [CrossRef]

21. Yang, C.; Yang, L.; Yang, Y.; Ouyang, Z. Rice Root Growth and Nutrient Uptake as Influenced by Organic Manure in Continuously and Alternately Flooded Paddy Soils. *Agric. Water Manag.* **2004**, *70*, 67–81. [CrossRef]

22. Jin, Z.; Shah, T.; Zhang, L.; Liu, H.; Peng, S.; Nie, L. Effect of Straw Returning on Soil Organic Carbon in Rice–Wheat Rotation System: A Review. *Food Energy Secur.* **2020**, *9*, e200. [CrossRef]

23. Štursová, M.; Baldrian, P. Effects of Soil Properties and Management on the Activity of Soil Organic Matter Transforming Enzymes and the Quantification of Soil-Bound and Free Activity. *Plant Soil* **2011**, *338*, 99–110. [CrossRef]

24. Wei, Q.; Xu, J.; Sun, L.; Wang, H.; Lv, Y.; Li, Y.; Hameed, F. Effects of Straw Returning on Rice Growth and Yield under Water-Saving Irrigation. *Chil. J. Agric. Res.* **2019**, *79*, 66–74. [CrossRef]

25. Wang, J.; Hussain, S.; Sun, X.; Zhang, P.; Javed, T.; Dessoky, E.S.; Ren, X.; Chen, X. Effects of Nitrogen Application Rate Under Straw Incorporation on Photosynthesis, Productivity and Nitrogen Use Efficiency in Winter Wheat. *Front. Plant Sci.* **2022**, *13*, 862088. [CrossRef] [PubMed]

26. Sui, P.; Tian, P.; Lian, H.; Wang, Z.; Ma, Z.; Qi, H.; Mei, N.; Sun, Y.; Wang, Y.; Su, Y.; et al. Straw Incorporation Management Affects Maize Grain Yield through Regulating Nitrogen Uptake, Water Use Efficiency, and Root Distribution. *Agronomy* **2020**, *10*, 324. [CrossRef]

27. Wang, J.; Qiu, Y.; Zhang, X.; Zhou, Z.; Han, X.; Zhou, Y.; Qin, L.; Liu, K.; Li, S.; Wang, W.; et al. Increasing Basal Nitrogen Fertilizer Rate Improves Grain Yield, Quality and 2-Acetyl-1-Pyrroline in Rice under Wheat Straw Returning. *Front. Plant Sci.* **2023**, *13*, 1099751. [CrossRef]

28. Zhou, Y.; Zhang, Y.; Tian, D.; Mu, Y. The Influence of Straw Returning on N$_2$O Emissions from a Maize-Wheat Field in the North China Plain. *Sci. Total Environ.* **2017**, *584–585*, 935–941. [CrossRef]

29. Zhou, L.-Y.; Zhu, Y.-H.; Kan, Z.-R.; Li, F.-M.; Zhang, F. The Impact of Crop Residue Return on the Food–Carbon–Water–Energy Nexus in a Rice–Wheat Rotation System under Climate Warming. *Sci. Total Environ.* **2023**, *894*, 164675. [CrossRef]
30. Liu, P.; Zhao, H.-J.; Wang, L.-L.; Liu, Z.-H.; Wei, J.-L.; Wang, Y.-Q.; Jiang, L.-H.; Dong, L.; Zhang, Y.-F. Analysis of Heavy Metal Sources for Vegetable Soils from Shandong Province, China. *Agric. Sci. China* **2011**, *10*, 109–119. [CrossRef]
31. Marschner, P.; Umar, S.; Baumann, K. The Microbial Community Composition Changes Rapidly in the Early Stages of Decomposition of Wheat Residue. *Soil Biol. Biochem.* **2011**, *43*, 445–451. [CrossRef]
32. Li, H.; Wang, L.; Peng, Y.; Zhang, S.; Lv, S.; Li, J.; Abdo, A.I.; Zhou, C.; Wang, L. Film Mulching, Residue Retention and N Fertilization Affect Ammonia Volatilization through Soil Labile N and C Pools. *Agric. Ecosyst. Environ.* **2021**, *308*, 107272. [CrossRef]
33. Yuan, L.; Gao, Y.; Mei, Y.; Liu, J.; Kalkhajeh, Y.K.; Hu, H.; Huang, J. Effects of Continuous Straw Returning on Bacterial Community Structure and Enzyme Activities in Rape-Rice Soil Aggregates. *Sci. Rep.* **2023**, *13*, 2357. [CrossRef] [PubMed]
34. Raina, A.; Khan, S. Field Assessment of Yield and Its Contributing Traits in Cowpea Treated with Lower, Intermediate, and Higher Doses of Gamma Rays and Sodium Azide. *Front. Plant Sci.* **2023**, *14*, 1188077. [CrossRef] [PubMed]
35. Shan, A.; Pan, J.; Kang, K.J.; Pan, M.; Wang, G.; Wang, M.; He, Z.; Yang, X. Effects of Straw Return with N Fertilizer Reduction on Crop Yield, Plant Diseases and Pests and Potential Heavy Metal Risk in a Chinese Rice Paddy: A Field Study of 2 Consecutive Wheat-Rice Cycles. *Environ. Pollut.* **2021**, *288*, 117741. [CrossRef]
36. Jin, S.; Huang, Y.; Dong, C.; Bai, Y.; Pan, H.; Hu, Z. Effects of Different Straw Returning Amounts and Fertilizer Conditions on Bacteria of Rice's Different Part in Rare Earth Mining Area. *Sci. Rep.* **2023**, *13*, 412. [CrossRef]
37. Xin, X.; Qin, S.; Zhang, J.; Zhu, A.; Yang, W.; Zhang, X. Yield, Phosphorus Use Efficiency and Balance Response to Substituting Long-Term Chemical Fertilizer Use with Organic Manure in a Wheat-Maize System. *Field Crops Res.* **2017**, *208*, 27–33. [CrossRef]
38. Lemanski, K.; Scheu, S. Incorporation of 13C Labelled Glucose into Soil Microorganisms of Grassland: Effects of Fertilizer Addition and Plant Functional Group Composition. *Soil Biol. Biochem.* **2014**, *69*, 38–45. [CrossRef]
39. Fan, W.; Yuan, J.; Wu, J.; Cai, H. Effects of Straw Maize on the Bacterial Community and Carbon Stability at Different Soil Depths. *Agriculture* **2023**, *13*, 1307. [CrossRef]
40. Yang, R.; Liu, K.; Geng, S.; Zhang, C.; Yin, L.; Wang, X. Comparison of Early Season Crop Types for Wheat Production and Nitrogen Use Efficiency in the Jianghan Plain in China. *PeerJ* **2021**, *9*, e11189. [CrossRef]
41. Yang, R.; Wang, Z.; Fahad, S.; Geng, S.; Zhang, C.; Harrison, M.T.; Adnan, M.; Saud, S.; Zhou, M.; Liu, K.; et al. Rice Paddies Reduce Subsequent Yields of Wheat Due to Physical and Chemical Soil Constraints. *Front. Plant Sci.* **2022**, *13*, 959784. [CrossRef]
42. Qi, D.; Pan, C. Responses of Shoot Biomass Accumulation, Distribution, and Nitrogen Use Efficiency of Maize to Nitrogen Application Rates under Waterlogging. *Agric. Water Manag.* **2022**, *261*, 107352. [CrossRef]
43. Dai, Y.; Shi, J.; Li, J.; Gao, Y.; Ma, H.; Wang, Y.; Wang, B.; Chen, J.; Cheng, P.; Ma, H. Transfer of the Resistance to Multiple Diseases from a Triticum-Secale-Thinopyrum Trigeneric Hybrid to Ningmai 13 and Yangmai 23 Wheat Using Specific Molecular Markers and GISH. *Genes* **2022**, *13*, 2345. [CrossRef] [PubMed]
44. Gao, R.; Ai, N.; Liu, G.; Liu, C.; Zhang, Z. Soil C:N:P Stoichiometric Characteristics and Soil Quality Evaluation under Different Restoration Modes in the Loess Region of Northern Shaanxi Province. *Forests* **2022**, *13*, 913. [CrossRef]
45. Zhao, X.; Wang, H.; Lu, D.; Chen, X.; Zhou, J. The Effects of Straw Return on Potassium Fertilization Rate and Time in the Rice–Wheat Rotation. *Soil Sci. Plant Nutr.* **2019**, *65*, 176–182. [CrossRef]
46. de Mendiburu, F.; Yaseen, M. *Agricolae: Statistical Procedures for Agricultural Research*; R Foundation for Statistical Computing: Vienna, Austria, 2000; pp. 3–19.
47. R Core Team. *R: A Language and Environment for Statistical Computing*; R Foundation for Statistical Computing: Vienna, Austria, 2020. Available online: https://www.R-project.org/ (accessed on 25 August 2023).
48. Kolde, R. *Pheatmap: Pretty Heatmaps*; R Foundation for Statistical Computing: Vienna, Austria, 2012.
49. Zhang, W.-Z.; Chen, X.-Q.; Wang, H.-Y.; Wei, W.-X.; Zhou, J.-M. Long-Term Straw Return Influenced Ammonium Ion Retention at the Soil Aggregate Scale in an Anthrosol with Rice-Wheat Rotations in China. *J. Integr. Agric.* **2022**, *21*, 521–531. [CrossRef]
50. Zhang, Z.; Liu, D.; Wu, M.; Xia, Y.; Zhang, F.; Fan, X. Long-Term Straw Returning Improve Soil K Balance and Potassium Supplying Ability under Rice and Wheat Cultivation. *Sci. Rep.* **2021**, *11*, 22260. [CrossRef] [PubMed]
51. Robertson, G.P.; Vitousek, P.M. Nitrogen in Agriculture: Balancing the Cost of an Essential Resource. *Annu. Rev. Environ. Resour.* **2009**, *34*, 97–125. [CrossRef]
52. Song, D.; Xi, X.; Zheng, Q.; Liang, G.; Zhou, W.; Wang, X. Soil Nutrient and Microbial Activity Responses to Two Years after Maize Straw Biochar Application in a Calcareous Soil. *Ecotoxicol. Environ. Saf.* **2019**, *180*, 348–356. [CrossRef]

Article

Conservation Tillage and Weed Management Influencing Weed Dynamics, Crop Performance, Soil Properties, and Profitability in a Rice–Wheat–Greengram System in the Eastern Indo-Gangetic Plain

Bushra Ahmed Alhammad [1], Dhirendra Kumar Roy [2,*], Shivani Ranjan [2], Smruti Ranjan Padhan [3], Sumit Sow [2], Dibyajyoti Nath [4], Mahmoud F. Seleiman [5,6,*] and Harun Gitari [7]

1 Biology Department, College of Science and Humanity Studies, Prince Sattam Bin Abdulaziz University, Al Kharj Box 292, Riyadh 11942, Saudi Arabia
2 Department of Agronomy, Dr. Rajendra Prasad Central Agricultural University, Pusa, Samastipur 848125, India
3 Division of Agronomy, ICAR-Indian Agricultural Research Institute, Pusa Campus, New Delhi 110012, India
4 Department of Soil Science, Dr. Rajendra Prasad Central Agricultural University, Pusa, Samastipur 848125, India
5 Plant Production, College of Food and Agriculture Sciences, King Saud University, Riyadh 11451, Saudi Arabia
6 Department of Crop Sciences, Faculty of Agriculture, Menoufia University, Shibin El-Kom 32514, Egypt
7 Department of Agricultural Science and Technology, School of Agriculture and Environmental Sciences, Kenyatta University, Nairobi P.O. Box 43844-00100, Kenya
* Correspondence: dr_dhirendra_krroy@yahoo.com (D.K.R.); mseleiman@ksu.edu.sa (M.F.S.)

Citation: Alhammad, B.A.; Roy, D.K.; Ranjan, S.; Padhan, S.R.; Sow, S.; Nath, D.; Seleiman, M.F.; Gitari, H. Conservation Tillage and Weed Management Influencing Weed Dynamics, Crop Performance, Soil Properties, and Profitability in a Rice–Wheat–Greengram System in the Eastern Indo-Gangetic Plain. *Agronomy* 2023, 13, 1953. https://doi.org/10.3390/agronomy13071953

Academic Editors: Ewa Szpunar-Krok, Mariola Staniak and Małgorzata Szostek

Received: 24 June 2023
Revised: 14 July 2023
Accepted: 20 July 2023
Published: 24 July 2023

Abstract: A three-year field experiment was carried out to assess the efficacy of various tillage and residue management practices, as well as weed management approaches, in a rice–wheat–green gram rotation. The treatments included: conventional till transplanted rice–conventional till wheat–fallow (T_1); conventional till transplanted rice–zero-till wheat–zero-till green gram (T_2); conventional till direct-seeded rice—conventional-till wheat—zero-till green gram (T_3); zero-till direct-seeded rice—zero-till wheat—zero-till green gram (T_4); zero-till direct-seeded rice + residue zero-till wheat + residue zero-till green gram (T_5). In weed management, three treatments are as follows: recommended herbicides (W_1); integrated weed management (W_2); and unweeded (W_3). The integrated weed management treatment had the lowest weed biomass, which was 44.3, 45.3, and 33.7% lower than the treatment W_3 at 30 and 60 days after sowing and harvest, respectively. T_1 grain and straw yielded more than T_2 in the early years than in subsequent years. The conventional till transplanted rice–zero-till wheat–zero-till green gram system produced 33.6, 37.6, and 27.7% greater net returns than the zero-till direct-seeded rice—zero-till wheat—zero-till greengram system, respectively. Conventional till transplanted rice–conventional till wheat–fallow had the biggest reduction (0.41%) in soil organic carbon from the initial value. The findings of the study demonstrated that adopting the transplanting method for rice, followed by zero tillage for wheat and green gram, enhanced productivity and profitability, while simultaneously preserving soil health.

Keywords: productivity; profitability; rice–wheat–green gram; soil health; tillage

1. Introduction

The rice–wheat cropping system, which spans roughly 14 million hectares in South Asia's Indo-Gangetic Plain (IGP) [1], has several obstacles that have hampered its efficacy. South Asia has been dealing with difficulties such as diminishing soil health [2], groundwater depletion [3], growing climatic variability [4], air pollution because of residue burning [5], and shifting socioeconomic conditions. These difficulties have had a substantial influence on the region's rice–wheat farming systems.

The Eastern Indo-Gangetic Plain (EIGP) is primarily made up of marginal farmers, who account for roughly 90% of the population, with a per capita income per household per year of INR 62,631 (USD 835), which is significantly lower compared to the county's average of INR 94,130 (USD 1256) [6]. As the population in Eastern India grows, there is an urgent need to raise agricultural intensity to satisfy the region's food and nutritional needs [7]. Initiatives have been launched to kickstart a second Green Revolution in Eastern India to achieve food security. However, the rice–wheat cropping system (RWCS) on the EIGP has many difficulties. Long-duration paddy types dominate most rice fields in the region, resulting in late transplanting and delayed harvesting. As a result of the delayed planting of wheat in the rice–wheat cropping system, yields have been lowered and grain quality has been affected due to heat stress during the grain-filling stage [8–10]. Furthermore, the management of rice residues, which are frequently left loose and scattered after harvest, is a substantial impediment since they interfere with tillage operations and the sowing of the next wheat crop [11]. Due to its cost-effectiveness, burning the residual rice and wheat residues is a widespread practice among local farmers in the EIGP region [12]. However, this burning process results in substantial nutrient losses, including 5.5 kg of N, 2.3 kg of P, 25 kg of K, and 1.2 kg of S, as well as organic carbon [13]. Furthermore, according to Jain et al. [4], agricultural residue burning adds to air pollution with emissions of 8.77 Mt CO, 0.23 Mt NO, 141.15 Mt CO_2, and 0.12 Mt NH_3. A considerable fraction of the nutrients found in crop residues is lost by gas and particle emissions, including 25 percent phosphorus, 80 to 90 percent nitrogen, 50 percent sulfur, and 20 percent potassium. These emissions, coupled with carbonaceous matter, considerably contribute to air pollution and global warming [14]. Greenhouse gas emissions (6266 Gg per year) can be lowered and soil health enhanced by minimizing agricultural residue burning and integrating residues into the soil [15]. As a result, an alternative production system that addresses these challenges by lowering production costs, conserving natural resources, reducing labor and time requirements, effectively controlling weeds, increasing productivity, and protecting the environment is urgently needed [16].

Conservation tillage is being adopted by an increasing number of farmers in South Asia as an alternate way to address rising difficulties. CA entails reducing or eliminating soil disturbance and keeping agricultural leftovers on the soil surface [17]. This transition towards CA technologies not only enhances production and revenue, but also addresses many challenges, such as restricted land size, diminishing agricultural output, growing cultivation expenses, farming risk, and the issue of climate change [18]. These difficulties represent serious concerns for livelihood security, especially for small-scale farmers. Zero tillage with residue retention has yielded excellent results, including a 5.8% increase in yields, a 25.9% increase in net income, and a 12.33% decrease in global warming potential [18,19]. While large-scale mechanized farms in the Americas and Australia have effectively embraced CA systems [20,21], smallholder farmers' adoption has been slower [22]. Furthermore, different regions' CA practices and cropping systems differ from those in the Eastern Indo-Gangetic Plain (EIGP). Due to the limited availability of resources such as sowing tools and pesticides, as well as traditional agricultural ideas, the broad adoption of zero tillage (ZT) in the EIGP has been hampered [23,24]. Although some CA features, such as ZT and residue retention, have been largely implemented in various crops, there is still a long way to go [25]. Various problems must be solved to encourage the broad adoption of CA systems within farming communities, with specific concerns varying based on the local situation.

Soil microorganisms are important in sustaining soil ecosystem health because they regulate several biochemical cycles and contribute to overall soil quality. Conservation agricultural practices have a large influence on soil microbial populations, making them useful markers of soil health and ecosystem resilience. Phosphate-solubilizing microorganisms (PSMs) are particularly essential among these beneficial microbial groups because they have the potential to hydrolyze phosphorus, boosting plant nutrition and enriching the soil [26,27]. However, both the physicochemical soil properties and agronomic practices impact the availability and activity of PSMs [28]. Factors such as soil pH, poor soil structure, low levels of soil organic carbon (SOC), and fluctuations in nitrogen, phosphorus, and potassium availability can limit the presence and functioning of PSMs [29,30]. In contrast to traditional tillage, zero tillage with residue retention reduces soil disturbance, which helps soil microbial populations in a variety of ways. It encourages microbial variety, increases SOC accumulation, preserves fungal hyphae, maintains soil food webs, and creates specialized microsites and microbial niches, all of which stimulate microbial activity and proliferation [29,31]. Conservation agricultural practices generate favorable circumstances for microbial communities to flourish by protecting the integrity of the soil structure and retaining organic wastes on the soil surface.

The successful implementation of conservation tillage in the rice (*Oryza sativa*)–wheat (*Triticum aestivum*)–green gram (*Vigna radiata*) cropping system has become dependent on the deployment of integrated weed management [32]. Tillage procedures that uproot, disturb, and bury weeds deep in the soil, limiting their emergence, are used in conventional tillage practices to accomplish effective weed management [14]. Weed seeds, on the other hand, tend to collect on the soil surface under conservation tillage systems that minimize or eliminate tillage, leading to increased weed development. In California, lower tillage intensity and frequency correlate to increased weed infestation levels. Furthermore, shifting from conventional to conservation-based farming might cause a shift in the composition of weed species within the crop field [33]. Furthermore, crop residues on the soil surface might intercept and bind herbicides, decreasing their ability to reach the soil surface. Consequently, the use of post-emergence herbicides in conservation tillage has become critical for weed control. Zero tillage also avoids bringing weed seed back up from the subsoil, residue cover impedes weed growth, and crop rotation reduces weeds, so conservation tillage, if well managed, can reduce weed problems in the medium to long term [34]. Furthermore, weed control practices have been widely encouraged across various tillage systems to boost soil fertility and crop production [35].

The majority of research in the Eastern Indo-Gangetic Plain (EIGP) has focused on zero-tillage practices for specific crops under the rice–wheat cropping system (RWCS). Thus, the major goal of this study was to assess the best tillage and weed control practices for increasing rice–wheat–green gram system productivity, soil fertility, and profitability. We hypothesized that the adoption of conservation tillage and weed management practices would result in improved crop yields, enhanced soil chemical and microbial properties, increased net income, and reduced weed infestation in the field.

2. Materials and Methods

2.1. Study Site

The experiment was conducted in an agricultural research center in Pusa, Bihar, India, with precise coordinates of 85° 48′ E longitude, 25° 59′ N latitude, and 52.92 m above mean sea level (Figure 1). This study was carried out between 2013 and 2016 as part of the Project Directorate on Weed Research's research program in Jabalpur. The trial lasted 36 months, with rice (*Oryza sativa* L.) grown during the rainy season (July–November), wheat (*Triticum aestivum* L.) grown from November to April, and green gram (*Vigna radiata* (L.) Wilczek) grown from April to July during the dry season. The climate in the study region is subtropical hot and humid, with a mean annual rainfall of 1210 mm. The majority of the rainfall, 75–80%, falls between July and September. The coldest temperature in January is around 5 °C, while the highest temperature in May is around 40.5 °C. During the research period, there were two instances of high rainfall: 255 mm in August 2014 and 316.2 mm in August 2015. The wet rice crop was lodged as a result of the severe rains in 2014. Figure 2 depicts the average weekly temperature, relative humidity, and monthly rainfall data obtained throughout the research from 2013–2014 to 2015–2016.

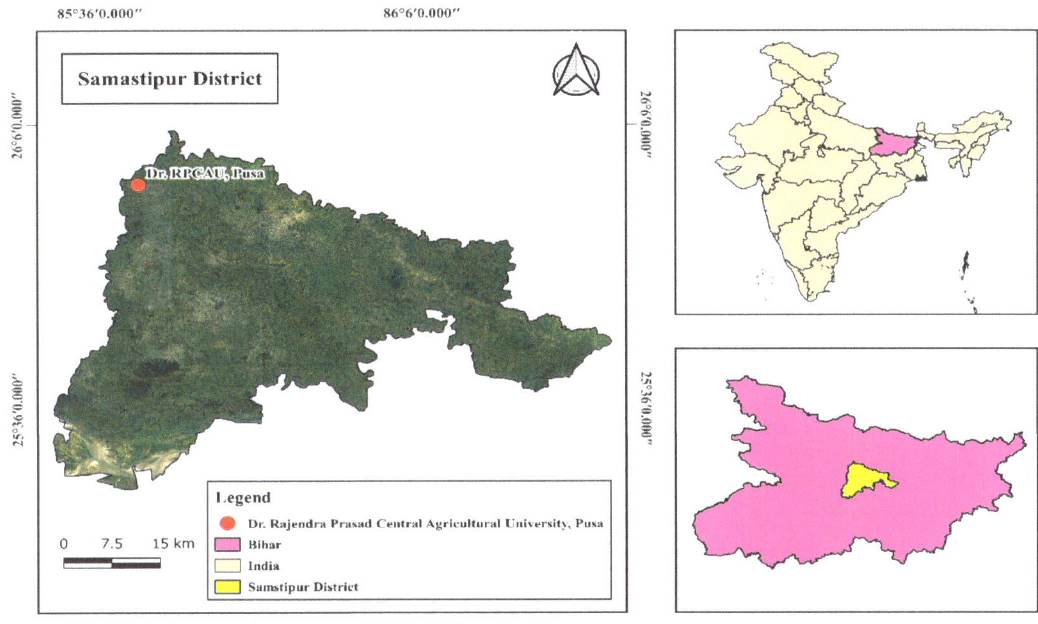

Figure 1. The geographical location map of the study area.

2.2. Weather Details

During the three years of the experiment (2013–2014 to 2015–2016), there were significant fluctuations in rainfall patterns, both in amount and distribution, raising concerns about rainfall uncertainty in the Eastern Indo-Gangetic Plain (EIGP). The mean maximum and minimum temperature, relative humidity, and rainfall received during the crop period are shown in Figure 2. Rainfall received from June to September accounted for 80–93% of the total yearly rainfall measured throughout the research period. During the rice season (July–November), the greatest reported rainfall was 811.7 mm in 2014, 160 mm in 2015, and 768.7 mm in 2016 (Figure 2). For the wheat season (November to April), the highest rainfall was 125.61 mm in 2013, followed by 44 mm in 2014, and 10.2 mm in 2015. In 2013–2014, 2014–2015, and 2015–2016, the average morning relative humidity was 81.48%, 87.32%, and 84.64%, respectively. During the same years, the average relative humidity in the evening was 50.42%, 55.36%, and 48% (Figure 2). Similarly, the mean weekly maximum and minimum temperatures for the summer green gram growth season varied between 36.8–40.7 °C and 21.1–23.2 °C, respectively, in all years. During the research years, the total rainfall was 157.2 mm (2014), 256 mm (2015), and 237.9 mm (2016).

2.3. Experimental Design and Treatment Details

The study employed a strip plot design, where each plot measured 20 m × 10 m. It consisted of a total of five tillage treatments in the main plots and three weed management treatments in subplots, replicated three times. The detailed treatment combinations are presented in Table 1.

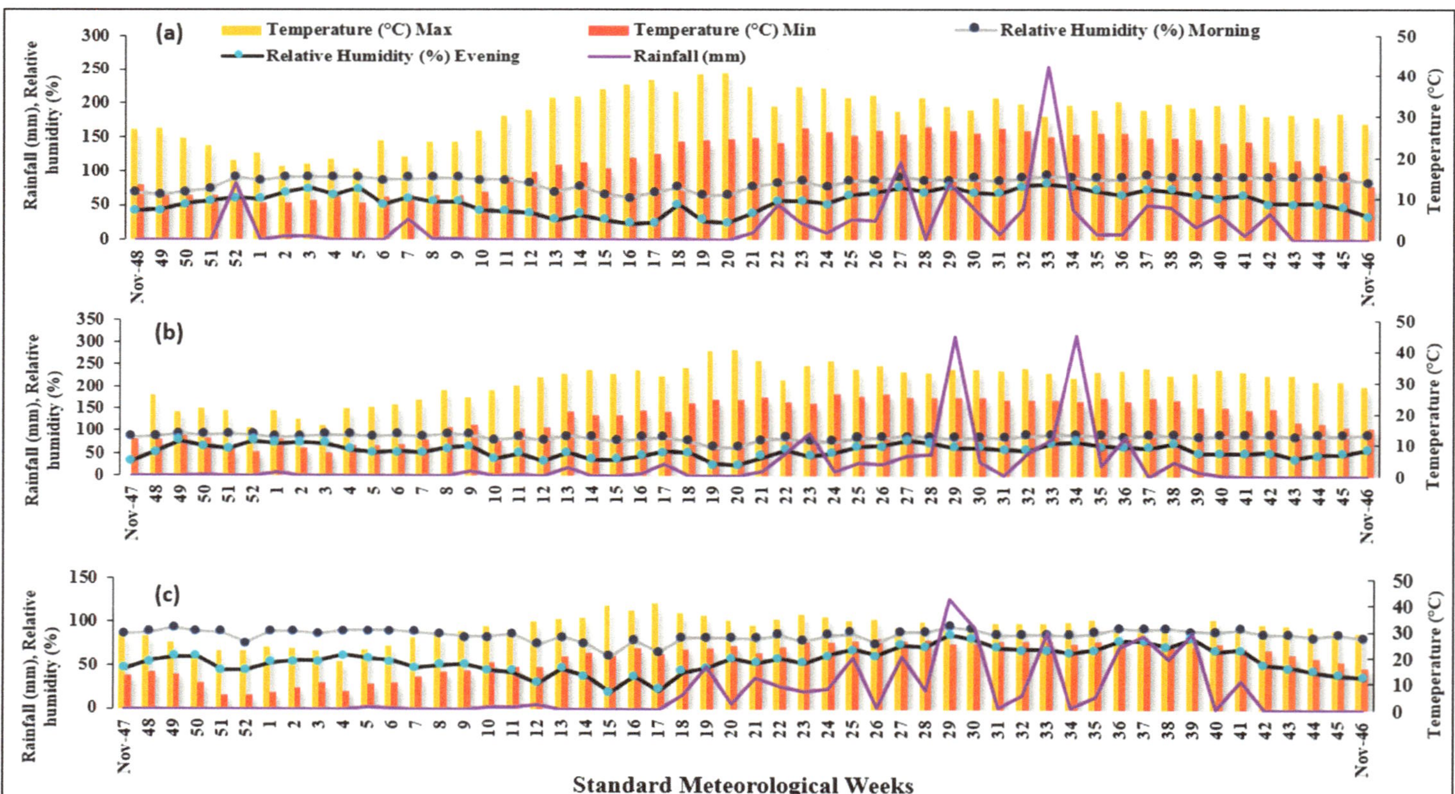

Figure 2. The mean weekly maximum temperature, minimum temperature, relative humidity in the morning, relative humidity in the evening, and rainfall for 2013–2014 (**a**), 2014–2015 (**b**), and 2015–2016 (**c**).

Table 1. Treatment details of the rice–wheat–green gram system.

Treatment	Rice	Wheat	Green Gram
Tillage and residue management			
T_1	CT (T)	CT	Fallow
T_2	CT (T)	ZT	ZT
T_3	CT (DS)	CT	ZT
T_4	ZT (DS)	ZT	ZT
T_5	ZT (DS) + R	ZT + R	ZT
Weed management			
W_1	Recommended herbicides		
W_2	Integrated weed management (IWM)		
W_3	Unweeded		

CT: Conventional tillage; ZT: Zero tillage; T: Transplanted; DS: Direct-seeded, R-Residue.

2.4. Crop Management

2.4.1. Rice

In the experiment, a medium-duration rice variety 'Rajendra Sweta (RAU710-99-22)' was used. For the direct-seeded rice (DSR) treatments, the rice seeds were sown using a Zero-Till Happy Seeder equipped with an inclined-plate seed metering system manufactured by Dasmesh, located in Malerkotla, Sangrur, Punjab, India. During the first to second week of July each year, sowing was performed in rows 22.5 cm apart. For all DSR treatments, the seed rate was set at 50 kg ha^{-1}, and a constant seeding depth of 3–4 cm was maintained using the seeder's depth control system. Nurseries were established in conventional transplanted rice (CTR) on the same day as DSR sowing, using the suggested package of practices outlined by Singh et al. [36]. The nurseries were established with a seed rate of 25 kg ha^{-1}. After 25 days, the resulting 2–3-week-old seedlings were manually transplanted into puddled fields. The transplantation process for CTR involved placing the seedlings at a spacing of 20 × 15 cm, with 2–3 seedlings per hill.

Rice was fertilized with a combination of urea and diammonium phosphate (DAP) at a rate of 120 kg of nitrogen (N) per hectare, along with 60 kg of phosphorous (P_2O_5) as DAP and 60 kg of potassium (K_2O) as muriate of potash (MOP), as described by Jat et al. [37]. During the sowing or transplanting stage, the entire amount of phosphorus and potassium fertilizers, as well as half of the recommended dose of nitrogen, were applied. The base dose of fertilizer was administered to transplanted rice during the final puddling phase, right before seedling transplantation. These fertilizers were drilled into the soil during the planting of direct-seeded rice (DSR) using the Zero-Till Happy Seeder. The remaining two-thirds of the nitrogen was supplied in two equal parts during the crop's mid-tillering and panicle initiation stages, as recommended by Kumar et al. [38]. To manage weeds, pre-sowing applications of pendimethalin at a rate of 1.0 kg a.i. ha^{-1} were applied in all the plots except the unweeded treatments. As per the recommended dose of herbicide treatments, pretilachlor was administered at a dosage of 0.75 kg a.i. ha^{-1} in the prescribed herbicide treatments at 20–25 days after sowing (DAS) in the CT/ZT/ZT + R-DSR treatments and 2–3 days after transplanting (DAT) in the CTR treatment. In the integrated weed management (IWM) treatments, two hand weeding sessions were performed, one at 40–45 DAS and another at 20–25 DAT, besides the pre-sowing application of pendimethalin. The Khurpi (trowel) was commonly used as a tool for weeding in the field.

Each herbicide was dissolved in 500 milliliters of water to make unique stock solutions for the herbicide solutions. Following that, the stock solutions were diluted with water to achieve a spray volume of 500 L per acre. A knapsack sprayer outfitted with a flat fan nozzle was used to apply the spray solution.

Weed density was assessed by counting weeds at three separate times: 30 days after sowing or transplanting (DAS/T), 60 DAS/T, and harvest. A 0.5 m × 0.5 m quadrate was set in each plot, except for the two boundary rows, and the number of weeds within the quadrate was counted. Weed biomass was measured by physically removing weeds from the sampling rows above ground with a sickle. The weeds were then sun-dried before being dried in a hot air oven at 60 °C until they reached a consistent dry weight.

To provide sufficient soil moisture for germination, direct-seeded rice (DSR) plots, including the ZT/ZT + R/CT treatments, were sown following the commencement of pre-monsoon showers. The land was prepped for transplanted rice by rotavator puddling with roughly 7 cm of water. Following the floodwater's receding, further irrigations were administered, and the irrigation schedule accounted for any rainfall occurrences. Depending on the amount and distribution of rainfall, the number of post-sowing irrigations ranged from 3 to 4. Various metrics were measured at regular intervals to determine rice growth and production. Plant height (cm) and effective tiller count (number per square meter) were measured at 30 and 60 DAS/T, as well as at harvest. Plant height was determined by randomly picking ten panicles from each plot. The number of effective tillers was obtained by counting them inside a 1 m^2 quadrate from four distinct sites within each plot and taking the average.

To assess the yield, a 10 m^2 area was allocated in the center of each plot for harvesting and measuring the grain and straw yields. In most treatments, the crop was picked manually with a sickle, around 15 cm above ground level. The crop was manually picked at a height of roughly 30 cm above ground level in the ZT + R treatments. The grain was sun-dried and manually threshed after harvest. To guarantee precise results, the grain yield was adjusted to a moisture content of 14%.

2.4.2. Wheat

Following the rice harvest, cultivation of the popular wheat variety 'HD 2967' was carried out. Wheat was planted in the fourth week of November. Traditional tillage practices, as reported by Singh et al. [36], were used in conventional tillage (CT) plots. In contrast, the zero-till (ZT) and zero-till with residue (ZT + R) plots were seeded directly onto the rice crop residue without any tillage, as shown in Table 1. The wheat plots, whether CT or ZT/ZT + R, were drilled with 22.5 cm rows using a Happy Seeder outfitted with an inclined-plate seed metering system. In all treatments, a seed rate of 100 kg/ha was used, and the seeds were uniformly planted at a depth of around 5 cm. Following the directions provided by Jat et al. [39], the wheat crop obtained 120 kg of nitrogen (N) in the form of urea and diammonium phosphate (DAP), 60 kg of phosphorous pentoxide (P$_2$O$_5$) as DAP, and 60 kg of potassium oxide (K$_2$O) as muriate of potash (MOP).

Using the Zero-Till Happy Seeder, the full amount of phosphorus and potash, as well as half of the recommended dose of nitrogen, were administered during wheat crop planting. The remaining two-thirds of nitrogen was applied in two equal treatments during the crop's crown root initiation (CRI) and maximum tillering stages. Pre-sowing irrigation was used to establish the wheat crop, followed by four further irrigations at important growth stages: CRI, tillering, blooming, and grain filling. Each irrigation required the application of around 5 cm^3 of water. Except for the unweeded treatments, weed control methods were applied, including the common spraying of glyphosate at a rate of 1.0 kg a.i. ha^{-1} before planting. A ready-mix solution of sulfosulfuron (75% WG) + metsulfuron-methyl (5% WG) at a combined rate of 32 (30 + 2) g a.i. ha^{-1} was treated 25 days after sowing (DAS) in the recommended herbicide treatments. In the integrated weed management (IWM) treatments, fenoxaprop ethyl 100 g a.i. ha^{-1} was applied at 25 DAS, and one manual weeding session was conducted at 40–45 DAS.

Several growth and yield characteristics of wheat were recorded at the maturity stage, including the number of effective tillers per square meter, grains per earhead, and test weight (1000-grain weight) at 12% moisture content. Ten spikes were randomly picked from each plot to assess plant height, number of earheads per square meter, number of grains per earhead, and test weight (in grams). Every year, the wheat harvest takes place in the latter week of April. A specified area of 5 m × 2 m (10 m^2) positioned in the center of each plot was harvested to measure grain and straw yield. Manual harvesting using a sickle was carried out in all treatments, with the plants being cut approximately 15 cm above ground level. Subsequently, the harvested wheat underwent manual threshing after appropriate sun drying. The recorded grain yield was adjusted to a moisture content of 12%.

2.4.3. Green Gram

The cultivation of a short-duration green gram cultivar, 'SML 668', developed by Punjab Agricultural University in Ludhiana, Punjab, India, was carried out during the study. Following the harvest of wheat, pre-sowing irrigation was applied to all plots, and green gram seeds were sown without any tillage into the wheat crop residue during the final week of April. The sowing process involved using a Zero-till Happy Seeder equipped with an inclined-plate seed metering system, with rows spaced 22.5 cm apart. A seed rate of 20 kg per hectare was utilized, and the seeds were sown at a depth of approximately 5 cm in all treatments, with the seeder's depth control system ensuring uniformity. In all the plots, the pre-sowing application of paraquat herbicide was applied to manage existing weeds except for unweeded treatments. Furthermore, a pre-emergence herbicide, pendimethalin, was applied at a rate of 1 kg a.i. per hectare one day after seeding (except in the unweeded treatments) to control subsequent weed emergence. In the integrated weed management (IWM) treatments, an additional hand weeding session was conducted 20–25 days after sowing (DAS), besides the application of pendimethalin one day after seeding to further control weed growth.

The green gram crop received a fertilization treatment consisting of 18 kg of nitrogen (N) and 46 kg of phosphorus pentoxide (P_2O_5) per hectare, which was supplied through diammonium phosphate (DAP) fertilizer. The entire dose of nitrogen and phosphorus was applied at the time of sowing using the Happy Seeder. Two additional irrigations were provided to all plots, with one at 25 days after sowing (DAS) and another at 45 DAS. To determine the grain yield, a designated area measuring 5 m × 2 m (10 m^2) located at the center of each plot was harvested. The recorded grain yield was reported at a moisture content of 12%. Harvesting of the mature pods was undertaken manually using a sickle, ensuring a consistent cutting height of approximately 15 cm above ground level in all treatments, except for the ZT + R treatments, where the crop was manually harvested at a cutting height of approximately 30 cm above ground level.

For the study of weed seed banks, soil samples were collected before the sowing/transplanting of the crop and were consistently irrigated. The number of germinated weed seedlings was regularly recorded at three intervals: 15 days (first flush), 25 days (second flush), and 40 days (third flush).

2.5. Soil Sampling and Analysis

Table 2 presents the initial physicochemical properties of the soil at the study site in 2013, before wheat sowing. After the experiment in 2016, soil samples were collected from all plots at a depth of 0–15 cm following the harvest of rice. To ensure representative soil samples, "V"-shaped slices were created, and five random samples were collected. These samples were thoroughly mixed, and approximately 500 g of soil was taken for analysis of the parameters listed in Table 3.

2.6. Economic Analysis

An economic analysis was conducted to assess the rice–wheat–green gram cropping system during the cropping year. The analysis involved calculating various economic indicators using the prevailing market prices of inputs and outputs. The cost of cultivation was determined by considering variable costs, excluding land rent. This included expenses for seeds, fertilizers, pesticides, human labor, machinery, irrigation, and other relevant activities. Fixed costs were not taken into account in the analysis.

The labor cost associated with various field activities was assessed based on person-days per hectare, with 8 h being comparable to 1 person-day under Indian labor regulations. The labor cost was estimated by multiplying the labor used in all processes by the government-mandated minimum pay rate stated in the Minimum Pay Act of 1948.

Table 2. Initial physicochemical properties of the soil at the study site.

Parameter	Value	Method Used	Reference
Sand (%)	24.72		
Silt (%)	48.85	Bouyoucos hydrometer	Piper [40]
Clay (%)	25.77		
Texture	Silty clay loam	Textural diagram	Black [41]
Bulk density (g cm^{-3})	1.43	Core sampler	Black [41]
Soil pH (1:2.5 soil water suspension)	8.04	Potentiometric	Jackson [42]
Electrical conductivity (dS m^{-1})	0.48	Potentiometric	Jackson [42]
Organic carbon (%)	0.51	Walkley and Black's rapid titration method	Jackson [42]
Available nitrogen (N) (kg ha^{-1})	247.64	Alkaline KMnO$_4$	Subbiah and Asija [43]
Available phosphorus (P$_2$O$_5$) (kg ha^{-1})	36.38	Olsen's method	Olsen et al. [44]
Available potash (K$_2$O) (kg ha^{-1})	249.48	1 N neutral ammonium acetate method	Jackson [42]

Table 3. Methods used in soil sample analysis.

Parameter	Method Used	Reference
Soil pH (1:2.5 soil water suspension)	Potentiometric	Jackson [42]
Organic carbon (%)	Walkley and Black's rapid titration method	Jackson [42]
Available nitrogen	Alkaline KMnO$_4$	Subbiah and Asija [43]
Available phosphate	Olsen's method	Olsen et al. [44]
Available potash	0.01 N neutral ammonium acetate method	Jackson [42]
Azotobacter (10^4 cfu g^{-1} soil)		
Total Pseudomonas (10^5 cfu g^{-1} soil)		
Total PSB (10^5 cfu g^{-1} soil)		
% of P solubilized by Pseudomonas	-	Schmidt and Coldwell [45]
Bacillus (10^5 cfu g^{-1} soil)		
% of P solubilized by Bacillus		
CO$_2$ evolution (mg kg^{-1})		Zibilske [46]

Gross returns (GR) were determined by multiplying the grain yield of each crop (in tons per hectare) by the minimum support price (MSP) offered by the Government of India for the respective years of 2013–2014, 2014–2015, and 2015–2016. The value of the straw was calculated using prevailing local market rates.

Net returns (NR) were calculated as the difference between gross returns and the cost of cultivation (CC) (NR = GR − CC). The benefit-to-cost ratio (B:C ratio) was computed by dividing the gross returns by the cost of cultivation (B:C ratio = GR/CC).

For the economic analysis, exchange rates of 1 USD = INR 60.99 (in 2014), INR 64.13 (in 2015), and INR 67.18 (in 2016) were considered, based on the average exchange rate for the period of 2014–2016 (source: https://www.exchangerates.org.uk (accessed on 12 April 2023)).

2.7. Statistical Analysis

Before doing statistical analysis, weed population data were square root transformed ((x + 1)). The analysis was carried out using the CPCS-1 statistical program developed by Punjab Agricultural University, Ludhiana [47]. Analysis of variance (ANOVA) was carried out with a strip plot design on all data relating to weed dynamics, growth, yield characteristics, yield, and economics using the Statistix 8.1 statistical tool (Analytical Software, Tallahassee, FL, USA) [48]. The significance of the treatment effect was determined using an F-test at a 5% level of significance. The least significant difference (LSD) or critical difference (CD) approach was used to assess differences between treatment means [49].

3. Results

3.1. Weed Dynamics

3.1.1. Weed Seed Bank and Its Dynamics in Soil

The major weed species that emerged from the soil during the three flushes in rice during the grow-out tests conducted on soil samples from the permanent tillage trial before the wet season of 2016 (after three years of completing the trial) were *Echinochloa crusgalli*, *Leptochloa chinensis*, *Cyperus difformis*, *Ammania baccifera*, and *Dactyloctenium aegyptium*. In the first flush, second flush, and third flush, the number of weed seeds that emerged was higher in the ZT(DS)–ZT–ZT (rice–wheat–green gram) treatment compared to all other tillage treatments. The unweeded treatment exhibited the highest weed density, while the IWM (W_2) treatment recorded a lower weed density compared to the RDH (W_1) treatment in all three flushes (Figure 3a).

A grow-out test was undertaken on soil samples obtained from different soil depths under various treatments before wheat sowing during the dry season of 2015 (after three years of the study). The findings revealed that Phalaris minor, C. album, and M. indica were the most common weed species across all treatments. The ZT(DS)–ZT–ZT treatment had the greatest weed density, followed by ZT(DS) + R–ZT + R–ZT, CT(DS)–CT–ZT, CT(T)–ZT–ZT, and CT(T)–CT–fallow treatments. The weed density was uniform over all three flushes (Figure 3b). Likewise, the unweeded treatment had the highest weed density.

Another grow-out test was undertaken on soil samples collected from different soil depths under different treatments before green gram seeding in the summer of 2016 (after three years of the study). In all treatments, *Euphorbia hirta*, *Amaranthus viridis*, *Celosia argentena*, *Chloris barbata*, and *Trianthema portulacum* predominated. In all three flushes, the ZT(DS)–ZT–ZT treatment had the highest weed density. Among the weed management treatments, W_1 with ZT(DS)–ZT–ZT had the highest weed density in the first, second, and third flushes, with 29.6%, 33.3%, and 16.7%, respectively (Figure 3c).

3.1.2. Weed Density

The highest weed density was observed in the rice–wheat–green gram system when zero tillage was performed without any residue retention [ZT(DS)-ZT–ZT], followed by the T5, T3, and T1 tillage systems. The lowest weed density was observed in conventional tillage with transplanting in rice and zero tillage in wheat and green gram under the CT(T)–ZT–ZT tillage system (Table 4). Weed density studies were conducted 30 days after sowing/transplanting (DAS/T), 60 days after sowing/transplanting (DAS/T), and at harvest during all three cropping systems.

3.1.3. Weed Biomass

Over the three-year experimental period, conservation tillage strategies considerably decreased weed biomass in the rice–wheat–green gram system (Table 5). In rice, T_2 treatment (CT(T)–ZT–ZT) had the lowest weed dry biomass (averaged across three seasons) at 30 and 60 DAS/T and at harvest, which was statistically similar to T_1 treatment (CT(T)–CT–fallow) and significantly better than the other conservation tillage and residue management treatments. Similarly, during the three–year trial, the lowest weed biomass was reported at 30 and 60 DAS/T with T_1 treatment. At harvest, CT(T)–ZT–ZT (T_2) had the lowest weed biomass (averaged throughout 2013–2014, 2014–2015, and 2015–2016), with values of 6.04, 6.88, and 5.98 g m^{-2}, respectively. Additionally, T2 treatment exhibited the lowest weed biomass at 30 and 60 DAS as well as at harvest in green gram during all three years of the study.

Across all years, the unweeded treatment (W_3) consistently had the largest weed biomass at 30 and 60 DAS, as well as at harvest. In contrast, the treatment W_2 consistently had the lowest weed biomass, which was 44.3%, 45.3%, and 33.7% lower than the biomass of W_3 at 30 and 60 DAS, as well as at harvest. Throughout the three-year trial, this tendency was found in both wheat and green gram crops.

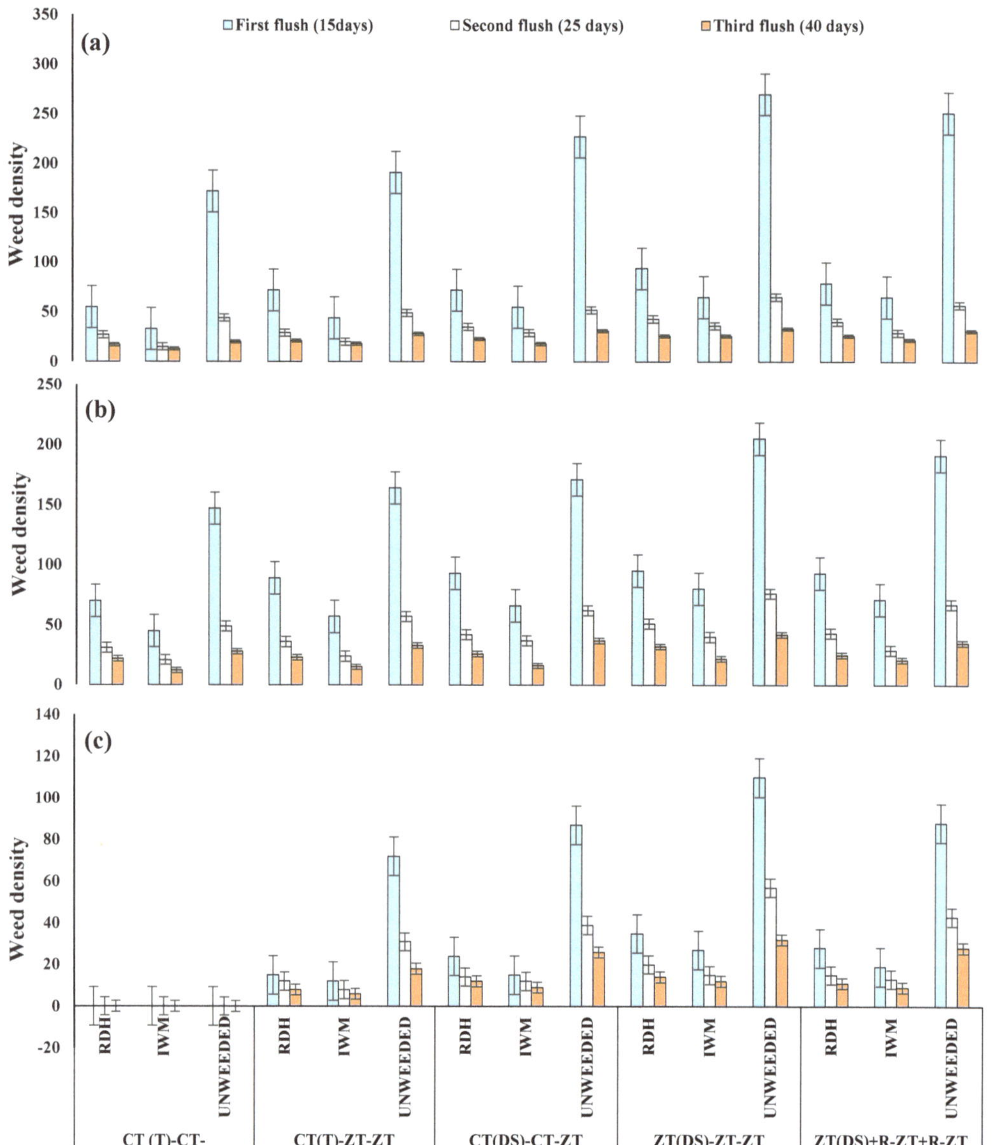

Figure 3. Weed seed emergence at different fluxes as affected by different tillage, residue management, and weed management methods in the wet season of 2016 (**a**), dry season of 2015 (**b**), and summer season of 2016 (**c**).

3.2. Crop Yield Attributes and Yield

3.2.1. Rice

Tillage and residue management had a substantial influence on yield characteristics and the yield of rice over the three-year experimental period, as shown in Table 6. The CT(DS)–CT–ZT treatment consistently had the highest number of effective tillers per square meter, with values of 44.71, 44.81, and 43.21 throughout the three years. Among the various weed control treatments, the IWM (W_2) treatment had the highest number of effective tillers per square meter in 2014 and 2015, which was substantially higher than the other treatments. It was equivalent to W_1 therapy in 2016. The number of effective tillers is an essential statistic since it shows plant development and vigor. It can be influenced by various factors such as tillage practices, residue management, herbicides, and their application timing.

Table 4. Weed density (number m^{-2}) for the years 2013–2014, 2014–2015, and 2015–2016 as affected by conservation tillage and different weed management practices.

Treatment *	Rice (Number m^{-2})			Wheat (Number m^{-2})			Green Gram (Number m^{-2})		
	30 DAS/T	60 DAS/T	At Harvest	30 DAS	60 DAS	At Harvest	20 DAS	40 DAS	At Harvest
				2013–2014					
			Tillage and residue management						
T_1	9.84 [c**]	13.12 [c]	8.16 [c]	6.45 [e]	24.50 [e]	6.57 [c]	-	-	-
T_2	8.96 [c]	11.86 [c]	7.49 [c]	7.93 [d]	27.96 [d]	6.04 [c]	6.58 [d]	35.06 [c]	7.81 [d]
T_3	15.12 [b]	20.81 [b]	12.85 [b]	11.26 [c]	31.68 [c]	11.23 [b]	9.09 [c]	44.34 [b]	10.63 [c]
T_4	17.65 [a]	25.65 [a]	14.46 [a]	13.92 [a]	40.79 [a]	13.58 [a]	11.72 [a]	49.94 [a]	12.92 [a]
T_5	15.85 [b]	22.05 [b]	13.58 [b]	12.21 [b]	35.44 [b]	11.07 [b]	10.35 [b]	46.76 [b]	11.58 [b]
			Weed management						
W_1	11.43 [b]	15.21 [b]	9.48 [b]	7.52 b	22.29 [b]	8.85 [b]	7.25 [b]	30.53 [b]	8.76 [b]
W_2	9.89 [b]	13.85 [b]	8.32 [b]	6.83 b	14.34 [c]	7.29 [c]	6.52 [c]	13.10 [c]	6.92 [c]
W_3	18.97 [a]	28.11 [a]	16.72 [a]	16.15 a	59.59 [a]	15.04 [a]	16.11 [a]	88.45 [a]	17.11 [a]
				2014–2015					
			Tillage and residue management						
T_1	10.23 [c]	12.65 [c]	9.41 [c]	7.62 [e]	23.68 [d]	7.12 [c]	-	-	-
T_2	9.31 [c]	10.45 [c]	8.25 [c]	9.05 [d]	26.75 [c]	6.88 [c]	7.25 [d]	33.85 [c]	6.21 [c]
T_3	17.84 [b]	21.67 [b]	10.47 [b]	12.47 [b]	31.39 [b]	10.85 [b]	11.38 [a]	43.21 [b]	11.35 [b]
T_4	18.17 [a]	27.04 [a]	12.68 [a]	14.03 [a]	40.86 [a]	12.50 [a]	10.07 [b]	48.34 [a]	13.42 [a]
T_5	17.07 [b]	22.95 [b]	11.42 [b]	11.19 [c]	31.57 [b]	10.62 [b]	9.23 [c]	44.74 [b]	12.69 [a]
			Weed management						
W_1	12.05 [b]	14.41 b	8.54 [b]	10.05 [b]	19.85 [b]	9.07 [b]	7.62 [b]	28.62 [b]	7.95 [b]
W_2	10.06 [b]	13.99 [b]	7.79 [b]	8.79 [c]	14.32 [c]	7.95 [c]	5.93 [c]	12.75 [c]	5.37 [c]
W_3	21.45 [a]	28.45 [a]	15.02 [a]	13.77 [a]	58.38 [a]	11.75 [a]	14.89 [a]	86.24 [a]	19.41 [a]
				2015–2016					
			Tillage and residue management						
T_1	9.65 [c]	12.65 [c]	8.25 [c]	6.41 [e]	23.61 [d]	6.73 [c]	-	-	-
T_2	8.72 [c]	10.45 [c]	7.44 [c]	8.02 [d]	25.43 [c]	5.98 [c]	6.39 [d]	33.37 [c]	7.65 [d]
T_3	14.98 [b]	21.67 [b]	12.69 [b]	11.24 [c]	30.13 [b]	11.35 [b]	8.93 [c]	43.00 [b]	10.59 [c]
T_4	17.61 [a]	27.04 [a]	14.32 [a]	14.05 [a]	41.04 [a]	13.48 [a]	11.65 [a]	47.23 [a]	12.81 [a]
T_5	15.72 [b]	22.95 [b]	13.55 [a]	12.18 [b]	30.21 [b]	10.98 [b]	10.21 [b]	43.26 [b]	11.55 [b]
			Weed management						
W_1	10.35 [b]	14.41 [b]	9.39 [b]	7.45 [b]	19.67 [b]	8.73 [b]	7.19 [b]	27.40 [b]	8.86 [b]
W_2	9.69 [b]	13.99 [b]	8.42 [b]	6.91 [b]	13.62 [c]	7.15 [c]	6.48 [b]	11.44 [c]	7.01 [c]
W_3	17.89 [a]	28.45 [a]	15.98 [a]	15.89 [a]	56.97 [a]	14.95 [a]	15.88 [a]	86.31 [a]	16.99 [a]
			Mean weed density of all the three years						
			Tillage and residue management						
T_1	9.90	12.80	8.60	6.82	23.93	6.80	6.74	34.09	7.22
T_2	8.99	10.92	7.72	8.33	26.71	6.30	9.80	43.51	10.85
T_3	15.98	21.38	12.00	11.65	31.06	11.14	11.14	48.50	13.05
T_4	17.81	26.57	13.82	14.00	40.89	13.18	9.93	44.92	11.94
T_5	16.21	22.65	12.85	11.86	32.40	10.89	6.74	34.09	7.22
			Weed management						
W_1	11.27	14.67	9.13	8.34	20.60	8.88	7.35	28.85	8.52
W_2	9.88	13.94	8.17	7.51	14.09	7.46	6.31	12.43	6.43
W_3	19.43	28.33	15.90	15.27	58.31	13.91	15.62	87.00	17.83

* Refer to Table 1 for treatment details. ** The means with similar letters down the column (per either tillage residue management or weed management) do not differ significantly at $p \leq 0.05$.

Table 5. Weed biomass ($g\ m^{-2}$) for the years 2013–2014, 2014–2015, and 2015–2016 as affected by conservation tillage and different weed management practices.

Treatment *	Rice ($g\ m^{-2}$)			Wheat ($g\ m^{-2}$)			Green Gram ($g\ m^{-2}$)		
	30 DAS/T	60 DAS/T	At Harvest	30 DAS	60 DAS	At Harvest	20 DAS	40 DAS	At Harvest
	2013–2014								
	Tillage and residue management								
T_1	9.84 [c**]	13.12 [c]	8.16 [c]	10.82 [d]	13.65 [d]	12.02 [c]	-	-	-
T_2	8.96 [c]	11.86 [c]	7.49 [c]	13.65 [c]	15.96 [c]	10.21 [d]	9.81 [d]	20.23 [c]	10.65 [d]
T_3	15.12 [b]	20.81 [b]	12.85 [b]	17.46 [b]	17.18 [b]	14.48 [b]	15.63 [c]	23.29 [b]	16.12 [c]
T_4	17.65 [a]	25.65 [a]	14.46 [a]	21.58 [a]	21.81 [a]	16.75 [a]	20.25 [a]	27.61 [a]	20.22 [a]
T_5	15.85 [b]	22.05 [b]	13.58 [b]	20.15 [a]	18.41 [b]	15.37 [b]	18.17 [b]	23.92 [b]	18.91 [b]
	Weed management								
W_1	17.05 [b]	30.18 [b]	13.42 [b]	13.25 [b]	11.77 [b]	12.46 [b]	11.67 [b]	13.29 [b]	11.72 [b]
W_2	15.82 [b]	27.23 [b]	12.65 [b]	12.68 [b]	7.73 [c]	11.07 [c]	9.81 [c]	5.70 [c]	9.65 [c]
W_3	28.11 [a]	52.35 [a]	18.62 [a]	23.96 [a]	32.52 [a]	16.37 [a]	22.18 [a]	52.31 [a]	21.11 [a]
	2014–2015								
	Tillage and residue management								
T_1	17.32 [c]	23.45 [c]	12.47 [b]	9.29 [d]	12.48 [d]	11.63 [c]	-	-	-
T_2	15.48 [c]	20.12 [c]	9.62 [b]	11.42 [c]	17.22 [c]	9.72 [d]	8.49 [d]	18.19 [d]	9.15 [d]
T_3	23.69 [b]	38.69 [b]	14.78 [a]	15.75 [b]	18.35 [b]	13.42 [b]	13.65 [c]	20.09 [c]	14.62 [c]
T_4	27.15 [a]	45.38 [a]	16.39 [a]	18.83 [a]	21.64 [a]	15.57 [a]	17.82 [a]	25.30 [a]	18.75 [a]
T_5	28.42 [a]	44.29 [a]	15.22 [a]	19.48 [a]	19.16 [b]	14.19 [b]	15.07 [b]	24.56 [b]	17.18 [b]
	Weed management								
W_1	19.47 [b]	29.62 [b]	19.47 [b]	11.47 [b]	12.32 [b]	11.12 [b]	10.63 [b]	12.48 [b]	10.74 [b]
W_2	16.88 [b]	26.45 [b]	16.88 [b]	9.85 [c]	8.94 [c]	9.83 [c]	8.45 [c]	4.82 [c]	8.36 [c]
W_3	30.88 [a]	47.10 [a]	30.88 [a]	22.53 [a]	32.05 [a]	17.78 [a]	22.17 [a]	49.35 [a]	25.66 [a]
	2015–2016								
	Tillage and residue management								
T_1	9.65 [c]	12.65 [c]	8.25 [c]	6.41 [e]	23.61 [d]	6.73 [c]	-	-	-
T_2	8.72 [c]	10.45 [c]	7.44 [c]	8.02 [d]	25.43 [c]	5.98 [c]	6.39 [d]	33.37 [c]	7.65 [d]
T_3	14.98 [b]	21.67 [b]	12.69 [b]	11.24 [c]	30.13 [b]	11.35 [b]	8.93 [c]	43.00 [b]	10.59 [c]
T_4	17.61 [a]	27.04 [a]	14.32 [a]	14.05 [a]	41.04 [a]	13.48 [a]	11.65 [a]	47.23 [a]	12.81 [a]
T_5	15.72 [b]	22.95 [b]	13.55 [a]	12.18 [b]	30.21 [b]	10.98 [b]	10.21 [b]	43.26 [b]	11.55 [b]
	Weed management								
W_1	16.88 [b]	29.62 [b]	13.62 [b]	12.98 [b]	12.98 [b]	11.98 [b]	11.72 [b]	12.48 [b]	11.69 [b]
W_2	15.69 [b]	26.45 [b]	12.78 [b]	12.41 [b]	9.33 [c]	10.87 [b]	9.72 [c]	5.07 [c]	9.56 [c]
W_3	27.85 [a]	47.10 [a]	17.98 [a]	22.69 [a]	31.71 [a]	15.95 [a]	21.36 [a]	48.89 [a]	20.88 [a]
	Mean weed biomass across the three years								
	Tillage and residue management								
T_1	12.27	16.40	9.62	8.84	16.58	10.12	-	-	-
T_2	11.05	14.14	8.18	11.03	19.53	8.63	8.23	23.93	9.15
T_3	17.93	27.05	13.44	14.81	21.88	13.08	12.73	28.79	13.77
T_4	20.80	32.69	15.05	18.15	28.16	15.26	16.57	33.38	17.26
T_5	19.9	29.76	14.11	17.27	22.59	13.51	14.48	30.58	15.88
	Weed management								
W_1	17.80	29.80	15.50	12.56	12.35	11.85	11.34	12.75	11.38
W_2	16.13	26.71	14.10	11.64	8.66	10.59	9.32	5.19	9.19
W_3	28.94	48.85	22.49	23.06	32.09	16.70	21.90	50.18	22.55

* Refer to Table 1 for treatment details. ** The means with similar letters down the column (per either tillage residue management or weed management) do not differ significantly at $p \leq 0.05$.

However, the T_1 treatment, which represents CT(T)–CT–fallow, consistently produced the maximum grain yield of rice (4.76, 4.81, and 4.73 tonnes per hectare) and straw yield (6.37, 6.40, and 6.40 tonnes per hectare). In terms of grain and straw yields, this treatment was statistically equivalent to the T_2 treatment, CT(T)–ZT–ZT, in 2014, 2015, and 2016, respectively. The T_4 treatment, on the other hand, resulted in the lowest rice grain production. The lowest grain and straw yields were obtained in the unweeded plots, where no weed control techniques were employed, among the various weed management treatments. The plots that received integrated weed management (IWM) followed by W_1 treatment showed the highest grain yields (4.65, 4.66, and 4.58 tons per hectare). These treatments exhibited a significant increase of 34.9%, 38%, 32.1%, and 13.6% in grain and straw yields, respectively, compared to the unweeded plots (Table 6).

Table 6. Yield attributes and yield of rice as influenced by conservation tillage and different weed management practices.

Treatments *	2014			2015			2016			Mean of All the Three Years		
	Effective Tillers (No m^{-2})	Grain Yield (t ha^{-1})	Straw Yield (t ha^{-1})	Effective Tillers (No m^{-2})	Grain Yield (t ha^{-1})	Straw Yield (t ha^{-1})	Effective Tillers (No m^{-2})	Grain Yield (t ha^{-1})	Straw Yield (t ha^{-1})	Effective Tillers (No m^{-2})	Grain Yield (t ha^{-1})	Straw Yield (t ha^{-1})
Tillage and residue management												
T_1	32.50 [b**]	4.76 [a]	6.37 [a]	32.58 [d]	4.81 [a]	6.40 [a]	34.98 [b]	4.73 [a]	6.40 [a]	33.35	4.76	6.39
T_2	37.35 [b]	4.66 [b]	6.07 [b]	37.84 [c]	4.69 [a]	6.12 [a]	39.54 [a]	4.62 [a]	6.15 [b]	38.24	4.65	6.11
T_3	44.71 [a]	4.04 [c]	5.66 [c]	44.81 [a]	4.05 [b]	5.69 [b]	43.21 [a]	4.02 [b]	5.63 [c]	44.24	4.03	5.66
T_4	37.09 [b]	3.49 [d]	4.81 [e]	37.12 [c]	3.48 [c]	4.80 [c]	39.63 [a]	3.49 [c]	4.82 [d]	37.94	3.48	4.81
T_5	41.60 [a]	4.03 [c]	4.92 [d]	41.84 [b]	4.05 [b]	4.93 [c]	41.28 [a]	4.02 [b]	4.95 [d]	41.57	4.03	4.93
Weed management												
W_1	38.36 [b]	4.52 [a]	5.39 [a]	38.80 [b]	4.55 [b]	5.41 [b]	40.82 [a]	4.52 [a]	5.44 [b]	39.32	4.53	5.41
W_2	44.13 [a]	4.65 [a]	6.56 [a]	44.20 [a]	4.66 [a]	6.60 [a]	43.92 [a]	4.58 [a]	6.56 [a]	44.08	4.63	6.57
W_3	33.46 [c]	3.43 [b]	4.75 [b]	33.50 [c]	3.44 [c]	4.76 [c]	34.44 [b]	3.43 [b]	4.77 [c]	33.80	3.43	4.76

* Refer to Table 1 for treatment details. ** The means with similar letters down the column (per either tillage residue management and Weed management) do not differ significantly at $p \leq 0.05$.

3.2.2. Wheat

The yield attributes of wheat were significantly influenced by conservation tillage and weed management treatments during all the years of the study (Tables 7 and 8). The number of earheads per square meter and grains per earhead were found to be the lowest in the T_4 treatment, with values that were 8.3% and 4.7% lower, respectively, compared to the T_1 treatment in the 2013–2014 season. This trend was consistently observed in the following two years of the study. Test weight, which serves as an indicator of grain quality, was highest in the T_1 treatment, with values of 47.68 g, 46.84 g, and 48.09 g in the three years of the study, respectively. Among the weed management treatments, the W_2 treatment had the highest yield qualities, including grains per earhead, the number of earheads per square meter, and test weight, whereas the unweeded plot (W_3) had the lowest values throughout all years (Table 7). The improved yield attributes in the weed control treatments may be attributed to the better photosynthetic efficiency of the crop, which allowed for effective weed control without causing crop damage. The weedy check plot had the lowest production characteristics due to intense crop–weed competition during the growing period. When compared to treatments with poor weed control, integrated weed management (IWM) treatments showed superior weed control and were able to obtain higher values for earheads per square meter and grains per earhead. Wheat crops with better weed control exhibited superior yield attributes due to reduced weed density. The availability of ample space, light, and nutrients for optimal crop growth and development, along with minimal interspecies competition, facilitated by the conventional tillage method in wheat, contributed to the superior yield attributes observed.

Table 7. Yield attributes and yield of wheat as affected by conservation tillage and different weed management practices.

Treatment *	2013–2014					2014–2015					2015–2016				
	Earhead (No m^{-2})	Grains Earhead^{-1}	Test Weight (g)	Grain Yield (t ha^{-1})	Straw Yield (t ha^{-1})	Earhead (No m^{-2})	Grains Earhead^{-1}	Test Weight (g)	Grain Yield (t ha^{-1})	Straw Yield (t ha^{-1})	Earhead (No m^{-2})	Grains Earhead^{-1}	Test Weight (g)	Grain Yield (t ha^{-1})	Straw Yield (t ha^{-1})
					Tillage and residue management										
T$_1$	314.85 [a**]	42.95 [b]	47.68 [a]	4.65 [a]	5.42 [a]	309.19 [a]	41.62 [b]	46.84 [a]	4.73 [a]	5.24 [a]	314.92 [a]	43.79 [a]	48.09 [a]	4.72 [a]	5.49 [a]
T$_2$	313.41 [a]	44.35 [a]	46.33 [a]	4.49 [b]	5.21 [b]	315.43 [a]	42.96 [a]	45.98 [b]	4.46 [b]	5.05 [b]	314.15 [a]	44.22 [a]	46.70 [a]	4.44 [b]	5.22 [b]
T$_3$	311.77 [a]	42.35 [b]	45.02 [a]	4.42 [b]	5.19 [b]	308.38 [a]	40.05 [b]	43.44 [c]	4.39 [b]	4.98 [b]	312.63 [a]	42.79 [b]	45.44 [a]	4.36 [c]	5.23 [b]
T$_4$	288.90 [b]	40.90 [c]	41.36 [b]	4.02 [d]	4.76 [d]	299.43 [b]	39.08 [c]	41.40 [d]	3.99 [c]	4.63 [c]	290.25 [b]	41.11 [c]	41.61 [b]	4.16 [e]	4.79 [d]
T$_5$	306.28 [a]	41.20 [c]	43.75 [a]	4.20 [c]	4.99 [c]	308.02 [a]	38.59 [c]	43.88 [c]	4.14 [c]	4.76 [c]	307.02 [a]	41.71 [c]	43.97 [a]	4.24 [d]	5.01 [c]
					Weed management										
W$_1$	311.46 [a]	43.51 [a]	45.14 [a]	4.68 [b]	5.58 [b]	317.11 [a]	40.77 [b]	44.17 [b]	4.75 [b]	5.47 [a]	312.93 [a]	43.63 [a]	45.32 [a]	4.74 [a]	5.64 [a]
W$_2$	311.20 [a]	44.50 [a]	47.96 [a]	4.89 [a]	5.78 [a]	315.86 [a]	42.58 [a]	47.74 [a]	4.91 [a]	5.48 [a]	310.99 [a]	44.97 [a]	48.50 [a]	4.91 [a]	5.83 [a]
W$_3$	298.46 [a]	39.05 [b]	41.38 [b]	3.49 [c]	3.99 [c]	291.29 [b]	38.04 [c]	41.02 [c]	3.37 [c]	3.84 [b]	299.46 [b]	39.58 [b]	41.66 [b]	3.50 [b]	3.98 [b]

* Refer to Table 1 for treatment details. ** The means with similar letters down the column (per either tillage residue management and weed management) do not differ significantly at $p \leq 0.05$.

Table 8. Mean of yield attributes and yield of wheat for the three years as influenced by conservation tillage and different weed management practices.

Treatment *	Earhead (No m^{-2})	Grains Earhead^{-1}	Test Weight (g)	Grain Yield (t ha^{-1})	Straw Yield (t ha^{-1})
		Tillage and residue management			
T$_1$	312.98	42.78	47.53	4.70	5.38
T$_2$	314.33	43.84	46.33	4.46	5.16
T$_3$	310.92	41.73	44.63	4.39	5.13
T$_4$	292.86	40.36	41.45	4.05	4.72
T$_5$	307.10	40.50	43.86	4.19	4.92
		Weed management			
W$_1$	313.83	42.63	44.87	4.72	5.56
W$_2$	312.68	44.01	48.06	4.90	5.69
W$_3$	296.40	38.89	41.35	3.45	3.93

* Refer to Table 1 for treatment details.

The T_1 treatment consistently produced the highest grain production (4.65 t/ha, 4.73 t/ha, and 4.72 t/ha averaged over three years), which was considerably superior to the other treatments. The T_4 treatment, on the other hand, had the lowest grain yield. This tendency was mirrored in straw production, with T_1 yielding 13.9% and 9.3% more straw (averaged over three years) than T_4 and T_5, respectively (Table 7). The lower yields in the zero-tillage treatments can be attributed to a lack of available growth resources as a result of increased weed competition, which hampered wheat crop growth and development, resulting in poorer yield-attributing traits. In the year 2013–2014, the greatest grain yield (4.89 t/ha) and straw yield (5.78 t/ha) was recorded in W_2 (4.68 t/ha, 5.58 t/ha), which was considerably superior to the other weed management treatments (Table 7). The greatest grain production of 4.91 t/ha was observed in the second year of the trial under the IWM treatment, which was considerably superior to all other weed management regimens. W_2 had the highest straw yield (5.48 t/ha), which was comparable to W_1 (5.47 t/ha) and much higher than W_3 (3.84 t/ha). Similarly, W_2 had the highest grain and straw yield, which was comparable to W_1 but much higher than W_3.

3.2.3. Green Gram

Yield attributes and yield of summer green gram were significantly influenced by conservation tillage and various weed management practices (Tables 9 and 10). The findings revealed that there was no significant difference in the number of pods per plant in 2014, while the T_2 treatment had the maximum number of pods per plant in 2014 and 2015, with values of 22.86 and 22.61 pods per plant, respectively. Furthermore, the CT(DS)–CT–ZT (T_3) treatment consistently produced the most seeds per pod for three years. In the first and second years of the study, there was no significant difference in test weight among the different tillage and residue management treatments. However, in the third year (2016), the T_5 treatment recorded the highest test weight, which was comparable to all treatments except T_4. Among the weed management treatments, the highest values for yield-attributing characteristics were observed in the W_2 treatment, while the lowest values were found in the W_3 treatment.

The T_3 treatment, which included CT(DS)–CT–ZT, produced the greatest grain yield of green gram (1.48 t/ha), and it was considerably superior to the other treatments. Similarly, T_3 produced the maximum straw yield of 2.60 t/ha, whereas T_4 produced the lowest straw yield (averaged over three years) of 1.85 t/ha. Throughout the years, the IWM (W_2) herbicidal treatment consistently produced the highest grain and straw yields. W_3 treatment, on the other hand, had a 32.0% and 32.7% lower grain and straw yield (averaged over three years) than W_1 treatment (Table 9). These differences in yield can be attributed to variations in the production of pods per plant and seeds per pod, which ultimately contributed to an increased overall yield. The significant reduction in weed competition, achieved through effective weed management, played a crucial role in promoting the overall growth of the crop at all stages of observation, thus positively impacting the final yield. The average yield of rice, wheat, and green gram for all the years is presented in Figure 4.

Table 9. Yield attributes and yield of green gram as affected by conservation tillage and different weed management practices.

Treatment *	2013–2014					2014–2015					2015–2016				
	Number of Pods Plants^{-1}	Number of Seeds Pod^{-1}	Test Weight (g)	Grain Yield (t ha^{-1})	Straw Yield (t ha^{-1})	Number of Pods Plants^{-1}	Number of Seeds Pod^{-1}	Test Weight (g)	Grain Yield (t ha^{-1})	Straw Yield (t ha^{-1})	Number of Pods Plants^{-1}	Number of Seeds Pod^{-1}	Test Weight (g)	Grain Yield (t ha^{-1})	Straw Yield (t ha^{-1})
						Tillage and residue management									
T$_1$	-	-	-	-	-	-	-	-	-	-	-	-	-	-	-
T$_2$	22.36 [a**]	8.22 [b]	54.25 [a]	1.52 [b]	2.34 [a]	22.86 [a]	8.65 [b]	53.36 [a]	1.14 [b]	2.35 [b]	22.61 [a]	8.36 [b]	54.15 [b]	1.17 [c]	2.40 [b]
T$_3$	22.22 [a]	9.53 [a]	54.01 [a]	1.82 [a]	2.57 [a]	21.92 [b]	10.04 [a]	53.16 [a]	1.31 [a]	2.60 [a]	22.44 [a]	8.90 [a]	54.39 [b]	1.31 [a]	2.64 [a]
T$_4$	20.54 [b]	7.61 [b]	51.43 [b]	1.35 [c]	1.84 [b]	20.56 [c]	7.75 [c]	50.10 [b]	0.94 [c]	1.88 [c]	20.57 [b]	7.68 [c]	51.56 [c]	0.97 [d]	1.85 [c]
T$_5$	21.59 [a]	8.51 [b]	54.75 [a]	1.64 [b]	2.56 [a]	21.48 [b]	8.70 [b]	52.53 [a]	1.30 [a]	2.53 [a]	21.68 [a]	8.63 [a]	54.76 [a]	1.24 [b]	2.57 [b]
						Weed management									
W$_1$	21.71 [b]	8.46 [b]	53.51 [b]	1.55 [b]	2.45 [b]	21.82 [b]	8.92 [b]	51.59 [b]	1.24 [a]	2.42 [b]	21.84 [b]	8.68 [a]	53.72 [b]	1.25 [a]	2.53 [b]
W$_2$	24.12 [a]	9.30 [a]	56.65 [a]	1.83 [a]	2.71 [a]	24.01 [a]	9.64 [a]	55.67 [a]	1.37 [a]	2.74 [a]	24.29 [a]	8.78 [a]	56.92 [a]	1.39 [a]	2.73 [a]
W$_3$	19.20 [c]	7.64 [c]	50.67 [c]	1.27 [c]	1.83 [c]	19.14 [c]	7.79 [c]	49.60 [c]	0.91 [b]	1.85 [c]	19.35 [c]	7.73 [b]	50.51 [c]	0.93 [b]	1.82 [c]

* Refer to Table 1 for treatment details. ** The means with similar letters down the column (per either tillage residue management and weed management) do not differ significantly at $p \leq 0.05$.

Table 10. Mean of yield attributes and yield of green gram across the three years as affected by conservation tillage and different weed management practices.

Treatment *	Number of Pods Plants^{-1}	Number of Seeds Pod^{-1}	Test Weight (g)	Grain Yield (t ha^{-1})	Straw Yield (t ha^{-1})
	Tillage and residue management				
T$_1$	-	-	-	-	-
T$_2$	22.61	8.41	53.92	1.27	2.36
T$_3$	22.19	9.49	53.85	1.48	2.60
T$_4$	20.55	7.68	51.03	1.08	1.85
T$_5$	21.58	8.61	54.01	1.39	2.55
	Weed management				
W$_1$	21.79	8.68	52.94	1.34	2.46
W$_2$	24.14	9.24	56.41	1.53	2.72
W$_3$	19.23	7.72	50.26	1.03	1.83

* Refer to Table 1 for treatment details.

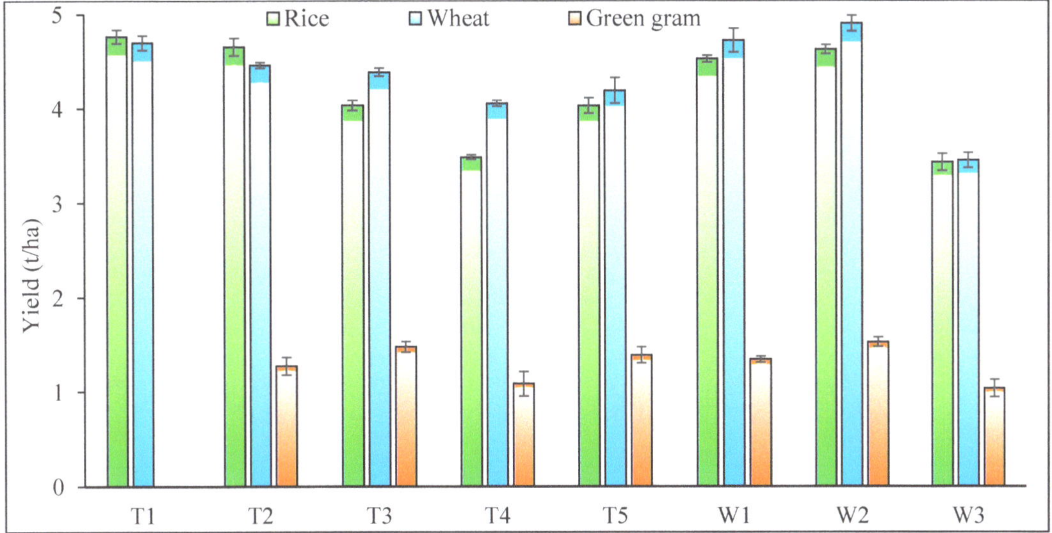

Figure 4. Mean yield (3 years) of rice, wheat, and green gram.

3.3. Soil Health Parameters

3.3.1. Soil Chemical Properties

There were significant differences in soil pH due to the adoption of various tillage and residue management practices. The maximum reduction in pH (7.32) from the initial value (8.04) was observed in the T_5 treatment (ZT(DS) + R–ZT + R–ZT). Conversely, the highest increase in pH from the initial value was found in T_1 treatment (CT(T)–CT–fallow) (Table 11). Among the weed management treatments, the unweeded treatment (W_3) resulted in the maximum reduction in pH (7.91).

Table 11. Chemical properties of the post-harvest soil under the rice–wheat–green gram cropping system as influenced by conservation tillage and different weed management practices.

Treatment *	pH	Organic Carbon (%)	Available N (kg ha^{-1})	Available P$_2$O$_5$ (kg ha^{-1})	Available K$_2$O (kg ha^{-1})
		Tillage and residue management			
T_1	8.85 [a]**	0.46 [c]	250.52 [a]	43.87 [a]	281.34 [a]
T_2	7.50 [b]	0.53 [b]	233.09 [b]	44.95 [a]	284.63 [a]
T_3	8.19 [a]	0.58 [a]	251.15 [a]	50.56 [a]	279.58 [a]
T_4	8.17 [a]	0.54 [b]	250.89 [a]	47.88 [a]	283.25 [a]
T_5	7.32 [b]	0.59 [a]	251.18 [a]	50.05 [a]	266.69 [b]
		Weed management			
W_1	7.93 [a]	0.47 [b]	240.05 [a]	49.47 [a]	284.35 [a]
W_2	8.34 [a]	0.52 [a]	241.22 [a]	50.38 [a]	286.59 [a]
W_3	7.91 [a]	0.55 [a]	239.35 [a]	44.05 [b]	268.87 [a]

* Refer to Table 1 for treatment details. ** The means with similar letters down the column (per either tillage residue management or weed management) do not differ significantly at $p \leq 0.05$.

The organic carbon (%) content of the soil after harvest was also measured and presented in Table 11. The treatments had a considerable impact on the soil's organic carbon concentration. At the start of the trial in 2013 (before wheat seeding), the original soil organic carbon concentration was 0.51%. The T_5 treatment showed the greatest increase in soil organic carbon content (0.59%) above the original value. The T_1 treatment (CT(T)–CT–fallow) had the greatest loss in soil organic carbon (0.41%) from the starting value.

In terms of weed management treatments, the W_3 treatment had the highest increase in organic carbon (0.55%), which was equivalent to the W_2 treatment (0.52%) and 17% greater than the W_1 treatment.

The analysis of available N, P, and K (kg ha^{-1}) was conducted after the harvest of the rice–wheat–green gram cropping system in the year 2016, specifically after the green gram harvest, and the results are presented in Table 11. Among the different tillage and residue management treatments, T_5 had the highest available N content of 251.18 kg ha^{-1}, representing a 2.5% increase over the original value. P_2O_5 and K_2O levels were greatest in T_3 and T_2, with values of 50.56 kg ha^{-1} and 284.63 kg ha^{-1}, respectively. Conversely, the minimum available N content was recorded in T_2 (233.09 kg ha^{-1}), while the lowest available P_2O_5 and K_2O levels were found in T1 and T5, with values of 40.65 kg ha^{-1} and 266.69 kg ha^{-1}, respectively (Table 11). Notably, T5 exhibited a 7.7% higher available N content compared to T_2. Additionally, T_3 and T_2 demonstrated a 15.2% and 6.7% increase in available P_2O_5 and K_2O, respectively, over T_1 and T_5 after the three-year study period. Similarly, the weed management treatment W_2 resulted in the maximum N, P_2O_5, and K_2O content in the post-harvest soil, measuring 241.22 kg ha^{-1}, 50.38 kg ha^{-1}, and 286.59 kg ha^{-1}, respectively, and it was significantly superior to the other weed management treatments.

3.3.2. Soil Biological Properties

Among the tillage and residue management treatments, soil biological properties differed significantly. The highest number of Azotobacter (104 cfu g^{-1} soil), with a value of 3.95, was found in the CT(DS)–CT–ZT treatment, which was 32.5% higher compared to the CT(T)–CT–fallow treatment. Similarly, T_5 exhibited the maximum values for Total Pseudomonas (105 cfu g^{-1} soil), Total phosphate solubilizing bacteria (PSB) (105 cfu g^{-1} soil), percentage of P solubilized by Pseudomonas and Bacillus (105 cfu g^{-1} soil), and CO_2 evolution (mg kg^{-1}), with increases of 40.4%, 34.3%, 32.7%, 49.1%, 54%, and 56.4%, respectively, compared to T_1 (Table 12). The biological properties of the soil vary significantly among the weed management treatments. The highest numbers of Azotobacter, total Pseudomonas, total PSB, and Bacillus (3.81, 6.54, 9.42, and 4.95, respectively) were found in W_3. Additionally, the percentage of P solubilized by Pseudomonas and by Bacillus was 14.5% and 19.9% higher, respectively, in W3 compared to W2 (Table 12). The lowest CO_2 evolution (mg kg^{-1}) was observed in W_1 (74.87), followed by W_2 (75.29) and W_3 (82.88).

Table 12. Biological properties of post-harvest soil under the rice–wheat–green gram cropping system as affected by conservation tillage and different weed management practices.

Treatment *	Azotobacter (10^4 cfu g^{-1} Soil)	Total Pseudomonas (10^5 cfu g^{-1} Soil)	Total PSB (10^5 cfu g^{-1} Soil)	% of P Solubilized by Pseudomonas	Bacillus (10^5 cfu g^{-1} Soil)	% of P Solubilized by Bacillus	CO_2 Evolution (mg kg^{-1})
			Tillage and residue management				
T_1	2.98 [b**]	4.63 [b]	7.11 [a]	19.53 [b]	4.05 [a]	15.68 [b]	63.78 [d]
T_2	3.31 [a]	4.95 [b]	8.22 [a]	23.75 [a]	4.13 [a]	21.32 [a]	69.95 [c]
T_3	3.95 [a]	5.57 [b]	8.89 [a]	23.12 [a]	4.59 [a]	18.26 [b]	69.52 [c]
T_4	3.45 [a]	5.03 [b]	9.39 [a]	23.51 [a]	4.35 [a]	20.89 [a]	83.56 [b]
T_5	3.93 [a]	8.68 [a]	9.55 [a]	25.92 [a]	6.24 [a]	23.38 [a]	99.78 [a]
			Weed management				
W_1	3.31 [a]	5.58 [b]	7.88 [b]	22.40 [a]	4.46 [a]	19.19 [a]	74.87 [b]
W_2	3.43 [a]	5.26 [b]	8.57 [a]	22.04 [a]	4.18 [a]	18.59 [b]	75.29 [b]
W_3	3.81 [a]	6.54 [a]	9.42 [a]	25.25 [a]	4.95 [a]	22.29 [a]	82.88 [a]

* Refer to Table 1 for treatment details. ** The means with similar letters down the column (per either tillage residue management or weed management) do not differ significantly at $p \leq 0.05$.

3.4. Economics

The selection of tillage, residue, and weed management practices is influenced by the economic returns they offer, as farmers prioritize higher returns per unit area, time, and investment. The cost of cultivation varied significantly depending on the tillage, residue, and weed management methods. In the rice–wheat–green gram cropping system, the total cost of production followed this order: $T_5 > T_2 > T_4 > T_3 > T_1$; $W_2 > W_1 > W_3$ (Table 13). At the system level, the highest cost of production was incurred in the ZT(DS) + R-ZT + R-ZT treatment (INR 68397 or USD 1121.4, 1066.5, 1019 ha^{-1} during the three years, respectively) among the tillage and residue management treatments, and in the IWM treatment (INR 77664 or USD 1273.4, 1211, 1157 ha^{-1} during the three years, respectively) among the weed management treatments.

The highest gross returns (INR 188841 or USD 3096.2, INR 184231 or USD 2872.7, INR 220322 or USD 3279.5 ha^{-1} during the three years, respectively) were recorded under the T2 treatment, CT(T)–ZT–ZT, in tillage and residue management, which were 19%, 21%, and 17.7% higher compared to T4 during the 2013–2014, 2014–2015, and 2015–2016 rice–wheat–green gram system, respectively (Table 13). Among the weed management practices, the highest gross return was recorded under the W_2 treatment during all the years.

The highest net return (INR 120628 or USD 1977.8, INR 116017 or USD 1809, INR 152109 or USD 2266.2 ha^{-1} during the three years, respectively) was also recorded under the T_2 treatment, CT(T)–ZT–ZT, in tillage and residue management, which were 33.6%, 37.6%, and 27.7% higher compared to T4 during the 2013–2014, 2014–2015, and 2015–2016 rice–wheat–green gram system, respectively (Table 13). Among the weed management practices, the highest net return was recorded under the W_2 treatment during all the years.

In the rice–wheat–green gram production system, the order of the benefit-to-cost (B:C) ratio among the tillage and residue management treatments was T_2 (2.90) = T_3 (2.88) > T_1 (2.79) > T_5 (2.77) > T_4 (2.44) (averaged over three years). Similarly, the highest B:C ratio was found under the W_1 treatment, which involved recommended herbicides. The higher economic yield under W_1 was attributed to the reduction in competition from weeds during the most critical stages of the crop-weed competition.

Table 13. Economics of the rice–wheat–green gram cropping system as affected by conservation tillage and different weed management practices.

Treatment *	2013–2014				2014–2015				2015–2016			
	Cost of Cultivation (INR ha^{-1})	Gross Returns (INR ha^{-1})	Net Returns (INR ha^{-1})	B:C Ratio	Cost of Cultivation (INR ha^{-1})	Gross Returns (INR ha^{-1})	Net Returns (INR ha^{-1})	B:C Ratio	Cost of Cultivation (INR ha^{-1})	Gross Returns (INR ha^{-1})	Net Returns (INR ha^{-1})	B:C Ratio
				Tillage and residue management								
T_1	55820	145504 [d]**	94384 [c]	2.75 [a]	55820	146852 [c]	95732 [c]	2.63 [b]	55820	166956 [d]	115836 [d]	2.99 [c]
T_2	68213	188841 [a]	120628 [a]	2.77 [a]	68213	184231 [a]	116017 [a]	2.70 [a]	68213	220322 [a]	152109 [a]	3.24 [a]
T_3	67597	184905 [a]	117308 [a]	2.76 [a]	67597	180782 [a]	113186 [a]	2.67 [a]	67597	216816 [b]	149219 [a]	3.23 [a]
T_4	67897	158616 [c]	90270 [c]	2.34 [c]	67897	152200 [c]	84304 [d]	2.24 [d]	67897	187035 [d]	119139 [c]	2.76 [d]
T_5	68397	179312 [b]	110916 [b]	2.64 [b]	68397	175290 [b]	106894 [b]	2.56 [c]	68397	211774 [c]	143378 [b]	3.11 [b]
				Weed management								
W_1	60620	183591 [b]	122971 [a]	3.04 [a]	60620	180771 [b]	120151 [a]	2.98 [a]	60620	214635 [b]	154015 [a]	3.55 [a]
W_2	77664	193351 [a]	115687 [b]	2.51 [b]	77664	189933 [a]	112269 [b]	2.45 [b]	77664	226149 [a]	148485 [a]	2.92 [b]
W_3	55650	137366 [c]	81716 [c]	2.47 [b]	55650	132909 [c]	77259 [c]	2.39 [b]	55650	160959 [c]	105309 [b]	2.90 [b]

* Refer to Table 1 for treatment details. ** The means with similar letters down the column (per either tillage residue management or weed management) do not differ significantly at $p \leq 0.05$.

4. Discussion

4.1. Weed Dynamics

Numerous variables influence how much tillage affects the amount of the weed seed bank [50]. Since tillage has a diminutive effect [51] on reducing [52] or increasing [53] weed seed bank density, empirical investigations produce contradictory consequences. The results of plentiful research indicate that the weed species governs how the weed seed bank reacts to tillage [32]. These studies also specified that the complex interplay between weather, the span of the experiment, and long-term field history affects how the weed seed bank responds to tillage. Although it can be time-consuming and challenging to assess, the initial state and distribution of the weed seed bank have a significant impact on study outcomes. Tillage reallocates seeds all over the soil profile, regardless of the texture and structure of the soil [54]. A higher percentage of ZT seedbanks will germinate than CT regime seedbanks [55], which have generalized patterns of seed distribution, because ZT seeds infiltrate the soil via very slow processes through thin cracks, diversified macro-fauna and freeze-dry cycles [56], resulting in an accumulation of weed seeds (60–90%) in the top 5 cm of the soil [57]. A higher density of weeds in zero-till rice was also reported by Nichols et al. [58]. ZT involves minimal or no soil disturbance, leaving the weed seeds undisturbed and closer to the soil surface, allowing the weed seeds to remain viable and readily available for germination, leading to higher weed densities [58]. Also, weed seeds are not buried as deeply into the soil in ZT as they would be with CT, and this shallow seed placement provides favorable conditions for weed seed germination and emergence [50,59], contributing to increased weed density. It might also be possible that CT can cause physical damage to weed seeds, resulting in decreased viability and germination potential, whereas ZT practices typically do not subject weed seeds to the same level of physical disturbance, allowing a higher percentage of seeds to remain viable, leading to increased weed densities.

Residue-laden treatments have shown better weed control over clean cultivation owing to the prevention of weeds from germinating and may facilitate higher seed predation due to favorable conditions for soil macro-fauna, such as ants and beetles [60,61]. Residue retention is beneficial for reducing weed populations in crop fields because it acts as a physical barrier [62], preventing weed seeds from reaching the soil surface and germinating. At the same time, the residue layer hinders weed seedling emergence by limiting light penetration and creating an unfavorable environment for weed growth. Additionally, the decaying crop residues release allelochemicals that possess herbicidal properties, inhibiting weed germination and growth [63,64]. By retaining crop residues, weed populations can be effectively suppressed, leading to improved weed control and higher yields in rice cultivation.

The continuous use of single herbicides or single weed management practices will lead to undesirable phenomena like herbicide resistance, weed shift, and many more issues [65]. Combining cultural practices such as crop rotation, like in the case of the present study involving summer green gram [66], and residue management with mechanical methods such as intercropping and manual weeding effectively suppresses weeds, optimizing resource utilization by crops [67]. This has ultimately resulted in a reduction in weed density as well as weed biomass in IWM treatment.

4.2. Crop Yield

Based on the findings of the present study, it can be inferred that the yield in ZT rice plots was comparatively lower than in CT, which aligns with the earlier research conducted by Alam et al. [68]. The decrease in yield observed in direct-seeded rice (DSR) can be attributed to various factors. These include soil sickness caused by nutrient unavailability compared to conventional tillage (CT), vigorous weed growth favored by alternating wet and dry cycles and the absence of standing water, moisture stress due to higher percolation rates, potential stress from nematodes and rice mealybugs, and increased spikelet sterility, all of which pose significant challenges to achieving high grain yields in rice cultivation. However, effective management of both biotic and abiotic stresses, such as controlling weed growth, nematode infestation, and leaf miner attacks, can greatly alleviate these yield losses in DSR. On the other hand, in CT rice cultivation, the practice of puddling provides several advantages. It enhances weed control by creating an unfavorable environment for their growth, reduces water and nutrient loss through deep percolation, facilitates the rapid establishment of rice seedlings, and improves nutrient availability by utilizing the redox potential phenomenon in waterlogged soil.

Being an integral part of conservation agricultural practices, ZT has gained popularity in wheat cultivation due to its potential benefits such as soil moisture conservation, reduced erosion, and cost savings [37,69]. However, despite these advantages, zero-tilled wheat systems sometimes experience a decline in yield compared to conventional tillage methods, attributed to numerous factors, including soil compaction, increased weed competition [32], challenges associated with residue management [62], nutrient imbalances [70], and disease and pest pressure. ZT often leads to increased soil compaction due to the absence of tillage operations that help alleviate compaction, hindering root growth, reducing nutrient uptake, and limiting water infiltration, resulting in decreased plant growth and, ultimately, lowering yields. These factors aggravated each other further because ZT experienced greater weed pressure as compared to CT methods. The absence of tillage disrupts weed seed burial and exposes them to favorable germination conditions, leading to amplified weed competition for resources such as nutrients, light, and water that can significantly impact the wheat yield [55]. On the contrary, CT provides better weed control due to tillage activity at the time of crop establishment, leading to better yield-attributing characteristics and yield [59]. Moreover, in ZT, crop residues are left on the soil surface, which might create challenges for the germination of emerging wheat crops by impeding seed-to-soil contact [71], hindering seedling emergence, and reducing early plant vigor, leading to a lesser number of tillers and other yield-attributing characteristics and, in turn, affecting the yield. ZT may also experience imbalances in nutrient availability and uptake [72], inferred due to the accumulation of crop residues on the soil surface leading to nutrient immobilization [73,74], making essential nutrients like N and P less accessible to the growing wheat plants [75]. Additionally, without tillage, nutrient stratification may occur [76], with nutrients concentrated in the surface layers and limiting their availability in the lower root zone. Also, in many instances, the residue was the chief source of promulgation of certain diseases and pests by acting as an alternate host and habitat for varied plant pathogens and pests [77], potentially leading to higher disease pressure and insect infestations and resulting in reduced stand establishment, poor plant health, and yield losses [78].

The higher yield in summer green gram following CT rice and wheat can be attributed to a combination of interconnected factors encompassing residue decomposition and nutrient availability, effective weed suppression, enhanced soil aeration and root penetration, better pest and disease management, and soil moisture conservation [79,80]. CT incorporates crop residues, promoting their decomposition and releasing nutrients that are readily available for the subsequent crop, enhancing the growth, yield, and yield-attributing characteristics of summer green gram [81]. Additionally, CT aids in effective weed control by burying weed seeds and disrupting their germination, reducing competition and providing the summer green gram crop with a competitive advantage [50,59]. Improved soil aeration and root penetration achieved through CT practices further support nutrient uptake and overall crop performance. Furthermore, CT disrupts pest life cycles, reduces disease incidence, and buries pests, pathogens, and infected crop residues, thereby minimizing pest and disease pressure and safeguarding the yield potential of further crops in sequence [77]. As CT operations break up surface soil crusts, facilitating better water infiltration and ensuring optimal moisture availability for summer green gram germination, establishment, and growth [82]. In contrast, ZT systems exhibit slower residue decomposition rates, potentially leading to delayed nutrient release, resulting in limited nutrient availability and reduced crop productivity [83]. Moreover, the presence of undisturbed crop residues in ZT systems can foster weed growth, increase weed competition, and hinder the yield potential of subsequent crops. Also, the compact soil layers in ZT can impede root growth, nutrient accessibility, and overall crop performance.

Conservation tillage in conjugation with integrated weed management (IWM) practices, i.e., W_2 in the rice–wheat–green gram sequence, provides scientifically substantiated benefits, including higher yields. This might be due to low weed density during the initial crop growth stages. The timely application of herbicides and further control of later germinated weeds by the supplemented intercultural operation followed by hand weeding caused a reduction in weed competition, leading to increased photosynthetic activity, and biomass accumulation [84,85]. The correlation among different parameters of rice and wheat across the years has been presented in Figures 5 and 6.

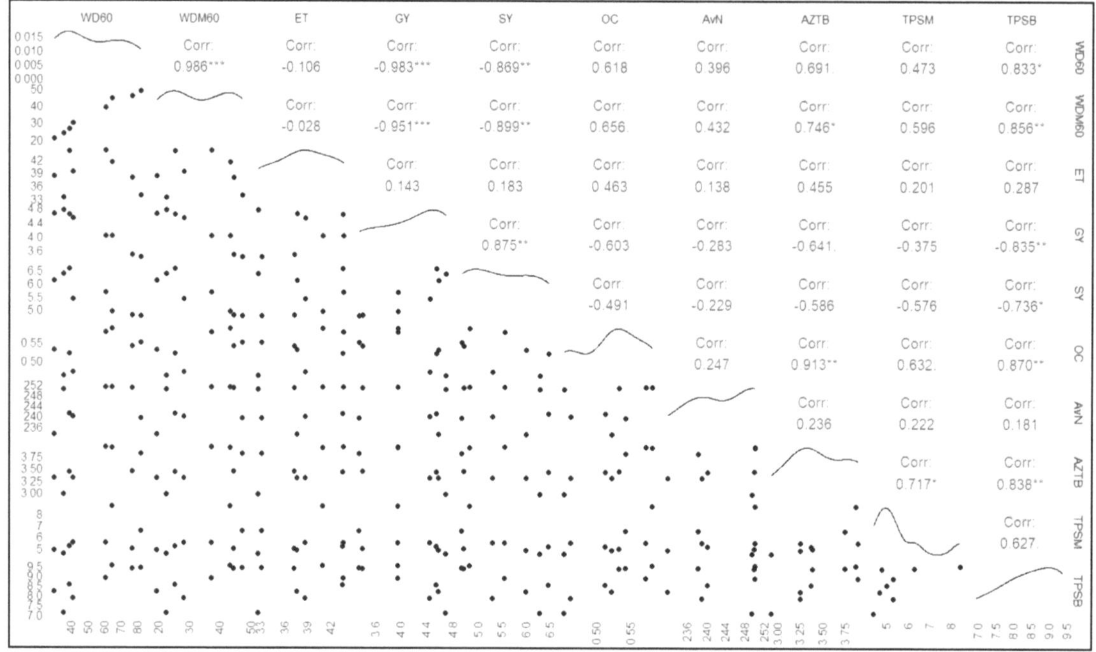

Figure 5. Correlation panel graph among different parameters of rice cultivation across the years (mean of 3 years, WD60: weed density at 60 days after sowing (60DAS), WDM60: weed dry matter at 60DAS, ET: number of effective tillers, GY: grain yield, SY: Straw yield, OC: organic carbon in soil, AvN: available nitrogen in soil, AZTB: Azotobacter population, TPSM: total Pseudomonas population, TPSB: total phosphate solubilizing bacteria population. * Significance at $p < 0.05$, ** significance at $p < 0.01$, *** significance at $p < 0.001$).

4.3. Soil Properties

The maximum upsurge in SOC content (0.59%) from the initial value was observed in T_5, where the ZT practices offer several interlinked reasons for higher OC content and increased nutrient availability in comparison to CT soil. Firstly, ZT minimizes soil disturbance, preserving soil organic matter (SOM) and preventing its oxidation, resulting in higher OC levels [86,87]. The presence of crop residues on the soil surface in ZT systems further contributes to OC accumulation by providing continuous organic material that gradually decomposes, which enhances the SOC content and positively influences soil health and nutrient availability [88,89]. Moreover, ZT promotes nutrient retention by reducing leaching through the physical barrier created by crop residues, improving nutrient availability, mainly N, P, and K, for plant uptake [90,91]. Additionally, the undisturbed soil structure and increased OC content in ZT soil foster a diverse and active microbial community that plays a vital role in nutrient cycling and mineralization [92], converting SOM into plant-available forms ([93]. The improved soil structure and stability of soil aggregates in ZT systems protect SOM and nutrients from degradation [94] and promote root exploration and nutrient uptake, enhancing nutrient availability [95]. Furthermore, ZT practices reduce erosion by maintaining soil surface cover, which prevents the loss of SOM and nutrients through wind or water erosion [96]. The increased water-holding capacity of ZT soil supports soil microbial activity, organic matter decomposition, and nutrient mineralization [76]. Additionally, accelerated carbon sequestration in ZT soil helps mitigate climate change while improving soil fertility.

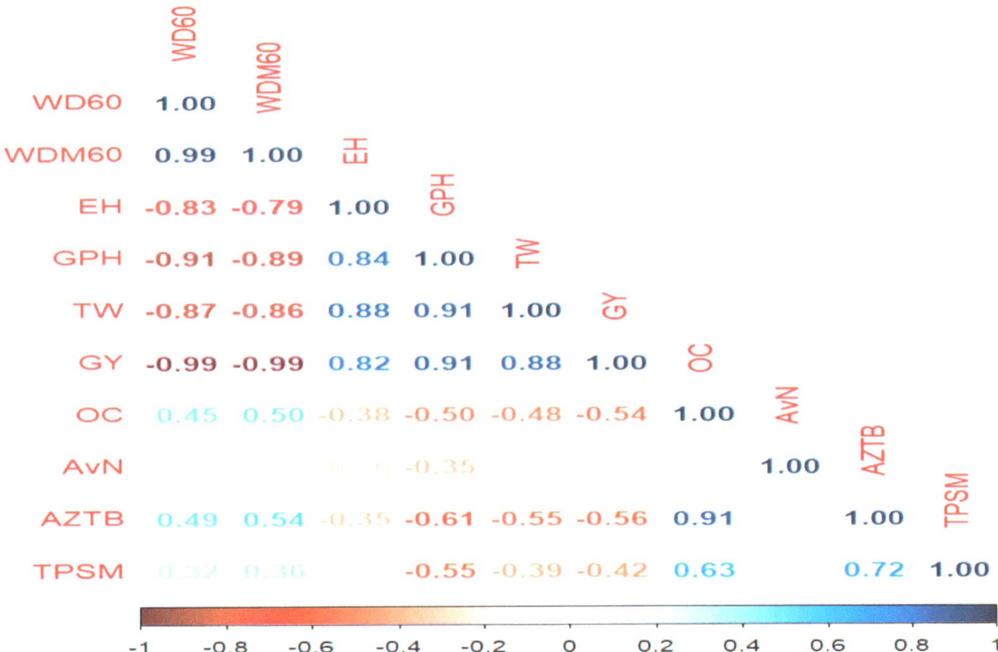

Figure 6. Correlation plot among different parameters of wheat cultivation across the years (mean of 3 years, WD60: weed density at 60 days after sowing (60DAS), WDM60: weed dry matter at 60DAS, EH: number of earheads, GPH: grains per earhead, TW: 1000 grain weight, GY: grain yield, OC: organic carbon in soil, AvN: available nitrogen in soil, AZTB: Azotobacter population, TPSM: total Pseudomonas population, TPSB: total phosphate solubilizing bacteria population).

Apart from the nutrient availability status of the soil, ZT promotes advanced microbial populations, including Azotobacter, Pseudomonas, and Bacillus, through a synergistic interplay of various factors involving the preservation of soil structure in T_5 involving ZT in all three crops in cropping sequence along with residue, creating protected microenvironments [97] and aggregates that serve as favorable niches for these microbial populations to establish and thrive [98,99]. Furthermore, ZT enables the accumulation of SOM, leading to a nutrient-rich soil environment that supports microbial growth and activity as well [100–102]. The improved soil aggregation and reduced disturbance in ZT systems enhance microbial diversity, while also safeguarding these microbes from environmental stresses [93,94]. Additionally, the enhanced water retention capacity and reduced soil erosion in ZT soil provide conducive conditions for the proliferation of Azotobacter, Pseudomonas, and Bacillus [103]. Moreover, beneficial interactions between microorganisms and plants, such as nitrogen fixation by Azotobacter and phosphate solubilization by Pseudomonas and Bacillus, are fostered in the undisturbed soil environment of ZT. Finally, the minimized chemical disturbances in ZT practices further support the growth and persistence of these microbial populations [101].

The incorporation of crop residues and the adoption of crop rotation in IWM enhance soil organic matter content [102], promoting the growth of diverse microbial communities. These beneficial microorganisms play vital roles in nutrient cycling, disease suppression, and soil health improvement, ultimately enhancing soil structure and nutrient availability [103]. Furthermore, IWM reduces the reliance on herbicides, minimizing potential negative effects on soil nutrient availability [104]. By effectively suppressing weeds, IWM allows crops to access and utilize available nutrients more efficiently [105]. The incorporation of organic matter through crop residues and green manuring practices in IWM further enhances soil fertility and nutrient content [106], nutrient mineralization [107], and soil biological properties [97]. The combined effects of reduced weed competition, improved soil biological properties, and enhanced nutrient cycling dynamics create a favorable environment for crop growth [108].

4.4. Economic Benefits

The higher cost of cultivation in ZT rice–wheat systems compared to conventionally tilled systems can be attributed to several interconnected factors. Firstly, the operational cost of specialized equipment, including seed drills and precision planters, adds to the overall expenses [83]. Additionally, the reliance on high-quality hybrid or certified seeds and the need for treated seeds for successful germination further increase seed costs. ZT practices often require greater inputs of fertilizers, herbicides, and pesticides to manage weeds and pests without tillage operations, contributing to higher production costs [37,109]. The retention of crop residues on the soil surface necessitates additional investment in machinery or labor for effective residue management. Moreover, acquiring specialized knowledge and training through workshops or consultants incurs educational costs [110]. Finally, the need for risk management strategies such as crop insurance to mitigate risks associated with diseases, pests, and adverse weather conditions adds to the overall cost of cultivation [111]. While ZT offers long-term soil conservation benefits [112], careful consideration of these cost factors is essential for farmers. CT systems exhibit higher yield potential due to favorable seedbed preparation, which promotes better seed germination and reduces weed competition [91]. Effective weed control and pest management in conventionally tilled systems contribute to improved crop growth and yield, minimizing yield losses [107]. Additionally, CT offers better opportunities for pest and disease management [78], reducing the risk of damage to crops, and precise nutrient management through tillage operations enhances nutrient availability and uptake [92], leading to higher crop yields [113]. Lower input costs in CT systems, including reduced reliance on specialized equipment and inputs, contribute to higher net returns. Considering these factors collectively provides insights into the economic advantages of CT rice–wheat systems, highlighting the need for further research to bridge the yield gap and improve the economic viability of ZT [114,115]. Farmers can make informed decisions based on these considerations to optimize their production practices and maximize profitability.

Additionally, the gross returns in IWM are higher, while the net returns and B:C ratio are higher in W1 [116] due to the higher cultivation costs. With appropriate location-specific optimization of the available techniques, IWM can be recommended with regard to environmental aspects and issues such as herbicide resistance and weed shift [91,117].

5. Conclusions

The research findings highlight the potential of conservation tillage and integrated weed management practices to promote sustainable agriculture in the rice–wheat–green gram system. The results revealed that CT–rice along with ZT–wheat significantly reduced weed emergence, distribution, and biomass, which was ascribed to the puddling destroying the weed habitat followed by no tillage that does not allow underground weed seed to come to the surface. IWM gives superior results in controlling weed flora due to the proper combination and additive effect of weed control measures. This reduction in weed pressure positively influenced crop performance, leading to better yield-attributing characteristics for all the crops in the sequence and, in turn, increased grain and straw yield. Additionally, conservation tillage involving ZT in all crops along with residue retention significantly enhanced soil chemical and biological properties in terms of enhancement in the soil organic carbon content and nutrient availability through an augmentation in the microbial population such as *Azotobacter*, *Pseudomonas* and *Bacillus* through providing better habitat for them along with substrates through residue incorporation, thus contributing to improved soil health and fertility. However, CT–rice, followed by ZT in wheat and green gram, gives more monetary remuneration in terms of net and gross return with a superior B:C ratio and vigorous crop growth with lesser weed pressure. Furthermore, integrated weed management practices demonstrated their ability to further suppress weed growth and enhance crop productivity. Therefore, by reducing weed pressure, improving soil chemical and biological health, and enhancing crop performance, these improved management practices offer promising avenues for farmers in the Eastern Indo-Gangetic Plain and similar agro-ecologies to achieve higher yields and profitability.

Author Contributions: Conceptualization, D.K.R.; methodology, D.K.R.; software, D.K.R., validation, D.K.R.; formal analysis, D.K.R.; investigation, D.K.R.; resources, D.K.R.; data curation, B.A.A., S.R., S.R.P. and S.S.; writing—original draft preparation, B.A.A., S.R., S.R.P. and S.S.; writing—review and editing, B.A.A., S.R., S.R.P, S.S., D.N. and H.G.; visualization, D.K.R. and M.F.S.; supervision, H.G. All authors have read and agreed to the published version of the manuscript.

Funding: The authors extend their appreciation to Prince Sattam bin Abdulaziz University for funding this research work through project number PSAU/2023/01/233562.

Data Availability Statement: Publicly available datasets were analyzed in this study. Some parts of the data can be found at: https://aicrp.icar.gov.in/wm/publication/annual-reports/ (accesed on 12 April 2023).

Conflicts of Interest: The authors declare no conflict of interest.

References

1. Alam, M.K.; Biswas, W.K.; Bell, R.W. Greenhouse gas implications of novel and conventional rice production technologies in the Eastern-Gangetic plains. *J. Clean. Prod.* **2016**, *112*, 3977–3987. [CrossRef]
2. Mondal, S.; Poonia, S.P.; Mishra, J.S.; Bhatt, B.P.; Karnena, K.R.; Saurabh, K.; Kumar, R.; Chakraborty, D. Short-term (5 years) impact of conservation agriculture on soil physical properties and organic carbon in a rice-wheat rotation in the Indo-Gangetic plains of Bihar. *Eur. J. Soil Sci.* **2020**, *71*, 1076–1089. [CrossRef]
3. Bhatt, R.; Kukal, S.S.; Busari, M.A.; Arora, S.; Yadav, M. Sustainability issues on rice-wheat cropping system. *Int. Soil Water Conserv. Res.* **2016**, *4*, 64–74. [CrossRef]
4. Jain, N.; Dubey, R.; Dubey, D.S.; Singh, J.; Khanna, M.; Pathak, H.; Bhatia, A. Mitigation of greenhouse gas emission with system of rice intensification in the Indo-Gangetic Plains. *Paddy Water Environ.* **2014**, *12*, 355–363. [CrossRef]
5. Shyamsundar, P.; Springer, N.P.; Tallis, H.; Polasky, S.; Jat, M.L.; Sidhu, H.S.; Krishnapriya, P.P.; Skiba, N.; Ginn, W.; Ahuja, V.; et al. Fields on fire: Alternatives to crop residue burning in India. *Science* **2019**, *365*, 6453. [CrossRef]
6. Mishra, J.S.; Bhatt, B.P.; Arunachalam, A.; Jat, M.L. *Conservation Agriculture for Sustainable Intensification in Eastern India*; Policy Brief, Indian Council of Agricultural Research and National Academy of Agricultural Sciences. New Delhi, India, 2020, 8p.
7. Upadhaya, B.; Kishor, K.; Kumar, V.; Kumar, N.; Kumar, S.; Yadav, V.K.; Kumar, R.; Gaber, A.; Laing, A.M.; Brestic, M.; et al. Diversification of Rice-Based Cropping System for Improving System Productivity and Soil Health in Eastern Gangetic Plains of India. *Agronomy* **2022**, *12*, 2393. [CrossRef]
8. Majeed, Y.; Fiaz, S.; Teng, W.; Rasheed, A.; Gillani, S.F.A.; Zhu, X.; Seleiman, M.F.; Diatt, A.A. Evaluation of twenty genotypes of wheat (*Triticum aestivum* L.) grown) grown under heat stress during germination stage. *Not. Bot. Horti Agrobot. Cluj-Napoca* **2023**, *51*, 13207. [CrossRef]
9. Jain, M.; Singh, B.; Srivastava, A.A.K.; Malik, R.K.; McDonald, A.J.; Lobell, D.B. Using satellite data to identify the causes of and potential solutions for yield gaps in India's Wheat Belt. *Environ. Res. Lett.* **2017**, *12*, 094011. [CrossRef]
10. Goher, R.; Alkharabsheh, H.M.; Seleiman, M.F.; Diatta, A.A.; Gitari, H.; Wasonga, D.O.; Khan, G.R.; Akmal, M. Impacts of heat shock on productivity and quality of *Triticum aestivum* L. at different growth stages. *Not. Bot. Horti Agrobot. Cluj-Napoca* **2023**, *51*, 13090. [CrossRef]
11. Sidhu, H.S.; Humphreys, E.; Dhillon, S.S.; Blackwell, J.; Bector, V. The Happy Seeder enables direct drilling of wheat into rice stubble. *Aust. J. Exp. Agric.* **2007**, *47*, 844–854. [CrossRef]
12. Gadde, B.; Menke, C.; Wassmann, R. Rice straw as a renewable energy source in India, Thailand, and the Philippines: Overall potential and limitations for energy contribution and greenhouse gas mitigation. *Biomass Bioenergy* **2009**, *33*, 1532–1546. [CrossRef]
13. NPMCR. National Policy for Management of Crop Residues. 2014. Available online: http://agricoop.nic.in/sites/default/files/NPMCR_1.pdf) (accessed on 13 April 2023).
14. Kisaka, M.O.; Shisanya, C.; Cournac, L.; Manlay, J.R.; Gitari, H.; Muriuki, J. Integrating no-tillage with agroforestry augments soil quality indicators in Kenya's dry-land agroecosystems. *Soil Tillage Res.* **2023**, *227*, 105586. [CrossRef]
15. Benbi, D.K. Carbon footprint and agricultural sustainability nexus in an intensively cultivated region of Indo-Gangetic Plains. *Sci. Total Environ.* **2018**, *644*, 611–623. [CrossRef]
16. Saharawat, Y.S.; Ladha, J.K.; Pathak, H.; Gathala, M.; Chaudhary, N.; Jat, M.L. Simulation of resource-conserving technologies on productivity, income and greenhouse gas emission in rice-wheat system. *J. Soil Sci. Environ. Manag.* **2012**, *3*, 9–22.

17. Gangwar, K.S.; Singh, K.K.; Sharma, S.K.; Tomar, O.K. Alternate tillage and crop residue management in wheat after rice in sandy loam soil of Indo-Gangetic plains. *Soil Tillage Res.* **2006**, *88*, 242–252. [CrossRef]
18. Nyawade, S.; Gitari, H.I.; Karanja, N.N.; Gachene, C.K.K.; Schulte-Geldermann, E.; Parker, M. Yield and evapotranspiration characteristics of potato-legume intercropping simulated using a dual coefficient approach in a tropical highland. *Field Crops Res.* **2021**, *274*, 108327. [CrossRef]
19. Jat, M.L.; Chakraborty, D.; Ladha, J.K.; Rana, D.S.; Gathala, M.K.; McDonald, A.; Gerard, B. Conservation agriculture for sustainable intensification in South Asia. *Nat. Sustain.* **2020**, *3*, 336–343. [CrossRef]
20. Bolliger, A.; Magid, J.; Amadon, T.C.; Neto, F.S.; Ribeiro, M.D.D.; Calegari, A.; Ralisch, R.; de Neergaard, A. Taking stock of the Brazilian "zero-till revolution": A review of landmark research and farmers' practice. *Adv. Agron.* **2006**, *91*, 47–100.
21. Derpsch, R. Making conservation tillage conventional, building a future on 25 years of research: Research and extension perspective. In Proceedings of the 25th Annual Southern Conservation Tillage Conference for Sustainable Agriculture, Auburn, AL, USA, 24–26 June 2002.
22. Derpsch, R.; Friedrich, T.; Kassam, A.; Hongwen, L. Current status of adoption of no-till farming in the world and some of its main benefits. *Int. J. Biol. Eng.* **2010**, *3*, 1–25.
23. Keil, A.; D'souza, A.; McDonald, A. Zero-tillage as a pathway for sustainable wheat intensification in the Eastern Indo-Gangetic Plains: Does it work in farmers' fields? *Food Secur.* **2015**, *7*, 983–1001. [CrossRef]
24. Lal, R. Carbon management in agricultural soils. *Mitig. Adapt. Strateg. Glob. Change* **2007**, *12*, 303–322. [CrossRef]
25. Maitra, S.; Hossain, A.; Brestic, M.; Skalicky, M.; Ondrisik, P.; Brahmachari, K.; Shankar, T.; Bhadra, P.; Palai, J.B.; Jena, J.; et al. Intercropping system—A low input agricultural strategy for food and environmental security. *Agronomy* **2020**, *11*, 343. [CrossRef]
26. Liu, W.; Zhang, Y.; Jiang, S.; Deng, Y.; Christie, P.; Murray, P.J.; Li, X.; Zhang, J. Arbuscular mycorrhizal fungi in soil and roots respond differently to phosphorus inputs in an intensively managed calcareous agricultural soil. *Sci. Rep.* **2016**, *6*, 24902. [CrossRef] [PubMed]
27. Kalayu, G. Phosphate solubilizing microorganisms: Promising approach as biofertilizers. *Int. J. Agron.* **2019**, *2019*, 4917256. [CrossRef]
28. Das, B.B.; Dkhar, M.S. Rhizosphere microbial populations and physicochemical properties as affected by organic and inorganic farming practices. *Am.-Eurasian J. Agri. Environ. Sci.* **2011**, *10*, 140–150.
29. Choi, S.; Song, H.; Tripathi, B.M.; Kerfahi, D.; Kim, H.; Adams, J.M. Effect of experimental soil disturbance and recovery on structure and function of soil community: A metagenomic and metagenetic approach. *Sci. Rep.* **2017**, *7*, 2260. [CrossRef] [PubMed]
30. Szoboszlay, M.; Dohrmann, A.B.; Poeplau, C.; Don, A.; Tebbe, C.C. Impact of land-use change and soil organic carbon quality on microbial diversity in soils across Europe. *FEMS Microbiol. Ecol.* **2017**, *93*, fix146. [CrossRef] [PubMed]
31. Martensson, L.M.; Olsson, P.A. Reductions in microbial biomass along disturbance gradients in a semi-natural grassland. *Appl. Soil Ecol.* **2012**, *62*, 8–13. [CrossRef]
32. Farooq, M.; Flower, K.C.; Jabran, K.; Wahid, A.; Siddique, K.H.M. Crop yield and weed management in rainfed conservation agriculture. *Soil Tillage Res.* **2011**, *117*, 172–183. [CrossRef]
33. Chauhan, B.S.; Johnson, D.E. Influence of tillage systems on weed seedling emergence pattern in rainfed rice. *Soil Tillage Res.* **2009**, *106*, 15–21. [CrossRef]
34. Fonteyne, S.; Singh, R.G.; Govaerts, B.; Verhulst, N. Rotation, Mulch and Zero Tillage Reduce Weeds in a Long-Term Conservation Agriculture Trial. *Agronomy* **2020**, *10*, 962. [CrossRef]
35. Ghimire, B.; Ghimire, R.; VanLeeuwen, D.; Mesbah, A. Cover crop residue amount and quality effects on soil organic carbon mineralization. *Sustainability* **2017**, *9*, 2316. [CrossRef]
36. Singh, M.; Kumar, P.; Kumar, V.; Solanki, I.S.; McDonald, A.J.; Kumar, A.; Poonia, S.P.; Kumar, V.; Ajay, A.; Kumar, A.; et al. Intercomparison of crop establishment methods for improving yield and profitability in the rice-wheat system of Eastern India. *Field Crops Res.* **2020**, *250*, 107776. [CrossRef]
37. Jat, R.K.; Sapkota, T.B.; Singh, R.G.; Jat, M.; Kumar, M.; Gupta, R.K. Seven years of conservation agriculture in a rice-wheat rotation of Eastern Gangetic Plains of South Asia: Yield trends and economic profitability. *Field Crops Res.* **2014**, *164*, 199–210. [CrossRef]
38. Kumar, V.; Jat, H.S.; Sharma, P.C.; Gathala, M.K.; Malik, R.K.; Kamboj, B.R.; Yadav, A.K.; Ladha, J.K.; Raman, A.; Sharma, D.K.; et al. Can productivity and profitability be enhanced in intensively managed cereal systems while reducing the environmental footprint of production? Assessing sustainable intensification options in the breadbasket of India. *Agric. Ecosyst. Environ.* **2018**, *252*, 132–147. [CrossRef]
39. Jat, R.K.; Singh, R.G.; Kumar, M.; Jat, M.L.; Parihar, C.M.; Bijarniya, D.; Sutaliya, J.M.; Jat, M.K.; Parihar, M.D.; Kakraliya, S.K.; et al. Ten years of conservation agriculture in a rice-maize rotation of Eastern Gangetic Plains of India: Yield trends, water productivity and economic profitability. *Field Crops Res.* **2019**, *232*, 1–10. [CrossRef]
40. Piper, C.S. *Soil and Plant Analysis*; Academic Press: New York, NY, USA, 1966; pp. 47–77.
41. Black, C.A. *Methods of Soil Analysis, Part 1 and 2*; Agronomy Monograph No. 3 in the Series "Agronomy"; American Society of Agronomy, Inc.: Madison, WI, USA, 1965; Volume 148.
42. Jackson, M.L. *Soil Chemical Analysis*; Prentice Hall of India Pvt. Ltd.: New Delhi, India, 1973.
43. Subbiah, B.; Asija, G.L. A rapid procedure for estimation of available nitrogen in soils. *Curr. Sci.* **1956**, *25*, 259–260.

44. Olsen, S.R.; Cole, C.V.; Watanabe, F.S. *Estimation of Available Phosphorus in Soils by Extraction with Sodium Bicarbonate*; USDA Circular, No. 939; U.S. Government Printing Office: Washington, DC, USA, 1954.
45. Schmidt, E.L.; Caldwell, A.C. *A Practical Manual of Soil Microbiology Laboratory Methods*; Food and Agriculture Organization of the United Nations: Rome, Italy, 1967; pp. 72–75.
46. Zibilske, L.M. Carbon Mineralization. In *Methods of Soil Analysis: Part 2 Microbiological and Biochemical Properties*; SSSA Book Series; Soil Science Society of America, Inc.: Madison, WI, USA, 2018; pp. 835–863. [CrossRef]
47. Cheema, H.S.; Singh, B. *Software Statistical Package CPCS-1*; Department of Statistics, PAU: Ludhiana, India, 1991.
48. Luo, H.; He, L.; Du, B.; Pan, S.; Mo, Z.; Meiyang, D.; Tian, H.; Tang, X. Biofortification with chelating selenium in fragrant rice: Effects on photosynthetic rates, aroma, grain quality and yield formation. *Field Crops Res.* **2020**, *255*, 107909. [CrossRef]
49. Gomez, K.A.; Gomez, A.A. *Statistical Procedures for Agricultural Research*; John Wiley & Sons: Hoboken, NJ, USA, 1984.
50. Mohler, C.L. A Model of the Effects of Tillage on Emergence of Weed Seedlings. *Ecol. Appl.* **1993**, *3*, 53–73. [CrossRef]
51. Bàrberi, P.; Bonari, E.; Mazzoncini, M.; García-Torres, L.; Benites, J.; Martínez-Vilela, A. Weed density and composition in winter wheat as influenced by tillage systems. Conservation agriculture, a worldwide challenge. In Proceedings of the First World Congress on Conservation Agriculture, Madrid, Spain, 1–5 October 2001; pp. 451–455.
52. Murphy, S.D.; Clements, D.R.; Belaoussoff, S.; Kevan, P.G.; Swanton, C.J. Promotion of weed species diversity and reduction of weed seedbanks with conservation tillage and crop rotation. *Weed Sci.* **2006**, *54*, 69–77. [CrossRef]
53. Sosnoskie, L.M.; Herms, C.P.; Cardina, J. Weed seedbank community composition in a 35-yr-old tillage and rotation experiment. *Weed Sci.* **2006**, *54*, 263–273. [CrossRef]
54. Yenish, J.P.; Doll, J.D.; Buhler, D.D. Effects of tillage on vertical distribution and viability of weed seed in soil. *Weed Sci.* **1992**, *40*, 429–433. [CrossRef]
55. Gallandt, E.R.; Fuerst, E.P.; Kennedy, A.C. Effect of tillage, fungicide seed treatment, and soil fumigation on seed bank dynamics of wild oat (*Avena fatua*). *Weed Sci.* **2004**, *52*, 597–604. [CrossRef]
56. Dorado, J.; Del Monte, J.; Lopez-Fando, C. Weed seedbank response to crop rotation and tillage in semiarid agroecosystems. *Weed Sci.* **1999**, *47*, 67–73. [CrossRef]
57. Hoffman, M.L.; Owen, M.D.; Buhler, D.D. Effects of crop and weed management on density and vertical distribution of weed seeds in soil. *Agron. J.* **1998**, *90*, 793–799. [CrossRef]
58. Nichols, V.; Verhulst, N.; Cox, R.; Govaerts, B. Weed dynamics and conservation agriculture principles: A review. *Field Crops Res.* **2015**, *183*, 56–68. [CrossRef]
59. Anderson, R.L. A Multi-Tactic Approach to Manage Weed Population Dynamics in Crop Rotations. *Agron. J.* **2005**, *97*, 1579–1583. [CrossRef]
60. Blubaugh, C.K.; Kaplan, I. Tillage compromises weed seed predator activity across developmental stages. *Biol. Control* **2015**, *81*, 76–82. [CrossRef]
61. Trichard, A.; Ricci, B.; Ducourtieux, C.; Petit, S. The spatio-temporal distribution of weed seed predation differs between conservation agriculture and conventional tillage. *Agric. Ecosyst. Environ.* **2014**, *188*, 40–47. [CrossRef]
62. Teasdale, J.; Mohler, C. The quantitative relationship between weed emergence and the physical properties of mulches. *Weed Sci.* **2000**, *48*, 385–392. [CrossRef]
63. Putnam, A.R.; DeFrank, J. Use of phytotoxic plant residues for selective weed control. *Crop Prot.* **1983**, *2*, 173–181. [CrossRef]
64. Prati, D.; Bossdorf, O. Allelopathic inhibition of germination by *Alliaria petiolata* (Brassicaceae). *Am. J. Bot.* **2004**, *91*, 285–288. [CrossRef] [PubMed]
65. Pratap, V.; Verma, S.; Dass, A. Weed growth, nutrient removal and yield of direct-seeded rice as influenced by establishment methods and chemical-cum-mechanical weed management practices. *Crop Prot.* **2023**, *163*, 106100. [CrossRef]
66. Singh, V.P.; Barman, K.K.; Singh, P.K.; Singh, R.; Dixit, A. Managing weeds in rice (*Oryza sativa*)-wheat (*Triticum aestivum*)-greengram (*Vigna radiata*) system under conservation agriculture in black cotton soils. *Indian J. Agric. Sci.* **2017**, *87*, 739–745. [CrossRef]
67. Chauhan, B.S.; Mahajan, G.; Sardana, V.; Timsina, J.; Jat, M.L. Productivity and Sustainability of the Rice-Wheat Cropping System in the Indo-Gangetic Plains of the Indian subcontinent: Problems, Opportunities, and Strategies. *Adv. Agron.* **2012**, *117*, 315–369. [CrossRef]
68. Alam, M.J.; Humphreys, E.; Sarkar, M. Comparison of dry seeded and puddled transplanted rainy season rice on the High Ganges River Floodplain of Bangladesh. *Eur. J. Agron.* **2018**, *96*, 120–130. [CrossRef]
69. Sapkota, T.B.; Majumdar, K.; Jat, M.L.; Kumar, A.; Bishnoi, D.K.; McDonald, A.J.; Pampolino, M. Precision nutrient management in conservation agriculture-based wheat production of Northwest India: Profitability, nutrient use efficiency and environmental footprint. *Field Crops Res.* **2014**, *155*, 233–244. [CrossRef]
70. Heydarzadeh, S.; Arena, C.; Vitale, E.; Rahimi, A.; Mirzapour, M.; Nasar, J.; Kisaka, O.; Sow, S.; Ranjan, S.; Gitari, H. Impact of different fertilizer sources under supplemental irrigation and rain-fed conditions on eco-physiological responses and yield characteristics of dragon's head (*Lallemantia iberica*). *Plants* **2023**, *12*, 1693. [CrossRef]
71. Hobbs, P.R.; Gupta, R.K. Resource-Conserving Technologies for Wheat in the Rice-Wheat System. In *Improving the Productivity and Sustainability of Rice-Wheat Systems: Issues and Impacts*; American Society of Agronomy, Inc.; Crop Science Society of America, Inc.; Soil Science Society of America, Inc.: Madison, WI, USA, 2001; pp. 149–171.

72. Johnson, S.E.; Angeles, O.R.; Brar, D.S.; Buresh, R.J. Faster anaerobic decomposition of a brittle straw rice mutant: Implications for residue management. *Soil Biol. Biochem.* **2006**, *38*, 1880–1892. [CrossRef]
73. Liao, P.; Huang, S.; van Gestel, N.C.; Zeng, Y.; Wu, Z.; van Groenigen, K.J. Liming and straw retention interact to increase nitrogen uptake and grain yield in a double rice-cropping system. *Field Crops Res.* **2018**, *216*, 217–224. [CrossRef]
74. Mandal, K.G.; Misra, A.K.; Hati, K.M.; Bandyopadhyay, K.K.; Ghosh, P.K.; Mohanty, M. Rice residue-management options and effects on soil properties and crop productivity. *J. Food Agric. Environ.* **2004**, *2*, 224–231.
75. Zheng, C.; Jiang, Y.; Chen, C.; Sun, Y.; Feng, J.; Deng, A.; Song, Z.; Zhang, W. The impacts of conservation agriculture on crop yield in China depend on specific practices, crops and cropping regions. *Crop J.* **2014**, *2*, 289–296. [CrossRef]
76. Hadas, A.; Kautsky, L.; Goek, M.; Erman Kara, E. Rates of decomposition of plant residues and available nitrogen in soil, related to residue composition through simulation of carbon and nitrogen turnover. *Soil Biol. Biochem.* **2004**, *36*, 255–266. [CrossRef]
77. Singh, Y.; Sidhu, H.S. Management of cereal crop residues for sustainable rice-wheat production system in the Indo-Gangetic plains of India. *Proc. Indian Natl. Sci. Acad.* **2014**, *80*, 95–114. [CrossRef]
78. Kaur, M.; Malik, D.P.; Malhi, G.S.; Sardana, V.; Bolan, N.S.; Lal, R.; Siddique, K.H. Rice residue management in the Indo-Gangetic Plains for climate and food security. A review. *Agron. Sustain. Dev.* **2022**, *42*, 92. [CrossRef]
79. Gautam, P.; Lal, B.; Panda, B.; Bihari, P.; Chatterjee, D.; Singh, T.; Nayak, P.; Nayak, A. Alteration in agronomic practices to utilize rice fallows for higher system productivity and sustainability. *Field Crops Res.* **2021**, *260*, 108005. [CrossRef]
80. Suryavanshi, T.; Sharma, A.R.; Nandeha, K.L.; Lal, S.; Porte, S.S. Effect of tillage, residue and weed management on soil properties, and crop productivity in greengram (*Vigna radiata* L.) under conservation agriculture. *J. Pharmacogn. Phytochem.* **2018**, *7*, 2022–2026.
81. Choudhary, M.; Patel, B.A.; Meena, V.S.; Yadav, R.P.; Ghasal, P.C. Seed bio-priming of green gram with Rhizobium and levels of nitrogen and sulphur fertilization under sustainable agriculture. *Legume Res.-Int. J.* **2019**, *42*, 205–210. [CrossRef]
82. Meena, J.; Behera, U.K.; Chakraborty, D.; Sharma, A. Tillage and residue management effect on soil properties, crop performance and energy relations in greengram (*Vigna radiata* L.) under maize-based cropping systems. *Int. Soil Water Conserv. Res.* **2015**, *3*, 261–272. [CrossRef]
83. Parihar, C.; Jat, S.; Singh, A.; Ghosh, A.; Rathore, N.; Kumar, B.; Pradhan, S.; Majumdar, K.; Satyanarayana, T.; Jat, M.; et al. Effects of precision conservation agriculture in a maize-wheat-mungbean rotation on crop yield, water-use and radiation conversion under a semiarid agro-ecosystem. *Agric. Water Manag.* **2017**, *192*, 306–319. [CrossRef]
84. Bajwa, A.A.; Walsh, M.; Chauhan, B.S. Weed management using crop competition in Australia. *Crop Prot.* **2017**, *95*, 8–13. [CrossRef]
85. Sapre, N.; Kewat, M.L.; Sharma, A.R.; Singh, P. Effect of tillage and weed management on weed dynamics and yield of rice in rice-wheat-greengram cropping system in vertisols of central India. *Indian J. Weed Sci.* **2022**, *54*, 233–239. [CrossRef]
86. Ghosh, S.; Das, T.; Rana, K.; Biswas, D.; Das, D.; Singh, G.; Bhattacharyya, R.; Datta, D.; Rathi, N.; Bhatia, A. Energy budgeting and carbon footprint of contrasting tillage and residue management scenarios in rice-wheat cropping system. *Soil Tillage Res.* **2022**, *223*, 105445. [CrossRef]
87. Raj, R.; Das, T.; Chakraborty, D.; Bhattacharyya, R.; Babu, S.; Govindasamy, P.; Kumar, V.; Ekka, U.; Sen, S.; Ghosh, S.; et al. Soil physical environment and active carbon pool in rice-wheat system of South Asia: Impact of long-term conservation agriculture practices. *Environ. Technol. Innov.* **2023**, *29*, 102966. [CrossRef]
88. Bhattacharyya, R.; Das, T.; Sudhishri, S.; Dudwal, B.; Sharma, A.; Bhatia, A.; Singh, G. Conservation agriculture effects on soil organic carbon accumulation and crop productivity under a rice-wheat cropping system in the western Indo-Gangetic Plains. *Eur. J. Agron.* **2015**, *70*, 11–21. [CrossRef]
89. Dutta, A.; Bhattacharyya, R.; Chaudhary, V.P.; Sharma, C.; Nath, C.P.; Kumar, S.N.; Parmar, B. Impact of long-term residue burning versus retention on soil organic carbon sequestration under a rice-wheat cropping system. *Soil Tillage Res.* **2022**, *221*, 105421. [CrossRef]
90. Kumar, K.; Goh, K. Crop Residues and Management Practices: Effects on Soil Quality, Soil Nitrogen Dynamics, Crop Yield, and Nitrogen Recovery. *Adv. Agron.* **1999**, *68*, 197–319. [CrossRef]
91. Tripathi, S.; Chander, S.; Meena, R.P.; Venkatesh, K.; Verma, A. Incorporation of rice residue and green gram cultivation saves nitrogen, improve soil health and sustainability of rice-wheat system. *Field Crops Res.* **2021**, *271*, 108248. [CrossRef]
92. Dey, A.; Dwivedi, B.S.; Meena, M.C.; Datta, S.P. Dynamics of soil carbon and nitrogen under conservation agriculture in rice-wheat cropping system. *Indian J. Fertil.* **2018**, *14*, 12–26.
93. Zhang, H.; Tang, X.; Hou, Q.; Zhu, Y.; Ren, Z.; Xie, H.; Liao, Y.; Wang, W.; Wen, X. Combining conservation tillage with nitrogen fertilizer measures promotes maize straw decomposition by regulating microbial community and enzyme activities. *Pedosphere* **2023**, *in press*. [CrossRef]
94. Guo, L.; Zheng, S.; Cao, C.; Li, C. Tillage practices and straw-returning methods affect topsoil bacterial community and organic C under a rice-wheat cropping system in central China. *Sci. Rep.* **2016**, *6*, 33155. [CrossRef] [PubMed]
95. Shao, Z.; Mwakidoshi, E.R.; Muindi, E.M.; Soratto, R.P.; Ranjan, S.; Padhan, S.R.; Wamukota, A.W.; Sow, S.; Alhammad, B.A.; Wasonga, D.O.; et al. Synthetic Fertilizer Application Coupled with Bioslurry Optimizes Potato (*Solanum tuberosum*) Growth and Yield. *Agronomy* **2023**, *13*, 1470.
96. Huang, S.; Zeng, Y.; Wu, J.; Shi, Q.; Pan, X. Effect of crop residue retention on rice yield in China: A meta-analysis. *Field Crops Res.* **2013**, *154*, 188–194. [CrossRef]

97. Nyawade, S.O.; Karanja, N.N.; Gachene, C.K.K.; Gitari, H.I.; Schulte-Geldermann, E.; Parker, M.L. Short-term dynamics of soil organic matter fractions and microbial activity in smallholder legume intercropping systems. *Appl. Soil Ecol.* **2019**, *142*, 123–135. [CrossRef]
98. Choudhary, M.; Sharma, P.C.; Jat, H.S.; Nehra, V.; McDonald, A.J.; Garg, N. Crop residue degradation by fungi isolated from conservation agriculture fields under rice-wheat system of North-West India. *Int. J. Recycl. Org. Waste Agric.* **2016**, *5*, 349–360. [CrossRef]
99. Li, T.; Xie, H.; Ren, Z.; Hou, Y.; Zhao, D.; Wang, W.; Wang, Z.; Liu, Y.; Wen, X.; Han, J.; et al. Soil tillage rather than crop rotation determines assembly of the wheat rhizobacterial communities. *Soil Tillage Res.* **2023**, *226*, 105588. [CrossRef]
100. Choudhary, M.; Datta, A.; Jat, H.S.; Yadav, A.K.; Gathala, M.K.; Sapkota, T.B.; Ladha, J.K. Changes in soil biology under conservation agriculture based sustainable intensification of cereal systems in Indo-Gangetic Plains. *Geoderma* **2018**, *313*, 193–204. [CrossRef]
101. Johnston, A.E.; Poulton, P.R.; Coleman, K. Soil Organic Matter: Its Importance in Sustainable Agriculture and Carbon Dioxide Fluxes. *Adv. Agron.* **2009**, *101*, 1–57. [CrossRef]
102. Modak, K.; Ghosh, A.; Bhattacharyya, R.; Biswas, D.R.; Das, T.K.; Das, S.; Singh, G. Response of oxidative stability of aggregate-associated soil organic carbon and deep soil carbon sequestration to zero-tillage in subtropical India. *Soil Tillage Res.* **2019**, *195*, 104370. [CrossRef]
103. Baghel, J.K.; Das, T.K.; Rana, D.S.; Paul, S. Effect of weed control on weed competition, soil microbial activity and rice productivity in conservation agriculture-based direct-seeded rice (*Oryza sativa*)-wheat (*Triticum aestivum*) cropping system. *Indian J. Agron.* **2018**, *63*, 129–136.
104. Usman, K.; Ullah, I.; Khan, S.M.; Khan, M.U.; Ghulam, S.; Khan, M.A. Integrated Weed Management Through Tillage and Herbicides for Wheat Production in Rice-Wheat Cropping System in Northwestern Pakistan. *J. Integr. Agric.* **2012**, *11*, 946–953. [CrossRef]
105. Seleiman, M.F.; Aslam, M.T.; Alhammad, B.A.; Hassan, M.U.; Maqbool, R.; Chattha, M.U.; Khan, I.; Gitari, H.I.; Uslu, O.S.; Roy, R.; et al. Salinity Stress in Wheat: Effects, Mechanisms and Management Strategies. *Phyton-Int. J. Exp. Bot.* **2013**, *91*, 667–694. [CrossRef]
106. Muoni, T.; Rusinamhodzi, L.; Thierfelder, C. Weed control in conservation agriculture systems of Zimbabwe: Identifying economical best strategies. *Crop Prot.* **2013**, *53*, 23–28. [CrossRef]
107. Mishra, J.; Kumar, R.; Mondal, S.; Poonia, S.; Rao, K.; Dubey, R.; Raman, R.K.; Dwivedi, S.; Kumar, R.; Saurabh, K.; et al. Tillage and crop establishment effects on weeds and productivity of a rice-wheat-mungbean rotation. *Field Crops Res.* **2022**, *284*, 108577. [CrossRef] [PubMed]
108. Chauhan, B.S.; Singh, R.G.; Mahajan, G. Ecology and management of weeds under conservation agriculture: A review. *Crop Prot.* **2012**, *38*, 57–65. [CrossRef]
109. Biswakarma, N.; Pooniya, V.; Zhiipao, R.; Kumar, D.; Verma, A.; Shivay, Y.; Lama, A.; Choudhary, A.; Meena, M.; Bana, R.; et al. Five years integrated crop management in direct seeded rice-zero till wheat rotation of north-western India: Effects on soil carbon dynamics, crop yields, water productivity and economic profitability. *Agric. Ecosyst. Environ.* **2021**, *318*, 107492. [CrossRef]
110. Parihar, C.; Jat, S.; Singh, A.; Majumdar, K.; Jat, M.; Saharawat, Y.; Pradhan, S.; Kuri, B. Bio-energy, water-use efficiency and economics of maize-wheat-mungbean system under precision-conservation agriculture in semi-arid agro-ecosystem. *Energy* **2017**, *119*, 245–256. [CrossRef]
111. Downing, A.S.; Kumar, M.; Andersson, A.; Causevic, A.; Gustafsson, Ö.; Joshi, N.U.; Krishnamurthy, C.K.B.; Scholtens, B.; Crona, B. Unlocking the unsustainable rice-wheat system of Indian Punjab: Assessing alternatives to crop-residue burning from a systems perspective. *Ecol. Econ.* **2022**, *195*, 107364. [CrossRef]
112. Magar, S.T.; Timsina, J.; Devkota, K.P.; Weili, L.; Rajbhandari, N. Conservation agriculture for increasing productivity, profitability and water productivity in rice-wheat system of the Eastern Gangetic Plain. *Environ. Chall.* **2022**, *7*, 100468. [CrossRef]
113. Wang, D.; Feng, H.; Li, Y.; Zhang, T.; Dyck, M.; Wu, F. Energy input-output, water use efficiency and economics of winter wheat under gravel mulching in Northwest China. *Agric. Water Manag.* **2019**, *222*, 354–366. [CrossRef]
114. Jat, S.; Parihar, C.; Singh, A.; Nayak, H.; Meena, B.; Kumar, B.; Parihar, M.; Jat, M. Differential response from nitrogen sources with and without residue management under conservation agriculture on crop yields, water-use and economics in maize-based rotations. *Field Crops Res.* **2019**, *236*, 96–110. [CrossRef]
115. Pooniya, V.; Biswakarma, N.; Parihar, C.; Swarnalakshmi, K.; Lama, A.; Zhiipao, R.; Nath, A.; Pal, M.; Jat, S.; Satyanarayana, T.; et al. Six years of conservation agriculture and nutrient management in maize–mustard rotation: Impact on soil properties, system productivity and profitability. *Field Crops Res.* **2021**, *260*, 108002. [CrossRef]
116. Ghosh, D.; Brahmachari, K.; Sarkar, S.; Dinda, N.K.; Das, A.; Moulick, D. Impact of nutrient management in rice-maize-greengram cropping system and integrated weed management treatments on summer greengram productivity. *Indian J. Weed Sci.* **2022**, *54*, 25–30. [CrossRef]
117. Chhokar, R.; Das, T.K.; Choudhary, V.; Chaudhary, A.; Raj, R.; Vishwakarma, A.; Biswas, A.; Singh, G.; Chaudhari, S. Weed Dynamics and Management in Conservation Agriculture. *J. Agric. Phys.* **2021**, *21*, 222–246.

Article

Effects of Ecotypes and Reduced N Fertilization on Root Growth and Aboveground Development of Ratooning Sorghum × Sudangrass Hybrids

Nayoung Choi [1], Miri Choi [2], Sora Lee [2], Chaelin Jo [2], Gamgon Kim [3], Yonghyun Jeong [4], Jihyeon Lee [5] and Chaein Na [2,6,*]

1   Future Agriculture Center, Kyung Nong Corporation, Gimje 54338, Republic of Korea; nychoi@dongoh.co.kr
2   Division of Applied Life Science, Gyeongsang National University, Jinju 52828, Republic of Korea; chlalfl0321@gnu.ac.kr (M.C.); dlthfk010111@gnu.ac.kr (S.L.); jcl5220@gnu.ac.kr (C.J.)
3   Future Technology Research Center, KT&G, Daejeon 34128, Republic of Korea; goney@ktng.com
4   Department of Seed Service, Korea Agriculture Technology Promotion Agency, Iksan 54667, Republic of Korea; yhjeong93@koat.or.kr
5   Crop Production and Physiology Division, National Institute of Crop Science, Rural Development Administration, Wanju 55365, Republic of Korea; dltla1264@korea.kr
6   Institute of Agriculture and Life Science, Gyeongsang National University, Jinju 52828, Republic of Korea
*   Correspondence: nachaein@gnu.ac.kr

Abstract: Reduced N input while maintaining biomass production of sorghum × sudangrass hybrids (*Sorghum bicolor* L. × *Sorghum sudanense*; SSG) is essential; however, its effects on root sustainability and photosynthetic capacity during the ratooning period are not well defined in a multiple harvests system. The physiological response and root morphology of SSG were investigated under different N application levels during the ratooning period in a two-year field experiment. Treatments were all combinations of two ecotypes (late-flowering, Greenstar; early-flowering, Honeychew) and four N levels (0, 50, 100, 150 kg N ha^{-1}). The total root length, surface area, volume, tips, and dry matter (DM) were significantly influenced by both ecotype and N level, with Greenstar outperforming Honeychew. Specifically, Greenstar's root length increased by up to three times with reduced N application (50 kg N ha^{-1}), while Honeychew showed significant root length increases only at higher N levels (100 and 150 kg N ha^{-1}). Our data support the conclusion that a low level of N (50–100 kg N ha^{-1}) was the optimal rate for ratooning root sustainability. The findings highlight the critical role of root development in sustaining biomass production and suggest that the late-flowering ecotype, Greenstar, is more suitable for a multiple harvests system with a robust root system.

Keywords: multiple harvests system; root sustainability; additional nitrogen; ecotype; photosynthetic capacity

Citation: Choi, N.; Choi, M.; Lee, S.; Jo, C.; Kim, G.; Jeong, Y.; Lee, J.; Na, C. Effects of Ecotypes and Reduced N Fertilization on Root Growth and Aboveground Development of Ratooning Sorghum × Sudangrass Hybrids. *Agronomy* **2024**, *14*, 2073. https://doi.org/10.3390/agronomy14092073

Academic Editor: Jorge Teixeira

Received: 16 August 2024
Revised: 3 September 2024
Accepted: 8 September 2024
Published: 10 September 2024

1. Introduction

Sorghum × sudangrass hybrids (*Sorghum bicolor* L. × *Sorghum sudanense*) are a multi-purpose crop for use as forage or dedicated cellulosic biofuels [1,2]. Sorghum × sudangrass hybrids have high biomass yields and tolerance to various environmental stresses such as high temperature, drought, and soil nutrient deficiency while requiring low agricultural input. For instance, forage sorghums generally have heat and low moisture tolerance and can be productive in areas of annual rainfall as low as 400–650 mm [3], including Southern USA and other semiarid subtropical environments [4]. Previous research indicated that hybrid forage sorghums and sorghum × sudangrass hybrids have the potential for increased biomass yields even under stressful growing conditions in temperate regions [5,6]. The physiological basis for the superior performance of sorghum hybrids has been ascribed to a sustained greater carbon exchange rate over wider environmental conditions with minimum management, but that may be contingent upon the parental genetic background more than heterosis [7].

Unlike the distinctive one-time harvest of annual row crops such as maize, wheat, rice, or soybean, multiple harvests are possible during the vegetative period (ratoon harvests) when sorghum hybrids are used for livestock feed [8–10]. To achieve a multiple harvests farming system, photoperiod-sensitive sorghums that remain vegetative for long periods of time are available [6,11–14]. This photosensitivity response to day-length delays heading until the day-length is from less than 13.2 h [12] to 12.2 h [13].

Even though the nutrient requirements for forage sorghum are relatively low compared with maize, adequate nutrient management is needed to sustain biomass production in multiple harvest conditions where N removal occurs. After the first harvest is conducted, the ratooning crop must distribute captured mineral nutrients to the newly growing above- and belowground parts. This will impact the subsequent pace of capturing N, or the uptake of resources from the soil. Previous research from our research group revealed that N-agronomic efficiency (NAE) was greatest at 50 kg N ha^{-1} after the first summer harvest and then decreased with higher N application rates for a sorghum $\times$ sudangrass hybrid [6]. There is limited research on the effects of multiple harvests on bioenergy crop growth and development throughout the entire ratooning season with reduced N fertilization, as the major focus of multiple harvest research has been on perennial forage grasses [15–17].

While plant roots play an essential role in anchoring the plant and the acquisition of nutrients and water, the corresponding trait responses to environmental conditions are complex and not well defined. This is due to overall difficulties in root phenotype monitoring and the limitations of existing methods for evaluating the characteristics of roots and environmental conditions [18]. Studies focusing on roots and their role in nutrient uptake are essential to support the development of management strategies to increase crop production while improving nutrient-use efficiency [19]. There has been some research describing the sorghum crown root angle effect on water use efficiency and yield [20,21]; however, there is no previous research evaluating ratooning effects on existing roots for sorghum $\times$ sudangrass hybrid. There have been some efforts to achieve a better understanding of root development under field conditions. For instance, the Winrhizo system (Regents Instruments Inc., QC, Canada) has the capability of quantifying the root length, diameter, surface area, and total volume of the root, which are acquired from field conditions [22–24]. The current research on root morphological changes has been conducted primarily in controlled environments with seedlings and pot experiments [25,26]. However, root growth in controlled conditions is often different than that observed under field conditions due to several abiotic and biotic factors that vary widely [26–30]. Thus, to better understand overall plant response to summer harvest and following regrowth, it is necessary to investigate the seasonal change of sorghum $\times$ sudangrass root morphology during the ratooning period in field conditions.

To understand the difference in the agronomic responses of ratooning sorghum $\times$ sudangrass hybrids, our research group previously investigated ratooning sorghum yield over 15 weeks for 3 years with early and late flowering ecotypes. We found that the practical range for additional N application rate can be reduced as much as between 50 and 100 kg ha^{-1}, and late-flowering cultivars tend to have high aboveground biomass yield and agronomic performance [6]. However, physiological response and root sustainability have not been reported for ratooning sorghum plants in response to additional N applications. Thus, the objectives of the current study were to evaluate the overall changes of root morphological traits and photosynthetic capacity in response to reduced N applications during the ratooning of early and late flowering sorghum $\times$ sudangrass hybrids.

2. Materials and Methods

2.1. Research Site and Field Management

The experiment was carried out in the Gyeongsang National University Research Farm in Jinju, Korea (35°14′39″ N 128°09′19″ E) for 2 years (2018–2019). Historically, the research site had been used for lowland rice production and was converted to upland for the current research. Soil samples were collected at a 30-cm depth using a soil auger (diameter 2.54 cm) prior to planting each year. Up to 20 samples were collected in a "W" pattern throughout the field and then composited. The chemical properties of soil are presented in Table 1. Overall soil organic matter and mineral nutrient concentrations were fairly poor, including the total N concentration. These data indicate that the fertilizer application did not affect the overall soil N, P, and K pools, except for available P_2O_5 in 2019.

Table 1. Soil chemical properties of the research site from 2018 to 2019.

Year	pH	EC	OM	Total N	Av. P_2O_5	K	Ca	Mg
	(1:5)	(dS m^{-1})	(g kg^{-1})	(g kg^{-1})	(mg kg^{-1})	(cmol$^+$ kg^{-1})		
2018	6.3	0.20	7.0	0.37	104	0.22	2.84	0.49
2019	6.3	0.22	10.9	0.34	37	0.21	3.49	0.40

At the beginning of the growing season, the field was plowed and then rotovated multiple times each year. A pre-planting application of 100 kg ha^{-1} of N was made using mixed fertilizer (21-17-17) that was manually broadcasted and then incorporated into the soil using a tractor-mounted rotovator. Then, 30 kg ha^{-1} of sorghum × sudangrass hybrid seeds (1000-seed weight of 25 g) were planted using a hand-pushed two-row manual seeder (AP-2, Agritecno Yazaki Korea, Cheongju, Republic of Korea) with 15-cm inter-row and 5.5-cm intra-row spacing. To investigate ecotype differences for root development during ratooning, two distinctive ecotypes were tested in both years: the early-flowering type Honeychew and the late-flowering type Greenstar. Major field management operations and the dates of flowering for each ecotype are shown in Table 2. There were no significant weed issues due to the good early germination and fast-growing characteristics of sorghum × sudangrass hybrids. Irrigation was not required due to sufficient rainfall during the season.

Table 2. Major field management operation schedule and developmental stages for two years.

Operations and Developmental Stages		2018	2019
Plowing, leveling		9 May	7 May
Preplant fertilizer application		9 May	7 May
Sowing		11 May	8 May
Summer harvest [a]		2 August	22 July
Additional N fertilizer application		10 August	26 July
Root sampling	Early	-	19 August (DAS [b] 28)
	Middle	17 October (DAS 76)	25 September (DAS 65)
	Late	23 November (DAS 113)	30 October (DAS 100)
Boot stage (Greenstar/Honeychew)		N.D. [c]/4 October	21 October/20 September
Full flowering stage (Greenstar/Honeychew)		N.D./17 October	28 October/4 October

[a] The yield data can be found in a previously published paper [6]. [b] DAS, Days after summer harvest. [c] N.D., Not Detected.

The weather data during the growing season were collected from the Automated Weather System station, operated by the Korea Meteorological Agency (https://data.kma.go.kr/, accessed on 1 February 2020). Rainfall events during the early regrowth period were distinctively different for the 2 years, especially from the summer harvest through the first 4-week ratooning period (341 vs. 11 mm in 2018 and 2019, respectively). The experimental site received a similar amount of rainfall during the overall regrowth period; a total of 664 (2018) and 597 (2019) mm (Figure 1). Monthly mean temperature and GDD data indicate that 2018 had a warmer spring and summer, while 2019 had a warmer fall.

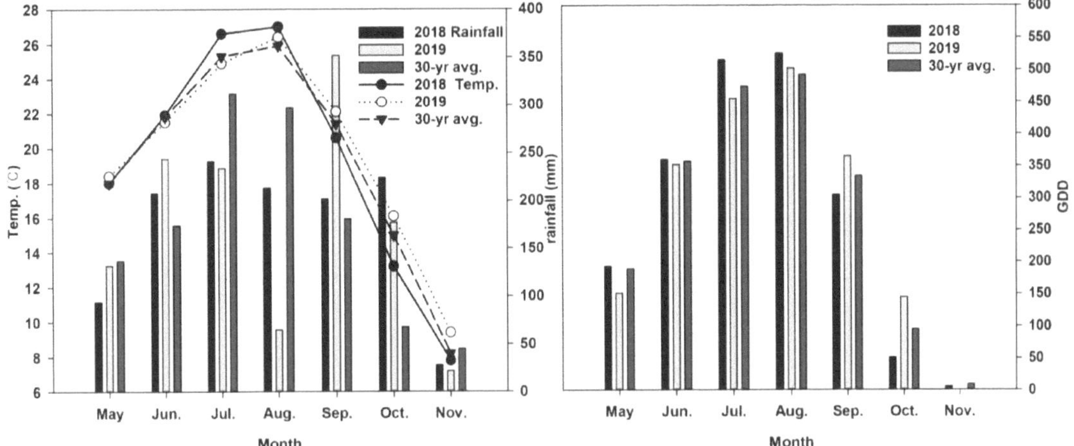

Figure 1. Monthly total rainfall (**left**), average temperature (**left**), and growing degree days (GDD, **right**) for two growing periods (2018–2019) and the 30-year average for the region.

2.2. Summer Harvest Management and N Treatments

The experimental design was a randomized complete block (RCBD) with a split-plot limitation to randomization with four replications; the main plots were the ecotypes and the subplots were the N levels. The size of each subplot was 6 m × 7 m. The entire field was summer harvested (Table 2) using a gas-powered handheld brush cutter (KG350S, Honda, Tokyo, Japan), leaving a target stubble height of 5 cm; it is recommended to avoid picking up soil during machine harvest. Cut biomass was removed from the field immediately. To implement N level treatments during ratooning, 0 (control), 50, 100, and 150 kg ha^{-1} of N fertilizer (urea) were broadcasted with a handheld spreader (GE-US 18 Li, Einhell, Landau an der Isar, Germany) within a week after the summer harvest in each year.

2.3. Ratooning Chlorophyll Content and LAI

Chlorophyll content was measured on fully developed uppermost leaves, excluding the flag leaf, of 10 plants per plot using a CCM-300 (Opti-Sciences Inc., Hudson, NH, USA), which records chlorophyll content on a surface basis (mg m^{-2}). Chlorophyll content data were collected at approximately 15-day intervals from days after summer harvest (DAS) 30 to 75 in 2018, and it extended from DAS 15 to 90 in 2019. Leaf area index (LAI) measurements were obtained from the center rows to minimize the border effect. The LAI of each plot was estimated using a LAI2200-C canopy analyzer (Li-Cor BioSciences, Lincoln, NE, USA) with a 90° view angle cap and six under-canopy readings in a "W" pattern. LAI data were collected at approximately 15-day intervals from DAS 30 to 75 in 2018, and from DAS 15 to 75 in 2019.

2.4. Ratooning Root Sampling

In 2018, after the different levels of N were applied, root sampling occurred at two different times (17 October and 23 November) while focused on the fully developed root morphology of the ratooning crop. After the first year of root characteristics analysis in 2018, we decided to further assess root morphological change during the 2019 growing season. Specifically, to describe root development before summer harvest, root sampling occurred at three different times (3 June, 2 July, and 22 July), and to extend our understanding of the root development of ratooning plants, regrowth plant root sampling was carried out at three different times (19 August, 25 September, and 30 October).

To obtain representative roots from each plot, we first made a circle (r = 15 cm) around the targeted plant, showing representative aboveground growth in the middle rows. Then, three shovels were inserted simultaneously at a 90-degree angle into the soil to a depth of 30 cm to collect roots in the targeted area (Figure 2). A total of 7 plant samples were collected from each plot, and the 5 most representative samples were chosen for further lab analysis. Each plant was separated into shoots and roots, and the roots were separated from the soil by adding tap water and straining the root out of the soil with a fine strainer (500 μm) [20]. The washed roots were stored in plastic zipper bags that were partially filled with water and stored at 4 °C for a maximum of 3 days. Prepared root samples were placed on a 30 cm × 40 cm acrylic tray, and enough water was added to submerge the samples to achieve clear resolution during the scanning. To obtain root images, samples were scanned (Epson Expression 12000XL, Seiko-Epson Corp. Suwa, Japan) at 400 dpi. The scanned root images were analyzed using the Winrhizo Pro ver. 2017a software program (Regent Instruments Inc., Sainte-Foy, QC, Canada), obtaining the root length, root surface area, root volume, and tips; tips are the number of root endings. The workflow of root morphological analysis is described in Figure 2.

Figure 2. Field-scale root morphological analysis workflows with the Winrhizo systems.

2.5. Statistical Methods

The PROC UNIVARIATE function of SAS 9.4 software (SAS Institute Inc., Cary, NC, USA) was used to assess the normality of data distribution. In 2018 and 2019, after summer harvest measurements (ratooning period chlorophyll content, LAI, total root length, total root surface area, total root volume, tips and root DM) were analyzed with ANOVA using the PROC MIXED model. N levels and ecotype were treated as fixed effects, and DAS was treated as a repeated measure. The block and block × ecotype were treated as random effects. The year was not included in the statistical analysis because sorghum × sudangrass is an annual crop and randomization occurred in each year. N level treatments were compared using polynomial contrast [6]. The least significant difference (LSD) test (p = 0.05) was utilized for mean separation.

3. Results

3.1. Chlorophyll Content and LAI

To evaluate the effects of N fertilization on the ratooning sorghum × sudangrass hybrid, chlorophyll content and LAI were measured throughout the ratooning period. There were significant DAS × N level interactions ($p = 0.001$, <0.001; 2018, 2019, respectively) for both years (Table 3). There were linear and quadratic responses of the chlorophyll content to the N level in 2018 and 2019, respectively, except for DAS 75 in 2019 (Table 4). In 2018, the control (0 N) consistently had the lowest chlorophyll content throughout the season compared to plots receiving the other N treatments (Table 4). At DAS 75, 50 N had a relatively lower chlorophyll content than 100 N and 150 N (466 vs. 492 and 503 mg m^{-2}). To identify early and late changes in chlorophyll content, measurements at DAS 15 and 90 were added in 2019. Overall trends in 2019 were similar to the previous year, with the control having the lowest chlorophyll content throughout the season. Additionally, 50 N had a relatively lower chlorophyll content than other N treatments, and no further change occurred when the N application level reached 100 N. For both years, plots receiving N had greater chlorophyll content in the early season.

Table 3. Source of variation and levels of probability (p) for chlorophyll content and LAI during the ratooning period.

Source of Variation	Chlorophyll Content		LAI	
	2018	2019	2018	2019
N level (N)	**<0.001** [a]	**<0.001**	**<0.001**	**<0.001**
Ecotype (E)	0.091	0.334	0.085	0.173
Days after summer harvest (DAS)	**<0.001**	**<0.001**	**<0.001**	**<0.001**
N × E	0.321	0.054	**0.012**	0.397
N × DAS	**0.001**	**<0.001**	**<0.001**	**<0.001**
E × DAS	0.968	**<0.001**	0.929	**<0.001**
N × E × DAS	0.460	0.807	0.865	0.075

[a] Significance evaluated at 0.05 probability level and bolded.

Table 4. Seasonal changes of chlorophyll content (mg m^{-2}) for two years (2018–2019) by N level (0, 50, 100, and 150 kg N ha^{-1}).

Year	Treatments		Ratooning Period					
			DAS [a] 15	DAS 30	DAS 45	DAS 60	DAS 75	DAS 90
2018	N level	0 N	-	438 a [b] B [c]	419 aB	395 bB	423 aC	-
		50 N	-	527 aA	449 bcA	439 cA	466 bB	-
		100 N	-	543 aA	450 cA	432 cA	492 bAB	-
		150 N	-	537 aA	462 cA	450 cA	503 bA	-
	Polynomial contrast [d]			L ***Q ***	L **	L *	L ***	
2019	N level	0 N	418 cB	418 cC	468 aB	462 aB	440 bB	425 bcB
		50 N	495 aA	456 bcB	467 bB	475 abB	474 bA	445 cAB
		100 N	507 aA	497 abA	480 bAB	479 bB	485 bA	459 cA
		150 N	500 aA	497 aA	495 aA	507 aA	489 aA	460 bA
	Polynomial contrast		L ***Q ***	L ***Q *	L *	L **	L **	NS

[a] DAS, Days after summer harvest. [b] Means followed by different lowercase letters within a row differ significantly (LSD, $p < 0.05$). [c] Means followed by different uppercase letters within a column differ significantly (LSD, $p < 0.05$). [d] Polynomial contrast was used to compare N level (L, linear; Q, quadratic; *** $p < 0.001$; ** $p < 0.01$; * $p < 0.05$; NS, statistically not significant).

The LAI of ratooning sorghum × sudangrass hybrids showed that there were significant DAS × N level interactions ($p < 0.001$) for both years (Table 3). Similar to chlorophyll content in 2018, the control continuously had the lowest LAI throughout the ratooning seasons (Figure 3). DAS 30 showed the highest LAI across the N levels, then it decreased throughout the season. At DAS 30, 50 N treated had relatively lower LAI than 100 N and 150 N (4.7 vs. 5.4 and 5.7, respectively). To identify the early change of LAI, a measurement at DAS 15 was added in 2019. In 2019, except for DAS 15, the control continuously had the lowest LAI throughout the ratooning season. Additionally, 100 N and 150 N had higher LAI than 50 N at DAS 30 to 45.

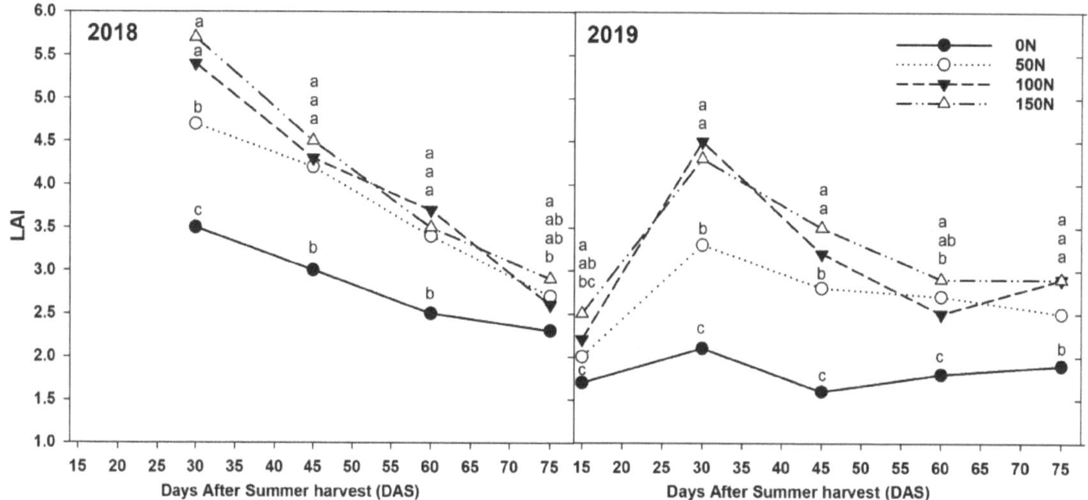

Figure 3. Seasonal changes of LAI during the ratooning period for two years (2018–2019) by N level (0, 50, 100, and 150 kg N ha^{-1}). Means followed by different lowercase letters within a DAS differ significantly (LSD, $p < 0.05$).

3.2. Root Morphology and Dry Matter

The total root length data showed that there were three-way interactions of N × E × DAS ($p = 0.011$) in 2018, while there were two-way interactions of N × E ($p = 0.002$) and E × DAS ($p < 0.001$) in 2019 (Table 5). Thus, data were presented showing the N × E and E × DAS interactions to explore the plant root responses after defoliation over the 2-year period. There were linear and quadratic effects of N level in 2018, but there was a linear response of total root length to the N level in 2019 (Table 6). Table 6 shows that, in 2018, Greenstar (late-flowering ecotype) had a greater increase in total root length in response to N fertilizer compared with Honeychew (early-flowering ecotype). Specifically, in comparison with the control (0 N), Greenstar showed a 3-fold increase, while Honeychew receiving 50 N was not different than the control. Moreover, Greenstar maintained a high root length regardless of N fertilization levels, whereas the Honeychew root length increased for 100 N and 150 N compared with 50 N. A similar pattern was observed in 2019. Greenstar showed a significant increase in root length with 50 N or greater, while Honeychew demonstrated an increase in total root length only at 100 N or greater. These results indicate that the two ecotypes responded differently to N levels across the 2 years, with Greenstar generally maintaining higher root length development even at relatively low levels of N fertilization.

When the summer defoliation responses of total root length in the later season were compared in 2018, both Greenstar and Honeychew showed a decrease between DAS 76 and 113, but the degree of difference varied. For instance, Greenstar only had a 16% decrease, while Honeychew showed a 32% decrease in total root length. In 2019, three different dates were compared (DAS 28, 65, and 100) for the total root length after summer defoliation, including the earlier determination of the total root length. In 2019, for the early regrowth season (DAS 28), Greenstar had a 61% greater total root length than Honeychew, and this difference was maintained throughout the season.

Table 5. Source of variation and levels of probability (p) for total root length, total root surface area, total root volume, tips, and root DM during the ratooning period.

Source of Variation	Total Root Length		Total Root Surface Area		Total Root Volume		Tips		Root DM
	2018	2019	2018	2019	2018	2019	2018	2019	2019
N level (N)	**<0.001** [a]	**<0.001**	**<0.001**	**<0.001**	**<0.001**	**<0.001**	**<0.001**	**<0.001**	**<0.001**
Ecotype (E)	0.381	**0.003**	0.773	**0.005**	0.893	**0.006**	0.627	**0.002**	**0.005**
Days after summer harvest (DAS)	**<0.001**	**<0.001**	**0.001**	**<0.001**	**0.019**	**<0.001**	**0.002**	**0.003**	**<0.001**
N × E	0.427	**0.002**	0.168	**0.005**	0.092	**0.032**	0.414	**<0.001**	0.077
N × DAS	0.401	0.770	0.443	0.837	0.348	0.560	0.450	0.585	**0.038**
E × DAS	0.213	**<0.001**	0.119	**<0.001**	0.066	**<0.001**	0.201	**0.005**	**<0.001**
N × E × DAS	**0.011**	0.267	**0.009**	0.504	**0.019**	0.790	**0.043**	0.193	0.872

[a] Significance evaluated at 0.05 probability level and bolded.

Table 6. Total root length, surface area, volume, and tips affected by N level (0, 50, 100, and 150 kg N ha^{-1}), ratooning periods (early, middle, and late seasons), and ecotype (late-flowering Greenstar, early-flowering Honeychew) for two growing seasons (2018–2019).

Year	Treatments		Total Root Length		Total Root Surface Area		Total Root Volume		Tips	
			cm plant^{-1}		cm^2 plant^{-1}		cm^3 plant^{-1}		plant^{-1}	
			Greenstar	Honeychew	Greenstar	Honeychew	Greenstar	Honeychew	Greenstar	Honeychew
2018	N level	0 N	377 a [b] B [c]	526 aB	76 aB	120 aB	1.73 aB	2.25 aC	1780 aB	2470 aB
		50 N	1063 aA	737 aAB	234 aA	165 aAB	4.22 aA	2.99 bBC	4630 aA	3380 aAB
		100 N	1079 aA	996 aA	232 aA	223 aA	4.05 aA	4.07 aAB	4490 aA	4490 aA
		150 N	1123 aA	982 aA	215 aA	226 aA	3.35 aA	4.21 aA	4690 aA	4380 aA
	Polynomial contrast [d]		L **Q *	L **	L **Q **	L **	L *Q ***	L **	L **Q *	L *
	Ratooning period	DAS [a] 76	992 aA	964 aA	201 aA	216 aA	3.41 aA	3.93 aA	4180 aA	4340 aA
		DAS 113	829 aB	657 aB	177 aA	151 aB	3.27 aA	2.83 aB	3620 aA	3020 aB
2019	N level	0 N	318 aC	226 bB	98 aC	65 bB	2.52 aC	1.52 bB	1240 aC	800 bB
		50 N	530 aB	246 bB	179 aB	80 bB	4.93 aB	2.12 bB	2110 aAB	880 bB
		100 N	512 aB	399 bA	186 aAB	136 bA	5.51 aAB	3.80 bA	1920 aB	1460 bA
		150 N	606 aA	422 bA	209 aA	141 bA	5.84 aA	3.80 bA	2370 aA	1570 bA
	Polynomial contrast		L **	L ***	L **	L ***	L **	L ***	L **	L ***
	Ratooning period	DAS 28	319 aC	198 bB	106 aC	67 bC	2.91 aC	1.86 bC	1270 aC	740 bB
		DAS 65	478 aB	377 bA	157 aB	113 bB	4.23 aB	2.70 bB	2020 aB	1510 bA
		DAS 100	677 aA	394 bA	241 aA	137 bA	6.95 aA	3.86 bA	2440 aA	1280 bA

[a] DAS, Days after summer harvest. [b] Means followed by different lowercase letters within a row differ significantly (LSD, $p < 0.05$). [c] Means followed by different uppercase letters within a column differ significantly (LSD, $p < 0.05$). [d] Polynomial contrast was used to compare N level (L, linear; Q, quadratic, *** $p < 0.001$, ** $p < 0.01$, * $p < 0.05$).

Similar to the total root length, the root surface area was affected by a three-way interaction of N × E × DAS ($p = 0.009$) in 2018, while there were two-way interactions of N × E ($p = 0.005$) and E × DAS ($p < 0.001$) in 2019 (Table 5). There were linear and quadratic responses of the total root surface area to the N level in 2018, but there was a linear effect of the N level in 2019 (Table 6). Plots fertilized with N exhibited a greater root surface area than the control for both Greenstar and Honeychew in 2018 (Table 6); however, no differences were observed across N levels. Similarly, in 2019, Greenstar showed an increased root surface area with 50 N and above, while Honeychew demonstrated an increase in total root surface area at 100 N and above. The total root surface area in the later ratooning season was compared in 2018, and both Greenstar and Honeychew tended to decrease between DAS 76 and 113, but the magnitude of the difference differed (16.4% vs. 31.8%). In the early ratooning season of 2019 (DAS 28),

Greenstar had a 58% greater total root surface area than Honeychew, a difference maintained throughout the season (21.1% and 41.8% for DAS 65 and 100, respectively).

The total root volume was affected by a three-way interaction of N $\times$ E $\times$ DAS ($p = 0.019$) in 2018, while there were two-way interactions of N $\times$ E ($p = 0.032$) and E $\times$ DAS ($p < 0.001$) in 2019 (Table 5). Similar to the total root length and surface area, there were linear and quadratic responses of the total root volume to the N level in 2018, but there was a linear effect of the N level in 2019 (Table 6). Table 6 demonstrates that Greenstar had a more pronounced increase in total root volume in response to N levels compared to Honeychew in 2018. Specifically, compared with the control, Greenstar at 50 N exhibited a two-fold increase, while Honeychew showed no difference. Furthermore, Greenstar maintained a high total root volume with all N fertilization levels, while Honeychew had a significant increase only at 100 N and 150 N in 2018. A similar trend was observed in 2019, where the 50 N treatment had a two-fold increase compared to the control for Greenstar, but there was no difference between these treatments for Honeychew. Greenstar had greater root volume compared to Honeychew across the N levels in 2019, while the difference showed only for 50 N in 2018. In 2018, for DAS 76 vs. 113, there was no difference for Greenstar, while Honeychew showed less total root volume at DAS 113. Conversely, when the comparison was conducted earlier in the regrowth period in 2019 (DAS 28), significant differences in root development were observed for both ecotypes, while Greenstar consistently showed greater total root volume than Honeychew.

Root tips data showed that there was a three-way interaction of N $\times$ E $\times$ DAS ($p = 0.043$) in 2018, while there were two-way interactions of N $\times$ E ($p < 0.001$) and E $\times$ DAS ($p = 0.005$) in 2019 (Table 5). There was a difference among N level treatments, but ecotype did not affect the response in 2018 (Table 6). There were linear and quadratic effects of the N level in 2018, while there was a linear response of root tips to the N level in 2019 (Table 6). Plots treated with N had greater numbers of root tips compared with the control, but there were no differences among the N-treated plots. In 2019, Greenstar had a greater number of tips compared to the control for N applications of 50 N and above, while it was 100 N and above for Honeychew. Additionally, similar to other root-associated parameters, Greenstar had greater tip number development than that of Honeychew. When later seasons were compared in 2018, both Greenstar and Honeychew exhibited a decrease in the root tip number between DAS 76 and 113, but to different extents. For instance, Greenstar had only a 13% decrease, while Honeychew exhibited a 30% decrease in tip numbers. At DAS 28, 65, and 100 in 2019, Greenstar had 71%, 33%, and 91% greater tip numbers than Honeychew, respectively, indicating that Greenstar develops root tips faster in the early ratooning season and maintains them better into the later season compared to Honeychew.

Root dry matter (DM) data showed that there were two-way interactions of N $\times$ DAS ($p = 0.038$) and E $\times$ DAS ($p < 0.001$) in 2019 (Table 5). The polynomial contrast result showed that there was a linear effect of the N level in 2019 (Table 7). On average, the control had the lowest root DM (0.97 g plant^{-1}), while 100 N and 150 N had the highest root DM (2.30 and 2.47 g plant^{-1}, respectively, Table 7). At DAS 65 and 100, Greenstar had 47% and 92% greater root DM than Honeychew, and the difference was greatest in late season (DAS 100; 4.19 vs. 2.18 g plant^{-1}). The control (0 N) consistently recorded the lowest root DM throughout the regrowth period. Both 100 N and 150 N resulted in up to three-fold greater root DM compared with the control at DAS 28, and the trend was maintained throughout the regrowth season. At DAS 65 and 100, any level of N applied increased root DM over that of the control. The 150 N had demonstrated the highest root DM values (4.16 g plant^{-1}) at DAS 100. Among N treatments, averages of 100 N and 150 N were 30% and 41% greater than those of 50 N, respectively.

Comparing root morphological characteristics over time before summer defoliation, there was an E $\times$ DAP (days after planting) interaction for total root length, total root surface area, and root DM, while the main effects of DAP and ecotype were noted for total root volume and tips (Table 8). Notable differences in the development of total root length were evident among the ecotypes. For instance, the total root length of Greenstar reached 1210 cm per plant at DAP 55, and then it plateaued (Table 8). Greenstar exhibited 64% and 89% greater root length development than Honeychew at DAP 55 and DAP 75, respectively.

The trend for the total root surface area was also similar. Specifically, Greenstar exhibited 47% and 104% more total root surface area than Honeychew, and both cultivars displayed rapid root development until DAP 55, after which they plateaued. In terms of total root volume, the average for Greenstar was 3.21 cm^3, while it was 2.17 cm^3 for Honeychew. The tip number for Greenstar was 3240 per plant, while it was only 2030 per plant for Honeychew. Unlike other root parameters, there were no statistical differences between ecotypes at DAP 55 for root DM. At DAP 75, Greenstar had 90% greater root DM than Honeychew (2.10 vs. 1.10 g plant^{-1}, respectively).

Table 7. Root dry matter (DM) affected by additional N level (0, 50, 100, and 150 kg N ha^{-1}), ratooning periods (early, middle, and late seasons), and ecotype (late-flowering Greenstar, early-flowering Honeychew) in 2019.

Treatments		Ecotype		
		g plant^{-1}		
		Greenstar	Honeychew	Avg.
N level	0 N	1.22	0.72	0.97 C [b]
	50 N	2.45	1.03	1.74 B
	100 N	2.78	1.82	2.30 A
	150 N	2.99	1.95	2.47 A
	Avg.	2.36 a [c]	1.38 b	
	Polynomial contrast [d]	L *	L ***	
Ratooning period	DAS [a] 28	0.92 aC	0.62 aC	
	DAS 65	1.98 aB	1.35 bB	
	DAS 100	4.19 aA	2.18 bA	
		Ratooning period		
		DAS 28	DAS 65	DAS 100
N level	0 N	0.29 bB	0.74 bC	1.89 aC
	50 N	0.67 cAB	1.55 bB	3.01 aB
	100 N	1.02 cA	2.22 bA	3.66 aA
	150 N	1.11 cA	2.14 bAB	4.16 aA

[a] DAS, Days after summer harvest. [b] Means followed by different uppercase letters within a column differ significantly (LSD, $p < 0.05$). [c] Means followed by different lowercase letters within a row differ significantly (LSD, $p < 0.05$). [d] Polynomial contrast was used to compare N level (L, linear; *** $p < 0.001$; * $p < 0.05$).

Table 8. Total root length, surface area, and volume in response to ecotype (late-flowering Greenstar, early-flowering Honeychew) before summer defoliation in 2019 and source of variation and levels of probability (p).

Treatments	Total Root Length		Total Root Surface Area		Total Root Volume			Tips			Root DM	
	cm plant^{-1}		cm^2 plant^{-1}		cm^3 plant^{-1}			plant^{-1}			g plant^{-1}	
	Greenstar	Honeychew	Greenstar	Honeychew	Greenstar	Honeychew	Avg.	Greenstar	Honeychew	Avg.	Greenstar	Honeychew
DAP [a] 26	217 a [b] B [c]	229 aB	50 aB	53 aB	0.95	1.00	0.98 B	680	740	710 B	0.23 aC	0.27 aB
DAP 55	1209 aA	738 bA	243 aA	165 bA	3.93	2.99	3.46 A	4520	2580	3550 A	1.14 aB	1.08 aA
DAP 75	1326 aA	701 bA	302 aA	148 bA	4.74	2.51	3.63 A	4520	2780	3650 A	2.10 aA	1.10 bA
Avg.					3.21 a	2.17 b		3240 a	2030 b			
Source of variation												
Ecotype (E)	<0.001 [d]		<0.001		0.039			0.002			0.031	
DAP	<0.001		<0.001		<0.001			<0.001			<0.001	
E × DAP	0.020		0.036		0.137			0.064			0.041	

[a] DAP, Days after planting. [b] Means followed by different lowercase letters within a row differ significantly (LSD, $p < 0.05$). [c] Means followed by different uppercase letters within a column differ significantly (LSD, $p < 0.05$). [d] Significance evaluated at 0.05 probability level and bolded.

4. Discussion

4.1. Chlorophyll Content and LAI Responses

The current study investigated seasonal changes in the above- and belowground development of two different ecotypes of sorghum × sudangrass hybrids, exploring the impact of different N application levels following summer defoliation. Aboveground development after summer defoliation in the current study indicated that N application linearly increased the leaf chlorophyll content of ratooning plants, and higher chlorophyll content was maintained until later in the growing season, which might have contributed to LAI and root development differences. Specifically, plants receiving 100 N (Avg. 7%) and 150 N (Avg. 9%) sustained higher chlorophyll content than other N applications throughout the ratooning season for both 2018 and 2019. Leaf nitrogen and photosynthesis are connected, as most of the N in leaves are associated with photosynthetic machinery [31]. In addition, it is known that leaf senescence is related to the photosynthetic capacity and chlorophyll content [32]. Previous research indicated that the supply of N increased the photosynthetic apparatus and chlorophyll content in the plant leaves, while leaf senescence was delayed in grain sorghum [33]. Similar results were reported in other C4 plants, such as sugarcane [34]. The stay-green trait delayed leaf senescence, but more importantly, it maintained leaf photosynthesis following the flowering period and increased grain yields in maize [35]. After DAS 45, there were no significant differences of LAI observed in terms of N input in 2018. This was due to measurements being taken after the plant canopy had already fully developed. Hence, the early growth period (DAS 15) in 2019 showed LAI development differences by N level began at the initiation of regrowth. A similar trend was reported in that 150 kg N ha^{-1}-applied plots had a greater LAI than 0 or 75 kg N ha^{-1} applied plots during the heading stage of irrigated grain sorghum [36].

4.2. Root Morphology and Dry Matter Responses

Throughout this study, N fertilization after summer defoliation linearly increased total root length during ratooning. However, two ecotypes exhibited different degrees of belowground development after summer defoliation in both 2018 and 2019. Greenstar continued to increase its total root length as the regrowth season progressed, while Honeychew maintained its total root length after DAS 65. These results indicated that late-flowering ecotypes such as Greenstar tend to increase total root length until the later season and maintain this growth until the end of the season. In contrast, early-flowering ecotype Honeychew appeared to cease total root length development earlier in the season and exhibited a greater decrease later in the season. Thus, the two ecotypes exhibited distinct responses in total root length under varying N levels across the two-year regrowth period, with Greenstar displaying more consistent growth and maintenance compared to Honeychew. Thus, the late-flowering ecotype is a better option for multiple-harvest management as it has the superior ability to extend its root system, which is essential for sustaining water and nutrient supply throughout the ratooning season. It might also help to maintain a well-developed plant canopy during ratooning with a relatively low level of N fertilization.

To the best of our knowledge, there has been no previous research testing the effect of N fertilization on ratooning root development for sorghum. In the ratooning sugarcane study, there were genotype differences in root length density after 90 and 270 days of ratooning growth [37]. In the study, sugarcane genotypes under drought stress had high root length density in lower soil layers. However, under moderate drought conditions, plants tend to increase root systems in order to take up more water from inadequate soil moisture environments [38]. Previous research also suggested that plants are believed to be capable of modifying root growth to meet water demands during droughts [39]. In terms of the total root surface area and volume for both Greenstar and Honeychew, values continued to increase linearly as the regrowth season progressed, while Greenstar continued to have a greater root surface area and volume than Honeychew. Chung et al. (2020) proposed that an increase in root tip number represents the accumulation of lateral roots or an increase in roots emerging from the lateral root [40]. Thus, the greater tip number for Greenstar in the current study suggests that the late-flowering ecotype has the capability to increase new roots and branching of roots. Additionally, Greenstar is able to maintain existing roots for longer periods during ratooning. Nitrogen fertilization is required as it increases both above- and belowground plant development. However, requirements vary, as Greenstar showed higher growth under low levels of N fertilization (50 N).

Unexpectedly, although differences in aboveground growth were evident in the early stages of ratooning, the variation in growth due to treatment was not observed after the mid-growth phase. This implies that in the later stages of regeneration, there are no significant differences attributable to ecotype or N application. However, this inference is contradicted when the results of both aboveground and belowground growth are considered. The root development of the ratooning crop exhibited differences until the late growth phase, influenced by N application and ecotype. Therefore, N supplementation following the summer harvest not only directly enhances the growth of the ratooning crop but also facilitates nutrient uptake from the soil by promoting root development. The results obtained from this study demonstrate a similarity to the aboveground biomass yield of our research [6]. Especially even under unfavorable weather conditions in 2019, Greenstar exhibited greater development of root DM compared to Honeychew. When examining the cultivars before the summer harvest was executed, rapid development was observed up to DAP 55. Greenstar exhibited greater development in terms of root length, surface area, volume, and tips, indicating differing root development potential between the two cultivars. This might affect differences in plant regrowth after defoliation by ecotypes. Therefore, well-developed existing roots might be crucial for vigorous above- and belowground regrowth in SSG. Thus, under unfavorable weather conditions, the late-flowering ecotype Greenstar displayed a greater development and maintenance of the root system. As SSG is often grown in marginal fields where nutrients are scarce, choosing a late-flowering ecotype with a low N fertilization level is a practical option for growers.

5. Conclusions

We investigated changes in aboveground and belowground growth development among sorghum × sudangrass hybrid cultivars following post-cut fertilization. The results indicated no significant difference in chlorophyll content and LAI between the two ecotypes. While chlorophyll content varied with N application levels, LAI did not show significant variation. Different from aboveground parts, belowground development showed variation between ecotypes, particularly under the relatively favorable growing conditions of 2018, where differences were less pronounced compared to the unfavorable environment of 2019. In the latter years, differences in root growth among cultivars were evident from the early growth stages of ratooning, with Greenstar exhibiting rapid and vigorous root development. As overall root development did not increase at 100 kg N ha^{-1} or above, reduced post-cut N application rates of 50–100 kg ha^{-1} are recommended with a late-flowering ecotype, as excessive N input for a low-value crop does not align with economic goals and sustainability.

Author Contributions: Methodology, G.K., Y.J. and J.L.; Software, N.C.; Validation, M.C., S.L. and C.J.; Formal analysis, N.C.; Investigation, M.C., G.K. and J.L.; Resources, C.J. and Y.J.; Data curation, M.C. and S.L.; Writing—original draft, N.C.; Writing—review and editing, C.N.; Supervision, C.N.; Project administration, C.N.; Funding acquisition, C.N. All authors have read and agreed to the published version of the manuscript.

Funding: This research was supported by Basic Science Research Program through the National Research Foundation of Korea (NRF) funded by the Ministry of Education (2021R1I1A3040330).

Data Availability Statement: Dataset available on request from the authors.

Acknowledgments: The authors would like to thank to Gyeongsang National University Research Farm staff, Crop Management and Seed Science Laboratory team members for data collection and field management.

Conflicts of Interest: Author Nayoung Choi was employed by the company Future Agriculture Center, Kyung Nong Corporation. The remaining authors declare that the research was conducted in the absence of any commercial or financial relationships that could be construed as a potential conflict of interest.

References

1. Reddy, B.V.; Ramesh, S.; Reddy, P.S.; Ramaiah, B.; Salimath, M.; Kachapur, R. Sweet sorghum-a potential alternate raw material for bio-ethanol and bio-energy. *Int. Sorghum Millets Newsl.* **2005**, *46*, 79–86.
2. Erickson, J.E.; Woodard, K.R.; Sollenberger, L.E. Optimizing sweet sorghum production for biofuel in the southeastern USA through nitrogen fertilization and top removal. *Bioenergy Res.* **2012**, *5*, 86–94. [CrossRef]
3. Barnes, R.F.; Miller, D.A.; Nelson, C.J. *Forages*, 5th ed.; Iowa State University Press: Ames, IA, USA, 1995; pp. 121–135.
4. Lauriault, L.M.; Marsalis, M.A.; VanLeeuwen, D.M. Planting date affects rainfed sorghum forage yields in semiarid, subtropical environments. *Forage Grazinglands* **2012**, *10*, 1–7. [CrossRef]
5. Venuto, B.; Kindiger, B. Forage and biomass feedstock production from hybrid forage sorghum and sorghum–sudangrass hybrids. *Grassland Sci.* **2008**, *54*, 189–196. [CrossRef]
6. Choi, N.; Kim, G.; Park, W.; Jeong, Y.; Kim, Y.-H.; Na, C.-I. Additional N application and ecotype affect yield and quality of ratoon harvested sorghum x sudangrass hybrid for temperate regions. *Biomass Bioenergy* **2022**, *160*, 106423. [CrossRef]
7. Blum, A.; Ramaiah, S.; Kanemasu, E.; Paulsen, G. The physiology of heterosis in sorghum with respect to environmental stress. *Ann. Bot.* **1990**, *65*, 149–158. [CrossRef]
8. Plucknett, D.L.; Evenson, J.; Sanford, W.G. Ratoon cropping. *Adv. Agron.* **1970**, *22*, 285–330. [CrossRef]
9. Tarumoto, I. Studies on breeding forage sorghum by utilizing heterosis. *Bul. Chugoku Nat. Agr. Exp. Sta. A* **1971**, *19*, 21–138.
10. Escalada, R.G.; Plucknett, D.L. Ratoon Cropping of Sorghum: II. Effect of Daylength and Temperature on Tillering and Plant Development 1. *Agron. J.* **1975**, *67*, 479–484. [CrossRef]
11. Quinby, J. The maturity genes of sorghum. *Adv. Agron.* **1967**, *19*, 267–305. [CrossRef]
12. Miller, F.; Barnes, D.; Cruzado, H. Effect of Tropical Photoperiods on the Growth of Sorghum When Grown in 12 Monthly Plantings 1. *Crop Sci.* **1968**, *8*, 499–509. [CrossRef]
13. Rooney, W.L.; Aydin, S. Genetic control of a photoperiod-sensitive response in *Sorghum bicolor* (L.) Moench. *Crop Sci.* **1999**, *39*, 397–400. [CrossRef]
14. Tarumoto, I.; Yanase, M.; Kadowaki, H.; Yamada, T.; Kasuga, S. Inheritance of photoperiod-sensitivity genes controlling flower initiation in sorghum, *Sorghum bicolor* Moench. *Grassl. Sci.* **2005**, *51*, 55–61. [CrossRef]
15. Na, C.-I.; Sollenberger, L.E.; Erickson, J.E.; Woodard, K.R.; Vendramini, J.M.; Silveira, M.L. Management of perennial warm-season bioenergy grasses. I. Biomass harvested, nutrient removal, and persistence responses of elephantgrass and energycane to harvest frequency and timing. *Bioenergy Res.* **2015**, *8*, 581–589. [CrossRef]
16. Wang, Z.; Jot Smyth, T.; Crozier, C.R.; Gehl, R.J.; Heitman, A.J. Yield and nitrogen removal of bioenergy grasses as influenced by nitrogen rate and harvest management in the Coastal Plain Region of North Carolina. *BioEnergy Res.* **2018**, *11*, 44–53. [CrossRef]
17. Knoll, J.E.; Johnson, J.M.; Lee, R.D.; Anderson, W.F. Harvest Management of 'Tifton 85' Bermudagrass for Cellulosic Ethanol Production. *BioEnergy Res.* **2014**, *7*, 1112–1119. [CrossRef]
18. Lopez, G.; Ahmadi, S.H.; Amelung, W.; Athmann, M.; Ewert, F.; Gaiser, T.; Gocke, M.I.; Kautz, T.; Postma, J.; Rachmilevitch, S. Nutrient deficiency effects on root architecture and root-to-shoot ratio in arable crops. *Front. Plant Sci.* **2023**, *13*, 1067498. [CrossRef] [PubMed]
19. Gregory, P.J.; Atkinson, C.J.; Bengough, A.G.; Else, M.A.; Fernández-Fernández, F.; Harrison, R.J.; Schmidt, S. Contributions of roots and rootstocks to sustainable, intensified crop production. *J. Exp. Bot.* **2013**, *64*, 1209–1222. [CrossRef]
20. Adams, C.B.; Reyes-Cabrera, J.; Nielsen, J.; Erickson, J.E. Root system architecture in genetically diverse populations of grain sorghum compared with shallow and steeply rooted monocultures. *Crop Sci.* **2020**, *60*, 2709–2719. [CrossRef]
21. Reyes-Cabrera, J.; Adams, C.B.; Nielsen, J.; Erickson, J.E. Yield, nitrogen, and water-use efficiency of grain sorghum with diverse crown root angle. *Field Crops Res.* **2023**, *294*, 108878. [CrossRef]
22. Slota, M.; Maluszynski, M.; Szarejko, I. An automated, cost-effective and scalable, flood-and-drain based root phenotyping system for cereals. *Plant Methods* **2016**, *12*, 34. [CrossRef] [PubMed]
23. Kim, Y.; Chung, Y.S.; Lee, E.; Tripathi, P.; Heo, S.; Kim, K.-H. Root response to drought stress in rice (*Oryza sativa* L.). *Int. J. Mol. Sci.* **2020**, *21*, 1513. [CrossRef]
24. Choi, M.; Choi, N.; Lee, J.; Lee, S.; Kim, Y.; Na, C. Effects of Italian Ryegrass (*Lolium multiflorum*) Cultivation for Green Manure and Forage on Subsequent Above-and Below-Ground Growth and Yield of Soybean (*Glycine max*). *Agriculture* **2023**, *13*, 2038. [CrossRef]
25. López-Bucio, J.; Cruz-Ramírez, A.; Herrera-Estrella, L. The role of nutrient availability in regulating root architecture. *Curr. Opin. Plant Biol.* **2003**, *6*, 280–287. [CrossRef]
26. Rich, S.M.; Watt, M. Soil conditions and cereal root system architecture: Review and considerations for linking Darwin and Weaver. *J. Exp. Bot.* **2013**, *64*, 1193–1208. [CrossRef]

27. Watt, M.; Moosavi, S.; Cunningham, S.C.; Kirkegaard, J.; Rebetzke, G.; Richards, R. A rapid, controlled-environment seedling root screen for wheat correlates well with rooting depths at vegetative, but not reproductive, stages at two field sites. *Ann. Bot.* **2013**, *112*, 447–455. [CrossRef]
28. Heinze, J.; Sitte, M.; Schindhelm, A.; Wright, J.; Joshi, J. Plant-soil feedbacks: A comparative study on the relative importance of soil feedbacks in the greenhouse versus the field. *Oecologia* **2016**, *181*, 559–569. [CrossRef]
29. Schittko, C.; Runge, C.; Strupp, M.; Wolff, S.; Wurst, S. No evidence that plant-soil feedback effects of native and invasive plant species under glasshouse conditions are reflected in the field. *J. Ecol.* **2016**, *104*, 1243–1249. [CrossRef]
30. Rich, S.M.; Christopher, J.; Richards, R.; Watt, M. Root phenotypes of young wheat plants grown in controlled environments show inconsistent correlation with mature root traits in the field. *J. Exp. Bot.* **2020**, *71*, 4751–4762. [CrossRef]
31. Thornley, J. Acclimation of photosynthesis to light and canopy nitrogen distribution: An interpretation. *Ann. Bot.* **2004**, *93*, 473–475. [CrossRef]
32. Gong, X.; Liu, C.; Ferdinand, U.; Dang, K.; Zhao, G.; Yang, P.; Feng, B. Effect of intercropping on leaf senescence related to physiological metabolism in proso millet (*Panicum miliaceum* L.). *Photosynthetica* **2019**, *57*, 993–1006. [CrossRef]
33. Moi, P.E.; Kitonyo, O.M.; Chemining'wa, G.N.; Kinama, J.M. Intercropping and Nitrogen Fertilization Altered the Patterns of Leaf Senescence in Sorghum. *Int. J. Agron.* **2021**, *2021*, 1–14. [CrossRef]
34. Bassi, D.; Menossi, M.; Mattiello, L. Nitrogen supply influences photosynthesis establishment along the sugarcane leaf. *Sci. Rep.* **2018**, *8*, 2327. [CrossRef] [PubMed]
35. Lee, E.; Tollenaar, M. Physiological basis of successful breeding strategies for maize grain yield. *Crop Sci.* **2007**, *47*, S-202–S-215. [CrossRef]
36. Wang, Z.; Nie, T.; Lu, D.; Zhang, P.; Li, J.; Li, F.; Zhang, Z.; Chen, P.; Jiang, L.; Dai, C. Effects of Different Irrigation Management and Nitrogen Rate on Sorghum (*Sorghum bicolor* L.) Growth, Yield and Soil Nitrogen Accumulation with Drip Irrigation. *Agronomy* **2024**, *14*, 215. [CrossRef]
37. Chumphu, S.; Jongrungklang, N.; Songsri, P. Association of physiological responses and root distribution patterns of ratooning ability and yield of the second ratoon cane in sugarcane elite clones. *Agronomy* **2019**, *9*, 200. [CrossRef]
38. Jongrungklang, N.; Toomsan, B.; Vorasoot, N.; Jogloy, S.; Boote, K.; Hoogenboom, G.; Patanothai, A. Drought tolerance mechanisms for yield responses to pre-flowering drought stress of peanut genotypes with different drought tolerant levels. *Field Crops Res.* **2013**, *144*, 34–42. [CrossRef]
39. Kato, Y.; Okami, M. Root morphology, hydraulic conductivity and plant water relations of high-yielding rice grown under aerobic conditions. *Ann. Bot.* **2011**, *108*, 575–583. [CrossRef]
40. Chung, Y.S.; Kim, S.; Park, C.; Na, C.-I.; Kim, Y. Treatment with silicon fertilizer induces changes in root morphological traits in soybean (*Glycine max* L.) during early growth. *J. Crop Sci. Biotech.* **2020**, *23*, 445–451. [CrossRef]

Article

Impact of No Tillage and Low Emission N Fertilization on Durum Wheat Sustainability, Profitability and Quality

Michele Andrea De Santis *, Luigia Giuzio, Damiana Tozzi, Mario Soccio and Zina Flagella *

Department of Agriculture, Food, Natural Resources and Engineering (DAFNE), University of Foggia, 71122 Foggia, Italy
* Correspondence: michele.desantis@unifg.it (M.A.D.S.); zina.flagella@unifg.it (Z.F.)

Abstract: Mitigation practices for cereal systems, including conservation agriculture and low emission fertilization, are required to face global challenges of food security and climate change. The combination of these climate-smart approaches was investigated for durum wheat in a dry region of the Mediterranean basin in two crop seasons. The experimental design consisted in two different genotypes, Marco Aurelio (high protein content) and Saragolla (higher adaptability), subjected to no tillage (NT) vs. conventional tillage (CT) and to two fertilization strategies (standard vs. low emission plus an unfertilized control). Different environmental and economic sustainability parameters as well as two different technological and nutritional quality traits were evaluated. Saragolla showed a better environmental adaptability and a higher nitrogen use efficiency, evaluated as partial nutrient balance (+27%), and was associated with a lower protein content (14.5% vs. 15.6%). NT was associated with an improvement in yield (+15%) and quality, i.e., micronutrients (Fe, Zn) and antioxidant capacity (+15%), in the drier crop year. Low emission fertilization did not reduce crop performance and its combination with NT showed a higher economic net return. The combination of the two mitigation practices improved not only environmental and economic sustainability but also the health quality of durum wheat under water limited conditions.

Keywords: conservative agriculture; climate-smart crop production; micronutrients; AOX; gluten; durum wheat quality; slow-release fertilizers

check for
updates

Citation: De Santis, M.A.; Giuzio, L.; Tozzi, D.; Soccio, M.; Flagella, Z. Impact of No Tillage and Low Emission N Fertilization on Durum Wheat Sustainability, Profitability and Quality. *Agronomy* **2024**, *14*, 2794. https://doi.org/10.3390/agronomy14122794

Academic Editors: Ewa Szpunar-Krok, Mariola Staniak and Małgorzata Szostek

Received: 28 October 2024
Revised: 21 November 2024
Accepted: 22 November 2024
Published: 25 November 2024

1. Introduction

Agriculture faces the dual challenges of increasing productivity to meet global food demands while minimizing environmental impacts. This is particularly relevant in regions vulnerable to climate change, such as the Mediterranean basin, where water scarcity and soil degradation are significant concerns. Durum wheat (*Triticum turgidum* L. *durum*) is a staple crop with high importance in Mediterranean area, with Italy representing the second largest producer after Canada [1]; durum wheat is mainly used for the production of pasta and semolina bread, and its productivity and quality is generally influenced by environmental conditions, including terminal abiotic stresses that can cause severe yield loss [2–5]). The major stresses typically occurring in Mediterranean area are heat and drought, especially during the grain filling period, and their severity and duration determine the extent of the yield loss, which can be higher than 50% [6,7]. To address these challenges, the implementation of climate-smart agricultural practices, aimed to improve mitigation both by increasing carbon sequestration and reducing greenhouse gases emissions due to tillage and fertilization [8–10], is essential to ensure the sustainability, profitability, and quality of durum wheat production.

Conservation agriculture (CA), particularly no-tillage (NT) systems, has emerged as a viable approach for improving soil structure, water retention, and carbon sequestration, thereby enhancing the resilience of cropping systems under climate stress [11–13]. No-tillage practices contribute to the preservation of soil organic matter, reduce soil erosion, and promote biodiversity, factors that collectively support more sustainable agricultural ecosystems [14–16]. No till practices are generally reported to reduce crop yield by about 5%; however, in drier environments, wheat yields match those achieved under conventional tillage [17]. Complementing NT, low-emission nitrogen (N) fertilization strategies, including the use of slow-release fertilizers with urease and nitrification inhibitors, have been shown to reduce greenhouse gas emissions without having a negative influence on nitrogen use efficiency (NUE), which is crucial for lowering the environmental footprint of crop production [18–20]. The EU N expert Panel indicated that the recommended NUE values, in terms of N output/input balance, should be comprised within 50 to 90%, with a great influence of farm type, management and environmental conditions [21,22].

While the environmental benefits of these practices are well-documented, their effects on quality, particularly in durum wheat, require further investigation. Durum wheat quality is categorized by several traits that are essential for its market value, particularly in the pasta-making industry. Among these traits, protein content and composition are critical for determining the technological performance of the grain, influencing dough strength and elasticity [23]. Beyond protein quality, the micronutrient content of durum wheat, particularly iron (Fe) and zinc (Zn), is gaining attention due to its importance in human nutrition. Micronutrient deficiencies, also known as hidden hunger, remain a global health issue, and enhancing Fe and Zn content in staple crops like wheat can significantly contribute to addressing this problem [24]. Furthermore, the antioxidant capacity of durum wheat, linked to its content of phenolic compounds and flavonoids, adds a nutritional dimension to its value, as antioxidants are known to reduce oxidative stress and contribute to human health [25,26]. However, few investigations report the effects of no tillage on durum wheat quality and its influence requires further clarification. For instance, in a study conducted in the Mediterranean area [27], the authors reported that variations in protein content and high and low molecular weight glutenin subunits expression in no tillage systems depend on field growing conditions, highlighting the complexity of storage protein regulation. Furthermore, no tillage has been found to enhance the activity of antioxidant enzymes during grain filling, but the correlation with grain antioxidant activity was not explored [28].

This study aims to evaluate the hypothesis that the combination of no-tillage and low-emission nitrogen fertilizers might improve the sustainability of durum wheat production in the context of climate-smart agriculture in South Italy. Specifically, the effects of these practices were investigated in relation to (i) sustainability, in terms of nitrogen use efficiency, (ii) economic profitability for farmers, and (iii) grain quality traits, including protein content and composition, technological quality, micronutrient content (Fe, Zn), and antioxidant activity (AOX). By integrating agronomic, economic, and environmental assessments, this study seeks to contribute to the knowledge on sustainable cereal production in Mediterranean environments.

2. Materials and Methods

2.1. Description of the Study Site

Field experiments were carried out in the province of Foggia, at farm level in Bovino (Foggia, Italy, 41°17′22.8″ N 15°26′40.7″ E, at about 243 m a.s.l.) during 2019–2022 in two crop seasons (2021 and 2022, respectively). Weather data were recorded from a proximal weather station, with monthly temperatures and precipitations (P) reported in Figure 1; in addition, 10-day trends of growing degree days (GDD), cumulative P, and potential evapotranspiration (PET) calculated according to Hargreaves method [29] are reported in Appendix A (Table A1).

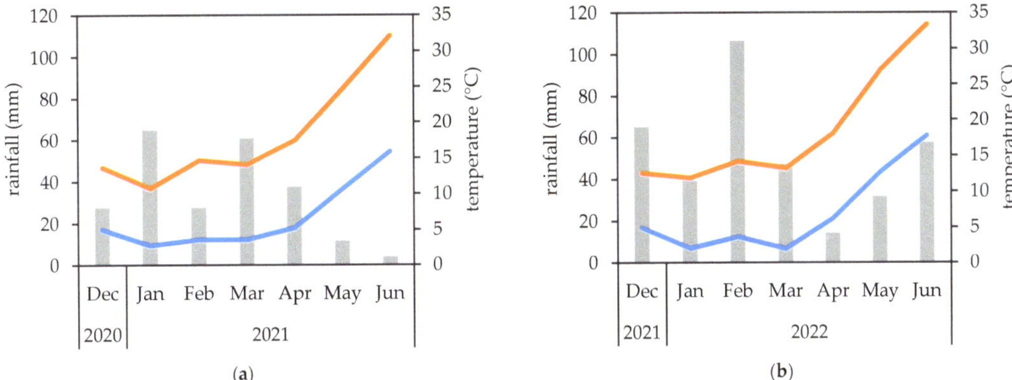

Figure 1. Weather conditions during experimental field trials, described in terms of monthly precipitation (grey histograms) and maximum (orange line) and minimum (blue line) temperatures during (**a**) 2021 and (**b**) 2022 crop years.

2.2. Experimental Design, Soil Sampling

Each plot was 18 m^2 (3 m × 6 m) in a split-split plot design with tillage as main plot, genotype as sub-plot, and fertilization as sub-sub plot, each with three replications, for a total of 36 plots in each experimental crop year. The soil for the experiments was a silty loam Vertisol, according to USDA classification, with 26.4% sand, 52.8% silt, and 15.1% clay, respectively, with 8.2 pH, 323 µs cm^{-1} of conductivity, low total nitrogen (N, 0.66 g/kg, by CHNS elemental analyzer), and good available P_2O_5 (43.1 ppm, determined by Olsen method) and soil organic carbon (SOC, 1.71%, by Walkley–Black method) as starting conditions and at the end of the experiment. Mean SOC stock variations (t/ha of C) between NT and CT were evaluated as following: soil C stock = (final SOC − initial SOC) × bulk density × soil depth (0.3 m).

2.3. Land Preparation, Fertilization, Sowing, Weeding, and Disease Control

Wheat was the preceding crop. Two soil tillage practices were compared within the field experiments, conventional tillage (CT) and no tillage (NT), as detailed in Supplementary Table S1. CT did not include straw incorporation. Two durum wheat genotypes largely cultivated in Italy were adopted, with Marco Aurelio, characterized by a higher protein content suitable for good technological performance, and Saragolla, characterized by a higher environmental adaptability, as detailed in Supplementary Material (Table S1).

Sowing occurred on 16 December 2020 and 20 December 2021 at a rate of 450 seeds per m^2. Herbicides and fungicides were adopted according to the local practices. A standard nitrogen fertilization, according to the local practice (T1), was compared to a low-emission strategy with the use of stabilized fertilizers with inhibitors of urease and nitrification (T2); an unfertilized control (T0) was included in the experiment to evaluate N use efficiency. Details of N rate, source, and timing are reported in Table 1. The N rates were defined on the basis of the indications for durum wheat in Italy for good quality targets [30]. Before sowing, 50 kg/ha of phosphorus was supplied as a single superphosphate. Before harvest, for each plot, plants from a square meter were collected for the determination of plant height (PH) and harvest index (HI). At maturity (197 and 194 days after sowing), grains were harvested by plot combine and grain yield (GY, t/ha) was determined and normalized at 12% moisture. An aliquot of 1 kg was collected for analyses, including grain weight (GW, mg), test weight (TW), and grain protein content (GPC, Foss Tecator 1241). Grain number per m^2 (GN) was calculated as the ratio between GY and GW.

Table 1. Details of nitrogen and sulfur fertilization strategies in terms of N source and timing.

Code		GS 23		GS 31	
		N (kg/ha)	N Source	N (kg/ha)	N Source
T0	unfertilized	0	-	0	-
T1	standard	90	U	50	AN
T2	low emission	80	UAS + NBPT	40	ASN + DMPP

N = nitrogen; S = sulfur; GS 23 = tillering; GS 31 = stem elongation; U = urea; AN = ammonium nitrate; UAS = urea ammonium sulfate; NBPT = N-(n-Butyl)thiophosphoric triamide, inhibitor of urease; ASN = ammonium sulfate nitrate; DMPP = 3,4-Dimethylpyrazole phosphate, inhibitor of nitrification.

2.4. Crop Physiological Measurements

Data on crop development were recorded and expressed as days after sowing (DAS) and GDD. At heading (GS 55, 137, and 141 days for 2021 and 2022), measurements of canopy spectral reflectance were carried using a field spectroradiometer (Apogee SS-110). Normalized difference vegetation index (NDVI) was calculated [31] according to the following formula:

$$NDVI = (\rho_{NIR} - \rho_{RED})/(\rho_{NIR} + \rho_{RED}) \tag{1}$$

with ρ_{NIR} as the crop reflectance at 800 nm and ρ_{RED} as the crop reflectance at 680 nm. At maturity, plant height (PH) was also measured. N uptake in grain was determined by multiplying grain yield (d.m.) to N concentration in grain. N use efficiency traits were calculated according to the formula:

$$NAE = (GY_x - GY_0)/N\ rate \tag{2}$$

$$ARE = (N\ uptake\ _x - N\ uptake\ _0)/N\ rate \tag{3}$$

$$PNB = N\ uptake/N\ rate \tag{4}$$

with NAE as N agronomic efficiency, ARE as apparent N recovery efficiency, and PNB as partial nutrient balance [32,33]; $_x$ and $_0$ referred to the N fertilization rates and unfertilized control, respectively.

2.5. Analysis of Durum Wheat Storage Protein Composition

Grains were milled by laboratory milling with a sieve of less than 1 mm (Bona 4RB, Monza, Italy) and flour was used for chemical analysis. Analysis of protein composition was carried out in order to evaluate differences in gliadin and glutenin content [34,35].

Briefly, 100 mg of flour was suspended in a 0.4 mL solution of KCl buffer (pH 7.8) and centrifuged at 4 °C at 10,000× g for 15 min to remove soluble proteins (albumins and globulins). The KCl-insoluble fraction was then suspended in 1-propanol solution (50% v/v) and centrifuged for 10 min at 4500× g (repeated twice) and gliadins were collected. Glutenins were extracted from the pellet by extraction solution (1-propanol 50% v/v, 1% DTT) after centrifugation at 10,000× g for 10 min (room temperature). Extracted glutenins and gliadins were quantified, and their subunits were separated by using a BioRad Mini Protean II system (Bio-Rad, Hercules, CA, USA) with precast acrylamide gels. Gels were stained with Coomassie Brilliant Blue G250 and digitally acquired (Epson Perfection V750pro). Molecular weight markers, from 10 to 250 kDa, were used (Bio-Rad Co., Hercules, CA, USA). Image analysis of gels was performed using ImageLab software (V6.1, Bio-Rad Co., Hercules, CA, USA).

Protein composition was reported in terms of gliadin to glutenin ratio (glia/glut) and HMW-GS to LMW-GS ratio (H/L).

2.6. Antioxidant Capacity

The antioxidant capacity (AOX) was evaluated using the Direct QUENCHER_ABTS Assay (QUick, Easy, New, CHEap, and Reproducible ABTS-2,2′-azinobis-3-ethylbenzothiazoline-6-sulfonic acid) [26]. This method is based on the direct reduction of ABTS radical cation

(ABTS$^{\bullet+}$) by the antioxidants present in the fine solid particles of the food sample, resulting in a decrease in absorbance at 734 nm (A$_{734}$). Measurements were carried out by reacting 10 mL of ABTS$^{\bullet+}$ solution with whole flour sample, ranging from 1 to 2 mg of dry weight, for 60 min. The (%) decrease of A$_{734}$ measured after sample incubation, compared to the A$_{734}$ of ABTS$^{\bullet+}$ solution, was calculated. A linear relationship of the (%) decrease of A$_{734}$ on sample amount was verified by linear regression analysis of the data. AOX was obtained by comparing the slope derived by linear regression analysis with that of the Trolox-derived calibration curve. Data are expressed as mmol Trolox equivalent per kg of dry weight.

2.7. Mineral Analysis

Grain micronutrients, i.e., iron (Fe) and zinc (Zn) content, were analyzed with an induced coupled plasma optical emission spectrometer (ICP-OES). Before the analytical process, samples (0.5 g d.w.) were dissolved in 10 mL HNO$_3$/H$_2$O$_2$ (3:1 v/v) by microwave assisted mineralization (CEM-Mars6). Digested samples were then diluted with Milli-Q water to 50 mL (US-EPA 1989, method 3050 B) and analyzed by ICP-OES. The data were expressed as ppm [35].

2.8. Economic Analysis

Economic analysis was carried out calculating economic net return (ENR, EUR/ha) as following:

$$\text{ENR} = \text{gross return} - \text{cost of cultivation} \tag{5}$$

with gross return as GY (t/ha, 12% moisture) multiplied per durum wheat market price (EUR/t) and cost of cultivation was determined based on the data reported in the Supplementary Materials. To this end, reference durum wheat prices were taken from the historical records of the weekly prices from the Chamber of Commerce of Foggia (https://www.fg.camcom.it, accessed on 10 July 2024), and three market price scenarios were considered due to the variability recorded in the last five years: 300 EUR/t, 400 EUR/t, and 500 EUR/t. According to the supply chain agreements largely present in Italy, a premium bonus on market price of 10 EUR/t is applied when GPC is higher than 13.5%, with a further bonus of 10 EUR/t with GPC of 14.5% or higher [36]. Supplementary EU, national, or regional conditions were not considered for the economic profitability.

2.9. Statistical Analysis

For each crop year, a standard least square regression model was conducted and means were separated by least significant difference according to Tukey's test as post hoc, with a level of significance of $p < 0.05$. Two separated Pearson's multiple regression analyses between the investigated parameters were carried out for samples from conventional tillage and no tillage. Statistical analysis was carried out by JMP (SAS Institute Inc., Cary, NC, USA, 2009).

3. Results

3.1. Effect of Weather Conditions and Agronomic Management on Crop Performances

The first crop year (2021) was characterized by higher rainfall during winter and vegetative growth stages with respect to the second year (2022). On the contrary, in 2022, warmer and wetter conditions were observed during spring, thus influencing grain filling duration, with 31 days in 2022 vs. 26 days in 2021. Thermal trend was comparable between the two years; in fact, anthesis was achieved at 137 and 141 days after sowing in the two years, respectively (about 1–2 days earlier for Saragolla, at about 1200 °C d).

Results of agronomic traits in the two crop years, as subjected to analysis of variance, are reported in Table 2. Spectral measurements carried out at heading (GS 55) only showed lower NDVI values under no tillage (NT) in 2021, which was characterized by a higher spring rainfall deficit, while no differences were observed due to genotype. In both years, significantly higher values due to fertilization were observed with respect to the unfertilized control (T0). The same response to N supply was observed in terms of plant height (PH).

Also, a lower PH was observed under NT only in 2022 (-14%), while a genotypic difference was found in 2021, with higher PH in Saragolla ($+9\%$).

Table 2. Effect of soil tillage, genotype, and N fertilization strategy and their interactions on durum wheat agronomic traits.

Year	Source of Variation	NDVI GS 55	PH cm	HI %	GW mg	GY t/ha	PNB kg/kg	NAE kg/kg	ARE kg/kg
2021	CT	0.70 a	69.1 a	27.7 b	32.9 b	2.0 b	0.49 a	6.0 a	25.2 a
	NT	0.65 b	71.2 a	29.7 a	36.1 a	2.3 a	0.55 a	5.9 a	23.5 a
	Marco Aurelio	0.67 a	67.1 b	25.8 b	33.8 a	1.6 b	0.39 b	3.6 b	17.4 b
	Saragolla	0.68 a	73.2 a	31.6 a	36.1 a	2.7 a	0.55 a	8.3 a	31.3 a
	T0	0.47 b	58.7 b	28.3 a	35.8 a	1.4 b	-	-	-
	T1	0.77 a	75.1 a	29.1 a	33.8 a	2.4 a	0.46 b	5.3 a	20.7 b
	T2	0.79 a	76.7 a	28.7 a	33.8 a	2.6 a	0.55 a	6.6 a	28.0 a
	TxG	ns	ns	ns	ns	ns	ns	*	*
	TxN	ns	ns	ns	ns	ns	ns	ns	ns
	GxN	ns	ns	*	ns	*	ns	ns	ns
	TxGxN	ns	ns	*	*	*	*	ns	*
2022	CT	0.72 a	67.0 a	31.6 a	44.0 a	2.1 a	0.51 a	3.6 a	20.2 a
	NT	0.69 a	58.7 b	31.7 a	42.9 a	2.2 a	0.51 a	4.5 a	16.9 a
	Marco Aurelio	0.70 a	61.6 a	29.4 b	43.7 a	1.9 b	0.48 a	3.2 a	17.5 a
	Saragolla	0.72 a	64.1 a	34.0 a	43.2 a	2.4 a	0.54 a	4.9 a	19.6 a
	T0	0.60 b	58.1 b	33.5 a	44.4 a	1.6 b	-	-	-
	T1	0.75 a	66.8 a	30.8 a	43.4 a	2.3 a	0.49 a	3.7 a	19.2 a
	T2	0.78 a	63.6 a	30.6 a	42.7 a	2.4 a	0.53 a	4.4 a	17.9 a
	TxG	ns	ns	ns	*	*	ns	*	ns
	TxN	ns	ns	ns	*	ns	ns	ns	ns
	GxN	ns	ns	ns	*	ns	ns	ns	ns
	TxGxN	ns	ns	ns	ns	ns	ns	ns	ns

Abbreviations: CT = conventional tillage; NT = no tillage; T0 = unfertilized control; T1 = standard fertilization; T2 = low emission fertilization; T = tillage; G = genotype; N = fertilization strategy; NDVI = normalized difference vegetation index; PH = plant height; HI = harvest index; GW = grain weight; GY = grain yield; PNB = partial nutrient balance; NAE = agronomic nitrogen use efficiency; ARE = apparent recovery efficiency. Different letters indicate significant differences according to Tukey's test; ns = not significant; * significant difference at $p < 0.05$. Level of significance: ns = not significant.

As regards the yield components, harvest index (HI) showed a higher general mean value in Saragolla; in addition, the TxGxN interaction was significant only in 2021, with no significant differences due to N treatment within each genotype (Figure 2). Grain weight (GW) was generally lower in the first year, which was characterized by a shorter and drier grain filling period. In the same year, NT showed a higher GW than wheat cultivated under CT. Genotype and N fertilization, in general, did not have a significant effect on GW (Figure 2). The observed behavior of the yield components influenced grain yield (GY), which were significantly higher under NT in 2021, characterized by higher water deficit during grain filling. In addition, a significant influence of genotype was found, with Saragolla generally more productive than Marco Aurelio (about +45%). This difference was marked in the N fertilized thesis; the highest GY was observed in Saragolla, under NT, in the T2 strategy (3.9 t ha^{-1}).

Different N use efficiency (NUE) traits were also calculated to assess the level of sustainability of the different agronomic practices. In general, Saragolla showed a significantly higher NUE in 2021 (0.55 vs. 0.39), while in 2022, no difference was found with Marco Aurelio (Table 2). In relation to N fertilization strategies, the low emission T2 showed higher efficiency in terms of PNB and ARE with respect to the standard strategy T1, but only in the first crop year (Table 2). The highest NUE (PNB) value (0.87) was achieved in 2021 with Saragolla fertilized under the T2 strategy in NT tillage (Figure 2).

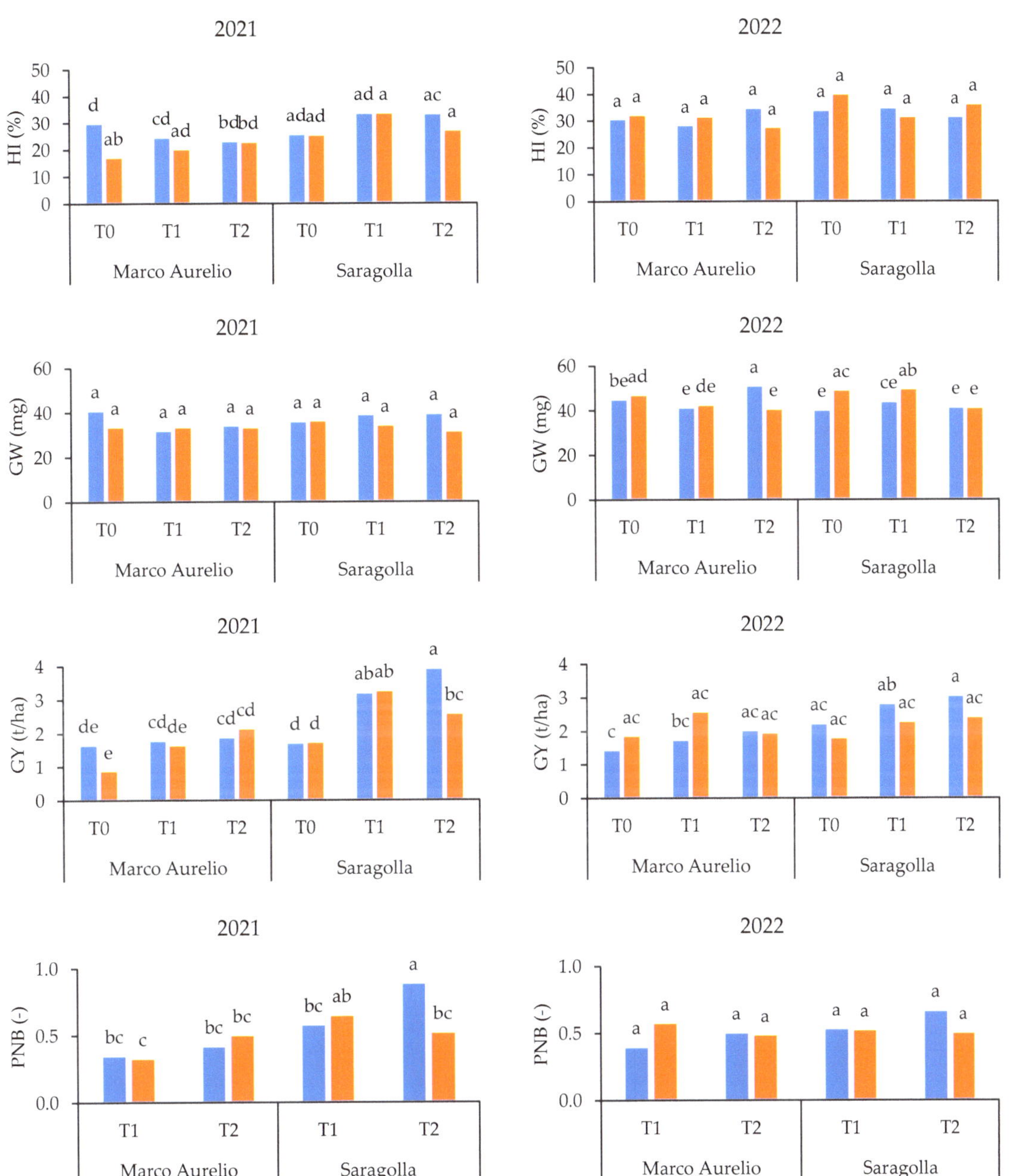

Figure 2. Response of two durum wheat genotypes subjected to two tillage and three N fertilization strategies in two consecutive growing seasons in terms of harvest index (HI), grain weight (GW), grain yield (GY), and partial N balance (PNB). Blue and orange histograms refer, respectively, to no tillage and conventional tillage samples. Different letters indicate significant differences according to Tukey's test.

3.2. Effects on Durum Wheat Technological and Health Quality

Test weight (TW), like GW, was higher in NT than CT only in 2021. In that year, higher values were observed for Saragolla and with T2 fertilization. As regards quality traits, in both crop years, Marco Aurelio showed a mean higher GPC than Saragolla (Table 3). In 2021, a mean higher GPC was observed (+5%) for NT compared to CT, in particular in the unfertilized control group T0 (Figure 3). N fertilization resulted in the highest source of variation for GPC, with a significant increase due to a higher N rate (140 kg-N/ha T1 > 120 kg-N/ha T2 > 0 kg-N/ha T0). However, both T1 and T2 fertilizations led to GPC values higher than the quality threshold (14.5%), which is satisfactory for the pasta industry, in both years.

Table 3. Effect of soil tillage, genotype, and N fertilization strategy and their interactions on durum wheat quality traits.

Year	Factor	TW kg/hl	GPC %	SSV Ml	glia/glut -	H/L -	AOX mmol/kg	Fe ppm	Zn ppm
2021	CT	74.5 b	14.3 b	33.9 a	1.16 a	0.22 b	47.6 b	53.9 b	35.9 b
	NT	75.6 a	15.0 a	33.4 a	1.12 b	0.30 a	54.9 a	59.0 a	43.4 a
	Marco Aurelio	73.8 b	15.0 a	37.0 a	1.13 a	0.30 a	51.5 a	60.1 a	44.9 a
	Saragolla	76.3 a	14.4 b	30.3 b	1.15 a	0.22 b	50.9 a	52.8 b	34.5 b
	T0	74.1 c	12.8 b	30.2 b	1.07 b	0.21 b	30.4 b	47.6 b	34.7 c
	T1	75.3 b	15.5 a	36.5 a	1.16 a	0.29 a	62.4 a	61.1 a	40.8 b
	T2	75.8 a	15.7 a	34.2 a	1.19 a	0.29 a	61.0 a	60.7 a	43.5 a
	TxG	*	**	*	*	ns	ns	*	*
	TxN	ns	***	ns	ns	ns	ns	*	*
	GxN	*	**	*	***	*	ns	*	*
	TxGxN	ns	*	ns	ns	ns	ns	*	*
2022	CT	76.4 a	15.2 a	38.2 a	1.26 a	0.08 b	33.6 a	31.1 b	25.4 b
	NT	73.8 b	15.5 a	37.9 a	1.19 b	0.12 a	32.5 a	37.1 a	39.8 a
	Marco Aurelio	75.1 a	16.2 a	38.3 a	1.16 b	0.10 a	35.7 a	27.7 b	27.1 b
	Saragolla	75.1 a	14.5 b	37.8 a	1.30 a	0.10 a	30.4 b	40.5 a	38.0 a
	T0	75.2 a	12.7 c	31.9 b	1.05 c	0.15 a	29.2 c	25.1 c	22.9 b
	T1	74.2 a	17.4 a	40.9 a	1.23 b	0.09 b	36.5 a	43.4 a	47.5 a
	T2	75.8 a	16.0 b	41.3 a	1.40 a	0.07 b	33.4 b	33.9 b	27.3 ab
	TxG	ns	*	ns	ns	*	ns	ns	ns
	TxN	ns	***	ns	*	ns	ns	ns	ns
	GxN	ns	ns	ns	*	*	*	*	*
	TxGxN	ns	*	ns	*	*	ns	*	ns

Abbreviations: CT = conventional tillage; NT = no tillage; T0 = unfertilized control; T1 = standard fertilization; T2 = low emission fertilization; T = tillage; G = genotype; N = fertilization strategy; GPC = grain protein content; SSV = sodium-dodecyl-sulfate sedimentation volume; glia/glut = gliadin-to-glutenin ratio; H/L = HMW-to-LMW glutenin subunits ratio; AOX = antioxidant capacity; Fe = iron content in grain; Zn = zinc content in grain. Different letters indicate significant differences according to Tukey's test; ns = not significant; * significant difference at $p < 0.05$. Level of significance: ns = not significant; ** = significant at $p < 0.01$; *** = significant at $p < 0.001$.

Technological quality was evaluated by SDS sedimentation volume (SSV). Of course, this was influenced by both protein content and composition. The higher GPC found in Marco Aurelio resulted in a better technological performance (SSV) only in 2021. Indeed, the same genotype showed a higher HMW-GS to LMW-GS ratio (H/L); this is due to a lower LMW-GS expression, which is generally responsible for good technological quality in durum wheat. No significant differences between the two N fertilization treatments (T1 and T2) were observed.

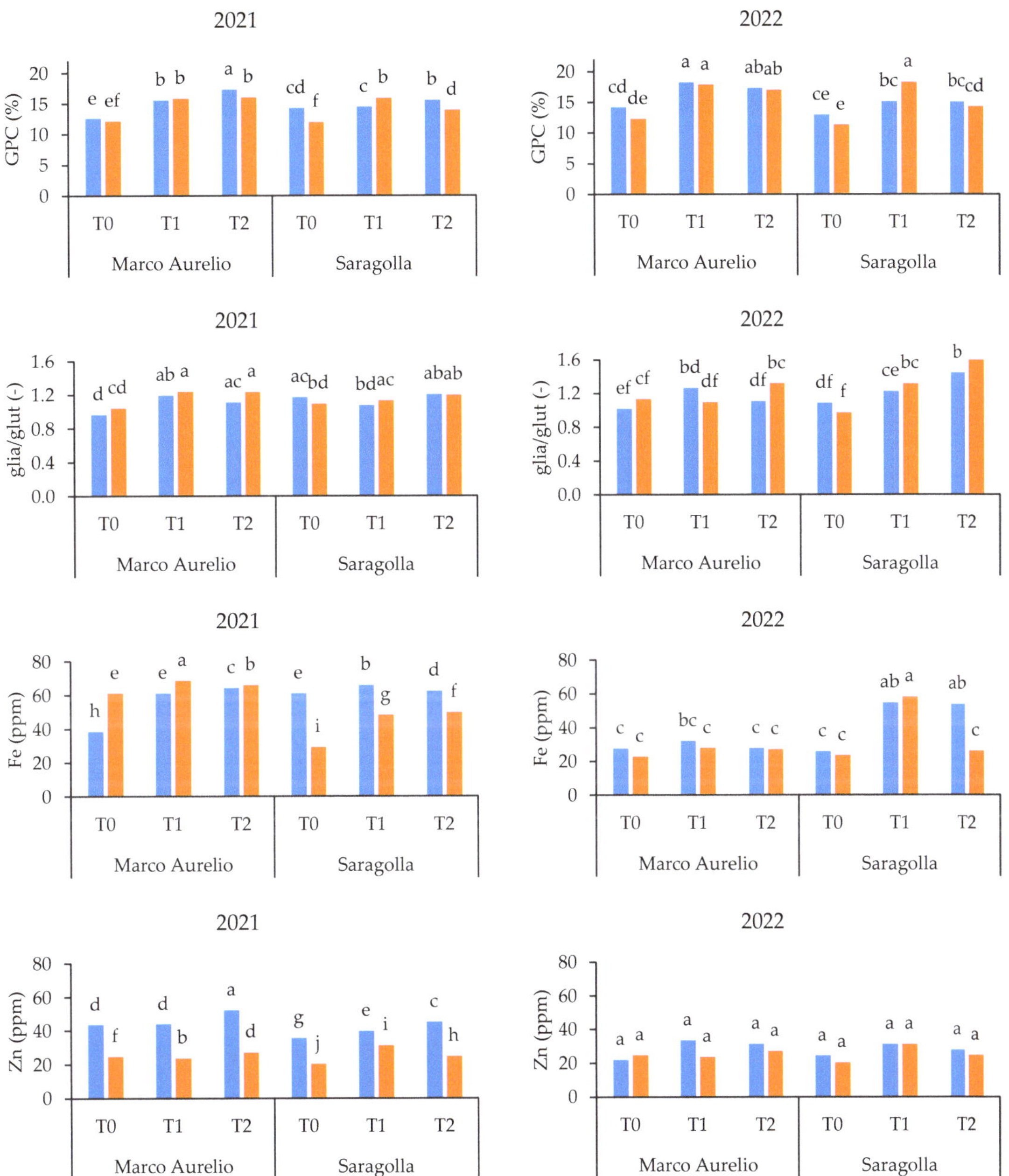

Figure 3. Response of two durum wheat genotypes subjected to two tillage and three N fertilization strategies in two consecutive growing seasons in terms of grain protein content (GPC), gliadin to glutenin ratio (glia/glut), iron (Fe), and zinc (Zn) content. Blue and orange histograms refer, respectively, to no tillage and conventional tillage samples. Different letters indicate significant differences according to Tukey's test.

Interesting results were observed for the investigated health-related quality traits, i.e., antioxidant capacity (AOX) and micronutrients content, showing higher values under NT (on average +15%, Table 3), even if only in 2021 for AOX. This trait was higher in Saragolla than Marco Aurelio only in 2022. Contrasting genotypic response in grain Fe (range within 23 to 64 ppm) and Zn (range within 22 to 52 ppm) content was observed, with higher concentrations in Marco Aurelio than Saragolla in 2021. Finally, higher N rates improved both AOX and micronutrient content, since unfertilized T0 generally showed lower values, except for Marco Aurelio in 2022 for Zn (Figure 3).

3.3. Multiple Regression Analysis

The analysis of the correlations between agronomic and quality traits is shown in Figure 4a (CT) and Figure 4b (NT). The trend of correlations was comparable between the two tillage systems. Also, the N rate showed a strong correlation with NDVI, yield, and quality traits, except H/L. The same NDVI measurements at heading showed a good correlation with both GY and GPC. As regards quality, a negative correlation between H/L and SSV was also found, as well as between GW and AOX and Fe and Zn, in both CT and NT.

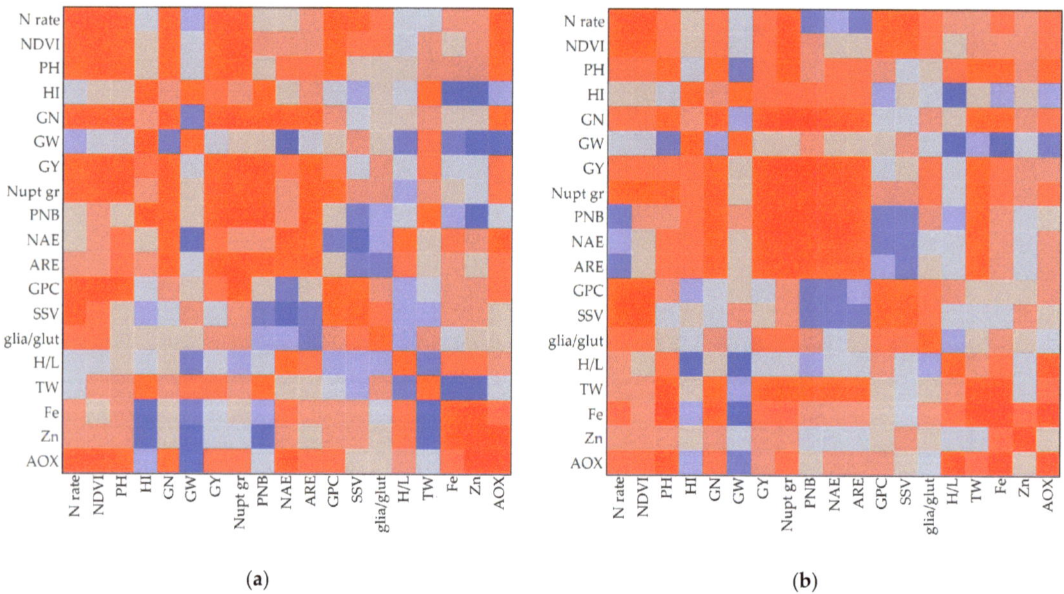

(**a**)　　　　　　　　　　　　　　(**b**)

Figure 4. Heatmap of Spearman's multiple regression analysis between the investigated agronomic and quality traits for durum wheat grown under (**a**) conventional tillage and (**b**) no tillage. R values are indicated in a scale of blue (negative) and red (positive). Abbreviations: NDVI = normalized difference vegetation index; PH = plant height; HI = harvest index; GW = grain weight; GY = grain yield; PNB = partial nutrient balance; NAE = agronomic nitrogen use efficiency; ARE = apparent recovery efficiency; GPC = grain protein content; SSV = sodium-dodecyl-sulfate sedimentation volume; glia/glut = gliadin-to-glutenin ratio; H/L = HMW-to-LMW glutenin subunits ratio; AOX = antioxidant capacity; Fe = iron content in grain; Zn = zinc content in grain.

3.4. Economic Profitability

The economic net return (ENR) of the different experimental combinations of durum wheat genotypes, soil tillage, and fertilization strategies in the current two-year field trial was predominantly influenced by the productive performance and, secondarily, by the achievement of quality targets and the different costs of cultivation. Wheat price, of course, resulted in one of the major factors influencing economic sustainability, falling into the 300 EUR/t scenario, generally with negative ENR (Table 4). Comparable ENR were observed between the two crop years. Conservative agriculture generally showed higher net return because of the lower costs (supplementary) without negatively influencing yield (from +36 to +176 EUR ha^{-1}). On the other hand, genotype was markedly relevant, with genotype Marco Aurelio characterized by a gross return that generally did not compensate for the costs, unless under the best price scenario (500 EUR/t). In general, fertilization benefited economic sustainability; in fact, the unfertilized control generally showed negative ENR values. Also, the low emission T2 strategy, −20 kg/ha with respect to the standard T1, was generally more economically efficient (Table 4) in all price scenarios.

Table 4. Economic analysis of the mean net return (ENR, EUR/ha) of durum wheat genotypes grown under different tillage and fertilization managements, with three different wheat price scenarios: 300 EUR/t, 400 EUR/t and 500 EUR/t.

Source of Variation	Factor	2021			2022		
		300 EUR/t	400 EUR/t	500 EUR/t	300 EUR/t	400 EUR/t	500 EUR/t
tillage	CT	−191	7	206	−159	51	260
	NT	−80	151	382	−123	93	310
genotype	Marco Aurelio	−296	−134	27	−215	−27	161
	Saragolla	24	292	560	−67	171	409
N fertilization	T0	−240	−96	49	−140	38	216
	T1	−102	140	382	−138	92	322
	T2	−65	193	450	−65	85	316

Abbreviations: CT = conventional tillage; NT = no tillage; T0 = unfertilized control; T1 = standard fertilization; T2 = low emission fertilization; T = tillage; G = genotype; N = fertilization strategy.

4. Discussion

The observations reported in the current study are within the framework of agronomic management suggested by the climate-smart crop production strategies (FAO) to promote mitigation and resource use efficiency [9,37]. This approach has been investigated for durum wheat, which represents one of the main species in Mediterranean cropping systems, especially in South Italy [5], by evaluating, at the same time, the effects of conservative agricultural practices (no tillage) under low emission fertilization in relation to both different qualitative durum wheat traits and the environmental and economic sustainability of the management systems.

4.1. The Influence of Agronomic Practices on Yield and Environmental Sustainability

Environmental influence on crop performance is generally significant in the Mediterranean area due to the considerable seasonal variability in temperature and rainfall distribution. In the current study, weather conditions in the two experimental years did not influence final grain yield, but an effect was observed in relation to yield components, with lower grain weight in the crop year characterized by drier and shorter grain filling [38]. In that year, no tillage showed an improvement in yield (grain weight) and protein content [39–41]. It is reported in the literature that the benefits of no tillage in wheat occur more frequently under dry conditions [17] due to the improvement of soil hydraulic properties [13] associated with an increase in soil organic matter [8,31], which reduces evapotranspiration losses [42,43]. Observations from the current study of 1 t/ha of SOC accumulated during the experiment under NT are in accordance with the well-known improvement in soil fertility due to conservation agriculture [12]. The adoption of slow-release fertilizers is reported to show comparable or improved N use efficiency with a lower greenhouse gas emission level [19,44–46]. The low moisture content of agricultural soils in Mediterranean environments favors nitrification activities; for this reason, the use of nitrification inhibitors, such as DMPP, is recommended for mitigation goals [47]. Under our experimental conditions, the combination of no tillage with the use of lower N rates of slow-release fertilizers does not negatively affect yield and quality traits. On the other hand, the contribution of both practices to mitigation resulted marked, in accordance with the indications of climate-smart crop production, with higher grain yield per unit of GWG [10,15,48]. It has been also reported that N fertilization reduces yield gaps between CT and NT, and that durum wheat yield is generally not influenced by tillage [4,5,40]; this trend was generally observed under our experimental conditions, since the interaction between tillage and fertilization was limited. On the other hand, the different response of the two genotypes suggests that varietal choice is a critical aspect for durum wheat cultivation in conservative agriculture, in particular on NUE [49]. Breeding activity for this target is in progress, reporting early vegetative growth as a key trait to improve yield under no tillage [50]. The better adaptation observed in Saragolla may be possibly related to differences in phenology, with good earliness and adaptation to the Mediterranean environment [51]. As regards the impact of fertilization on NUE traits, in general the low emission strategy showed a higher efficiency, with values within the optimal range indicated by the EU Nitrogen Expert Panel [22]. The standard fertilization adopted by farmers, in fact, is proposed for high quality targets; however, under low yield conditions, these could be associated with a low NUE, even if comprised within the EU panel's range, in terms of ratio between N output to N input, i.e., partial N balance [21]. The low emission fertilization strategy achieved higher NUE values, closer to the target (90%) proposed by EU N Expert Panel [22].

4.2. The Influence of Agronomic Practices on Quality and Economic Profitability

Grain filling is generally one of the most sensitive stages to abiotic stresses, especially water deficit, having a significant influence on quality [2,3]. The impact on protein accumulation was more marked in terms of protein composition, rather than content, with a higher gliadin value and its ratio with glutenin in the wetter crop year during grain filling. Few studies report the effect of tillage on wheat storage protein composition. In a study under transition to conservative agriculture, the authors reported differences of response to N fertilization of gliadin accumulation due to rainfall during grain filling [30]. The same group also reported a higher glutenin to gliadin ratio under no tillage [27], thus highlighting the complexity of storage protein regulation as affected by environmental conditions. However, the same trend was observed in the current study under Mediterranean climate conditions. The limited differences due to genotype in terms of gluten composition is possibly explained by the fact that the two modern durum wheat genotypes have been selected for their high technological quality target, and this resulted in a general low gliadin-to-glutenin ratio and a high expression of LMW-GS for good pasta making quality [34,51]. The investigated fertilizations showed, in general, an increase in gliadin-to-glutenin ratio with respect to the unfertilized control. Higher N supply tends, in fact, to increase this ratio due to the source-related dynamics of grain protein fractions accumulation [52].The low emission fertilization showed an increase in gliadin expression. This is possibly explained by the prolonged soil N availability granted by the controlled-release fertilizers during grain filling, which contributed the accumulation of the monomeric alcohol-soluble fractions [46,53]. With regard to the health-related quality traits, the shorter grain filling time, due to lower water availability, led to lower grain weight and was associated with a higher grain mineral concentration (Fe and Zn) and antioxidant capacity in accordance with other observations on durum wheat [3,54–56].

Studies on bread wheat reported the benefits of no tillage in terms of grain quality, i.e., protein and Zn [57], which is explained by the contribution of straw incorporation [58]. Indeed, the improvement in soil fertility due to SOC is reported to increase grain nutrient concentration [59,60] by influencing soil nutrient dynamics [61]. The influence of soil tillage on antioxidant capacity is less investigated in durum wheat. Under our conditions, the trend of higher AOX measured with no tillage was observed in the year with the grain filling period characterized by higher evapotranspiration deficit. A comparable response was also observed in the Mediterranean environment, supporting the hypothesis of complexity on the regulation of this health-related trait [62]. It is worth noting that the use of no tillage with straw mulching was recently associated with an improvement in antioxidant enzyme activity under grain filling stage in wheat; the improvement in antioxidant enzyme metabolism reported at leaf level might be responsible for the increase in antioxidant capacity in grain observed in our study [28]. With regard to the effect of N supply, health-related traits were also generally depressed in the unfertilized control, indicating how crop nutrient deficiency could affect not only technological but also health wheat quality, in terms of both micronutrients (Fe and Zn) and antioxidant capacity [30,63]. The combination of no tillage (conservative) and slow-release fertilization showed great benefits in terms of economic profitability, in agreement with other observations drawn from the Mediterranean environment [64,65]. Indeed, the reduction of costs for cultivation is suggested as the major factor in improving economic stability for farmers [66]. This aspect is critical in a market increasingly subjected to price instability and consumer demands in terms of health quality [67].

5. Conclusions

Under our experimental conditions, no tillage showed the best performance for both durum wheat grain yield and quality in the crop year characterized by water deficit during grain filling. Low emission fertilization, also carried out using controlled release fertilizers, did not compromise yield and quality but led to an advantage in terms of sustainability and economic profitability with respect to the standard fertilization. The outcomes of this study indicate that the combination of no tillage and low emission fertilization is the best practice among those investigated to counteract climate changes by reducing nitrogen supply while maintaining yield performance, with benefits for both farmers and consumers. The novel finding that has emerged from this study is the improvement of health-related quality traits, such as micronutrient content and antioxidant activity, which is consistent with GHG emission mitigation and economic benefits. Since this study was carried out in a South Italy environment, further experiments under different growing conditions, and those that also investigate a wider genotypic variability, are necessary to support the outcomes of this study in order to face the goals of food security and mitigation in Mediterranean environment in a climate change scenario.

Supplementary Materials: The following supporting information can be downloaded at: https://www.mdpi.com/article/10.3390/agronomy14122794/s1, Table S1: Details of the investigated durum wheat genotypes; Table S2: Details of agronomic management and economic analysis.

Author Contributions: Conceptualization, M.A.D.S. and Z.F.; methodology, M.A.D.S. and Z.F.; software, M.A.D.S.; formal analysis, M.A.D.S.; investigation, M.A.D.S., L.G., D.T. and M.S.; resources, Z.F.; data curation, M.A.D.S.; writing—original draft preparation, M.A.D.S.; writing—review and editing, Z.F.; visualization, M.A.D.S. and Z.F.; supervision, Z.F.; project administration, Z.F.; funding acquisition, Z.F. All authors have read and agreed to the published version of the manuscript.

Funding: This research was funded by Regione Puglia through the grant "Ottimizzazione delle pratiche di semina su sodo in frumento duro per migliorare la sostenibilità della cerealicoltura pugliese—SODOSOST" (P.S.R. Puglia 2014/2020-Misura 16.2)—CUP: B79J20000090009; DDS: 94250033811.

Data Availability Statement: Data are available on request.

Acknowledgments: We thank Maura Laus for her technical support.

Conflicts of Interest: The authors declare no conflicts of interest.

Appendix A

Table A1. Details of agrometeorological and phenology data of durum wheat grown in south Italy during two consecutive crop years.

Month	10-Day	Phenology	GDD °C d		P mm		PET mm		P-PET mm	
			2021	2022	2021	2022	2021	2022	2021	2022
January	I		175	190	77	37	18	20	59	18
	II		220	259	78	37	26	30	52	7
	III		315	313	93	46	39	43	54	4
February	I	tillering	424	394	101	62	54	57	47	4
	II		485	498	120	63	68	75	52	−12
	III		567	559	120	153	85	87	35	66
March	I		655	606	158	192	104	101	54	90
	II	stem elongation	731	674	181	192	124	124	57	68
	III		837	792	181	200	152	154	29	46
April	I		941	892	181	214	182	180	−1	34
	II		998	1005	202	214	201	214	2	0
	III	heading	1137	1152	215	215	240	253	−25	−38
May	I		1305	1300	217	246	286	292	−68	−46
	II	grain filling	1475	1503	225	246	331	348	−106	−102
	III		1680	1763	227	246	391	416	−165	−170
June	I		1882	2010	231	304	449	483	−218	−179
	II	maturity	2110	2244	231	304	515	545	−284	−241
	III		2396	2527	231	304	587	616	−357	−312

GS = growth stage; GDD = growing degree days; P = precipitation; PET = potential evapotranspiration; P-PET = rainfall deficit.

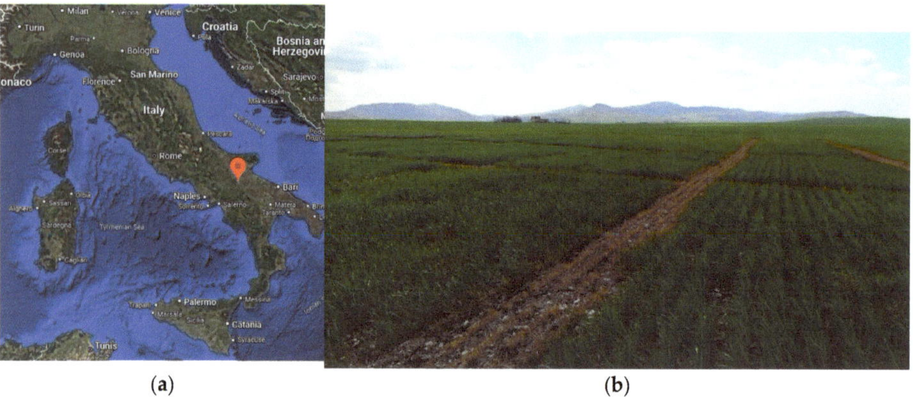

Figure A1. Geographic reference of the experimental site (**a**) and image of the crop plots (**b**).

References

1. Grosse-Heilmann, M.; Cristiano, E.; Deidda, R.; Viola, F. Durum Wheat Productivity Today and Tomorrow: A Review of Influencing Factors and Climate Change Effects. *Resour. Environ. Sustain.* **2024**, *17*, 100170. [CrossRef]
2. Flagella, Z.; Giuliani, M.M.; Giuzio, L.; Volpi, C.; Masci, S. Influence of Water Deficit on Durum Wheat Storage Protein Composition and Technological Quality. *Eur. J. Agron.* **2010**, *33*, 197–207. [CrossRef]
3. De Santis, M.A.; Soccio, M.; Laus, M.N.; Flagella, Z. Influence of Drought and Salt Stress on Durum Wheat Grain Quality and Composition: A Review. *Plants* **2021**, *10*, 2599. [CrossRef] [PubMed]
4. Cossani, C.M.; Slafer, G.A.; Savin, R. Nitrogen and Water Use Efficiencies of Wheat and Barley under a Mediterranean Environment in Catalonia. *Field Crops Res.* **2012**, *128*, 109–118. [CrossRef]
5. Ceglar, A.; Toreti, A.; Zampieri, M.; Royo, C. Global Loss of Climatically Suitable Areas for Durum Wheat Growth in the Future. *Environ. Res. Lett.* **2021**, *16*, 104049. [CrossRef]
6. Farooq, M.; Hussain, M.; Siddique, K.H.M. Drought Stress in Wheat during Flowering and Grain-Filling Periods. *Crit. Rev. Plant Sci.* **2014**, *33*, 331–349. [CrossRef]

7. Araus, J.L.; Slafer, G.A.; Royo, C.; Serret, M.D. Breeding for Yield Potential and Stress Adaptation in Cereals. *Crit. Rev. Plant Sci.* **2008**, *27*, 377–412. [CrossRef]
8. Raimondi, G.; Maucieri, C.; Toffanin, A.; Renella, G.; Borin, M. Smart Fertilizers: What Should We Mean and Where Should We Go? *Ital. J. Agron.* **2021**, *16*. [CrossRef]
9. Liu, Z.; Wang, N.; Lü, J.; Wang, L.; Li, G.; Ning, T. Climate-Smart Tillage Practices with Straw Return to Sustain Crop Productivity. *Agronomy* **2022**, *12*, 2452. [CrossRef]
10. Alhajj Ali, S.; Tedone, L.; Verdini, L.; De Mastro, G. Effect of Different Crop Management Systems on Rainfed Durum Wheat Greenhouse Gas Emissions and Carbon Footprint under Mediterranean Conditions. *J. Clean. Prod.* **2017**, *140*, 608–621. [CrossRef]
11. Seddaiu, G.; Iocola, I.; Farina, R.; Orsini, R.; Iezzi, G.; Roggero, P.P. Long Term Effects of Tillage Practices and N Fertilization in Rainfed Mediterranean Cropping Systems: Durum Wheat, Sunflower and Maize Grain Yield. *Eur. J. Agron.* **2016**, *77*, 166–178. [CrossRef]
12. Giambalvo, D.; Amato, G.; Badagliacca, G.; Ingraffia, R.; Di Miceli, G.; Frenda, A.S.; Plaia, A.; Venezia, G.; Ruisi, P. Switching from Conventional Tillage to No-Tillage: Soil N Availability, N Uptake, 15N Fertilizer Recovery, and Grain Yield of Durum Wheat. *Field Crops Res.* **2018**, *218*, 171–181. [CrossRef]
13. Castellini, M.; Fornaro, F.; Garofalo, P.; Giglio, L.; Rinaldi, M.; Ventrella, D.; Vitti, C.; Vonella, A.V. Effects of No-Tillage and Conventional Tillage on Physical and Hydraulic Properties of Fine Textured Soils under Winter Wheat. *Water* **2019**, *11*, 484. [CrossRef]
14. Puig-Sirera, À.; Acutis, M.; Bancheri, M.; Bonfante, A.; Botta, M.; De Mascellis, R.; Orefice, N.; Perego, A.; Russo, M.; Tedeschi, A.; et al. Zero-Tillage Effects on Durum Wheat Productivity and Soil-Related Variables in Future Climate Scenarios: A Modeling Analysis. *Agronomy* **2022**, *12*, 331. [CrossRef]
15. Meng, X.; Meng, F.; Chen, P.; Hou, D.; Zheng, E.; Xu, T. A Meta-Analysis of Conservation Tillage Management Effects on Soil Organic Carbon Sequestration and Soil Greenhouse Gas Flux. *Sci. Total Environ.* **2024**, *954*, 176315. [CrossRef]
16. Dong, W.; Liu, E.; Yan, C.; Tian, J.; Zhang, H.; Zhang, Y. Impact of No Tillage vs. Conventional Tillage on the Soil Bacterial Community Structure in a Winter Wheat Cropping Succession in Northern China. *Eur. J. Soil Biol.* **2017**, *80*, 35–42. [CrossRef]
17. Pittelkow, C.M.; Linquist, B.A.; Lundy, M.E.; Liang, X.; van Groenigen, K.J.; Lee, J.; van Gestel, N.; Six, J.; Venterea, R.T.; van Kessel, C. When Does No-till Yield More? A Global Meta-Analysis. *Field Crops Res.* **2015**, *183*, 156–168. [CrossRef]
18. Casolani, N.; Pattara, C.; Liberatore, L. Water and Carbon Footprint Perspective in Italian Durum Wheat Production. *Land Use Policy* **2016**, *58*, 394–402. [CrossRef]
19. Volpi, I.; Laville, P.; Bonari, E.; di Nasso, N.N.O.; Bosco, S. Improving the Management of Mineral Fertilizers for Nitrous Oxide Mitigation: The Effect of Nitrogen Fertilizer Type, Urease and Nitrification Inhibitors in Two Different Textured Soils. *Geoderma* **2017**, *307*, 181–188. [CrossRef]
20. De Santis, M.A.; Cammarano, D. Agronomic Management Factors Impacting Yield, Quality Stability, and Environmental Footprints of Barley in a Mediterranean Environment. *Field Crops Res.* **2024**, *309*, 109334. [CrossRef]
21. Tamagno, S.; Maaz, T.M.; van Kessel, C.; Linquist, B.A.; Ladha, J.K.; Lundy, M.E.; Maureira, F.; Pittelkow, C.M. Critical Assessment of Nitrogen Use Efficiency Indicators: Bridging New and Old Paradigms to Improve Sustainable Nitrogen Management. *Eur. J. Agron.* **2024**, *159*, 127231. [CrossRef]
22. EU Nitrogen Expert Panel. *Nitrogen Use Efficiency (NUE): An Indicator for the Utilization of Nitrogen in Agriculture and Food Systems*; Wageningen University: Wageningen, The Netherlands, 2015; pp. 1–47.
23. Lafiandra, D.; Shewry, P.R. Wheat Glutenin Polymers 2. The Role of Wheat Glutenin Subunits in Polymer Formation and Dough Quality. *J. Cereal Sci.* **2022**, *106*, 103487. [CrossRef]
24. Velu, G.; Singh, R.P.; Huerta, J.; Guzmán, C. Genetic Impact of Rht Dwarfing Genes on Grain Micronutrients Concentration in Wheat. *Field Crops Res.* **2017**, *214*, 373–377. [CrossRef] [PubMed]
25. Fu, B.X.; Chiremba, C.; Pozniak, C.J.; Wang, K.; Nam, S. Total Phenolic and Yellow Pigment Contents and Antioxidant Activities of Durum Wheat Milling Fractions. *Antioxidants* **2017**, *6*, 78. [CrossRef] [PubMed]
26. Laus, M.N.; Di Benedetto, N.A.; Caporizzi, R.; Tozzi, D.; Soccio, M.; Giuzio, L.; De Vita, P.; Flagella, Z.; Pastore, D. Evaluation of Phenolic Antioxidant Capacity in Grains of Modern and Old Durum Wheat Genotypes by the Novel QUENCHERABTS Approach. *Plant Foods Hum. Nutr.* **2015**, *70*, 207–214. [CrossRef]
27. Pagnani, G.; Galieni, A.; D'Egidio, S.; Visioli, G.; Stagnari, F.; Pisante, M. Effect of Soil Tillage and Crop Sequence on Grain Yield and Quality of Durum Wheat in Mediterranean Areas. *Agronomy* **2019**, *9*, 488. [CrossRef]
28. Li, P.; Yin, W.; Fan, Z.; Hu, F.; Zhao, L.; Fan, H.; He, W.; Chai, Q. Improving Crop Productivity by Optimizing Straw Returning Patterns to Delay Senescence of Wheat Leaves. *Eur. J. Agron.* **2024**, *159*, 127274. [CrossRef]
29. Hargreaves George, H.; Samani Zohrab, A. Estimating Potential Evapotranspiration. *J. Irrig. Drain. Div.* **1982**, *108*, 225–230. [CrossRef]
30. Galieni, A.; Stagnari, F.; Visioli, G.; Marmiroli, N.; Speca, S.; Angelozzi, G.; D'Egidio, S.; Pisante, M. Nitrogen Fertilisation of Durum Wheat: A Case Study in Mediterranean Area during Transition to Conservation Agriculture. *Ital. J. Agron.* **2016**, *10*, 12. [CrossRef]
31. De Santis, M.A.; Campaniello, D.; Tozzi, D.; Giuzio, L.; Corbo, M.R.; Bevilacqua, A.; Sinigaglia, M.; Flagella, Z. Agronomic Response to Irrigation and Biofertilizer of Peanut (*Arachis hypogea* L.) Grown under Mediterranean Environment. *Agronomy* **2023**, *13*, 1566. [CrossRef]

32. Guarda, G.; Padovan, S.; Delogu, G. Grain Yield, Nitrogen-Use Efficiency and Baking Quality of Old and Modern Italian Bread-Wheat Cultivars Grown at Different Nitrogen Levels. *Eur. J. Agron.* **2004**, *21*, 181–192. [CrossRef]
33. Lollato, R.P.; Figueiredo, B.M.; Dhillon, J.S.; Arnall, D.B.; Raun, W.R. Wheat Grain Yield and Grain-Nitrogen Relationships as Affected by N, P, and K Fertilization: A Synthesis of Long-Term Experiments. *Field Crops Res.* **2019**, *236*, 42–57. [CrossRef]
34. De Santis, M.A.; Giuliani, M.M.; Giuzio, L.; De Vita, P.; Flagella, Z. Assessment of Grain Protein Composition in Old and Modern Italian Durum Wheat Genotypes. *Ital. J. Agron.* **2018**, *11*, 40–43. [CrossRef]
35. De Santis, M.A.; Giuliani, M.M.; Flagella, Z.; Pellegrino, E.; Ercoli, L. Effect of Arbuscular Mycorrhizal Fungal Seed Coating on Grain Protein and Mineral Composition of Old and Modern Bread Wheat Genotypes. *Agronomy* **2022**, *12*, 2418. [CrossRef]
36. Ciliberti, S.; Stanco, M.; Frascarelli, A.; Marotta, G.; Martino, G.; Nazzaro, C. Sustainability Strategies and Contractual Arrangements in the Italian Pasta Supply Chain: An Analysis under the Neo Institutional Economics Lens. *Sustainability* **2022**, *14*, 8542. [CrossRef]
37. Guardia, G.; Aguilera, E.; Vallejo, A.; Sanz-Cobena, A.; Alonso-Ayuso, M.; Quemada, M. Effective Climate Change Mitigation through Cover Cropping and Integrated Fertilization: A Global Warming Potential Assessment from a 10-Year Field Experiment. *J. Clean. Prod.* **2019**, *241*, 118307. [CrossRef]
38. Garrido-Lestache, E.; López-Bellido, R.J.; López-Bellido, L. Durum Wheat Quality under Mediterranean Conditions as Affected by N Rate, Timing and Splitting, N Form and S Fertilization. *Eur. J. Agron.* **2005**, *23*, 265–278. [CrossRef]
39. De Vita, P.; Di Paolo, E.; Fecondo, G.; Di Fonzo, N.; Pisante, M. No-Tillage and Conventional Tillage Effects on Durum Wheat Yield, Grain Quality and Soil Moisture Content in Southern Italy. *Soil Tillage Res.* **2007**, *92*, 69–78. [CrossRef]
40. Dang, H.; Sun, R.; She, W.; Hou, S.; Li, X.; Chu, H.; Wang, T.; Huang, T.; Huang, Q.; Siddique, K.H.M.; et al. Updating Soil Organic Carbon for Wheat Production with High Yield and Grain Protein. *Field Crops Res.* **2024**, *317*, 109549. [CrossRef]
41. Baiamonte, G.; Novara, A.; Gristina, L.; D'Asaro, F. Durum Wheat Yield Uncertainty under Different Tillage Management Practices and Climatic Conditions. *Soil Tillage Res.* **2019**, *194*, 104346. [CrossRef]
42. Djouadi, K.; Mekliche, A.; Dahmani, S.; Ladjiar, N.I.; Abid, Y.; Silarbi, Z.; Hamadache, A.; Pisante, M. Durum Wheat Yield and Grain Quality in Early Transition from Conventional to Conservation Tillage in Semi-Arid Mediterranean Conditions. *Agriculture* **2021**, *11*, 711. [CrossRef]
43. Gandía, M.L.; Del Monte, J.P.; Tenorio, J.L.; Santín-Montanyá, M.I. The Influence of Rainfall and Tillage on Wheat Yield Parameters and Weed Population in Monoculture versus Rotation Systems. *Sci. Rep.* **2021**, *11*, 22138. [CrossRef] [PubMed]
44. Abalos, D.; Jeffery, S.; Sanz-Cobena, A.; Guardia, G.; Vallejo, A. Meta-Analysis of the Effect of Urease and Nitrification Inhibitors on Crop Productivity and Nitrogen Use Efficiency. *Agric. Ecosyst. Environ.* **2014**, *189*, 136–144. [CrossRef]
45. Rose, T.J.; Wood, R.H.; Rose, M.T.; Van Zwieten, L. A Re-Evaluation of the Agronomic Effectiveness of the Nitrification Inhibitors DCD and DMPP and the Urease Inhibitor NBPT. *Agric. Ecosyst. Environ.* **2018**, *252*, 69–73. [CrossRef]
46. De Santis, M.A.; Giuliani, M.M.; Flagella, Z.; Reyneri, A.; Blandino, M. Impact of Nitrogen Fertilisation Strategies on the Protein Content, Gluten Composition and Rheological Properties of Wheat for Biscuit Production. *Field Crops Res.* **2020**, *254*, 107829. [CrossRef]
47. Abalos, D.; Sanz-Cobena, A.; Andreu, G.; Vallejo, A. Rainfall Amount and Distribution Regulate DMPP Effects on Nitrous Oxide Emissions under Semiarid Mediterranean Conditions. *Agric. Ecosyst. Environ.* **2017**, *238*, 36–45. [CrossRef]
48. Plaza-Bonilla, D.; Álvaro-Fuentes, J.; Arrúe, J.L.; Cantero-Martínez, C. Tillage and Nitrogen Fertilization Effects on Nitrous Oxide Yield-Scaled Emissions in a Rainfed Mediterranean Area. *Agric. Ecosyst. Environ.* **2014**, *189*, 43–52. [CrossRef]
49. Ingraffia, R.; Lo Porto, A.; Ruisi, P.; Amato, G.; Giambalvo, D.; Frenda, A.S. Conventional Tillage versus No-Tillage: Nitrogen Use Efficiency Component Analysis of Contrasting Durum Wheat Genotypes Grown in a Mediterranean Environment. *Field Crops Res.* **2023**, *296*, 108904. [CrossRef]
50. Honsdorf, N.; Mulvaney, M.J.; Singh, R.P.; Ammar, K.; Govaerts, B.; Verhulst, N. Dataset of Historic and Modern Bread and Durum Wheat Cultivar Performance under Conventional and Reduced Tillage with Full and Reduced Irrigation. *Data Brief* **2022**, *43*, 108439. [CrossRef]
51. De Santis, M.A.; Giuliani, M.M.; Giuzio, L.; De Vita, P.; Lovegrove, A.; Shewry, P.R.; Flagella, Z. Differences in Gluten Protein Composition between Old and Modern Durum Wheat Genotypes in Relation to 20th Century Breeding in Italy. *Eur. J. Agron.* **2017**, *87*, 19–29. [CrossRef]
52. Ferrise, R.; Bindi, M.; Martre, P. Grain Filling Duration and Glutenin Polymerization under Variable Nitrogen Supply and Environmental Conditions for Durum Wheat. *Field Crops Res.* **2015**, *171*, 23–31. [CrossRef]
53. Rekowski, A.; Wimmer, M.A.; Hitzmann, B.; Hermannseder, B.; Hahn, H.; Zörb, C. Application of Urease Inhibitor Improves Protein Composition and Bread-Baking Quality of Urea Fertilized Winter Wheat. *J. Plant Nutr. Soil Sci.* **2020**, *183*, 260–270. [CrossRef]
54. Ficco, D.B.M.; Riefolo, C.; Nicastro, G.; De Simone, V.; Di Gesù, A.M.; Beleggia, R.; Platani, C.; Cattivelli, L.; De Vita, P. Phytate and Mineral Elements Concentration in a Collection of Italian Durum Wheat Cultivars. *Field Crops Res.* **2009**, *111*, 235–242. [CrossRef]
55. Magallanes-López, A.M.; Ammar, K.; Morales-Dorantes, A.; González-Santoyo, H.; Crossa, J.; Guzmán, C. Grain Quality Traits of Commercial Durum Wheat Varieties and Their Relationships with Drought Stress and Glutenins Composition. *J. Cereal Sci.* **2017**, *75*, 1–9. [CrossRef]

56. Martini, D.; Taddei, F.; Ciccoritti, R.; Pasquini, M.; Nicoletti, I.; Corradini, D.; D'Egidio, M.G. Variation of Total Antioxidant Activity and of Phenolic Acid, Total Phenolics and Yellow Coloured Pigments in Durum Wheat (*Triticum turgidum* L. Var. *durum*) as a Function of Genotype, Crop Year and Growing Area. *J. Cereal Sci.* **2015**, *65*, 175–185. [CrossRef]

57. Li, Y.; Hou, R.; Tao, F. Interactive Effects of Different Warming Levels and Tillage Managements on Winter Wheat Growth, Physiological Processes, Grain Yield and Quality in the North China Plain. *Agric. Ecosyst. Environ.* **2020**, *295*, 106923. [CrossRef]

58. Chen, Y.; Shi, J.; Dong, J.; Wu, Y.; Li, C.; Ye, Y.; Tian, X.; Wang, Y. Synergistic Improvement of Soil Organic Carbon Storage and Wheat Grain Zinc Bioavailability by Straw Return in Combination with Zn Application on the Loess Plateau of China. *Catena* **2021**, *197*, 104920. [CrossRef]

59. Adesanya, T.; Zvomuya, F.; Fernandez, M.R.; Luce, M.S. Crop Rotation Diversity and Tillage Effects on Soil and Wheat Grain Nutrient Concentration in an Organically-Managed System. *J. Agric. Food Res.* **2024**, *18*, 101411. [CrossRef]

60. Carrara, J.E.; Beelman, R.B.; Duiker, S.W.; Heller, W.P. Reduced Tillage Agriculture May Improve Plant Nutritional Quality through Increased Mycorrhizal Colonization and Uptake of the Antioxidant Ergothioneine. *Soil Tillage Res.* **2024**, *244*, 106283. [CrossRef]

61. Öztürk, F.; Ortaş, I. The Impact of Long-Term Tillage Systems on Soil Carbon and Nitrogen Dynamics and Other Nutrient Contents. *Int. J. Agron.* **2024**, *2024*, 8037593. [CrossRef]

62. Kerbouai, I.; M'hamed, H.C.; Jenfaoui, H.; Riahi, J.; Mokrani, K.; Jribi, S.; Arfaoui, S.; Sassi, K.; Ben Ismail, H. Long-Term Effect of Conservation Agriculture on the Composition and Nutritional Value of Durum Wheat Grains Grown over 2 Years in a Mediterranean Environment. *J. Sci. Food Agric.* **2022**, *102*, 7379–7386. [CrossRef] [PubMed]

63. Ma, D.; Sun, D.; Li, Y.; Wang, C.; Xie, Y.; Guo, T. Effect of Nitrogen Fertilisation and Irrigation on Phenolic Content, Phenolic Acid Composition, and Antioxidant Activity of Winter Wheat Grain. *J. Sci. Food Agric.* **2015**, *95*, 1039–1046. [CrossRef] [PubMed]

64. Colecchia, S.A.; De Vita, P.; Rinaldi, M. Effects of Tillage Systems in Durum Wheat under Rainfed Mediterranean Conditions. *Cereal Res. Commun.* **2015**, *43*, 704–716. [CrossRef]

65. Souissi, A.; Bahri, H.; Cheikh M'hamed, H.; Chakroun, M.; Benyoussef, S.; Frija, A.; Annabi, M. Effect of Tillage, Previous Crop, and N Fertilization on Agronomic and Economic Performances of Durum Wheat (*Triticum durum* Desf.) under Rainfed Semi-Arid Environment. *Agronomy* **2020**, *10*, 1161. [CrossRef]

66. Keil, A.; Mitra, A.; McDonald, A.; Malik, R.K. Zero-Tillage Wheat Provides Stable Yield and Economic Benefits under Diverse Growing Season Climates in the Eastern Indo-Gangetic Plains. *Int. J. Agric. Sustain.* **2020**, *18*, 567–593. [CrossRef]

67. Bimbo, F.; De Meo, E.; Carlucci, D. Hedonic Analysis of Dried Pasta Prices Using E-Commerce Data—An Explorative Study. *Foods* **2024**, *13*, 903. [CrossRef]

Article

Early-Stage Impacts of Irrigated Conservation Agriculture on Soil Physical Properties and Crop Performance in a French Mediterranean System

Juan David Dominguez-Bohorquez [1,2], Claire Wittling [1,*], Bruno Cheviron [1], Sami Bouarfa [1], Nicolas Urruty [2], Jean-Marie Lopez [1] and Cyril Dejean [1]

[1] G-EAU, INRAE, AgroParisTech, BRGM, CIRAD, Institute Agro, IRD, University of Montpellier, 34000 Montpellier, France; juan-david.dominguez-bohorquez@inrae.fr (J.D.D.-B.); bruno.cheviron@inrae.fr (B.C.); sami.bouarfa@inrae.fr (S.B.); jean-marie.lopez@cirad.fr (J.-M.L.); cyril.dejean@inrae.fr (C.D.)

[2] Société du Canal de Provence (SCP), Le Tholonet, 13182 Aix-en-Provence, France; nicolas.urruty@canal-de-provence.com

[*] Correspondence: claire.wittling@inrae.fr

Academic Editors: Mariola Staniak, Ewa Szpunar-Krok and Małgorzata Szostek

Received: 20 December 2024
Revised: 21 January 2025
Accepted: 23 January 2025
Published: 25 January 2025

Citation: Dominguez-Bohorquez, J.D.; Wittling, C.; Cheviron, B.; Bouarfa, S.; Urruty, N.; Lopez, J.-M.; Dejean, C. Early-Stage Impacts of Irrigated Conservation Agriculture on Soil Physical Properties and Crop Performance in a French Mediterranean System. *Agronomy* 2025, 15, 299. https://doi.org/10.3390/agronomy15020299

Abstract: The Mediterranean region faces intensified climate change effects, increasing irrigation demands to sustain crop yields and increasing pressure on water resources. Adaptive management strategies such as conservation agriculture (CA) offer potential benefits for soil quality and water use efficiency. However, there is limited research on the short-term effects of this farming system under irrigated Mediterranean climatic conditions. This study aimed to explore the short-term impacts of conservation agriculture (no tillage, cover crops and crop rotation) on the soil properties, water flows and crop and water productivity in a French Mediterranean agrosystem of irrigated field crops, using a multifactorial approach. From 2021 to 2023, maize, sorghum and soybean were grown successively under either conventional tillage (CT) or conservation agriculture (CA), combined with sprinkler irrigation, subsurface drip irrigation or non-irrigated conditions. The dynamics of the surface soil properties (bulk density, penetration resistance, soil temperature), water flows (infiltration, soil evaporation) and agronomic indicators (leaf area index, crop yield, water productivity) were measured across the three cropping seasons. In the pedoclimatic conditions of the study, CA was shown to clearly impact the soil properties, water flows and crop yields, from the first year of adoption. CA practices caused an increased bulk density and soil resistance penetration, leading to decreased quasi-steady ponded infiltration in the surface horizon, particularly in the CA–subsurface drip and CA–non-irrigated conditions. These effects were also reflected in the leaf area index, crop yield and water productivity, with CA showing lower values compared to CT. Crop residues in CA reduced soil evaporation, particularly under sprinkler irrigation. However, this benefit diminished as the residues decomposed, leading to soil evaporation rates comparable to those observed in CT. Agronomic indicators were better under sprinkler irrigation than under subsurface drip irrigation. Overall, compaction emerged as a significant challenge in the adoption of CA, considering its negative impact on crop yields.

Keywords: conservation agriculture; sprinkler irrigation; subsurface drip irrigation; short-term effects; soil compaction

1. Introduction

Climate change has led to significant increases in temperature on both global and regional scales, while also amplifying the spatial and temporal variability in the rain amounts and intensity at very local scales. The Mediterranean climate, characterized by warm and rainy winters and hot and dry summers [1], is particularly vulnerable to these impacts. Mediterranean regions experience frequent droughts and rising temperatures together with occasional rainstorms, which have overall negative effects on both water resource availability and soil conservation and quality [2]. In Montpellier (Southern France), the 30-year mean annual precipitation is 730 mm, while the mean potential evapotranspiration is 930 mm, resulting in an annual deficit of -200 mm. During the summer crop period (April to September), the mean deficit intensifies to -450 mm, leading to severe drought. Irrigation has become crucial in mitigating these adverse effects on crop yields overall and for summer crops in this context. However, it often leads to conflicts among different water users. In 2016, Southern France accounted for 46% of the total water withdrawn for irrigation in the country [3]. Projections suggest that, solely due to climate change, the irrigation demand in the Mediterranean region could rise by 4 to 18% by 2100 [4,5].

To address this issue and mitigate the impacts of climate change, it is essential to implement adaptive strategies in agrosystem management. One such approach is the adoption of conservation agriculture (CA), which focuses on preserving soil health and fertility [6,7] through practices encompassing three principles: (1) reduced or zero tillage, (2) crop residue or cover crop mulching and (3) diversified crop rotation [8]. The long-term adoption of CA has demonstrated numerous benefits, including an improved soil structure (better aggregate stability and continuity in microporosity), reduced erosion risks and increased water storage capacities (minimized soil evaporation and facilitated water infiltration and retention) [9,10].

Given its potential to improve the soil water dynamics by reducing runoff, enhancing the water use efficiency and mitigating drought conditions, CA is particularly relevant for Mediterranean temperate climates [11–13]. A meta-analysis conducted by Lee et al. [14] confirms that CA can lead to more efficient water use in Mediterranean regions, particularly in rainfed semi-arid contexts. As a result, CA has emerged as a realistic and effective tool to sustainably intensify agricultural production in these regions [15].

However, the effectiveness of CA can widely vary depending on the specific combination of CA practices implemented, the pedoclimatic conditions [16,17] and the duration after adoption. As presented in Table 1, the implemented CA practices differ from one system to another, with some systems adopting one, two or all three CA principles, while others adapt these principles to the local conditions (e.g., mulch till, strip till, occasional crop rotation). This diversity of practices, along with various pedoclimatic factors, results in variable outcomes regarding the soil properties, which affect soil functions and crop performance [13]. For example, depending on soil type, studies generally report an increase in bulk density (3–19%) and higher penetration resistance (25–56%) alongside reduced infiltration rates (13–40%) during the first three years of CA adoption [17–25]. Similar trends were observed even after 6 years of CA adoption [26]. Additionally, these studies often report a decrease in grain yield with considerable variability (15–72%). Some studies have found similar results even after more than ten years [11,27,28]. Over time, however, there is a tendency toward favorable trends, such as reduced penetration resistance, increased water infiltration and enhanced crop and water productivity by up to +94% [29–31]. The literature also indicates that CA systems reduce fluctuations in soil temperature, as well as mean soil temperatures, with a drop of up to 3 °C [17,32,33]. Notably, soil evaporation remains a relatively underexplored aspect.

Focusing on Mediterranean conditions, the short-term effects of CA can be highly variable. Some research shows positive outcomes, similar to the long term, including increased infiltration, higher yields and improved water productivity, despite rises in bulk density and penetration resistance [11,33–35]. In contrast, other studies describe reduced infiltration, an increased bulk density and penetration resistance, accompanied sometimes by declines in yield and water productivity [22,23,36]. These inconsistencies highlight the complexity of CA adoption and underscore the importance of considering local factors such as the soil type, climate and irrigation practices.

Despite extensive research on CA's effects on the soil properties and water dynamics [37–42], relatively few studies have focused on the short-term effects following CA adoption in irrigated Mediterranean agrosystems [43,44], with even fewer in the context of Southern France. Additionally, limited research has explored the interaction between CA and different irrigation systems. Furthermore, there is a lack of studies providing a comprehensive overview of the simultaneous evolution in the short term of the soil properties, water dynamics and agronomic performance, which may differ from the long term (as highlighted in Table 1) and can present real challenges for CA adoption. Evaluating these early-stage effects is crucial, as it can help farmers to anticipate potential obstacles during the transition and ensure that they realize the full benefits of CA over time.

The present study aims to fill the gap in the literature by assessing the effects of CA (three principles implemented simultaneously: zero tillage, cover crops, diversified crop rotation) on the soil hydrodynamic properties and crop productivity, during the short-term period following its adoption in a French Mediterranean irrigated agrosystem. It consists of the experimental assessment of the physical soil properties (bulk density, penetration resistance, temperature), water fluxes (quasi-steady ponded infiltration rate and soil evaporation), crop development (leaf area index) and crop performance (grain yield and water productivity). The investigation was conducted over a three-year period (2021 to 2023) to compare the outcomes of two farming approaches, conservation agriculture (CA) and conventional tillage (CT), both combined with sprinkler (S) irrigation, subsurface drip (SSD) irrigation or non-irrigated (NI) conditions. Although some properties have been evaluated at slightly greater depths, this study focused mainly on the surface soil layer, which is the most relevant as the surface part of the soil is the most affected by tillage and organic matter enrichment.

Table 1. Effects of conservation agriculture practices on some soil properties, water fluxes, grain yield and water productivity across diverse geographical contexts, compared to conventional tillage. Positioning of the results of the present study (grey line). "CA": conservation agriculture with the three principles (no tillage, cover crops and crop rotation). "ND": not determined. Symbols: ↑ increase, ↓ decrease, ≈ no significant difference.

Reference	Time of Practice	CA Practices	Climate (Country)	Soil Type (Texture)	Irrigation	Soil Temperature	Bulk Density	Penetration Resistance	Infiltration	Soil Evaporation	Grain Yield	Water Productivity
[24]	1 year	CA	Semi-arid continental (Spain)	Vertic Luvisol (Loam)	Sprinkler	↓ fluctuations, ↓ 1.9–2.5 °C	↑ 12–19%	↑ 25–33%	ND	ND	↓ 15.4%	↓ 15.4% (crop yield/irrigation applied)
[45]	1 year	CA	Sub-tropical (Cuba)	Red Ferralitic	Not irrigated	ND	↓ 7%	ND	↑ 20%	ND	ND	ND
[18]	2 years	CA	Sub-tropical (Nepal)	(Silt–Loam)	Not irrigated	ND	↑ 5%	ND	Soil sorptivity about three times slower	ND	≈ in the first year, ↓ 72% second year	ND
[21]	2 years	No tillage	Semi-arid (Iran)	(Silty Clay Loam)	Sprinkler	ND	↑ 6%	↑ 37%	ND	ND	↓ 18%	ND
[19]	2 or 5 years	No tillage and cover crop	Semi-arid (Southern Malawi)	Chromic Luvisols and Chromic Cambisols—(Clay Loams to Clay)	Not irrigated	ND	↑ 5%	ND	↓ 13% of hydraulic conductivity after 2 years, ≈ after 5 years	ND	ND	ND
This study	3 years	CA	Mediterranean (France)	Fluvisol- (Loam)	Sprinkler, subsurface drip and not irrigated	-	-	-	-	-	-	-
[22,23]	3 years	CA	Mediterranean (Turkey)	Haploxererts (Heavy Clay)	Sprinkler	ND	↑ 1–12%	↑ 7–56%	↓ 20–40% of saturated hydraulic conductivity	ND	ND	ND
[11]	3 years	CA	Mediterranean (Spain)	Xerofluvent, (Loamy Alluvial)	Sprinkler	ND	↑ 2–6%	↓ 3–41%	↑ soil water storage	ND	↑ 8–24%	↑ 11–31%
[32]	3 years	No tillage and cover crops	Semi-humid, monsoon (China)	Hepludolls (Clay Loam)	Not irrigated	↓ 0.5–0.9 °C	ND	ND	ND	ND	ND	ND

Table 1. *Cont.*

Reference	Time of Practice	CA Practices	Climate (Country)	Soil Type (Texture)	Irrigation	Soil Temperature	Bulk Density	Penetration Resistance	Infiltration	Soil Evaporation	Grain Yield	Water Productivity
[46]	3 years	Permanent beds, permanent beds with short-duration pulse crop	Semi-arid (India)	(Clay Loam)	Flood	ND	ND	ND	ND	ND	↑ 4.2–13.5%	↑ 28–40% (grain yield/crop evapotran-spiration)
[20]	3 years	CA	Tropical (Ghana)	Ferric Lixisols (Sandy with Low Clay Content)	Not irrigated	ND	↑ 10–15%	ND	ND	ND	↓ 23–37%	ND
[24]	3 years	No tillage and cover crop	Sub-humid mediter-ranean (Chile)	Ultic Palexerals (Sandy Loam and Clay)	Not irrigated	ND	ND	↑ 25% in the first 20 cm depth	ND	ND	↓ 29–35%	ND
[47]	3 years	No tillage and cover crop	Sub-humid (Italy)	Fluvi-Calcaric Cambisol (Silty Loam)	Not irrigated	ND	↑ 6.5%	↑ 31–36%	↑ more than twice	ND	ND	ND
[25]	3 to 5 years	No tillage, cover crop (mulch)	Semi-arid (Iran)	Haploxerepts (Silty Clay Loam)	Flood	ND	↑ 7–11%	ND	↓ 13% of cumulative water infiltration	ND	↓ 11–31%	↑ 8% (crop yield/total water applied)
[48]	3, 6 and 9 years	No tillage and cover crop	Sub-tropical continental monsoon (India)	Alluvium (Sandy Clay Loam and Clay Loam)	Flood	ND	↑ 7%, 2% and 3% after 3, 6 and 9 years, respec-tively	ND	ND	ND	ND	ND
[33]	3 and 6 years	No tillage and cover crop	Mediterranean (Spain)	Vertisol	Not irrigated	↓ of fluctuations	ND	ND	ND	ND	ND	ND
[34]	3 and 9 years	No tillage and cover crops	Mediterranean (Spain)	Stony District Luvisol (Loam)	Sprinkler	ND	ND	↓ 60%	↑ 26–44% of soil water content	ND	ND	ND

Table 1. *Cont.*

Reference	Time of Practice	CA Practices	Climate (Country)	Soil Type (Texture)	Irrigation	Soil Temperature	Bulk Density	Penetration Resistance	Infiltration	Soil Evaporation	Grain Yield	Water Productivity
[35]	3 and 7 years	No tillage and cover crops	Mediterranean (Spain)	Hydragic Anthrosol	Sprinkler and flood	ND	ND	↑ in the 0–25 cm layer ≈ for deeper layers	ND	ND	↑ 7–70% for the first 3 years and ↑ 50–170% after 7 years	↑ 10–70% for the first 3 years ↑ 50–170% after 7 years (grain yield/irrigation water)
[49]	4 years	No tillage, permanent beds, residue retention, crop rotation	Sub-tropical (India)	(Silty Loam)	Flood	ND	ND	ND	ND	ND	↑ 12–16% for maize and ↑ 5–9% for wheat	↑ 35–94% (grain yield/irrigation water applied)
[36]	5 to 10 years	CA	Mediterranean (Spain)	Xerofluvent (Sandy Clay Loam)	Not irrigated (winter season)	ND	ND	↑ 10–15 times	ND	ND	↓ 97%	ND
[28]	5 to 7 years	No tillage, organic amendments	Moist Mediterranean (Spain)	Clay-Skeletal, Kaolinitic, Acid, Thermic Plinthoc Palexerults	Not irrigated (winter season)	ND	↓ 12%	ND	↑ 56–59% of hydraulic conductivity	ND	↑ 3–4 times	ND
[50]	5 to 8 years	Mulch till and cover crop and crop rotation	Oceanic with both Atlantic and Mediterranean influences (Southwestern France)	Gleyic Luvisol (Loamy, Illuvial Clay and Alluvial Pebbly)	Center pivot	↓ 0.8–2.8 °C at sowing	ND	ND	ND	ND	ND	ND
[51]	5 to 7 years	CA	Semi-arid (India)	Alluvium (Sandy Clay Loam)	Flood	↓ 1.3 °C	ND	ND	ND	ND	↑ 12%	↑ 15% (grain yield/irrigation applied)
[52]	6 years	No tillage, cover crop and occasional crop rotation	Semi-arid (United States)	Abilene (Clay Loam)	Subsurface drip	ND	ND	ND	ND	ND	↑ 9%	↑ 9–11% (grain yield/irrigation applied)

Table 1. *Cont.*

Reference	Time of Practice	CA Practices	Climate (Country)	Soil Type (Texture)	Irrigation	Soil Temperature	Bulk Density	Penetration Resistance	Infiltration	Soil Evaporation	Grain Yield	Water Productivity
[26]	6 years	CA	Sub-humid and cool temperate (Canada)	Gray Luvisol	Not irrigated	ND	↑ 9%	↑ 10–79%	↓ 33%	ND	ND	ND
[31]	6 years	CA in monocropping and intercropping	Tropical wet and savanna (Malawi)	Ferric/Orthic Acrisol and Ferric Luvisol (Sandy Clay Loam)	Not irrigated	ND	ND	ND	↑ 18–42%	ND	↑ 30–133%	ND
[53]	7 years	CA	Semi-arid (India)	Typic Haplustept (Clay Loam)	Flood	ND	≈	↓ 20–41%	≈	ND	↑ 13–21%	ND
[27]	7 years	CA	Semi-arid (India)	Haplustept (Sandy Loam)	Flood	NS	↓ 4–7%	↓ 16–27%	↑ 11–12%	ND	ND	ND
[7]	7 years	No tillage and crop rotation	Mediterranean (Italy)	(Clay Loam)	Not irrigated (winter season)	↑ 0.3 °C	ND	ND	ND	ND	ND	ND
[54]	7 to 9 years	No tillage, crop rotation	Humid temperate (Argentina)	Argiaquoll (Clay Loam, Clayey, Silty Clay)	Not irrigated	ND	↑ 5%	ND	≈	ND	ND	ND
[28]	8 years	CA	Mediterranean (Spain)	Parexerults (Sandy Loam, Sandy Clay and Sandy Clay Loam)	Not irrigated (winter season)	ND	↓ 13%	ND	↑ 120% of mean infiltration ↑ 140% of hydraulic conductivity	ND	↑ 3 times	ND
[41]	9 to 28 years	No till, strip till, crop rotation	Oceanic (Southwestern France)	Calcisols, Umbrisols and Luvisols	Sprinkler and not irrigated	ND	ND	ND	↑ 1.5–3 times of mean hydraulic conductivity	ND	ND	ND

Table 1. *Cont.*

Reference	Time of Practice	CA Practices	Climate (Country)	Soil Type (Texture)	Irrigation	Soil Temperature	Bulk Density	Penetration Resistance	Infiltration	Soil Evaporation	Grain Yield	Water Productivity
[55]	10 years	No tillage and crop rotation	Temperate semi-humid continental monsoon (China)	Dark Loessial Soil (Middle Loam)	Not irrigated	ND	↓ 7.7%	ND	ND	ND	≈	ND
[56]	10–12 years	CA in monocropping and intercropping	Tropical wet and savanna (Malawi)	Chromic Luvisol, Haplic Lixisols (Sandy Clay Loam and Sandy Loam)	Not irrigated	ND	ND	ND	↑ 44–450% of hydraulic conductivity	ND	ND	ND
[57]	12 years	CA	Humid continental (Canada)	Dystric Gleysol (Loamy Sand)	Not irrigated	ND	↑ 10%	ND	≈	ND	ND	ND
[29]	14 years	CA	Humid sub-tropical (Zambia)	Ferric Lixisols	Not irrigated	ND	ND	ND	↑ 74% of mean infiltration rates	ND	↑ 64%	ND
[58]	15 years	No tillage and crop rotation	Semi-arid (India)	(Loam, Sandy Loam and Clay Loam)	Flood	ND	↑ on the topsoil and ≈ after 10 cm	ND	↑ 38–51% of saturated hydraulic conductivity ↑ 28% of water intake rate	ND	≈	ND
[59]	16 years	CA and dif-ferentiated fertilization	Mediterranean (Italy)	Typic Xerofluvent (Loam)	Not irrigated	ND	↓ 3% in the 0–10 cm layer ↑ 5% in the lower layers	ND	ND	ND	↓ 17% of total biomass yield	ND
[33]	20 years	No tillage and crop rotation	Mediterranean (Spain)	Vertisol	Not irrigated	↓ 0.7–2.6 °C	ND	ND	ND	ND	ND	ND

Table 1. *Cont.*

Reference	Time of Practice	CA Practices	Climate (Country)	Soil Type (Texture)	Irrigation	Soil Temperature	Bulk Density	Penetration Resistance	Infiltration	Soil Evaporation	Grain Yield	Water Productivity
[44]	20 years	No tillage and barley monocropping	Semiarid Mediterranean (Spain)	Xerofluvent	Solid set	ND	↑ 2–12%	↑ 17%	↑ 45%	ND	Slight ↑	ND
[60]	21 years	No tillage, differentiated fertilization	Humid sub-tropical (USA)	Typic Paleudalf—Maury (Silt Loam)	Not irrigated	ND	≈	ND	ND	ND	≈	ND
[30]	22 years	CA	Cold semi-arid (Mexico)	Feozem (Sandy Clay Loam)	Flood	ND	↓ 20%	↓ 92%	↑ 95%	ND	↑ 52%	ND
[61]	26 years	CA	Semiarid Mediterranean (Spain)	Xerofluvent (Silt Loam)	Solid set	ND	↑ 6%	ND	↑ 33%	ND	ND	ND

2. Materials and Methods

2.1. Site Description

This study was conducted in Southern France, in Montpellier, at the INRAE experimental platform (43.647° N, 3.871° E) over three years (2021–2023). Specific plots were allocated for the transition from 20 years of conventional homogeneous tillage, primarily for maize production, to conservation agriculture. The adoption began in autumn 2020 by implementing soil conservation practices with, simultaneously, (i) no tillage, (ii) direct seeded winter cover crops and (iii) diversified crop rotation. The soil was classified as a Fluvisol [62] of colluvio-alluvial origin, with a texture comprising 43% silt, 36% sand and 21% clay and displaying few spatial heterogeneities [63]. The pH was alkaline at approximately 8.5. The organic matter content for the 0–20 cm soil layer was 1.6%. The total available water capacity was 150 mm per 100 cm soil depth, determined using a soil water retention curve obtained with pressure chambers.

2.2. Climatic Data

Local climatic data were collected daily at a height of 1–2 m using a Cimel data acquisition system, model 516i, which transmitted the data via a GPRS link. The measured variables included the wind speed, which was measured with a digital wind vane (Cimel model CE 157 N, Cimel, Paris, France) and a cup anemometer (Cimel model CE 155 N). The air temperature was recorded using waterproof PT100 probes (Priggen Special Electronic GmbH, Wettringen, Germany) with a 4-wire system. The air humidity was assessed using a Vaisala capacitive hygrometer (HMP110, Vaisala, Vantaa, Finland). Precipitation was measured with a tipping bucket rain gauge from Precis Mécanique (Bezons, France). Global radiation was quantified with a CMP6 pyranometer. All measurements were conducted on a nearby plot approximately 100 m away from the experimental plot in the absence of any obstacles or significant disturbances of any kind.

The climatic conditions for each cropping season (from sowing to harvest) were determined by following the daily average values of four significant meteorological variables: the air temperature, global radiation, reference evapotranspiration (ETo) and rainfall (R) (Figure 1). In the 2021 cropping season, the mean air temperature ranged between 9.0 and 28.0 °C; in 2022, it was between 17.0 and 30.0 °C; and in 2023, it was between 16.2 and 29.9 °C. The average cumulative daily global radiation (Rg) recorded was 2600 $W \cdot m^{-2}$ in 2021, 2192 $W \cdot m^{-2}$ in 2022 and 2229 $W \cdot m^{-2}$ in 2023. The precipitation (R) during the cropping season totaled 390 mm in 2021, 115 mm in 2022 and 158 mm in 2023. Additionally, the cumulative reference evapotranspiration (ETo) for the same periods reached 726 mm in 2021, 642 mm in 2022 and 660 mm in 2023. The drought intensity was characterized by assessing the drought indicator calculated over the cropping season as cumulative ETo/cumulative R and comparing it with the same ratio calculated over the period of April to September from 1991 to 2023. Based on this assessment, the years 2021 (2.37) and 2023 (3.84) were categorized as "average", while 2022 (5.14) was categorized as "dry".

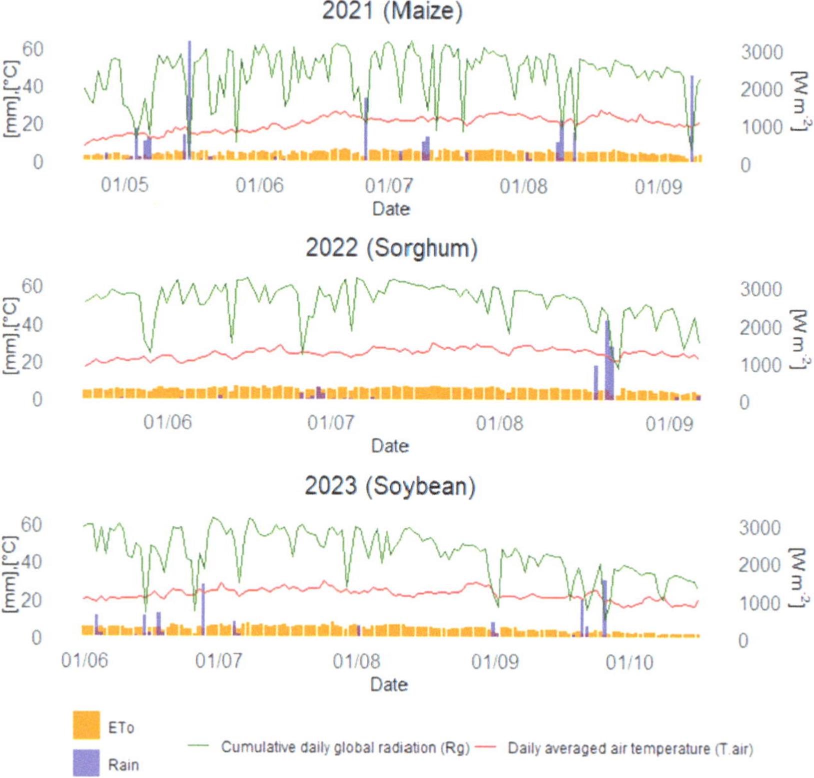

Figure 1. Daily meteorological variables observed during each cropping season from sowing to harvest. Averaged air temperature (°C), reference evapotranspiration ETo (mm) and rain (mm) on the left axis and cumulative daily global radiation (W·m^{-2}) on the right axis.

2.3. Experimental Design

Maize (*Zea mays* L., RAGT IXABEL), sorghum (*Sorghum bicolor* L., HUGGO) and soybean (*Glycine max* L., RGT SPEEDA) crops were cultivated under either conventional tillage (CT) or conservation agriculture (CA). In CA plots, (1) zero tillage, (2) winter cover cropping (direct seeded) with residue-covered soil and (3) crop rotation were simultaneously implemented. CT plots involved conventional tillage and bare soil in winter. Both CT and CA were followed under sprinkler (S) irrigation, subsurface drip (SSD) irrigation or no irrigation (NI), leading to an experimental agronomic set-up in large strips with six treatments in total, covering an area of 1.5 ha (Figure 2). SSD irrigation was installed in 2019 at a depth of 35 cm, which represented a constraint regarding the layout of the treatments from the start of the experiment in 2021. Consequently, it was not possible to randomize the treatments as shown Figure 2: SSD irrigation covered the eastern part of the plot, while S irrigation covered the western part. Previous studies, however, indicated few spatial heterogeneities within the top 30 cm of soil [63], supporting the suitability of this layout.

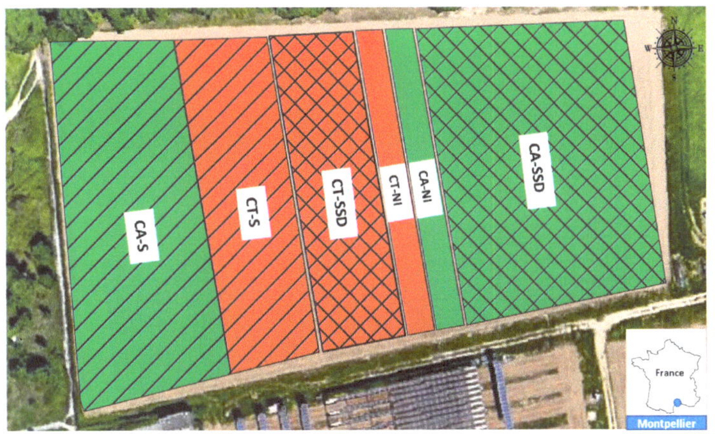

Figure 2. Treatments on the experimental site (Montpellier, France). From left to right: conservation agriculture irrigated by sprinkler irrigation (CA-S), conventional tillage irrigated by sprinkler irrigation (CT-S), conventional tillage irrigated by subsurface drip irrigation (CT-SSD), conventional tillage not irrigated (CT-NI), conservation agriculture not irrigated (CA-NI) and conservation agriculture irrigated by subsurface drip irrigation (CA-SSD).

2.4. Farming Practices

Table 2 lists the practices adopted during the three cropping seasons, for summer and winter cover crops, under CA and CT.

Table 2. Cultural practices implemented for the three cropping seasons.

Practice	2021	2022	2023
Winter cover crop sown in CA	Faba bean	Mixture of mustard, phacelia and vetch	Mixture of faba bean and oat
Sowing date of winter cover crop	12 September 2020	17 November 2021	20 September 2022
Dry biomass of winter cover crop in CA	3 t·ha^{-1}	6 t·ha^{-1}	6 t·ha^{-1}
Main crop: variety	Maize: IXABEL	Sorghum: HUGGO	Soybean: SPEEDA
Date of tillage in CT with plow Huard (30 cm)	10 October 2020 23 November 2021	22 October 2022	20 January 2024
Date of ground fertilizer application for CA and CT	18 March 2021 (K: 120U) 09 April 2021 (N: 180U)	11 April 2022 (N: 90U, P: 70U, K: 150U)	-
Date of seedbed preparation in CT (rotary harrow)	15 April 2021	9 May 2022	9 May 2023
Date of winter cover crop termination in CA (farming tool used)	15 April 2021 (FACA roller)	11 May 2022 (Roll' N' Sem ROLLS)	17 May 2023 (Roll' N' Sem ROLLS)
Date of main crop sowing (direct seeding in CA)	16 April 2021	12 May 2022	17 May 2023
Seeding density of main crop	8.3 seeds·m^{-2}	31.5 seeds·m^{-2}	33 seeds·m^{-2}
Date and type of weed control	27 May 2021: Hoeing on the CT plots 1 June 2021: Pass of the flail mower on CA plots 4 June 2021: Selective herbicide application on CA plots	10 May 2022: Herbicide application on CA plots 11 May 2022: Pass of "Roll N Sem" roller on CA plots 30 May 2022: Hoeing on the CT plots	27 June 2023: Hoeing on the CT plots
Date of main crop harvest	6 September 2021	3 September 2022	12 October 2023
Duration of main crop cycle (days)	143	113	139

2.5. Irrigation

The system employed for sprinkler (S) irrigation was the solid set for 2021 and the hose reel for 2022 and 2023. Irrigation inputs were monitored by reading rain gauges placed in the plot in a grid pattern. This allowed the determination of the actual irrigation amount reaching the soil in comparison to the scheduled setpoint dose fed to the system. The subsurface drip (SSD) irrigation system comprised buried drip lines, NAAN HYDRO PC (diameter 16 mm), spaced at intervals of 80 cm, with pressure-compensating drippers, for a nominal flow rate of 1.6 $L \cdot h^{-1}$, positioned at every 50 cm along the drip tapes. This system was positioned at a 35 cm soil depth to avoid damage-related problems due to tillage before CA adoption. The monitoring of the irrigation water supply with SSD was allowed by individual volumetric meters installed at the terminals. Initial irrigation was applied to ensure seed germination for all treatments using sprinkler irrigation in 2022 and 2023. The irrigation schedule was based on weekly doses programmed to ensure hydric comfort and provide the same amount of water for all treatments.

In 2021 (maize), ten irrigation applications were performed, with measured volumes of 210 mm for CA-S, 240 mm for CT-S and 250 mm for both the CA-SSD and CT-SSD treatments. In 2022 (sorghum), eight applications were performed, with CA-S receiving 240 mm, CA-SSD receiving 260 mm and both CT-S and CT-SSD receiving 260 mm. In 2023 (soybean), eight applications were performed, with CA-S receiving 300 mm, while CA-SSD, CT-S and CT-SSD all received 270 mm. As can be seen, some variability in the irrigation volumes was observed in sprinkler irrigation, primarily due to wind-induced drift losses during irrigation. However, soil moisture monitoring at different depths confirmed that the soil moisture levels remained consistently above the permanent wilting point throughout the growing season, ensuring adequate water availability for crop development.

2.6. Soil Monitoring

2.6.1. Soil Temperature

The soil temperature was assessed using T-type thermocouples, commonly known as Copper/Constantan, connected to a Campbell Scientific data logger (CR100) recording temperature values every 20 min. In the central area of each treatment, one thermocouple was placed at a depth of 3 cm and another one at a depth of 10 cm. This device allowed the evaluation of temperature fluctuations in the soil upper layer and the assessment of their correlations with soil evaporation.

2.6.2. Bulk Density

Soil sampling was realized using the cylindrical core method with stainless-steel cylinders (Eijkelkamp—E53 model https://www.royaleijkelkamp.com/products/augers-samplers/soil-samplers/undisturbed-core-samplers/soil-sampling-ring-kit-model-e53-heavy/ (accessed on 6 June 2024)). The cylinders were inserted into the soil to collect representative soil samples (three repetitions at each soil depth by treatment). After retrieval, they were immediately covered with plastic film to preserve their integrity.

In the laboratory, the collected soil samples were weighed to obtain the individual initial weights and subsequently placed in an oven at 105 °C for 48 h, ensuring complete drying regardless of their soil texture. After drying, the samples were weighed again to determine their total dry weight. The dry bulk density ρd $(g \cdot cm^{-3})$ was calculated as

$$\rho d = \frac{W_d - W_c}{V_c} \tag{1}$$

where W_d is the weight of the dry sample (g), W_c is the weight of the steel cylinder (g) and V_c is the volume of the cylinder (cm^3).

2.6.3. Soil Penetration Resistance

In 2021, the soil penetration resistance was evaluated using the Eijkelkamp IB model hand penetrometer (Giesbeek, The Netherlands) [64]. Subsequently, for 2022 and 2023, Field Scout's shaft-mounted digital probe SC900 (FieldScout) (Spectrum Technologies, Inc., Aurora, IL, USA) was used, aiming to enhance the measurement convenience and data accuracy. Random measurements with at least 10 repetitions per treatment were realized. Using a half-inch-diameter cone, the penetrometer provided continuous readings up to a depth of 40 cm for every 2.5 cm depth.

2.7. Soil Water Flux Monitoring

2.7.1. Soil Infiltration

Soil infiltration was characterized by measuring the quasi-steady ponded infiltration rate using the double Müntz ring method. This approach implies vertically infiltrating a uniform layer of water into the soil for a defined period until a constant infiltration rate is attained [65]. An automated device, based on the Müntz method developed by Garnier and Elamri [66], was employed to facilitate the systematic recording of infiltration data. Automation consisted of a controlled filling process for both the measurement and guard rings, ensuring a constant uniform water height during infiltration events. The system included sensors for precise water level management in the rings, enabling the filling of a specific ring from a designated reservoir. Throughout the measurement period, a Diver® probe, positioned at the base of the reservoir supplying the central cylinder, autonomously acquired the infiltrated water heights. In the central zone of each treatment, after sowing, four infiltration device replicates were used, spaced at 5 m intervals. The experiment spanned over two hours to ensure a steady-state infiltration rate. Following the approach defined by Vinatier et al. [67], the quasi-steady ponded infiltration rate was determined as the change in the infiltrated water volume per unit area of the measurement cylinder during the final 30 min of steady-state conditions.

2.7.2. Soil Evaporation

Soil evaporation was monitored using field mini-lysimeters adapted from some experimental designs [68–70], constructed on PVC material with dimensions of 20 cm in length and 10 cm in diameter, with a geotextile placed at the bottom. In this study, each mini-lysimeter was filled with undisturbed soil by sinking the mini-lysimeter directly into the soil to obtain representative soil conditions. The method implied monitoring the variation in the mini-lysimeter weight to assess the evaporation rate (ER) as follows:

$$ER = R - \frac{\Delta M}{S} \qquad (2)$$

where R denotes the amount of rainfall or irrigation (mm), ΔM represents the mass variation converted into water loss in mm^3 (e.g., grams of water associated with each cubic millimeter) and S represents the evaporating surface area of the lysimeter (mm^2). As indicated by Trambouze [69], this methodology has a limitation, because, when a water input occurs through rainfall or irrigation, the mini-lysimeter gains weight, making it difficult to differentiate the actual amount of water evaporated. Consequently, in our experiment, only the data collected during periods without rainfall or irrigation were considered.

In 2021, two mini-lysimeters with undisturbed soil were placed along the crop inter-row in the central area of each of the six treatments (Figure 3). To simulate the conditions of CA with mulch at the soil surface, cover crop residues (equivalent to 3 t·ha^{-1} in 2021 and 6 t·ha^{-1} in 2022 and 2023) were laid directly onto the surface of each mini-lysimeter. In 2022 and 2023, further improvements were made to this device. Net global radiation probes (pyranometers Skye, Campbell Scientific SP1110) were placed on the soil surface of each mini-lysimeter (Figure 4). These pyranometers were used to better understand soil evaporation by measuring the net radiation arriving at the soil surface. The pyranometers were connected to the same data logger used for the soil temperature measurements, recording data every 20 min. To prevent the loss of surface mulch and its impact on soil evaporation, plant residues equivalent to the total biomass of the winter crop (6 t·ha^{-1} annually) were placed in litterbags. These litterbags were positioned on the surface of each mini-lysimeter (Figures 3 and 4).

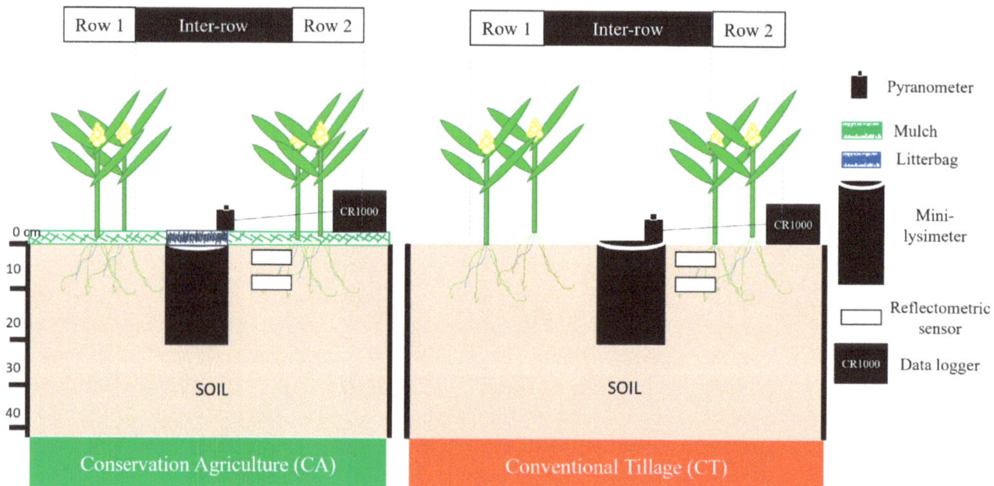

Figure 3. Spatial configuration of mini-lysimeters in conservation agriculture (CA) and conventional tillage (CT) field treatments.

Figure 4. Weighing process of mini-lysimeter without mulch (**left**); pyranometer and litterbag added to mini-lysimeter (**right**).

2.7.3. Surface Soil Water Content

The surface soil water content of each treatment was measured to evaluate its impact on soil evaporation, using CS650 reflectometric sensors (Campbell Scientific, https://www.campbellsci.fr/cs650 (accessed on 6 June 2024): Campbell Scientific France S.A.S. 41 Rue Périer, 92120 Montrouge, France), with one sensor positioned at a 3 cm depth and one placed at a 10 cm depth (Figure 3). These sensors were connected to the data logger

(CR1000) used for temperature assessment, with continuous measurements at 20 min intervals throughout the cropping season.

2.8. Crop Monitoring

2.8.1. Leaf Area Index (LAI)

The leaf area index (LAI) is an indicator expressed in square meters of leaves per square meter of soil, representing crop canopy development and thus the ability to intercept light and radiation use for photosynthesis [71]. Throughout the three years of experimentation, the vegetative growth of the main crops was assessed by weekly measurements of the LAI (ten spatial repetitions per treatment to assess potentially uneven distributions of the LAI values) using the LAI-2200C plant canopy analyzer (LI-COR, Inc., Lincoln, NE, USA) from the LI-COR system [72]. Measurements were conducted during the late hours of the day to avoid direct sunlight, which is known to impact the accuracy of such measurements. These non-destructive measurements were conducted from the initial stages of crop vegetative development until the beginning of the crop senescence phase, when a decrease in the LAI values was observed, shortly after the maximum LAI values (LAImax) were reached.

2.8.2. Grain Yield (GY)

The dry grain yield was determined at harvest by calculating the average of five random samples taken from the aboveground components of the plants, including the leaves, stems and cobs, within an area of 1.44 m^2. The plants were meticulously separated into individual vegetative components, weighed and subsequently dried in laboratory ovens at a constant temperature of 65 °C for 48 h, until the grain reached stable moisture content of 12–15%. Following the drying process, the kernels were separated from the corncobs, sorghum panicles or soybean pods to estimate the grain yield.

2.9. Total Water Productivity and Irrigation Water Productivity

The total water productivity and irrigation water productivity were calculated for each cropping season as practical, straightforward indicators of the water use efficiency in crop production. The total water productivity (TWP), expressed in kg·m^{-3}, can be defined as follows [73]:

$$TWP = \frac{GY}{I + R} \tag{3}$$

where GY is the grain yield in kg·ha^{-1}, and I and R denote the water application depths from irrigation and rainfall, respectively, measured in m^3·ha^{-1}.

The irrigation water productivity (IWP), also expressed in kg·m^{-3}, can be defined [74] as

$$IWP = \frac{GY_i - GY_r}{I} \tag{4}$$

where GY_i and GY_r are the grain yields in kg·ha^{-1} under irrigated and rainfed conditions, respectively.

2.10. Dates of Measurement

Table 3 shows the measurement dates for the period 2021–2023. All measurements were performed in the middle of each treatment plot.

Table 3. Summary of measurement dates for variables across three crop cycles.

Measurement	2021	2022	2023
Bulk density	April 22nd	June 15th	September 8th
Soil temperature	From May 5th to September 3rd	From June 10th to August 29th	From June 6th to October 19th
Soil penetration resistance	April 22nd	September 27th	December 7th
Quasi-steady ponded infiltration rate	May 6th	May 30th	June 28th
Soil evaporation	From June 21st to September 3rd	From June 14th to August 30th	From May 26th to August 18th
Leaf area index (LAI)	From June 10th to September 2nd	From July 1st to August 18th	From June 20th to September 18th
Grain yield and water productivity	September 6th	September 3rd	October 12th

2.11. Statistical Analysis

This study employed a factorial design to investigate the effects of two categorical variables ("practice" and "irrigation system") on a set of continuous responses. Practice was divided into two levels (CA and CT) and irrigation system into three levels (S, SSD and NI), resulting in a total of six experimental treatments. To assess the main effects of each variable independently, the Kruskal–Wallis test was employed to ensure robustness against violations of parametric assumptions (normality and variance homogeneity). To evaluate interactions, two linear models were constructed to assess the joint and individual effects of the practice and irrigation system and their interactions on a continuous response variable. The first model exclusively considered the main individual effects, expressed as

$$Y = \beta_0 + \beta_1 X_1 + \beta_2 X_2 + \epsilon \tag{5}$$

where Y is the so-called "dependent variable" (the effect), and X_1 and X_2 denote the "categorical independent variables" (practice and irrigation system, respectively). Here, β_0 is the "intercept", which represents the residual value of Y when all independent variables are zero—in other words, the part of the effect that is unexplained by the variables accounted for. The coefficients β_1 and β_2 are the "regression coefficients" denoting the strength of the effects associated with X_1 and X_2, respectively, and ϵ represents the error. The second model considered both the main individual effects and interactions:

$$Y = \beta_0 + \beta_1 X_1 + \beta_2 X_2 + \beta_3 (X_1 \times X_2) + \epsilon \tag{6}$$

where β_3 is the coefficient associated with the interaction term $X_1 \times X_2$. The significance of the interaction term was assessed using an ANOVA or analysis of variance. The null hypothesis stated that the inclusion of the interaction terms did not significantly improve the overall fit of the model. A p-value below the pre-specified significance level (typically 0.05) indicated a significant interaction effect. This analytical procedure was applied to a larger set of response variables, handled separately, allowing for a systematic assessment of the impact of the tested factors. To conduct these statistical analyses, the R package (r-stats) version 2.0 was employed, ensuring the accuracy and reproducibility of the results.

A principal component analysis (PCA) was conducted specifically for soil evaporation assessment, to evaluate the correlation between soil evaporation and other variables, including the soil moisture (taking the average value between a 3 and 10 cm soil depth), net global radiation at the soil surface, LAI and soil temperature (also taking the average value between a 3 and 10 cm soil depth), which are also thought to control evaporation. The analysis was limited to the days on which all variables were measured in the field, during the 2022 and 2023 cropping seasons.

3. Results

3.1. Soil Temperature

The seasonal mean soil temperatures, between a 3 and 10 cm depth (horizontal black line in each sketch in Figure 5), presented comparable values for most of the CA and CT treatments, with the differences remaining below 5%. However, significant variations appeared between the irrigation systems, notably in 2022, where treatments with S irrigation resulted in mean soil temperatures that were up to 4 °C lower than in the SSD and NI treatments.

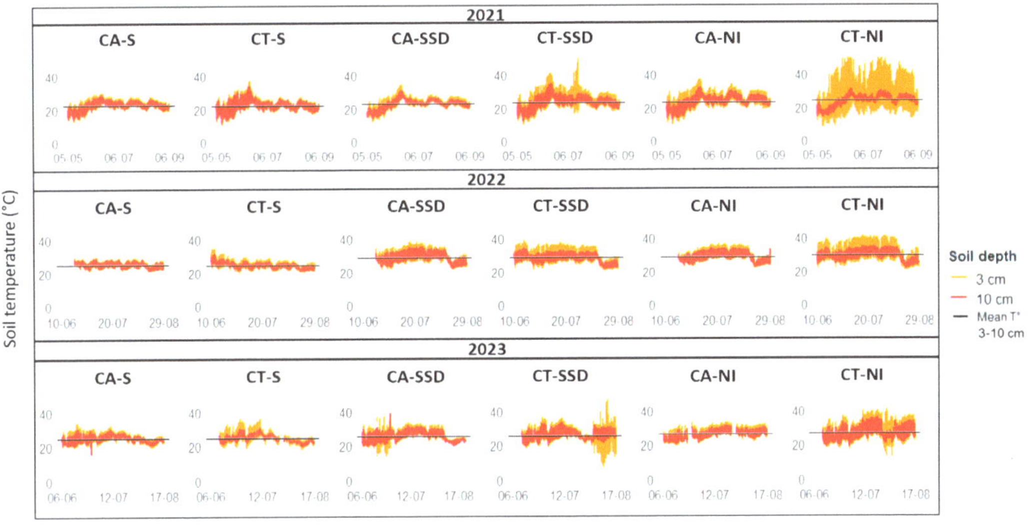

Figure 5. Temporal variation in soil temperature across different depths and mean soil temperature (3 and 10 cm) for each crop cycle under different agricultural practices (conservation agriculture—CA and conventional tillage—CT) combined with different irrigation systems (non-irrigated—NI, sprinkler—S and subsurface drip—SSD).

In CA, the soil temperature fluctuations were notably damped in comparison with CT (the envelopes of the red and orange curves are narrower for CA), especially during periods of high temperatures early in the crop cycle, when residues remained on the soil surface (Figure 5). Across all cropping cycles, CA consistently led to lower maximum soil temperatures and higher minimum soil temperatures than CT. For example, during the S cropping cycle in 2021, the seasonal mean of the maximum soil temperatures at the 3 and 10 cm depths was 6 °C lower under CA, while the seasonal mean of the minimum soil temperatures was 2 °C higher. This trend persisted across the years and cropping cycles, highlighting the insulating effect of residues under CA practices.

3.2. Bulk Density

Figure 6 presents the soil bulk density profiles across the three experimental years. In 2021, measurements were only performed on the CA-S, CA-SSD and CT-SSD treatments, for the 0–20 cm horizon. The bulk density ranged from approximately 1.35 g·cm^{-3} for CT to 1.55 g·cm^{-3}, with CA exceeding CT by 11–14%. During 2022, the bulk density evaluations were extended to the 0–60 cm depth for all treatments, with values ranging from 1.48 g·cm^{-3} to 1.75 g·cm^{-3}. Treatments irrigated with SSD showed the highest density values, particularly beyond the 15 cm depth, surpassing 1.70 g·cm^{-3}. While the mean bulk density varied between CA and CT at different soil depths, the differences remained below 6% across the 0–60 horizon, except for the S treatments at 5 cm (15%). In 2023, the measurements were limited to the 0–30 cm horizon, with the mean values ranging from 1.4 g·cm^{-3} to 1.68 g·cm^{-3}. The CA mean values surpassed those of CT, especially at 5 cm (15%) and 15 cm (up to 10%), with minimal differences at 25 cm (<5%).

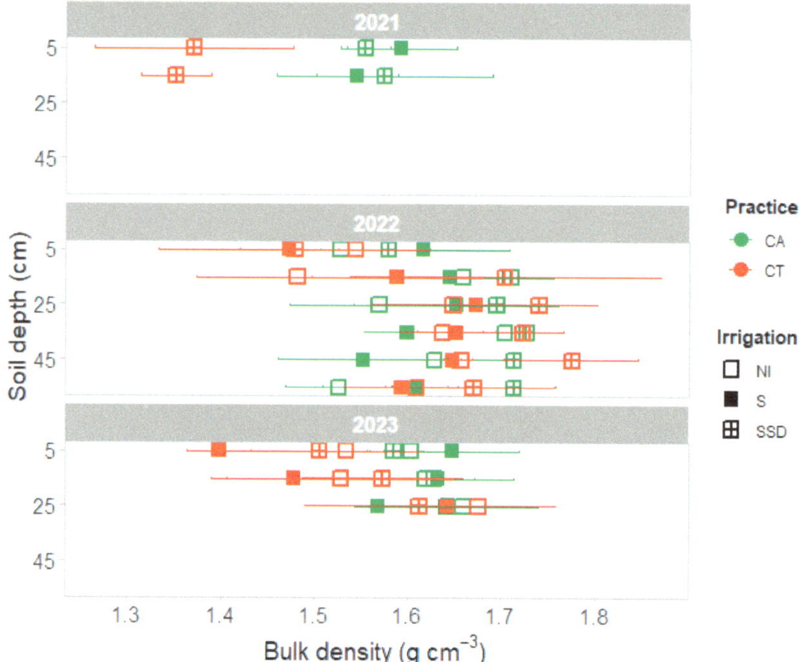

Figure 6. Soil bulk density profile for different agricultural practices (conservation agriculture—CA and conventional tillage—CT) combined with different irrigation systems (not irrigated—NI, sprinkler—S and subsurface drip—SSD). Error bars indicate standard deviations.

Between 2021 and 2022, there was a 4.5% increase in the mean bulk density (between all depths) for CA and a 20% increase for CT. However, these values declined in 2023, by 2.5% for CA and 4.8% for CT, in comparison to 2022. The bulk densities for CA appeared roughly stable across all depths for all years, while the bulk density values for CT decreased near the soil surface as a direct consequence of ploughing. Notably, CT showed lower values in 2021 compared to subsequent years, likely due to the measurements being taken shortly after seedbed preparation, in contrast to 2022 (1 month later) and 2023 (3 months later), allowing more time for natural soil consolidation and the formation of surface crusts as a result of rainfall and sprinkler irrigation.

3.3. Soil Penetration Resistance

Measurements performed in 2021 suggested that CA exhibited higher penetration resistance compared to CT. However, the manual method employed, lacking precision, was finally discarded. An improved methodology was implemented from 2022 onwards, using a digital penetrometer. In 2022 and 2023, along the soil profile, with soil volumetric water content of approximately 20%, the penetration resistance tended to be higher in CA by 33–47% compared to CT, and there was a common trend toward increasing penetration resistance with increasing depth (Figure 7). Specifically, the CA-NI and CA-SSD treatments showed penetration resistance values exceeding 2500 kPa below 10 cm, which then further increased with the depth. In 2023, these treatments resulted in increased resistance by more than 50%, particularly below 20 cm. In contrast, the CA-S treatment demonstrated lower and stable values, around 1500 kPa for both years. Finally, CT showed low penetration resistance at the surface (<1000 kPa), which increased abruptly for the SSD and NI treatments below 30 cm, surpassing 2000 kPa.

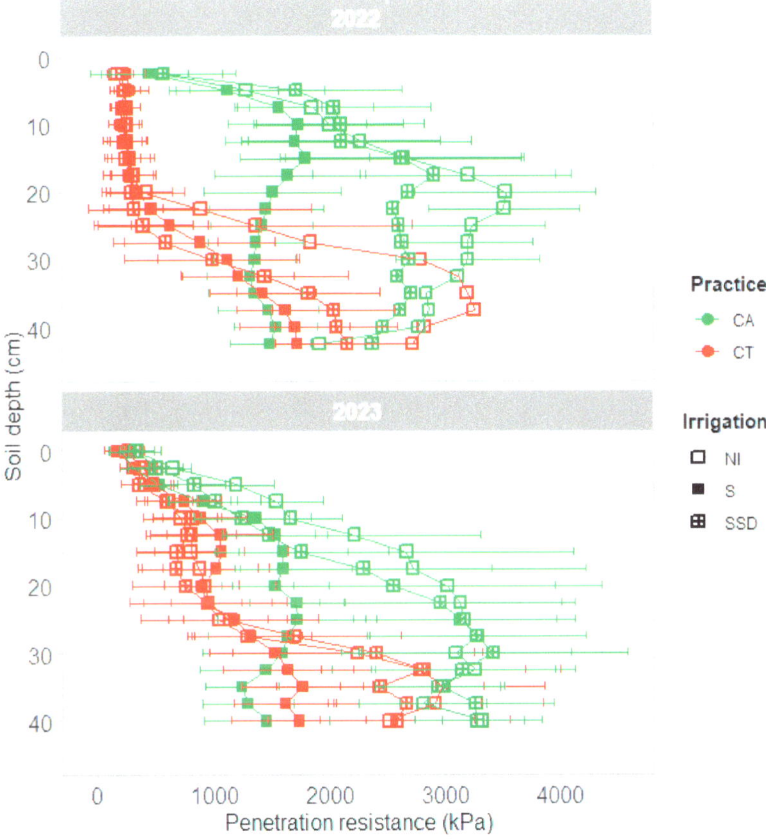

Figure 7. Soil penetration resistance profile in 2022 and 2023 under different agricultural practices (conservation agriculture—CA and conventional tillage—CT) combined with different irrigation systems (not irrigated—NI, sprinkler—S and subsurface drip—SSD). Error bars indicate standard deviations. Measurements were realized after the sowing date of the main crop (CT and CA) and before the tillage date (CT).

3.4. Quasi-Steady Ponded Infiltration

The quasi-steady ponded infiltration values exhibited clear differences between CA and CT (Figure 8). In 2021, the standard deviations for CA ranged from 23 mm·h^{-1} to 96 mm·h^{-1}, while, for CT, they ranged from 86 mm·h^{-1} to 274 mm·h^{-1}. These deviations decreased in 2022, with CA showing values ranging from 3 mm·h^{-1} to 13 mm·h^{-1} and CT from 40 mm·h^{-1} to 58 mm·h^{-1}. By 2023, the standard deviations were further reduced to 7 mm·h^{-1} to 14 mm·h^{-1} for CA and 15 mm·h^{-1} to 90 mm·h^{-1} for CT. These variations highlight the small-scale heterogeneity in the soil properties despite the relatively short distance between the measurement points (5 m).

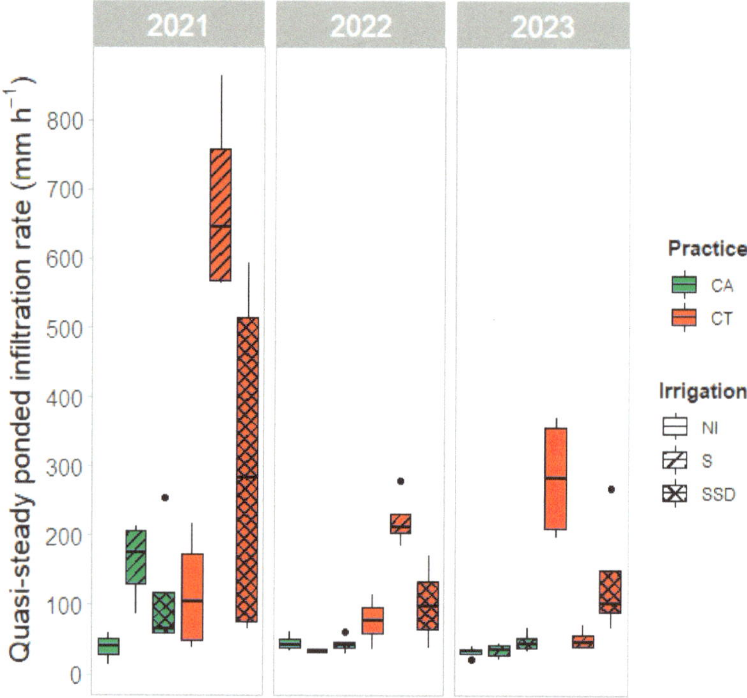

Figure 8. Quasi-steady ponded infiltration rates under different agricultural practices (conservation agriculture—CA and conventional tillage—CT) combined with different irrigation systems (not irrigated—NI, sprinkler—S and subsurface drip—SSD). Error bars indicate standard deviations.

Throughout all three years, CT consistently displayed higher infiltration rates compared to CA, regardless of the irrigation system. In 2021, CT-S gave the highest value at 680 mm·h^{-1}, followed by CT-SSD at 306 mm·h^{-1}. CA-S followed with 162 mm·h^{-1}, while CT-NI and CA-SSD were close at 117 mm·h^{-1} and 111 mm·h^{-1}, respectively. The lowest value was observed for CA-NI at 38 mm·h^{-1}. In 2022, the infiltration rates decreased across all treatments, compared to 2021. CT-S led with 223 mm·h^{-1}, followed by CT-SSD at 101 mm·h^{-1} and CT-NI at 77 mm·h^{-1}; then, CA-SSD, CA-NI and CA-S followed at 46 mm·h^{-1}, 44 mm·h^{-1} and 34 mm·h^{-1}, respectively. In 2023, CT-NI gave the highest rate at 284 mm·h^{-1}, followed by CT-SSD at 136 mm·h^{-1} and CT-S at 51 mm·h^{-1}. CA showed lower rates between 33 and 48 mm·h^{-1}.

Moreover, within CA, the infiltration rates decreased with the time that had elapsed since the last tillage (spring 2020). Additionally, CA exhibited less temporal variability in the infiltration rates compared to CT.

3.5. Soil Evaporation

The cumulative soil evaporation curves represent measurements taken outside of rainy or irrigated periods (Figure 9). The soil evaporation measurements began 69 days after sowing in 2021, 34 days in 2022 and 28 days in 2023. In 2021, CA exhibited lower cumulative soil evaporation values compared to CT, with differences of −8%, −2% and −8% for S, SSD and NI, respectively. However, in 2022, CA showed higher values, with differences of +28%, +6% and +24% for S, SSD and NI, respectively. In 2023, CA demonstrated lower values than CT for S irrigation, with a difference of −22%, while presenting higher values with differences of +4% and +6% for SSD and NI, respectively. Hence, all trends were present, meaning that multiple factors likely intervene and may be challenging to identify and even more so to isolate. Nevertheless, clear variations were noted between the irrigation systems: the S treatments displayed higher cumulative soil evaporation values compared to SSD and NI, with differences of 36% and 17% for 2021, 42% and 39% for 2022 and 27% and 16% for 2023 for SSD and NI, respectively.

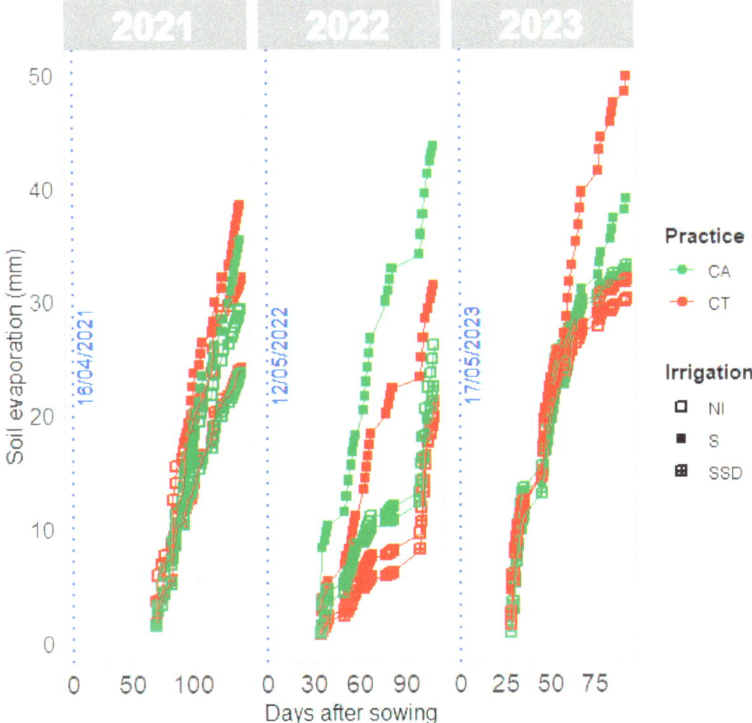

Figure 9. Cumulative soil evaporation for each cropping season (2021, 2022 and 2023) under different agricultural practices (conservation agriculture—CA and conventional tillage—CT) combined with different irrigation systems (not irrigated—NI, sprinkler—S and subsurface drip—SSD). Dotted blue lines indicate the sowing dates.

The impact of cover crop residues left on the soil surface in CA was not clearly discernible across all years of observation. This was due to the quick degradation (a few weeks at the very most) of most cover crop residues with a predominance of legumes during the soil evaporation study periods in 2021 and 2022, resulting in minimal global effects on the cumulative evaporation. Additionally, noticeable differences in vegetative growth, between treatments and during the same periods, were expected to have a strong influence on the net soil evaporation dynamics.

To clarify these points, a more detailed examination was conducted in 2023, focusing on three specific periods along the crop cycle, following rainfall or irrigation events for the SSD treatment under CA and CT (Figure 10). During the first period (May 31 to June 5), the LAI of the main crop was 0 $m^2 \cdot m^{-2}$ and the cover crop residues on the soil surface in CA had a mass of 421 g of dry matter per m^2. This resulted in lower evaporation rates for CA compared to CT. By the second period (June 14 to 21), the LAI was 0.2 $m^2 \cdot m^{-2}$ and the residue mass in CA had decreased to 248 g of dry matter per m^2, leading to slightly higher evaporation rates in CA compared to CT. As for the third period (July 4 to 10), the

LAI values increased to 1.1 m²·m⁻² for CT and 0.8 m²·m⁻² for CA, with the residues in CA further degraded to 200 g of dry matter per m². At this point, the evaporation rates between the two treatments were nearly identical, with less than a 0.5 mm difference, suggesting that both the residue mass and LAI influenced soil evaporation. However, by this stage of crop growth, additional factors may also have contributed to the observed evaporation rates.

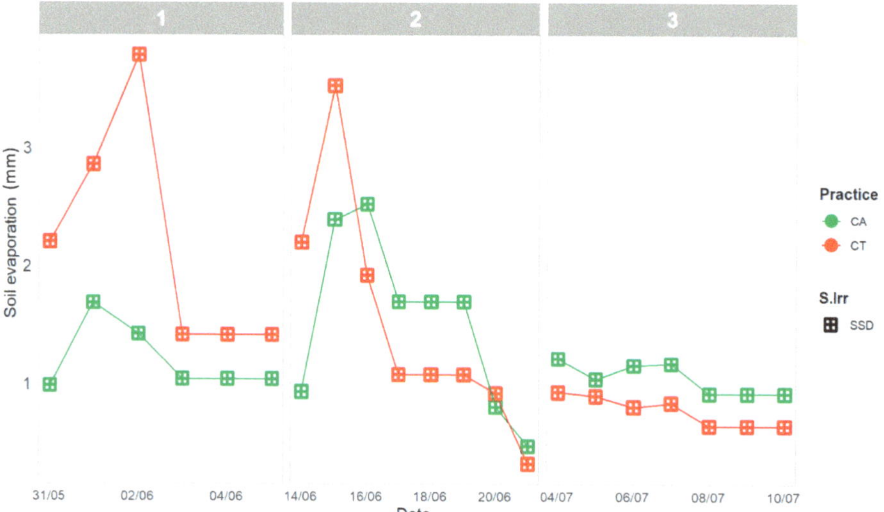

Figure 10. Daily soil evaporation observations periods in 2023 under different agricultural practices (conservation agriculture—CA and conventional tillage—CT) irrigated by subsurface drip (SSD): (1) May 31 to June 5, (2) June 14 to 21 and (3) July 4 to 10.

To further investigate the multiple determinants of the soil evaporation dynamics, a principal component analysis (PCA) was conducted using data from 2022 and 2023 (Figure 11). This dataset included variables such as the LAI, net global radiation at the soil surface, soil surface temperature (averaged from 3 to 10 cm) and soil surface moisture (averaged from 3 to 10 cm).

The analysis reveals that the first two principal components together capture approximately 80% of the total variance in the data (Figure 11a), providing a comprehensive understanding of the factors driving soil evaporation. Dimension 1 (Dim 1, 53.9% of the variance) highlights soil moisture as the dominant factor controlling soil evaporation, as demonstrated by their strong positive correlation. This indicates that water availability is the primary constraint on evaporation within the dataset. Conversely, the soil temperature exhibits a negative correlation with evaporation, suggesting that limited moisture availability may inhibit evaporation, even when the soil temperature is elevated. This inverse relationship implies that the temperature alone is not a sufficient driver of evaporation without enough soil moisture. Additionally, the LAI shows a negative correlation with the net global radiation, indicating that the increased interception of solar radiation by the canopy reduces the radiation reaching the soil surface. This shading effect plays a critical role in moderating both the soil temperature and moisture retention.

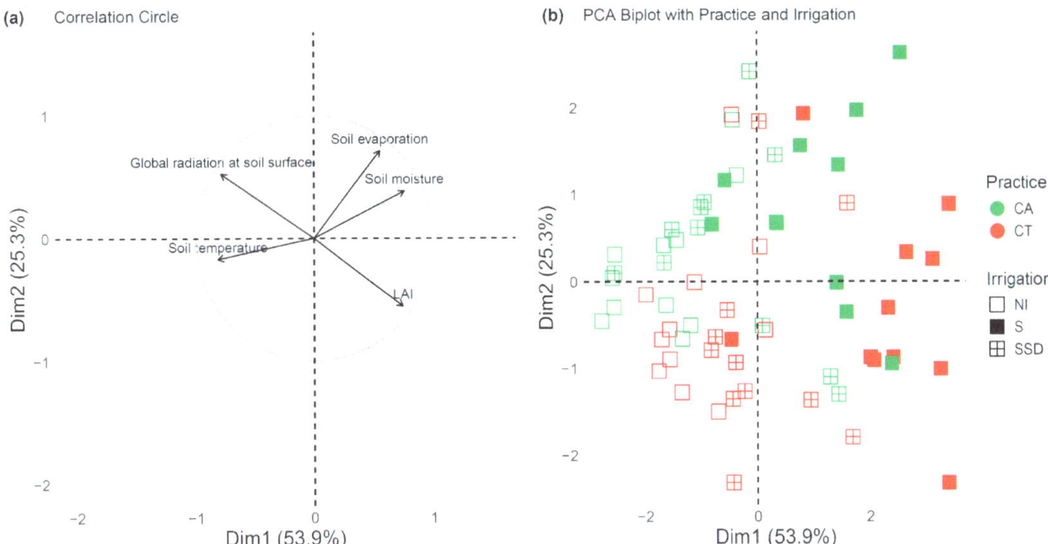

Figure 11. Principal component analysis (PCA) of the 2022 and 2023 dataset for soil evaporation assessment. (**a**) Correlation circle displaying vectors for explanatory variable scores; (**b**) graph of individuals illustrating the distribution of treatments. Each point represents an agricultural practice (conservation agriculture—CA and conventional tillage—CT) and an irrigation system (not irrigated—NI, sprinkler—S and subsurface drip—SSD).

Dimension 1 clearly separates the treatments based on their moisture availability and evaporation patterns (Figure 11b). The S treatments are clustered on one side of Dim1, reflecting higher water availability and evaporation rates, while the NI and SSD treatments are grouped on the opposite side, indicating lower moisture levels and reduced evaporation. In contrast, Dimension 2 (Dim 2, 25.3% of the variance) reveals only slight differences between CA and CT. In CA systems, soil evaporation appears to be more influenced by the interaction between soil moisture and global radiation, whereas, in CT systems, factors such as the LAI and soil temperature seem to exert a stronger influence on the evaporation dynamics.

3.6. Leaf Area Index (LAI) and Grain Yield (GY)

The maximum leaf area index (LAImax) and total grain yield (GY) values are indicated in Figure 12. They followed a similar trend throughout the three years, with CA consistently showing lower values than CT, regardless of the irrigation method, indicating reduced crop development in CA. In 2021, the LAImax values for CA were 31%, 14% and 24% lower than CT for S, SSD and NI, respectively. The differences became more pronounced in 2022, with reductions of 44%, 68% and 56% for S, SSD and NI, respectively, and with 49%, 54% and 29% reductions in 2023 for S, SSD and NI, respectively. Among the irrigation methods, S consistently produced the highest LAImax, followed by SSD and NI.

Similarly, in 2021, the GY was lower in CA by 28%, 15% and 9% for S, SSD and NI, respectively, compared to CT. These differences widened significantly in subsequent years. In 2022, the GY in CA was 65%, 98% and 88% lower than in CT for S, SSD and NI, respectively. In 2023, the reductions were 65%, 76% and 95%. Across all years, the S irrigation method always resulted in the highest GY, followed by SSD irrigation and NI.

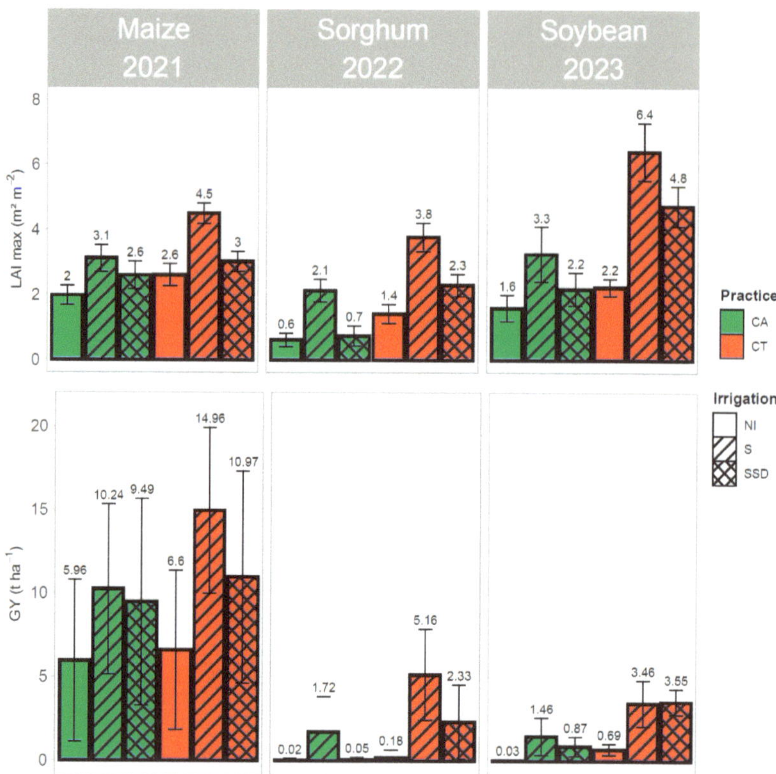

Figure 12. Maximum leaf area index (LAImax) and grain yield (GY) values across cropping seasons (2021, 2022, 2023) under different agricultural practices (conservation agriculture—CA and conventional tillage—CT) combined with different irrigation systems (not irrigated—NI, sprinkler—S and subsurface drip—SSD). Error bars indicate standard deviations.

3.7. Water Productivity

The trends in the total water productivity (TWP) and irrigation water productivity (IWP) matched the observed grain yield patterns over the three years (Table 4). Specifically, higher values of TWP and IWP were constantly observed in CT compared to CA, and the values for S-irrigated treatments were higher compared to the SSD and NI treatments. This was expected as the water productivity is closely linked to the yield. In 2021, the TWP in CA was reduced by 40%, 16% and 11% compared to CT for the S, SSD and NI treatments, respectively. These differences became more pronounced in 2022, where the reductions in TWP reached 65%, 97% and 89% for S, SSD and NI, respectively. In 2023, the gaps changed, with reductions of 61%, 75% and 96% for S, SSD and NI, respectively.

A similar trend was observed for IWP. In 2021, the IWP in CA was lower than that in CT by 42% and 20% for the S and SSD irrigation treatments, respectively. These gaps increased significantly in 2022, reaching reductions of 63% and 99% for S and SSD. By 2023, the IWP reductions were 60% and 72% for S and SSD, respectively. Across most scenarios, IWP consistently outperformed TWP, except in the CA-SSD treatment in 2021.

Table 4. Summary of real irrigation volumes (I), total water input (rainfall R + irrigation I), total water productivity (TWP) and irrigation water productivity (IWP) across cropping seasons under different agricultural practices (conservation agriculture—CA and conventional tillage—CT) combined with different irrigation systems (not irrigated—NI, sprinkler—S and subsurface drip—SSD).

	2021 (Maize)				2022 (Sorghum)				2023 (Soybean)			
Treatment	I ($m^3 \cdot ha^{-1}$)	R + I ($m^3 \cdot ha^{-1}$)	TWP ($kg \cdot m^{-3}$)	IWP ($kg \cdot m^{-3}$)	I ($m^3 \cdot ha^{-1}$)	R + I ($m^3 \cdot ha^{-1}$)	TWP ($kg \cdot m^{-3}$)	IWP ($kg \cdot m^{-3}$)	I ($m^3 \cdot ha^{-1}$)	R + I ($m^3 \cdot ha^{-1}$)	TWP ($kg \cdot m^{-3}$)	IWP ($kg \cdot m^{-3}$)
CA-S	2100	6000	1.70	2.02	2400	3550	0.48	0.71	3000	4570	0.32	0.48
CA-SSD	2500	6400	1.48	1.41	2600	3750	0.01	0.01	2700	3870	0.20	0.31
CA-NI	-	3900	1.53	-	-	1150	0.02	-	-	1570	0.02	-
CT-S	2400	6300	2.37	3.48	2600	3750	1.38	1.92	2700	4270	0.81	1.02
CT-SSD	2500	6400	1.71	1.75	2600	3750	0.62	0.83	2700	4270	0.83	1.06
CT-NI	-	3900	1.69	-	-	1150	0.16	-	-	1570	0.44	-

3.8. Effect of Practice and Irrigation System on Dependent Variables

Table 5 presents the statistical results for each cropping season, examining the effect of the "practice" (CT or CA), "irrigation system" (S, SSD, or NI) or their interaction on various dependent variables. This section includes the Kruskal–Wallis p-values. Subsequently, p-values derived from linear models for the interactions are also provided, indicating the statistical significance of the factors.

Table 5. Statistical comparison of effects of practice (CA and CT), irrigation system (S, SSD and NI) and their interaction on variables measured for all cropping seasons. Kruskal–Wallis p-values for main effects "practice" and "irrigation system" and linear model p-values for interaction are shown. '***' indicates highly significant ($p < 0.001$). '**' indicates moderately significant ($0.001 \leq p < 0.01$). '*' indicates marginally significant ($0.01 \leq p < 0.05$). (ns) indicates not statistically significant ($p \geq 0.05$). '-' indicates that p-values were not assessed.

	2021			2022			2023		
Variable	Practice (CA, CT)	Irrigation System (S, SSD, NI)	Practice and Irrigation System	Practice (CA, CT)	Irrigation System (S, SSD, NI)	Practice and Irrigation System	Practice (CA, CT)	Irrigation System (S, SSD, NI)	Practice and Irrigation System
LAI	*	***	ns	***	**	ns	ns	ns	ns
Grain Yield	***	***	*	***	***	***	***	***	*
Bulk density	ns	ns	ns	ns	**	ns	ns	ns	ns
Soil temperature (mean between 3 and 10 cm depth)	*	***	***	*	***	***	***	***	**
Soil penetration resistance	-	-	-	***	*	ns	***	ns	ns
Quasi-steady ponded infiltration	*	*	*	***	ns	***	***	ns	**
Soil evaporation	ns	***	ns	*	***	ns	ns	***	ns

In 2021, the practice significantly influenced most variables, except soil evaporation and the bulk density. Similarly, the irrigation systems had notable effects on the variables, except the bulk density. The interaction between these factors notably affected the soil temperature and marginally impacted the quasi-steady ponded infiltration and grain yield. In 2022, the practice continued to exert a significant influence, impacting most variables except the bulk density. The irrigation system also showed substantial effects across the variables, except the quasi-steady ponded infiltration. The interaction effects were significant, notably on the grain yield, soil temperature and quasi-steady ponded infiltration. By 2023, the practice significantly influenced the grain yield, soil temperature, soil penetration resistance and quasi-steady ponded infiltration. The irrigation system also had substantial impacts, particularly on the grain yield, soil temperature and soil evaporation. The interaction effects were notable, especially on the soil temperature and quasi-steady ponded infiltration, with marginal significance for the grain yield.

The grain yield, soil temperature and quasi-steady ponded infiltration rate remained constantly sensitive variables affected by interactions between the practice and the irrigation system across all years. The soil penetration resistance was significantly influenced by the agricultural practices, while the irrigation system notably affected soil evaporation.

4. Discussion

4.1. Effects of CA and Irrigation System on Soil Physical Properties

In the present study, significant effects of both the agricultural practice (CA or CT) and the irrigation system (S, SSD or NI) were observed from the first year of the experiment.

4.1.1. Temperature

In CA, smaller temperature fluctuations were noted (Figure 5), attributed to the buffering effect of cover crop residues. Even in low quantities, these residues reduce the direct radiation on the soil surface, leading to more stable temperatures [75]. Similar findings have been reported in several studies (Table 1), including those by Shen et al. [32], Alletto et al. [50] and Meena et al. [51]. The irrigation method also played a role in temperature regulation. Sprinkler (S) irrigation resulted in soil surface temperatures that were up to 4 °C lower than in SSD and NI. On the one hand, S treatments are characterized by higher surface humidity, contributing to lower temperatures by cooling the soil surface, especially during hot days under a strong evaporative demand. This cooling effect occurs as liquid vapor absorbs heat from the surrounding air and evaporates, lowering the temperatures via phase changes [76]; however, it is generally limited to the top few centimeters of the soil, where the evaporative front takes place. A few centimeters below this, another effect takes over: the infiltration of water at temperatures lower than the soil's has a direct cooling effect. Moreover, the higher the water content, the more efficiently the cooling effect propagates downwards; this is due to convection but also because the thermal conductivity of water is at least twice that of soil particles [77]. On the other hand, the SSD and NI treatments, characterized by dry surface conditions, displayed stronger temperatures due to reduced moisture levels, which limit the soil's capacity to absorb and retain heat.

These temperature fluctuations are influenced not only by crop residues and the soil characteristics but also by the developmental stage of the crops [78]. While the sun's warming effect on the soil surface is prone to increasing from spring to summer, the growing canopy intercepts progressively more radiation.

The higher LAI values observed for S irrigation indicate better radiation interception, leading to less radiation reaching the soil surface, in comparison with SSD and NI, exhibiting lower LAI values. In Mediterranean climates, reduced soil temperatures may help to mitigate the effects of wider daily temperature fluctuations and increased drought frequencies [33,79]. Moreover, cooler soil temperatures in summer can create a more stable environment for crops, enhancing water retention, nutrient availability and microbial activity, as lower temperatures reduce evaporation and help to maintain the optimal conditions for soil organisms, which collectively support crop growth.

4.1.2. Bulk Density

The bulk density, an indicator of soil health and compaction and affected by the soil structure, texture, climatic conditions and agricultural practices [80], showed no significant differences across the years (Figure 6). In general, in CT, the bulk density values were lower than those in CA (5–15%) due to the effect of tillage in loosening soil packing and particle arrangement [81]. The values found in this study are similar to those also observed by [47] for a silty soil in CT. Particularly high values (>1.7 g·cm^{-3}) were observed in CA. According to the USDA guidelines (2019) for loam soils, bulk densities exceeding 1.7 g·cm^{-3} may hinder root development. Nevertheless, the "ideal" bulk density for optimal plant growth (i.e., adequate pore space, optimal water retention, good support for root development, sufficient nutrient availability, etc.) varies depending on the soil type and crop variety [82]. Research suggests that adopting CA can lead to an increase in bulk density in many cases [83,84], which often implies a reduction in micro-, meso- and macroporosity, thereby reducing the hydraulic conductivity [85].

4.1.3. Penetration Resistance

By contrast, the soil penetration resistance, which is a proxy for soil compaction [86] and represents the difficulty with which the roots will develop throughout the crop cycle [82], was significantly influenced by the farming practice (Table 5), with higher values observed in CA compared to CT, from the first year (Figure 7). The difference between CA and CT was observed in the upper soil layers and could be mainly attributed to the tillage practice in CT, which mechanically allows the loosening and restoration of the initial soil conditions for root development. Conversely, in CA, the absence of tillage, combined with the repeated use of agricultural machinery for sowing, weed control and harvesting [87], likely contributed to higher surface compaction.

In CT, however, the penetration resistance values significantly increased at depths of 25 to 30 cm, exceeding the critical threshold of 2000 kPa recommended for optimal root growth in the literature [87,88]. This increase suggests the presence of a plough pan, a compacted layer formed during tillage that can raise the bulk density in the deeper soil layers, negatively impacting root penetration and overall crop development [89]. In CA, the penetration resistance values also generally exceeded 2000 kPa, indicating surface compaction issues. Similar trends have already been observed in different short-term studies, as shown in Table 1 [17,21–23,36]. Additionally, Peigné et al. [90] emphasize that, even with conservation practices, compaction is reinforced by the soil's susceptibility to compaction, particularly in soils with high silt and clay content, as observed in our study.

Notably, the highest resistance values in CA were observed in the NI treatments, followed by the SSD-irrigated treatments, with the lowest values recorded in the S-irrigated treatments. Since the soil moisture levels during penetration resistance measurement were equivalent across the treatments (approximately 20%), these resistance differences directly reflect the impacts of the irrigation systems on the soil structure. For the SSD and NI treatments, the drying cycles and localized water distribution patterns likely exacerbated surface compaction. In the SSD system, where the driplines were buried at a depth of 35 cm, the frequent dry conditions in the surface layers may have increased the friction between the soil particles, strengthened the soil structure and shrunk the pores [91], even though sufficient moisture was maintained at the depth of the driplines. Particularly in SSD, Lamm [92] has observed potential challenges, where driplines may encounter deformation due to vertical soil compaction or lateral overburden. This deformation can result in reduced flow rates and compromised distribution uniformity, affecting, in turn, the irrigation efficiency.

4.2. Effect of CA and Irrigation System on Water Flux

4.2.1. Infiltration

The quasi-steady ponded infiltration values were significantly lower in CA (three to four times) compared to CT (Figure 8), due to soil compaction in CA, as indicated by the elevated penetration resistance, especially in the SSD and NI treatments. This trend, related to the higher bulk density and penetration resistance in the topsoil under CA, is also noted in previous studies [26,93] (see Table 1). The difference is also attributed to greater small-scale soil heterogeneity in the conventionally tilled plots, in coherence with observations made by Vinatier et al. [67], who attribute this dynamic to preferential flows through cracks or macropores that are unevenly distributed. This soil heterogeneity was consistent across all CT repetitions, likely influenced by factors such as the soil structure, the presence of compacted layers or aggregates and prior tillage conditions [94]. However, some authors [41,61,95] indicate that the infiltration rates in tilled systems may decrease rapidly due to post-tillage soil reconsolidation. Finally, the infiltration rates in CA decreased after the first year, because of the progressive soil compaction with the passage of heavy machinery for agricultural operations, known to affect the soil structure and decrease the soil porosity [96]. Moreover, Esser and Roua et al. [97,98] suggest that CA practices may have different effects on the soil infiltration rates according to specific practices (minimum tillage, no tillage, etc.) and the initial or local soil conditions (soil texture, organic matter content, compaction level, etc.). Finally, numerous authors report that CA practices may lead, in the long term, to an improved soil structure, increased soil organic matter content and enhanced soil biota, which overall favor infiltration [29,41,56,58,99–102].

4.2.2. Evaporation

According to the literature, surface residues in CA practices reduce soil evaporation by creating a physical barrier, often referred to as mulch, which shields the soil surface underneath from direct exposure to sunlight and air, minimizing water loss [103]. The observations in 2023 (Figure 10) showed a more pronounced effect during the initial phases of the crop cycle, before the decomposition of the residues. Later on, when the surface gradually dries out over time and the residues naturally degrade, the decrease in evaporation in CA becomes less pronounced. In fact, water located deeper in the soil does not reach the surface rapidly enough to sustain the rate of evaporation from moist soil. As the surface soil surface dries out, it begins to impede the transport of water [102]. Additionally, the thickness of the mulch, which limits the lateral movement of humid air at the surface [104], decreases as the residues decompose, reducing the mulch's efficacy in dampening soil evaporation.

Treatments irrigated by S had significantly higher soil evaporation (Table 5) compared to SSD and NI, because, when the soil surface is wet, the evaporation rates are primarily driven by atmospheric energy. As the surface dries out, the evaporation rates are constrained by water movement from the soil to the surface. In the case of the SSD (buried at 35 cm) and NI treatments, the increase in soil surface moisture was only due to precipitation, which results in lower evaporation values when compared to S irrigation. This trend was corroborated by the principal component analysis (PCA) shown in Figure 11, highlighting the surface soil moisture as a principal variable affecting soil evaporation. However, the methodology used to evaluate soil evaporation may present a limitation. Rainwater may drain out quickly from lysimeters, preventing water redistribution through capillarity, which would normally occur. This likely results in lower evaporation from lysimeters [105] than from the soil volume that they aim to monitor.

4.3. Effect of CA and Irrigation System on LAI, GY, TWP and IWP

It was constantly observed that the LAI values remained lower in CA compared to CT, regardless of the irrigation system employed (Figure 12). Moreover, the S irrigation system consistently gave the highest values of all indicators. In 2022, the LAI and GY values for the CA-SSD and CA-NI treatments were far lower than those obtained for all other combinations. Several interdependent factors may have contributed to these poor results—most likely soil compaction, low infiltration rates and high bulk densities. These factors are closely linked to vegetative growth and, when unfavorable, they can hinder root development and water uptake, limiting plant growth and productivity. Indeed, Chen et al. [106] and Cárceles Rodríguez et al. [107] mentioned that the main challenge in CA is root growth limitation due to compaction and poor water infiltration. Some authors observed also lower yields in CA [21,36], attributed to poor seedling emergence due to soil compaction and low soil temperatures in the no-till system (see other examples in Table 1). In a meta-analysis, Van den Putte et al. [108] confirmed that no tillage results in lower yields under drier climatic conditions compared to CT and other conservation tillage techniques that do not involve soil inversion.

Regarding water productivity, similar patterns could be observed as in the crop yield. The treatments in CT exhibited higher values of total water productivity (TWP) and irrigation water productivity (IWP), indicating more efficient water use (rainfall and irrigation) by the crops under these conditions. On the contrary, CA may face challenges in maintaining water productivity, potentially due to factors such as soil compaction or changes in soil moisture dynamics that can affect root development and water uptake. Generally, S-irrigated systems allowed better TWP and IWP than SSD-irrigated systems for both CA and CT. In particular, the IWP values surpassed the TWP values, suggesting that using irrigation is more effective in enhancing crop yields than relying only on rainwater, although both remained relatively low for the CA-SSD treatments. For comparison, Rana et al. [109] reported that an SSD system, buried at a 0.15 m depth with spacing of 0.67 m under CA for 8 years, allowed two times less water use (increase of 1.8% in total water productivity) compared to flood irrigation under semi-arid to sub-humid conditions. Sánchez-Llerena et al. [35] reported an important increase in total water productivity, with water savings of up to 75% when using sprinkler irrigation with no tillage compared to flood irrigation under Mediterranean conditions.

It is important to note that the TWP and IWP have limitations, as they do not account for the timing of rainfall or irrigation, which affects the water availability. More detailed approaches, such as daily water balance models or models with a frequency sufficient to capture temporal variations, may be needed, especially in CA systems, where the soil structure and water retention can vary.

4.4. Challenges of CA Adoption

Our study identified significant challenges in the initial years following CA adoption, particularly high soil compaction, in irrigated systems under a Mediterranean climate. It is hypothesized that, before transitioning to CA (before distinguishing CA and CT plots prior to winter crop sowing in 2020 under CA), a "plough pan" was already established at our experimental site, down to a depth of about 30 cm, as evidenced in the literature and confirmed by the results regarding soil penetration resistance. Additionally, compaction in the shallower soil horizon may have been intensified by machinery passage and the impact of raindrops during the preceding crop period. After transitioning to CA, soil compaction, as indicated by the penetration resistance values, became more evident in CA compared to CT, primarily in the first 30 cm, owing to the absence of tillage. In CT, the tillage annually reconditions the soil, alleviating compaction. Therefore, reconditioning measures to address soil compaction in CA systems may be necessary, as persistent compaction can affect other critical soil physical properties, ultimately hindering crop performance and irrigation water productivity, even if the irrigation system is recognized as efficient. Martínez et al. [24] propose that adopting no tillage with subsoiling before sowing may be especially relevant in Mediterranean areas affected by soil compaction. Similarly, Salem et al. [17] recommend exploring alternative forms of conservation tillage, like reservoir and minimum tillage, in specific scenarios to mitigate compaction without sacrificing the benefits of CA. Special attention needs to be given to SSD-irrigated plots, ensuring that these measures are applied to an optimal depth, preferably not exceeding 30 cm, to prevent any damage to the irrigation system buried at a depth of 35 cm.

Despite the agronomic challenges associated with the adoption of CA, it is feasible to address them through the more tailored adaptation of CA principles to the local conditions, such as adapted machinery or agricultural practices adapted to the climatic conditions [10]. Moreover, as observed in Table 1, the temporal dimension of CA has diverse and varying effects on the observed properties [13], with expectations for their evolution to benefit crop productivity over the long term [110,111], underscoring the interest in monitoring these properties over the long term.

4.5. Subsurface Drip Irrigation in CA

In the Mediterranean climate, irrigation is essential to maintain productivity and enhance the regulating services improved by CA, especially during dry seasons [14]. This study aimed to explore how different irrigation systems influence the outcomes (and diagnostic) of CA, with a particular focus on subsurface drip irrigation (SSD), recognized for its efficiency in water delivery.

The combination of SSD with CA offers a compelling strategy, as it enables direct, localized water delivery to the root zone, minimizing soil evaporation [112,113]. However, despite the potential benefits of SSD irrigation in improving the water use efficiency and mitigating environmental impacts [114,115], this irrigation system may be unable to deliver water in the shallow soil layers. This issue is particularly critical during the early stages of crop growth, when insufficient surface moisture can lead to water stress, lessened seed emergence, reduced vegetative growth and lower yields, particularly in the absence of sufficient rainfall [116]. Consequently, SSD-irrigated systems may face disadvantages compared to sprinkler irrigation (S), which consistently maintains the soil moisture throughout the profile from the beginning of the crop cycle. To address SSD's limitations, supplementary irrigation with an alternative system, such as sprinklers, may be necessary during the initial growth stages to ensure proper crop establishment and root access to moisture provided by SSD at deeper levels [92]. This strategy could be especially crucial in Mediterranean climates, with low initial soil moisture and extended summer droughts. However, the success of this approach depends heavily on favorable soil conditions that support root growth and development [117]. Further research is needed to optimize SSD systems for Mediterranean crops under CA practices, including determining the ideal drip line depth, spacing and emitter discharge rates to ensure that the water delivery aligns with the crop needs and environmental conditions.

5. Conclusions

This study presents the impacts of conservation agriculture (CA), over a three-year period following adoption, and compares it to conventional tillage (CT), with different irrigation systems (sprinkler (S), subsurface drip (SSD)) and under non-irrigated conditions (NI). The physical soil properties, water fluxes, vegetative growth, crop yields and water productivity were studied in the Mediterranean French context, likely to be extrapolated to other Mediterranean systems.

The results show that the short-term effects of CA differ significantly from those seen in long-term studies reported in the literature, with initial challenges observed in the soil properties, water fluxes, crop development, yields and water productivity. CA treatments exhibited higher soil penetration resistance and lower infiltration rates, particularly the CA-SSD and CA-NI treatments, leading to reduced crop performance. Sprinkler irrigation produced the best crop results across all systems. In CA, mulch initially reduced soil evaporation, but this effect diminished over time as the residue decomposed, and crop development lagged. Soil temperature fluctuations were moderated by surface residues, especially in S-irrigated plots.

Addressing soil compaction is vital when implementing CA, especially in soils that are sensitive to compaction, as it can severely hinder crop yields and water productivity, even with efficient irrigation systems. Our findings suggest that mid- and long-term studies in irrigated Mediterranean systems are needed to assess whether the initial challenges of CA can be mitigated through improvements in the soil structure and system adaptation. Additionally, the water consumption of winter crops should be considered, as they help to maintain soil cover but also use rain or irrigation water (commonly applied in Mediterranean conditions), affecting the water reserves for summer crops. Future research should focus on the optimal irrigation strategies under varying water deficit scenarios and explore how combining conservation practices and irrigation systems can enhance the water use efficiency and provide other ecosystemic benefits.

Author Contributions: Conceptualization, J.D.D.-B. and C.W.; methodology, J.D.D.-B. and C.W.; validation, J.D.D.-B., C.W., S.B. and N.U.; formal analysis, J.D.D.-B. and C.W.; investigation, J.D.D.-B. and C.W.; resources, C.W. and N.U.; data curation, J.D.D.-B., C.W., J.-M.L. and C.D.; writing—original draft preparation, J.D.D.-B.; writing—review and editing, J.D.D.-B., C.W., S.B., N.U., J.-M.L., C.D. and B.C.; visualization, J.D.D.-B. and C.W.; supervision, C.W. and S.B.; project administration, C.W.; funding acquisition, C.W. and N.U. All authors have read and agreed to the published version of the manuscript.

Funding: This project is part of the PRIMA programme supported by the European Union, under grant agreement number PRIMA 2020 Section 1 Water IA 2022 MAGO. This work was also supported by INRAE and the Région Occitanie in the framework of the TETRAE program. Additional support was provided by ANRT (Association Nationale de la Recherche et de la Technologie) under convention 2021/0835 and the Société du Canal de Provence (SCP).

Data Availability Statement: The original contributions presented in the study are included in the article; further inquiries can be directed to the corresponding author.

Acknowledgments: The authors wish to extend their sincere appreciation to the PRESTI (G-EAU) experimental team, comprising Gaël Hermet, Marine Muffat-Jeandet and Geoffrey Froment, for their invaluable support in the installation and ongoing management of the field experiment. Special recognition goes to Milancha Babity, Lucas Schwartz, Julien Chabat, Feriel Gheboub and Rami Ben-Moussa for their dedicated contributions to the experimental follow-up during their internships. Gratitude is also extended to the LISAH experimental team for providing the necessary materials for this experimental setup. Finally, heartfelt thanks are extended to the reviewers for their meticulous proofreading of the manuscript.

Conflicts of Interest: Juan David Dominguez-Bohorquez and Nicolas Urruty were employed by the company Société du Canal de Provence (SCP). The remaining authors declare that the research was conducted in the absence of any commercial or financial relationships that could be construed as a potential conflict of interest.

Abbreviations

The following abbreviations are used in this manuscript:

CA	Conservation Agriculture
CT	Conventional Tillage
S	Sprinkler
SSD	Subsurface Drip

NI   No Irrigation
GY   Grain Yield
LAI   Leaf Area Index
TWP   Total Water Productivity
IWP   Irrigation Water Productivity

References

1. Morugán-Coronado, A.; Linares, C.; Gómez-López, M.D.; Faz, Á.; Zornoza, R. The Impact of Intercropping, Tillage and Fertilizer Type on Soil and Crop Yield in Fruit Orchards under Mediterranean Conditions: A Meta-Analysis of Field Studies. *Agric. Syst.* **2020**, *178*, 102736. [CrossRef]
2. Ferreira, C.S.S.; Seifollahi-Aghmiuni, S.; Destouni, G.; Ghajarnia, N.; Kalantari, Z. Soil Degradation in the European Mediterranean Region: Processes, Status and Consequences. *Sci. Total Environ.* **2022**, *805*, 150106. [CrossRef]
3. Eaufrance Bulletin N°5: Prélèvements Quantitatifs sur la Ressource en Eau. Available online: https://www.eaufrance.fr/publications/prelevements-quantitatifs-sur-la-ressource-en-eau-donnees-2016 (accessed on 15 November 2024).
4. Fader, M.; Shi, S.; von Bloh, W.; Bondeau, A.; Cramer, W. Mediterranean Irrigation under Climate Change: More Efficient Irrigation Needed to Compensate for Increases in Irrigation Water Requirements. *Hydrol. Earth Syst. Sci.* **2016**, *20*, 953–973. [CrossRef]
5. Galeotti, M. The Economic Impacts of Climate Change in the Mediterranean. In *IEMed Mediterranean Yearbook*; IEMed: Barcelona, Spain, 2020.
6. Busari, M.A.; Kukal, S.S.; Kaur, A.; Bhatt, R.; Dulazi, A.A. Conservation Tillage Impacts on Soil, Crop and the Environment. *Int. Soil Water Conserv. Res.* **2015**, *3*, 119–129. [CrossRef]
7. Stagnari, F.; Pagnani, G.; Galieni, A.; D'Egidio, S.; Matteucci, F.; Pisante, M. Effects of Conservation Agriculture Practices on Soil Quality Indicators: A Case-Study in a Wheat-Based Cropping Systems of Mediterranean Areas. *Soil Sci. Plant Nutr.* **2020**, *66*, 624–635. [CrossRef]
8. FAO Conservation Agriculture. Available online: https://www.fao.org/conservation-agriculture/en/ (accessed on 6 June 2024).
9. Jat, R.K.; Sapkota, T.B.; Singh, R.G.; Jat, M.L.; Kumar, M.; Gupta, R.K. Seven Years of Conservation Agriculture in a Rice-Wheat Rotation of Eastern Gangetic Plains of South Asia: Yield Trends and Economic Profitability. *Field Crops Res.* **2014**, *164*, 199–210. [CrossRef]
10. Page, K.L.; Dang, Y.P.; Dalal, R.C. The Ability of Conservation Agriculture to Conserve Soil Organic Carbon and the Subsequent Impact on Soil Physical, Chemical, and Biological Properties and Yield. *Front. Sustain. Food Syst.* **2020**, *4*, 31. [CrossRef]
11. Boulal, H.; Gómez-Macpherson, H.; Villalobos, F.J. Permanent Bed Planting in Irrigated Mediterranean Conditions: Short-Term Effects on Soil Quality, Crop Yield and Water Use Efficiency. *Field Crops Res.* **2012**, *130*, 120–127. [CrossRef]
12. Kassam, A.; Friedrich, T.; Derpsch, R.; Lahmar, R.; Mrabet, R.; Basch, G.; González-Sánchez, E.J.; Serraj, R. Conservation Agriculture in the Dry Mediterranean Climate. *Field Crops Res.* **2012**, *132*, 7–17. [CrossRef]
13. Ghaley, B.B.; Rusu, T.; Sandén, T.; Spiegel, H.; Menta, C.; Visioli, G.; O'Sullivan, L.; Gattin, I.T.; Delgado, A.; Liebig, M.A.; et al. Assessment of Benefits of Conservation Agriculture on Soil Functions in Arable Production Systems in Europe. *Sustainability* **2018**, *10*, 794. [CrossRef]
14. Lee, H.; Lautenbach, S.; García-Nieto, A.P.; Bondeau, A.; Cramer, W.; Geijzendorffer, I.; García-Nieto, A. The Impact of Conservation Farming Practices on Mediterranean Agro-Ecosystem Services Provisioning-a Meta-Analysis. *Reg. Environ. Change* **2019**, *19*, 2187–2202. [CrossRef]
15. Mastrorilli, M.; Zucaro, R. Towards Sustainable Use of Water in Rainfed and Irrigated Cropping Systems: Review of Some Technical and Policy Issues. *AIMS Agric. Food* **2016**, *1*, 294–314. [CrossRef]
16. Giller, K.E.; Witter, E.; Corbeels, M.; Tittonell, P. Conservation Agriculture and Smallholder Farming in Africa: The Heretics' View. *Field Crops Res.* **2009**, *114*, 23–34. [CrossRef]
17. Salem, H.M.; Valero, C.; Muñoz, M.Á.; Rodríguez, M.G.; Silva, L.L. Short-Term Effects of Four Tillage Practices on Soil Physical Properties, Soil Water Potential, and Maize Yield. *Geoderma* **2015**, *237–238*, 60–70. [CrossRef]
18. Laborde, J.P.; Wortmann, C.S.; Blanco-Canqui, H.; McDonald, A.J.; Baigorria, G.A.; Lindquist, J.L. Short-Term Impacts of Conservation Agriculture on Soil Physical Properties and Productivity in the Midhills of Nepal. *Agron. J.* **2019**, *111*, 2128–2139. [CrossRef]
19. Mloza-Banda, H.R.; Makwiza, C.N.; Mloza-Banda, M.L. Soil Properties after Conversion to Conservation Agriculture from Ridge Tillage in Southern Malawi. *J. Arid. Environ.* **2016**, *127*, 7–16. [CrossRef]
20. Naab, J.B.; Mahama, G.Y.; Yahaya, I.; Prasad, P.V.V. Conservation Agriculture Improves Soil Quality, Crop Yield, and Incomes of Smallholder Farmers in North Western Ghana. *Front. Plant Sci.* **2017**, *8*, 996. [CrossRef] [PubMed]

21. Afzalinia, S.; Zabihi, J. Soil Compaction Variation during Corn Growing Season under Conservation Tillage. *Soil Tillage Res.* **2014**, *137*, 1–6. [CrossRef]

22. Celik, I. Effects of Tillage Methods on Penetration Resistance, Bulk Density and Saturated Hydraulic Conductivity in a Clayey Soil Conditions. *Tarim Bilimleri Dergisi* **2011**, *17*, 143–156.

23. Celik, I.; Turgut, M.M.; Acir, N. Crop Rotation and Tillage Effects on Selected Soil Physical Properties of a Typic Haploxerert in an Irrigated Semi-Arid Mediterranean Region. *Int. J. Plant Prod.* **2012**, *6*, 157–480.

24. Martínez, I.; Ovalle, C.; Del Pozo, A.; Uribe, H.; Natalia Valderrama, V.; Prat, C.; Sandoval, M.; Fernández, F.; Zagal, E. Influence of Conservation Tillage and Soil Water Content on Crop Yield in Dryland Compacted Alfisol of Central Chile. *Chil. J. Agric. Res.* **2011**, *71*, 615–622. [CrossRef]

25. Khorami, S.S.; Kazemeini, S.A.; Afzalinia, S.; Gathala, M.K. Changes in Soil Properties and Productivity under Different Tillage Practices and Wheat Genotypes: A Short-Term Study in Iran. *Sustainability* **2018**, *10*, 3273. [CrossRef]

26. Singh, B.; Malhi, S.S. Response of Soil Physical Properties to Tillage and Residue Management on Two Soils in a Cool Temperate Environment. *Soil Tillage Res.* **2006**, *85*, 143–153. [CrossRef]

27. Parihar, C.M.; Yadav, M.R.; Jat, S.L.; Singh, A.K.; Kumar, B.; Pradhan, S.; Chakraborty, D.; Jat, M.L.; Jat, R.K.; Saharawat, Y.S.; et al. Long Term Effect of Conservation Agriculture in Maize Rotations on Total Organic Carbon, Physical and Biological Properties of a Sandy Loam Soil in North-Western Indo-Gangetic Plains. *Soil Tillage Res.* **2016**, *161*, 116–128. [CrossRef]

28. Gómez-Paccard, C.; Hontoria, C.; Mariscal-Sancho, I.; Pérez, J.; León, P.; González, P.; Espejo, R. Soil–Water Relationships in the Upper Soil Layer in a Mediterranean Palexerult as Affected by No-Tillage under Excess Water Conditions—Influence on Crop Yield. *Soil Tillage Res.* **2015**, *146*, 303–312. [CrossRef]

29. Mhlanga, B.; Thierfelder, C. Long-Term Conservation Agriculture Improves Water Properties and Crop Productivity in a Lixisol. *Geoderma* **2021**, *398*, 115107. [CrossRef]

30. Martínez, M.; Salvador, E.; Ceja, O.; Espinosa Ramírez, M. *Accumulated Impact of Conservation Agriculture on Soil Properties and Corn Yield*; Instituto Nacional de Investigaciones Forestales, Agrícolas y Pecuarias (INIFAP): Mexico City, Mexico, 2019; Volume 10.

31. Ngwira, A.R.; Thierfelder, C.; Lambert, D.M. Conservation Agriculture Systems for Malawian Smallholder Farmers: Long-Term Effects on Crop Productivity, Profitability and Soil Quality. *Renew. Agric. Food Syst.* **2013**, *28*, 350–363. [CrossRef]

32. Shen, Y.; McLaughlin, N.; Zhang, X.; Xu, M.; Liang, A. Effect of Tillage and Crop Residue on Soil Temperature Following Planting for a Black Soil in Northeast China. *Sci. Rep.* **2018**, *8*, 4500. [CrossRef]

33. Muñoz, V.; Lopez, L.; Lopez, R. Effect of Tillage System on Soil Temperature in a Rainfed Mediterranean Vertisol. *Int. Agrophys.* **2015**, *29*, 467–473. [CrossRef]

34. Muñoz, A.; López-Piñeiro, A.; Ramírez, M. Soil Quality Attributes of Conservation Management Regimes in a Semi-Arid Region of South Western Spain. *Soil Tillage Res.* **2007**, *95*, 255–265. [CrossRef]

35. Sánchez-Llerena, J.; López-Piñeiro, A.; Albarrán, Á.; Peña, D.; Becerra, D.; Rato-Nunes, J.M. Short and Long-Term Effects of Different Irrigation and Tillage Systems on Soil Properties and Rice Productivity under Mediterranean Conditions. *Eur. J. Agron.* **2016**, *77*, 101–110. [CrossRef]

36. López-Garrido, R.; Madejón, E.; León-Camacho, M.; Girón, I.; Moreno, F.; Murillo, J.M. Reduced Tillage as an Alternative to No-Tillage under Mediterranean Conditions: A Case Study. *Soil Tillage Res.* **2014**, *140*, 40–47. [CrossRef]

37. Gabriel, J.L.; Muñoz-Carpena, R.; Quemada, M. The Role of Cover Crops in Irrigated Systems: Water Balance, Nitrate Leaching and Soil Mineral Nitrogen Accumulation. *Agric. Ecosyst. Environ.* **2012**, *155*, 50–61. [CrossRef]

38. Aguilera, E.; Lassaletta, L.; Gattinger, A.; Gimeno, B. Managing Soil Carbon for Climate Change Mitigation and Adaptation in Mediterranean Cropping Systems: A Meta-Analysis. *Agric. Ecosyst. Environ.* **2013**, *168*, 25–36. [CrossRef]

39. Francaviglia, R.; Di Bene, C.; Farina, R.; Salvati, L. Soil Organic Carbon Sequestration and Tillage Systems in the Mediterranean Basin: A Data Mining Approach. *Nutr. Cycl. Agroecosyst.* **2017**, *107*, 125–137. [CrossRef]

40. Vastola, A.; Zdruli, P.; D'Amico, M.; Pappalardo, G.; Viccaro, M.; Di Napoli, F.; Cozzi, M.; Romano, S. A Comparative Multidimensional Evaluation of Conservation Agriculture Systems: A Case Study from a Mediterranean Area of Southern Italy. *Land. Use Policy* **2017**, *68*, 326–333. [CrossRef]

41. Alletto, L.; Cueff, S.; Bréchemier, J.; Lachaussée, M.; Derrouch, D.; Page, A.; Gleizes, B.; Perrin, P.; Bustillo, V. Physical Properties of Soils under Conservation Agriculture: A Multi-Site Experiment on Five Soil Types in South-Western France. *Geoderma* **2022**, *428*, 116228. [CrossRef]

42. Blanco-Canqui, H.; Ruis, S.J. No-Tillage and Soil Physical Environment. *Geoderma* **2018**, *326*, 164–200. [CrossRef]

43. Panettieri, M.; Carmona, I.; Melero, S.; Madejón, E.; Gómez-Macpherson, H. Effect of Permanent Bed Planting Combined with Controlled Traffic on Soil Chemical and Biochemical Properties in Irrigated Semi-Arid Mediterranean Conditions. *Catena* **2013**, *107*, 103–109. [CrossRef]

44. Pareja-Sánchez, E.; Plaza-Bonilla, D.; Ramos, M.C.; Lampurlanés, J.; Álvaro-Fuentes, J.; Cantero-Martínez, C. Long-Term No-till as a Means to Maintain Soil Surface Structure in an Agroecosystem Transformed into Irrigation. *Soil Tillage Res.* **2017**, *174*, 221–230. [CrossRef]

45. Rodríguez González, A.; Arcia Porrúa, J.; Antonio Martínez Cañizares, J.; Domínguez Vento, C.; Herrera Puebla, J.; Miranda Caballero, A.; González, R. Influence of Conservation Agriculture on Some Physical Properties of a Red Ferra-Litic Soil. *Int. J. Food Sci. Agric.* **2022**, *2022*, 320–326. [CrossRef]

46. Jat, R.D.; Jat, H.S.; Nanwal, R.K.; Yadav, A.K.; Bana, A.; Choudhary, K.M.; Kakraliya, S.K.; Sutaliya, J.M.; Sapkota, T.B.; Jat, M.L. Conservation Agriculture and Precision Nutrient Management Practices in Maize-Wheat System: Effects on Crop and Water Productivity and Economic Profitability. *Field Crops Res.* **2018**, *222*, 111–120. [CrossRef]

47. Sartori, F.; Piccoli, I.; Polese, R.; Berti, A. Transition to Conservation Agriculture: How Tillage Intensity and Covering Affect Soil Physical Parameters. *Soil* **2022**, *8*, 213–222. [CrossRef]

48. Bhattacharya, P.; Maity, P.P.; Mowrer, J.; Maity, A.; Ray, M.; Das, S.; Chakrabarti, B.; Ghosh, T.; Krishnan, P. Assessment of Soil Health Parameters and Application of the Sustainability Index to Fields under Conservation Agriculture for 3, 6, and 9 Years in India. *Heliyon* **2020**, *6*, e05640. [CrossRef] [PubMed]

49. Jat, H.S.; Kumar, V.; Datta, A.; Choudhary, M.; Yadvinder-Singh; Kakraliya, S.K.; Poonia, T.; McDonald, A.J.; Jat, M.L.; Sharma, P.C. Designing Profitable, Resource Use Efficient and Environmentally Sound Cereal Based Systems for the Western Indo-Gangetic Plains. *Sci. Rep.* **2020**, *10*, 19267. [CrossRef] [PubMed]

50. Alletto, L.; Coquet, Y.; Justes, E. Effects of Tillage and Fallow Period Management on Soil Physical Behaviour and Maize Development. *Agric. Water Manag.* **2011**, *102*, 74–85. [CrossRef]

51. Meena, R.; Vashisth, A.; Das, T.K. Effect of Conservation Agriculture on Soil Hydrothermal Regimes, Crop Evapo-Transpiration and Water Productivity of Wheat Crop in the Semi-Arid Region of North-Western India. *J. Soil Water Conserv.* **2021**, *20*, 424–430. [CrossRef]

52. DeLaune, P.B.; Mubvumba, P.; Ale, S.; Kimura, E. Impact of No-till, Cover Crop, and Irrigation on Cotton Yield. *Agric. Water Manag.* **2020**, *232*, 106038. [CrossRef]

53. Mondal, S.; Chakraborty, D.; Das, T.K.; Shrivastava, M.; Mishra, A.K.; Bandyopadhyay, K.K.; Aggarwal, P.; Chaudhari, S.K. Conservation Agriculture Had a Strong Impact on the Sub-Surface Soil Strength and Root Growth in Wheat after a 7-Year Transition Period. *Soil Tillage Res.* **2019**, *195*, 104385. [CrossRef]

54. Sokolowski, A.C.; Prack McCormick, B.; De Grazia, J.; Wolski, J.E.; Rodríguez, H.A.; Rodríguez-Frers, E.P.; Gagey, M.C.; Debelis, S.P.; Paladino, I.R.; Barrios, M.B. Tillage and No-Tillage Effects on Physical and Chemical Properties of an Argiaquoll Soil under Long-Term Crop Rotation in Buenos Aires, Argentina. *Int. Soil. Water Conserv. Res.* **2020**, *8*, 185–194. [CrossRef]

55. Zhang, Y.; Tan, C.; Wang, R.; Li, J.; Wang, X. Conservation Tillage Rotation Enhanced Soil Structure and Soil Nutrients in Long-Term Dryland Agriculture. *Eur. J. Agron.* **2021**, *131*, 126379. [CrossRef]

56. Eze, S.; Dougill, A.J.; Banwart, S.A.; Hermans, T.D.G.; Ligowe, I.S.; Thierfelder, C. Impacts of Conservation Agriculture on Soil Structure and Hydraulic Properties of Malawian Agricultural Systems. *Soil Tillage Res.* **2020**, *201*, 104639. [CrossRef] [PubMed]

57. Dam, R.F.; Mehdi, B.B.; Burgess, M.S.E.; Madramootoo, C.A.; Mehuys, G.R.; Callum, I.R. Soil Bulk Density and Crop Yield under Eleven Consecutive Years of Corn with Different Tillage and Residue Practices in a Sandy Loam Soil in Central Canada. *Soil Tillage Res.* **2005**, *84*, 41–53. [CrossRef]

58. Singh, A.; Phogat, V.K.; Dahiya, R.; Batra, S.D. Impact of Long-Term Zero till Wheat on Soil Physical Properties and Wheat Productivity under Rice–Wheat Cropping System. *Soil Tillage Res.* **2014**, *140*, 98–105. [CrossRef]

59. Mazzoncini, M.; Sapkota, T.B.; Bàrberi, P.; Antichi, D.; Risaliti, R. Long-Term Effect of Tillage, Nitrogen Fertilization and Cover Crops on Soil Organic Carbon and Total Nitrogen Content. *Soil Tillage Res.* **2011**, *114*, 165–174. [CrossRef]

60. Ismail, I.; Blevins, R.L.; Frye, W.W. Long-Term No-Tillage Effects on Soil Properties and Continuous Corn Yields. *Soil Sci. Soc. Am. J.* **1994**, *58*, 193–198. [CrossRef]

61. Talukder, R.; Plaza-Bonilla, D.; Cantero-Martínez, C.; Di Prima, S.; Lampurlanés, J. Spatio-Temporal Variation of Surface Soil Hydraulic Properties under Different Tillage and Maize-Based Crop Sequences in a Mediterranean Area. *Plant Soil* **2024**, *500*, 263–277. [CrossRef]

62. IUSS Working Group WRB International Soil Classification System for Naming Soils and Creating Legends for Soil Maps. *World Soil. Resour. Rep.* **2015**, *106*, 166–168.

63. Albasha, R.; Thiesson, J.; Buvat, S.; Lopez, J.M.; Cheviron, B. Prospection Géophysique Dans Le Cadre d'une Étude de La Variabilité Spatiale Des Rendements Agricoles. In Proceedings of the GEOFCAN, Orsay, France, 13–14 November 2014; UMR GEOPS: Orsay, France, 2014.

64. Eijkelkamp, R. *Hand-Penetrometer for Top Layers, Type IB Meet the Difference Operating Instructions*; Royal Eijkelkamp: Giesbeek, The Netherlands, 2022.

65. Yolcubal, I.; Brusseau, M.L.; Artiola, J.F.; Wierenga, P.J.; Wilson, L.G. Environmental physical properties and processes. *Environ. Monit. Charact.* **2004**, 207–239. [CrossRef]

66. Garnier, F.; Elamri, Y. *Le Cahier des Techniques de l'INRA 2014 (83) n°3 Automatisation du Dispositif de Müntz Pour la Détermination In Situ de la Conductivité Hydraulique à Saturation*, 84th ed.; 2014; Volume 83, 12p, hal-04632381.

67. Vinatier, F.; Rudi, G.; Coulouma, G.; Dagès, C.; Bailly, J.-S. Dynamics of Quasi-Steady Ponded Infiltration under Contrasting Plant Cover and Management Strategies. *Soil Tillage Res.* **2024**, *237*, 105985. [CrossRef]

68. Bhatt, R.; Singh, P. Evaporation Trends on Intervening Period for Different Wheat Establishments under Soils of Semi-Arid Tropics. *J. Soil Water Conserv.* **2018**, *17*, 41. [CrossRef]

69. Trambouze, W. *Caractérisation et Eléments de Modelisation de L'évapotranspiration Réelle de la Vigne à L'échelle de la Parcelle*; Ecole Nationale Supérieure Agronomique de Montpellier: Montpellier, France, 1996.

70. Flumignan, D.L.; de Faria, R.T.; Lena, B.P. Test of a Microlysimeter for Measurement of Soil Evaporation. *Eng. Agrícola* **2012**, *32*, 80–90. [CrossRef]

71. Fang, H.; Liang, S. Leaf Area Index Models. In *Encyclopedia of Ecology*; Elsevier: Oxford, UK, 2014. [CrossRef]

72. LI-COR LAI-2200C Plant Canopy Analyzer. Available online: https://www.licor.com/env/products/leaf-area/LAI-2200C/ (accessed on 15 November 2024).

73. Shaheen, T.; Riaz, M.S.; Zafar, Y. Soybean Production and Drought Stress. In *Abiotic and Biotic Stresses in Soybean Production*; Academic Press: San Diego, CA, USA, 2016; pp. 177–196. [CrossRef]

74. Mailhol, J.C.; Ruelle, P.; Walser, S.; Schütze, N.; Dejean, C. Analysis of AET and Yield Predictions under Surface and Buried Drip Irrigation Systems Using the Crop Model PILOTE and Hydrus-2D. *Agric Water Manag.* **2011**, *98*, 1033–1044. [CrossRef]

75. Bussière, F.; Cellier, P. Modification of the Soil Temperature and Water Content Regimes by a Crop Residue Mulch: Experiment and Modelling. *Agric. For. Meteorol.* **1994**, *68*, 1–28. [CrossRef]

76. Wang, D.; Shannon, M.C.; Grieve, C.M.; Yates, S.R. Soil Water and Temperature Regimes in Drip and Sprinkler Irrigation, and Implications to Soybean Emergence. *Agric. Water Manag.* **2000**, *43*, 15–28. [CrossRef]

77. Abu-Hamdeh, N.H. SW—Soil and Water: Measurement of the Thermal Conductivity of Sandy Loam and Clay Loam Soils Using Single and Dual Probes. *J. Agric. Eng. Res.* **2001**, *80*, 209–216. [CrossRef]

78. Sándor, R.; Fodor, N. Simulation of Soil Temperature Dynamics with Models Using Different Concepts. *Sci. World J.* **2012**, *2012*, 590287. [CrossRef]

79. Kassam, A.; Friedrich, T.; Shaxson, F.; Pretty, J. The Spread of Conservation Agriculture: Justification, Sustainability and Uptake. *Int. J. Agric. Sustain.* **2009**, *7*, 292–320. [CrossRef]

80. Özdemir, N.; Demir, Z.; Bülbül, E. Relationships between Some Soil Properties and Bulk Density under Different Land Use. *Soil Stud.* **2022**, *11*, 43–50. [CrossRef]

81. Rahimi, A.A.; Sepaskhah, A.R.; Ahmadi, S.H. Evaluation of Different Methods for the Prediction of Saturated Hydraulic Conductivity in Tilled and Untilled Soils. *Arch. Agron. Soil Sci.* **2011**, *57*, 899–914. [CrossRef]

82. Nyeki, A.; Milics, G.; Kovacs, A.J.; Neményi, M. Effects of Soil Compaction on Cereal Yield: A Review. *Cereal Res. Commun.* **2017**, *45*, 1–22. [CrossRef]

83. Zhou, H.; LÜ, Y.; YANG, Z.; LI, B. Influence of Conservation Tillage on Soil Aggregates Features in North China Plain. *Agric. Sci. China* **2007**, *6*, 1099–1106. [CrossRef]

84. Taser, O.; Metinoglu, F. Physical and Mechanical Properties of a Clayey Soil as Affected by Tillage Systems for Wheat Growth. *Acta Agric. Scand. Sect. B-Soil Plant Sci.* **2007**, *55*, 186–191. [CrossRef]

85. Indoria, A.K.; Sharma, K.L.; Reddy, K.S. Hydraulic Properties of Soil under Warming Climate. In *Climate Change and Soil Interactions*; Elsevier: Amsterdam, The Netherlands, 2020; Chapter 18; pp. 473–508. [CrossRef]

86. Souza, R.; Hartzell, S.; Freire Ferraz, A.P.; de Almeida, A.Q.; de Sousa Lima, J.R.; Dantas Antonino, A.C.; de Souza, E.S. Dynamics of Soil Penetration Resistance in Water-Controlled Environments. *Soil Tillage Res.* **2021**, *205*, 104768. [CrossRef]

87. Hamza, M.A.; Anderson, W.K. Soil Compaction in Cropping Systems: A Review of the Nature, Causes and Possible Solutions. *Soil Tillage Res.* **2005**, *82*, 121–145. [CrossRef]

88. Aase, J.K.; Bjorneberg, D.L.; Sojka, R.E. Zone-Subsoiling Relationships to Bulk Density and Cone Index on a Furrow-Irrigated Soil. *Trans. Am. Soc. Agric. Eng.* **2001**, *44*, 577. [CrossRef]

89. Jeřábek, J.; Zumr, D.; Dostál, T. Identifying the Plough Pan Position on Cultivated Soils by Measurements of Electrical Resistivity and Penetration Resistance. *Soil Tillage Res.* **2017**, *174*, 231–240. [CrossRef]

90. Peigné, J.; Vian, J.-F.; Payet, V.; Saby, N.P.A. Soil Fertility after 10 Years of Conservation Tillage in Organic Farming. *Soil Tillage Res.* **2018**, *175*, 194–204. [CrossRef]

91. da Silva, V.R.; Reinert, D.J.; Reichert, J.M. Resistência Mecânica Do Solo à Penetração Influenciada Pelo Tráfego de Uma Colhedora Em Dois Sistemas de Manejo Do Solo. *Ciência Rural* **2000**, *30*, 795–801. [CrossRef]

92. Lamm, F.R. Managing the Challenges of Subsurface Drip Irrigation. In Proceedings of the Irrigation Association Technical Conference, San Antonio, TX, USA, 2–5 December 2009.

93. Materechera, S.A.; Mloza-Banda, H.R. Soil Penetration Resistance, Root Growth and Yield of Maize as Influenced by Tillage System on Ridges in Malawi. *Soil Tillage Res.* **1997**, *41*, 13–24. [CrossRef]
94. Coulouma, G.; Boizard, H.; Trotoux, G.; Lagacherie, P.; Richard, G. Effect of Deep Tillage for Vineyard Establishment on Soil Structure: A Case Study in Southern France. *Soil Tillage Res.* **2006**, *88*, 132–143. [CrossRef]
95. Mubarak, I.; Mailhol, J.C.; Angulo-Jaramillo, R.; Ruelle, P.; Boivin, P.; Khaledian, M. Temporal Variability in Soil Hydraulic Properties under Drip Irrigation. *Geoderma* **2009**, *150*, 158–165. [CrossRef]
96. Horn, R.; Way, T.; Rostek, J. Effect of Repeated Tractor Wheeling on Stress/Strain Properties and Consequences on Physical Properties in Structured Arable Soils. *Soil Tillage Res.* **2003**, *73*, 101–106. [CrossRef]
97. Esser, K.B. Water Infiltration and Moisture in Soils under Conservation and Conventional Agriculture in Agro-Ecological Zone IIa, Zambia. *Agronomy* **2017**, *7*, 40. [CrossRef]
98. Roua, A.; Ibrahimi, K.; Sher, F.; Milham, P.; Ghazouani, H.; Chehaibi, S.; Hussain, Z.; Iqbal, H. Impacts of Different Tillage Practices on Soil Water Infiltration for Sustainable Agriculture. *Sustainability* **2021**, *13*, 3155. [CrossRef]
99. Azooz, R.H.; Arshad, M.A. Soil Infiltration and Hydraulic Conductivity under Long-Term No-Tillage and Conventional Tillage Systems. *Can. J. Soil Sci.* **1996**, *76*, 143–152. [CrossRef]
100. Lipiec, J.; Kuś, J.; Słowińska-Jurkiewicz, A.; Nosalewicz, A. Soil Porosity and Water Infiltration as Influenced by Tillage Methods. *Soil Tillage Res.* **2006**, *89*, 210–220. [CrossRef]
101. Thierfelder, C.; Wall, P.C. Effects of Conservation Agriculture Techniques on Infiltration and Soil Water Content in Zambia and Zimbabwe. *Soil Tillage Res.* **2009**, *105*, 217–227. [CrossRef]
102. Van Donk, S.J.; Martin, D.L.; Irmak, S.; Melvin, S.R.; Petersen, J.L.; Davison, D.R. Crop Residue Cover Effects on Evaporation, Soil Water Content, and Yield of Deficit-Irrigated Corn in West-Central Nebraska. *Trans. ASABE* **2010**, *53*, 1787–1797. [CrossRef]
103. Horton, R.; Bristow, K.L.; Kluitenberg, G.J.; Sauer, T.J. Crop Residue Effects on Surface Radiation and Energy Balance—Review. *Theor. Appl. Clim.* **1996**, *54*, 27–37. [CrossRef]
104. Fuchs, M.; Hadas, A. Mulch Resistance to Water Vapor Transport. *Agric. Water Manag.* **2011**, *98*, 990–998. [CrossRef]
105. Baalousha, H.M.; Ramasomanana, F.; Fahs, M.; Seers, T.D. Measuring and Validating the Actual Evaporation and Soil Moisture Dynamic in Arid Regions under Unirrigated Land Using Smart Field Lysimeters and Numerical Modeling. *Water* **2022**, *14*, 2787. [CrossRef]
106. Chen, Y.; Cavers, C.; Tessier, S.; Monero, F.; Lobb, D. Short-Term Tillage Effects on Soil Cone Index and Plant Development in a Poorly Drained, Heavy Clay Soil. *Soil Tillage Res.* **2005**, *82*, 161–171. [CrossRef]
107. Cárceles Rodríguez, B.; Durán-Zuazo, V.H.; Soriano Rodríguez, M.; García-Tejero, I.F.; Gálvez Ruiz, B.; Cuadros Tavira, S. Conservation Agriculture as a Sustainable System for Soil Health: A Review. *Soil Syst.* **2022**, *6*, 87. [CrossRef]
108. Van den Putte, A.; Govers, G.; Diels, J.; Gillijns, K.; Demuzere, M. Assessing the Effect of Soil Tillage on Crop Growth: A Meta-Regression Analysis on European Crop Yields under Conservation Agriculture. *Eur. J. Agron.* **2010**, *33*, 231–241. [CrossRef]
109. Rana, B.; Parihar, C.M.; Jat, M.L.; Patra, K.; Nayak, H.S.; Reddy, K.S.; Sarkar, A.; Anand, A.; Naguib, W.; Gupta, N.; et al. Combining Sub-Surface Fertigation with Conservation Agriculture in Intensively Irrigated Rice under Rice-Wheat System Can Be an Option for Sustainably Improving Water and Nitrogen Use-Efficiency. *Field Crops Res.* **2023**, *302*, 109074. [CrossRef]
110. Martínez Gamiño, M.Á.; Osuna Ceja, E.S.; Espinosa Ramírez, M.; Martínez-Gamiño, M.Á.; Osuna Ceja, E.S.; Espinosa Ramírez, M. Impacto Acumulado de La Agricultura de Conservación En Propiedades Del Suelo y Rendimiento de Maíz. *Rev. Mex. Cienc. Agric.* **2019**, *10*, 765–778. [CrossRef]
111. Abdallah, A.M.; Jat, H.S.; Choudhary, M.; Abdelaty, E.F.; Sharma, P.C.; Jat, M.L. Conservation Agriculture Effects on Soil Water Holding Capacity and Water-Saving Varied with Management Practices and Agroecological Conditions: A Review. *Agronomy* **2021**, *11*, 1681. [CrossRef]
112. Serra-Wittling, C.; Molle, B.; Cheviron, B. Plot Level Assessment of Irrigation Water Savings Due to the Shift from Sprinkler to Localized Irrigation Systems or to the Use of Soil Hydric Status Probes. Application in the French Context. *Agric. Water Manag.* **2019**, *223*, 105682. [CrossRef]
113. Umair, M.; Hussain, T.; Jiang, H.; Ahmad, A.; Yao, J.; Qi, Y.; Zhang, Y.; Min, L.; Shen, Y. Water-Saving Potential of Subsurface Drip Irrigation for Winter Wheat. *Sustainability* **2019**, *11*, 2978. [CrossRef]
114. Jacobsen, S.E.; Jensen, C.R.; Liu, F. Improving Crop Production in the Arid Mediterranean Climate. *Field Crops Res.* **2012**, *128*, 34–47. [CrossRef]
115. Sidhu, H.S.; Jat, M.L.; Singh, Y.; Sidhu, R.K.; Gupta, N.; Singh, P.; Singh, P.; Jat, H.S.; Gerard, B. Sub-Surface Drip Fertigation with Conservation Agriculture in a Rice-Wheat System: A Breakthrough for Addressing Water and Nitrogen Use Efficiency. *Agric. Water Manag.* **2019**, *216*, 273–283. [CrossRef]

116. Cakir, R. Effect of Water Stress at Different Development Stages on Vegetative and Reproductive Growth of Corn. *Field Crops Res.* **2004**, *89*, 1–16. [CrossRef]
117. Wang, H.; Wang, N.; Quan, H.; Zhang, F.; Fan, J.; Feng, H.; Cheng, M.; Liao, Z.; Wang, X.; Xiang, Y. Yield and Water Productivity of Crops, Vegetables and Fruits under Subsurface Drip Irrigation: A Global Meta-Analysis. *Agric. Water Manag.* **2022**, *269*, 107645. [CrossRef]

 *agronomy*

Article

Effects of No-Tillage on Field Microclimate and Yield of Winter Wheat

Zhiqiang Dong [1], Shuo Yang [2,3], Si Li [2,3], Pengfei Fan [2,3], Jianguo Wu [2], Yuxin Liu [2], Xiu Wang [2,3], Jingting Zhang [1,*] and Changyuan Zhai [2,3,*]

1. Institute of Cereal and Oil Crops, Hebei Academy of Agricultural and Forestry Sciences, Shijiazhuang 050035, China; woshidongzhiqiang81@126.com
2. Intelligent Equipment Research Center, Beijing Academy of Agricultural and Forestry Sciences, Beijing 100097, China; yangshuo@nercita.org.cn (S.Y.); lis@nercita.org.cn (S.L.); fanpf@nercita.org.cn (P.F.); wujg@nercita.org.cn (J.W.); 18684214617@163.com (Y.L.); wangx@nercita.org.cn (X.W.)
3. National Engineering Research Center of Intelligent Equipment for Agriculture, Beijing 100097, China
* Correspondence: jingting58@126.com (J.Z.); zhaicy@nercita.org.cn (C.Z.); Tel.: +86-0311-87670620 (J.Z.); +86-010-51503886 (C.Z.)

Abstract: Field studies were conducted in the North China Plain (NCP) during the 2023–2024 season to investigate the vertical microclimate, yield, and yield-related characteristics of winter wheat during the grain-filling stage under no-till direct seeding and conventional tillage. The aim was to compare the differences in microclimate between the two tillage methods in wheat fields and the impact of microclimate on yield. The results indicated that, compared to conventional tillage, no-till direct seeding reduced the air temperature and increased the relative humidity of the air at 20 cm and 100 cm above the ground during the wheat grain-filling period. The soil moisture content at 20 cm below the ground under no-till direct seeding was higher than under conventional tillage during the early grain-filling stage. Seven days before the wheat harvest, the dry weight per plant and the dry weight per spike were significantly greater under no-till direct seeding than under conventional tillage. Consequently, the thousand-grain weight of no-till direct seeding was significantly higher than that of conventional tillage, with an increase of 7.9%. The number of wheat sterile spikelets under no-till direct seeding was significantly lower than that under conventional tillage. Furthermore, the number of grains per spike was higher than that of conventional tillage. Although the number of harvested spikes under no-till direct seeding was 10.8% lower than under conventional tillage, the increase in thousand-grain weight and the number of grains per spike compensated for the reduced number of harvested spikes. As a result, the grain yield of winter wheat under no-till direct seeding was higher than that of conventional tillage, increasing by 2.7%. Therefore, adopting no-till direct seeding in the NCP is conducive to increasing winter wheat production and efficiency, as well as supporting sustainable agricultural development.

Keywords: winter wheat; no-tillage direct seeding; conventional tillage; air temperature; relative humidity of air; soil moisture content; yield

Citation: Dong, Z.; Yang, S.; Li, S.; Fan, P.; Wu, J.; Liu, Y.; Wang, X.; Zhang, J.; Zhai, C. Effects of No-Tillage on Field Microclimate and Yield of Winter Wheat. *Agronomy* **2024**, *14*, 3075. https://doi.org/10.3390/agronomy14123075

Academic Editor: Tie Cai

Received: 12 November 2024
Revised: 19 December 2024
Accepted: 20 December 2024
Published: 23 December 2024

1. Introduction

The NCP is one of the major grain producing areas in China, playing a crucial role in ensuring national food security. Double cropping of winter wheat and summer corn is the predominant grain planting method in this region [1]. During the planting of winter wheat in this area, there are issues, such as multiple agricultural machinery operations, low efficiency, and high production costs. Compared to traditional tillage, no-till has advantages, including reducing the number of agricultural machinery operations, improving production efficiency, protecting soil structure, increasing soil water retention capacity, and enhancing soil fertility. Additionally, no-till can prevent soil erosion, protect the agricultural ecological environment, and contribute to the sustainable development of agriculture [2–5].

In the Loess Plateau, no-till with mulch is considered one of the best tillage systems for increasing wheat water storage and yield, as well as improving water use efficiency and conserving energy [6]. However, compared to traditional tillage, no-till often faces controversy regarding its impact on crop yield and its adaptability in different agricultural environments [7–9], which limits its widespread application, especially in winter wheat production in the NCP [10].

Farmland microclimate refers to the climate within a small, localized area near the field surface, commonly represented by the values of agrometeorological elements, such as air temperature, air humidity, radiation, wind speed, and soil temperature and moisture in the near-surface layer of the farmland. The impact of soil tillage systems on the vertical distribution of the farmland microclimate has practical significance both theoretically and in practice, as it can alter the microclimate within the farmland ecosystem, thereby affecting soil moisture, nutrients, and crop yield [6]. The no-till sowing technology for wheat minimally disturbs the soil structure while retaining a large amount of above-ground crop residue, which has a certain impact on the farmland microclimate. Under no-till conditions, the undisturbed crop residue layer acts as a barrier to reduce soil evaporation and respiration. No-till improves water retention compared to conventional tillage, resulting in increased soil moisture storage [11]. Conservation tillage, which involves no or reduced soil disturbance (no-till or reduced tillage), is one of the effective ways to alleviate crop drought [12,13]. From the regreening stage to the maturity of winter wheat, the soil moisture content in each soil layer under no-till with straw mulch is higher than that under traditional tillage [14]. Most studies have shown that conservation tillage can improve crop water use efficiency [4–6,14–16].

No-till with straw mulch delays the growth process, prolongs the grain-filling duration, enhances the capacity for dry matter transport to grains, and increases the number of grains per spike and thousand-grain weight of winter wheat, compared to traditional tillage, thereby improving the grain yield of winter wheat [17]. Compared to traditional rotary tillage, no-till wheat has a slight increase in the number of ears, grains per ear, and thousand-grain weight, with a yield increase of 3.2% [18]. In the dryland areas of northern China, compared to conventional tillage, less tillage increases the yield of spring maize by 13–16% and winter wheat by 9–37%, while the yield under no-till is very close to that of conventional tillage [4]. Under ambient and warming conditions, conservation agriculture (permanent crop residue cover and no-till) maintains similar winter wheat yields to traditional agriculture (crop residue removal and annual tillage) [19]. Less tillage and no-till are effective measures to increase corn yields in dryland areas [20,21] and can also reduce the uncertainty of crop yields between years [22–24]. Guo et al. [25] demonstrated through a 12-year long-term conservation tillage experiment in Weibei Upland Plateau that in the winter wheat-fallow-spring maize rotation cycle, the average yield of crops under no-till was 1.6% higher than that under traditional tillage. In the NCP, long-term no-till for winter wheat can improve water use efficiency by reducing deep root distribution and water uptake, but continuous no-till can increase soil compaction, leading to a decrease in yield [26]. In southern China, a strategy based on no-till and rotary tillage (rotating one year of rotary tillage with three years of no-till) could reduce the negative impacts of continuous no-till and enhance profitability through increased yields, thereby maintaining the sustainability of rice production [27]. Pittelkow et al. [7] compared the yield differences between no-till and traditional tillage for 48 crops in 63 countries. Overall, no-till reduced crop yields; however, when combined with straw return and crop rotation, no-till could produce yields equivalent to or higher than traditional tillage. Of course, yield is only one component of the agricultural system, and there is an urgent need to optimize tillage methods based on environmental index [28].

Although numerous studies have been conducted on wheat yield and water use efficiency under no-till direct seeding by predecessors [4,6,17,18,26,27], there are fewer reports on the impact of tillage systems on the microclimate in wheat fields. Research on the effects of no-till practices on field microclimate and how the field microclimate influences wheat yield has not been reported. This study conducted research on no-till direct seeding technology for winter wheat in the double-cropping area of winter wheat and summer maize in the NCP. The purpose is to compare the changes in microclimate under two tillage methods and the impact of microclimate on wheat grain yield.

2. Materials and Methods

2.1. Overview of the Experimental Site

The experiment was conducted from 2023 to 2024 at the National Precision Agriculture Demonstration and Research Base in Xiaotangshan, Changping, Beijing (40°21′ N, 116°34′ E, altitude 36 m). The annual average temperature in the experimental area is 10–13 °C, with an average daily sunlight duration of 6.5–8.5 h and an annual average precipitation of 602 mm, with 80% of the precipitation concentrated from June to August. The soil type is fluvo-aquic soil, and the main physical and chemical properties of the soil in the 0–20 cm layer are as follows: pH value of 7.62, organic matter content ranging from 5.8 to 20.0 g kg^{-1}, total nitrogen is 1.0–1.2 g kg^{-1}, nitrate nitrogen is 3.16–14.82 mg kg^{-1}, available phosphorus is 3.14–21.18 mg kg^{-1}, and available potassium is 86.83–120.62 mg kg^{-1}.

2.2. Experimental Materials

2.2.1. Equipment for No-Till and Conventional Wheat Sowing and Fertilization

A navigation-based Internet of Things (IoT) wheat sowing and fertilization machine is used for no-till direct seeding operations (NT). This machine, as shown in Figure 1, based on the 2BMQF-6/12 no-till seeder (Luoyang Xinle Machinery Equipment Co., Ltd., Luoyang, China), is equipped with a wheat sowing and fertilization IoT monitoring controller (NERCITA, Beijing, China) for monitoring and recording information on sowing rates, fertilization amounts, and operation trajectories. The navigation controller (Nongxin Technology Co., Ltd., Beijing, China) is used for tractor navigation and operational control, while the sowing and fertilization fault alarm controller (Changzhou Huaiyu Electronics Co., Ltd., Changzhou, China) monitors for any failures in seed and fertilizer distribution across the 12-row sowing and 6-row fertilization systems and then alerts the operator. The main components of the no-till seeder platform include a disc sawtooth furrow opener, a hard residue shoe-type opener, a dual-chamber dual-tube seed distribution system, and a roller for soil compaction. This machine is capable of simultaneously performing residue removal, furrowing, fertilization, seeding, and packing operations in a single pass. It employs a wide-narrow row planting system, with wide rows at 22 cm and narrow rows at 13 cm, while the fertilization position is 3–5 cm below the seed placement. In contrast, conventional tillage involves two passes of rotary tillage followed by seeding and fertilization with a 2BFX-24 seeder (Shijiazhuang Agricultural Machinery Co., Ltd., Shijiazhuang, China) at a uniform row spacing of 15 cm. The seeding speed during these operations ranges from 5 to 8 km h^{-1}.

2.2.2. Environmental Monitoring System

An environmental monitoring system was set up to monitor changes in air temperature and humidity in the wheat field, as well as soil temperature and humidity, see Figure 2. The air temperature and humidity sensor (model HSTL-BXYWS, Beijing Huakong Xingye Technology Development Co., Ltd., Beijing, China) has a measurement range of 0% RH to 100% RH for humidity, with an accuracy of ±4.5% RH, and a measurement range of −40 °C to 120 °C for temperature, with an accuracy of ±0.5 °C. The output signal is 4–20 mA. The soil temperature and humidity sensor (model HSTL-102STR, Beijing Huakong Xingye Technology Development Co., Ltd., Beijing, China) has a measurement range of 0 to 100% for soil relative humidity, with an accuracy of ±2%, and a measurement range of −40 °C to 80 °C for soil temperature, with an accuracy of ±0.2 °C. The output signal is 4–20 mA. The data acquisition module (model DAM0888, Beijing Juying Aoxiang Electronics Co., Ltd., Beijing, China) collects sensor data through its current analog input ports and transmits the data via the GSM network to a cloud platform for storage. The environmental monitoring system is powered by a solar panel, enabling continuous data collection in the wheat field.

Figure 1. Navigation-based Internet of Things wheat seeding and fertilizer machine.

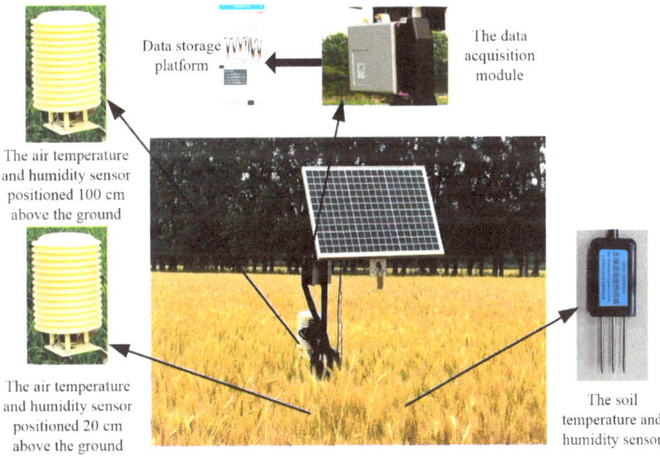

Figure 2. Environmental monitoring system.

2.3. Experimental Methods

2.3.1. Experimental Design

Treatment 1 was no-tillage sowing (NT), and treatment 2 was conventional tillage (CT). Sowing occurred on 6 October 2023, and harvesting took place on 12 June 2024. The preceding crop was silage summer maize. The plot size was 500 m^2 (50 m × 10 m), with three replicates in a randomized block design. The wheat variety tested was Jingdong 22 (Provided by the Wheat Research Institute of Beijing Academy of Agriculture and Forestry Sciences), with a base fertilizer of diammonium phosphate (N 15%, P$_2$O$_5$ 42%) applied at a rate of 375 kg ha^{-1}. Urea (N 46%) was applied at a rate of 225 kg ha^{-1} during the wheat jointing stage. The seeding rate for both treatments was 270 kg ha^{-1}, and the irrigation method used was a movable spray irrigation system. The total irrigation amount during the wheat growth period was 180 mm, including 22.5 mm after sowing, 45 mm for the pre-wintering stage, 22.5 mm for the erecting stage, 45 mm for the jointing stage, 22.5 mm for the anthesis stage, and 22.5 mm for the grain-filling stage. Other field management practices followed local production habits, and there were no severe pests or diseases during the wheat growth period.

2.3.2. Data Collection and Measurement Methods

Air temperature, humidity, and soil moisture content: An environmental monitoring system was utilized to measure air temperature and relative humidity at 20 cm and 100 cm above the wheat field surface, and soil moisture content at 20 cm below the ground. The data collection period for air temperature and relative humidity was from 6 May 2024, to 12 June 2024 (the wheat flowering time was on 5 May 2024), and for soil moisture content, it was from 6 May 2024, to 20 May 2024, with both data collection intervals set at 1 h. Plant moisture content and yield: Seven days before wheat harvest, approximately 20 representative single stems and 20 ears were collected from each plot. The samples underwent steam killing for 30 min at 105 °C, followed by drying at 75 °C until a constant weight was achieved. During harvest, three representative points with uniform growth were selected along the diagonal direction of each plot, and 3 m^2 of wheat was manually harvested at each sampling point.

2.4. Data Analysis

Data were organized using Microsoft Excel 2007, and statistical analysis was conducted using SPSS 25.0 software. The least significant difference (LSD) method was used to test for significant differences.

3. Results

3.1. Air Temperature at 20 cm and 100 cm Above the Ground During the Wheat Grain-Filling Stage (69–89)

As shown in Figure 3, the trends of the average air temperature at 20 cm above the ground during the wheat grain-filling stage were similar for both tillage methods: a gradual decline followed by a rapid increase, then a slower rise to a peak, after which the temperature gradually and rapidly decreased. The minimum values all occurred at 5:00 for both tillage methods, while the maximum values were observed at 14:00. During the wheat grain-filling stage, from 6:00 to 23:00, the average air temperature at 20 cm above the ground under no-till direct seeding was lower than that under conventional tillage, with reductions ranging from 0.4% to 10.4%. From 1:00 to 5:00, the average air temperature at 20 cm under no-till direct seeding was slightly higher than that under conventional tillage. The difference in the average air temperature at 20 cm between the two tillage methods was more significant from 8:00 to 18:00, with an average decrease of 8.1% under no-till direct seeding compared to conventional tillage.

The trend of the average air temperature changes at 100 cm above the ground (canopy) during the winter wheat grain-filling stage was basically the same as that at 20 cm for both no-till direct seeding and conventional tillage (Figure 4): a gradual and slow decrease from 0:00 to 5:00, with the lowest value at 5:00; a rapid increase from 6:00 to 10:00, a slow increase from 10:00 to 14:00, reaching the highest value at 14:00; and a slow decrease from 14:00 to 17:00, a rapid decrease from 17:00 to 20:00, and a slow decrease from 20:00 to 23:00. During the wheat grain-filling stage, the average canopy temperature under no-till direct seeding was lower than that under conventional tillage from 0:00 to 23:00, with a decrease of 0.5% to 8.1%. Notably, from 10:00 to 6:00, the canopy temperature under no-till direct seeding was significantly lower than that under conventional tillage.

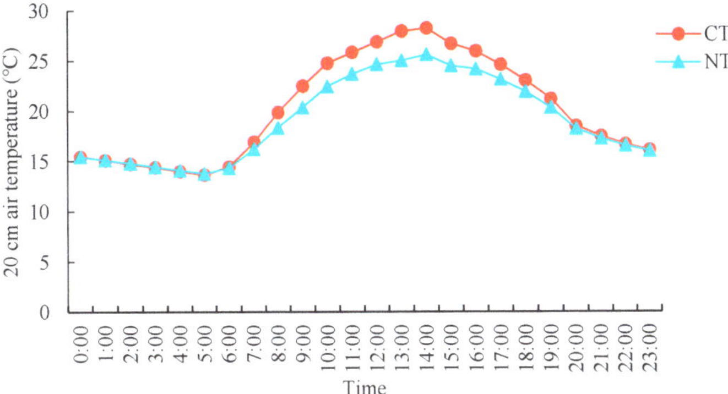

Figure 3. The average air temperature at 20 cm above the ground during the grain-filling stage (69–89) of winter wheat in the Beijing area.

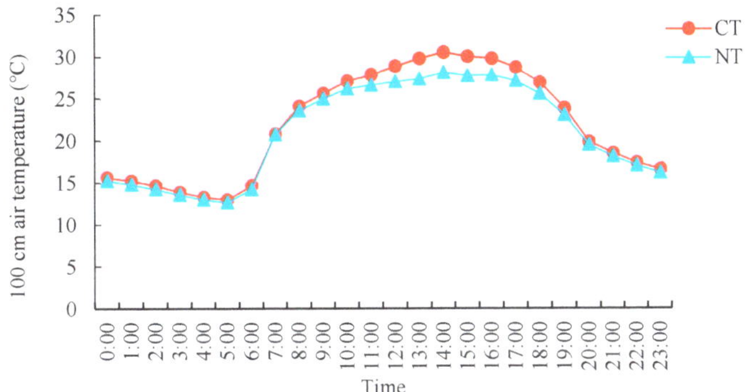

Figure 4. The average air temperature at 100 cm above the ground during the grain-filling stage (69–89) of winter wheat in the Beijing area.

3.2. Relative Humidity of the Air at 20 cm and 100 cm Above the Ground During the Wheat Grain-Filling Stage (69–89)

As shown in Figure 5, the trend of the average relative humidity of the air change at 20 cm above the ground during the grain-filling period of winter wheat under no-till direct seeding was basically the same as that under conventional tillage: it generally rises slowly from 0:00 to 7:00, with the maximum value occurring at 7:00. It drops rapidly from 7:00 to 12:00, and the rate of decline slows down after 12:00, with the minimum value appearing at 14:00, followed by a slow rise until 18:00, then it rises rapidly from 18:00 to 20:00, and afterwards, it continues to rise slowly. The overall trend of change in the average relative humidity of the air at 20 cm above the ground during the grain-filling period of wheat under both tillage methods was opposite to the trend of change in the average air temperature at 20 cm above the ground (Figure 3). The average relative humidity of the air at 20 cm above the ground during the grain-filling period of wheat under no-till direct seeding was greater than that under conventional tillage, especially from 8:00 to 19:00, when it was significantly higher than that of conventional tillage, increasing by 6.0% to 18.4%. During other time periods, the difference in the average relative humidity of the air at 20 cm above the ground between the two tillage methods was smaller.

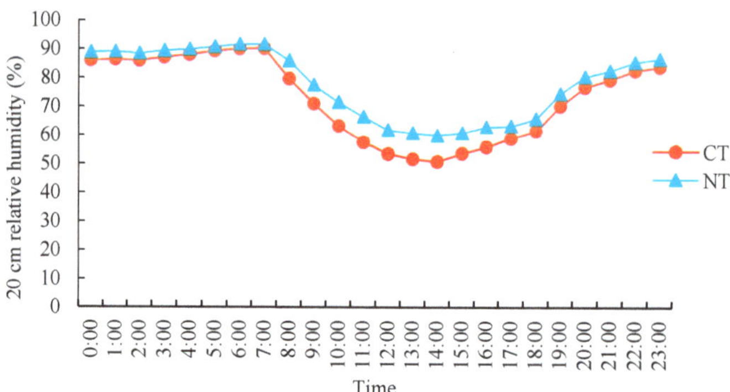

Figure 5. The average relative humidity of the air at 20 cm above the ground during the grain-filling stage (69–89) of winter wheat in the Beijing area.

The trend of the average air relative humidity of the air changes at 100 cm above the ground during the winter wheat grain-filling stage showed certain differences between no-till direct seeding and conventional tillage compared to the 20 cm level (Figure 6). Both tillage methods exhibited the following pattern: a gradual and slow increase from 0:00 to 6:00, followed by a slow decrease, with the maximum value occurring at 5:00, which was also the maximum value of the day; from 6:00 to 10:00, there was a rapid decrease, and then the rate of decrease gradually slowed down, reaching the minimum value at 14:00 or 15:00. After that, it slowly increased until 18:00, and then rapidly increased from 18:00 to 22:00. During the wheat grain-filling stage, the average relative humidity of the air at 100 cm above the ground under no-till direct seeding was higher than that under conventional tillage from 8:00 to 19:00, with an increase of 1.2% to 13.8%. Furthermore, from 10:00 to 18:00, the average relative humidity of the air under no-till direct seeding was significantly higher than under conventional tillage. From 0:00 to 8:00 and 19:00 to 23:00, the average relative humidity of the air at 100 cm above the ground under no-till direct seeding was slightly lower than that under conventional tillage.

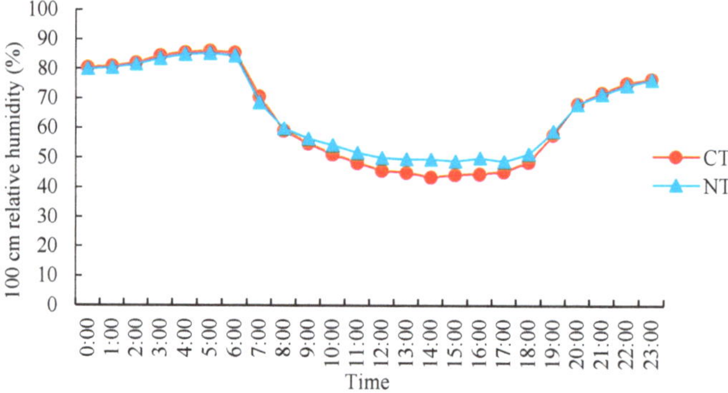

Figure 6. The average relative humidity of the air at 100 cm above the ground during the grain-filling stage (69–89) of winter wheat in the Beijing area.

3.3. Soil Moisture Content at 20 cm Below the Ground at the Early Stage of Wheat Grain-Filling (69–75)

During the early grain-filling period (15 days after flowering), in the absence of effective precipitation and artificial irrigation, the soil moisture content at 20 cm below the ground under both no-till direct seeding and conventional tillage methods showed a gradual decrease (Figure 7). On 14 May, the average soil moisture content at 20 cm below the ground decreased by 10.6% and 15.9%, respectively, compared to 6 May. The average soil moisture content at 20 cm below the ground under no-till direct seeding was greater than that under conventional tillage, increasing by 3.1% to 11.2%. From 6 May to 14 May and from 17 May to 20 May, the difference in the average soil moisture content at 20 cm below the ground between the two tillage methods showed a gradual increase. Compared to conventional tillage, the average soil moisture content of no-till direct seeding increased by 4.7%, 11.2%, 3.1% and 4.9% on 6 May, 14 May, 17 May and 20 May, respectively. The above results indicated that under the same irrigation period and amount of irrigation, no-till direct seeding was more conducive to maintaining soil moisture content in the tillage layer of wheat fields and reducing water evaporation.

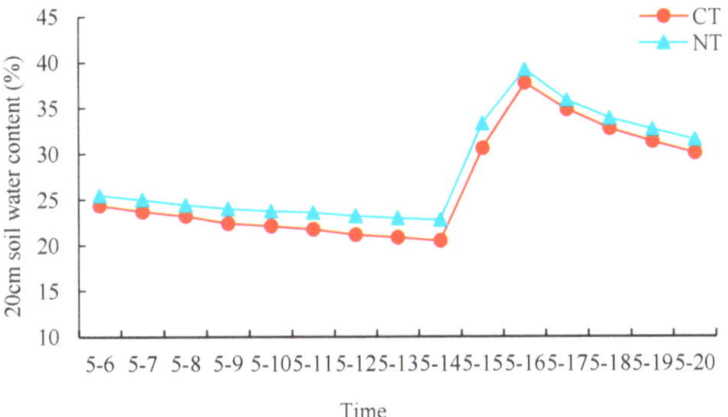

Figure 7. The average soil moisture content at 20 cm below the ground at the beginning of the grain-filling stage (69–75) of winter wheat in the Beijing area.

The watering began at 9:30 am on 15 May, and the irrigation amount was 30 mm.

3.4. Moisture Content of Wheat Plants and Spike Moisture Content Before Harvest (86)

The moisture content of wheat plants and spike moisture content were measured 7 days before harvest, and the results are shown in Table 1. The moisture content of wheat plants and spike moisture content under no-till direct seeding were significantly higher than those under conventional tillage. The dry weight of a single stem and single spike of wheat under no-till direct seeding were significantly greater than those under conventional tillage. In other words, at the end of the wheat grain-filling stage, the greenness retention and dry matter accumulation of wheat plants under no-till direct seeding were significantly better than those under conventional tillage, which is beneficial for grain-filling and, thus, increases grain weight.

Table 1. Moisture content of wheat plants and spike moisture content in the Beijing area.

Treatment	Plant Moisture Content (%)	Dry Weight per Stem (g)	Spike Moisture Content (%)	Dry Weight per Ear (g)
NT	49.35 ± 1.26 A	2.67 ± 0.04 a	42.02 ± 0.87 A	1.64 ± 0.04 a
CT	35.69 ± 0.85 B	2.49 ± 0.10 b	30.30 ± 1.03 B	1.55 ± 0.02 b

Different uppercase and lowercase letters in the same column indicate very significant ($p < 0.01$) and significant ($p < 0.05$) differences between treatments, respectively.

3.5. Wheat Yield and Yield Components

The grain yield of winter wheat under no-till direct seeding was higher than that under conventional tillage, although the difference was not significant (Table 2). The main reason that the wheat yield of the no-tillage direct seeding method was higher than that of conventional tillage was that the increase in thousand-grain weight and grain number per spike compensated for the yield reduction caused by the decrease in the number of harvested spikes. Specifically, the number of harvested spikes using no-till direct seeding was significantly lower than that of conventional tillage. In contrast, the thousand-grain weight was significantly higher under no-till direct seeding than that under conventional tillage, reflecting an increase of 7.9%, while the grain number of spike was higher than that under conventional tillage. The difference in the number of spikelets between the two tillage methods was very small. However, the number of sterile spikelets under no-till direct seeding was significantly lower than that under conventional tillage, with a decrease of 8.6%. This might be one of the reasons why no-till direct seeding had a higher number of grains per spike compared to conventional tillage.

Table 2. Wheat yield and yield components in the Beijing area.

Treatment	Spikelet Numbers	Numbers of Sterile Spikelet	Spike Numbers (10^4 ha^{-1})	Grain Numbers per Spike	Thousand-Grain Weight (g)	Grain Yield (kg ha^{-1})
NT	16.33 ± 0.55 a	2.45 ± 0.10 b	605 ± 19 b	29.98 ± 0.60 a	45.24 ± 0.18 a	7413 ± 233 a
CT	16.23 ± 0.40 a	2.68 ± 0.18 a	678 ± 14 a	29.08 ± 0.33 a	41.93 ± 0.44 b	7218 ± 232 a

Different lowercase letters in the same column indicate significant difference ($p < 0.05$) between treatments.

This study was carried out at one location throughout the course of a winter wheat growing season. Given that field trials are greatly affected by environmental and climatic conditions, further experiments are required to derive more conclusive and representative conclusions on the impact of no-till direct seeding and conventional tillage on the microclimate in wheat fields, as well as the influence of these microclimates on yield.

4. Discussion

4.1. Effects of Different Tillage Methods on Microclimate in Wheat Fields

The growth, development, and yield formation of wheat plants are closely related to the microclimate in the fields, and different tillage methods have varying impacts on the microclimate. During the later stages of winter wheat growth, straw mulching improves the above-ground microclimate, which is beneficial for increasing yield and water use efficiency [29]. In the later stages of maize growth, the air temperature in no-till maize fields is lower than that in conventionally tilled fields [30]. This study indicated that during the grain-filling stage of winter wheat, the average air temperature at 20 cm above the ground under no-till direct seeding was slightly higher than that under conventional tillage during the period from 1:00 to 5:00. However, it was lower than that of conventional tillage from 6:00 to 23:00, and especially significantly lower during the period from 8:00 to 18:00, with an average reduction of 8.1%. The average air temperature at 100 cm above the ground (canopy level) under no-till direct seeding was lower than that under conventional tilling, especially during the period from 10:00 to 18:00, when it was significantly lower than conventional tillage, with an average reduction of 6.1%. This was consistent with the findings of Irmak et al. [30], but contrary to the conclusion that no-till increases the air temperature above the canopy during the grain-filling stage of winter wheat compared to conventional tillage [31], which might be caused by differences in growth environments or climatic factors, especially differences in precipitation during the growth period of winter wheat. According to the research findings of Han et al. [31], the advantage of the no-till system is its ability to improve the vertical distribution of the population microclimate. Compared to conventional tillage, no-till practices enhance the relative humidity of the air in the upper, middle, and lower parts of the winter wheat canopy during the grain-filling period. In this study, the average relative humidity of the air at 20 cm above the ground under no-till direct seeding was higher than that under conventional tillage, especially during the period from 8:00 to 19:00, when it was significantly higher than conventional tillage, with an increase of 6.0% to 18.4%. The average relative humidity of the air at 100 cm above the ground under no-till direct seeding was slightly lower than that under conventional tillage during the period from 0:00 to 7:00 and from 20:00 to 23:00, while it was higher than conventional tillage from 8:00 to 19:00, with a significant increase from 10:00 to 18:00, with an average increase of 9.2%. This was consistent with the above research conclusions. This study also found that during the grain-filling stage of winter wheat, the variation range of the average air temperature and humidity at 20 cm and 100 cm above the ground under no-till direct seeding was smaller than that under conventional tillage (Figures 3–6). The variation trends of the average air temperature and humidity at 20 cm and 100 cm above the ground under the two tillage methods were basically the same, and the variation trends of the air temperature and relative humidity of the air were opposite, meaning that when the air temperature was high, the relative humidity of the air was low.

Straw mulching reduces the evapotranspiration of winter wheat from the seedling stage to the maturity stage, with the mulched treatment showing a reduction of 47.4 mm in evapotranspiration compared to the non-mulched treatment [29]. Compared to traditional tillage, no-till on the Loess Plateau drylands improves the efficiency of precipitation storage during the fallow period, and the soil available water during the wheat planting period is significantly higher under no-till direct seeding than under the traditional methods [6]. The soil moisture content in the tillage layer during the regreening stage, grain-filling stage, and harvest stage of winter wheat under less tillage and no-till increase by 4.49%, 6.86%, and 7.88%, respectively, compared to traditional tillage methods [32]. The results of this study indicated that under the same irrigation period and amount of irrigation, during the early grain-filling stages of winter wheat, the average soil moisture content at 20 cm below the ground under no-till direct seeding was higher than that under conventional tillage, with an increase of 3.1% to 11.2%. This was basically consistent with the above research conclusions and the findings of Feng et al. [14]. The average soil moisture content at 20 cm below the ground under both no-till direct seeding and conventional tillage methods showed a trend of gradual decrease, and the difference between the two methods showed a trend of gradual increase with the delay of the growth period, which was also confirmed by the research results of Du et al. [32]. The period from 10:00 to 18:00 is a critical time for the growth and development of wheat plants in a day. In this study, during the grain-filling stage, the average air temperature at 20 cm and 100 cm above the ground under no-till direct seeding was significantly lower than that under conventional tillage, while the average relative humidity of the air was significantly higher than that under conventional tillage. In addition, the soil moisture content at 20 cm below the ground under no-till direct seeding was higher than that under conventional tillage. In conclusion, the vertical microclimate in the field during the grain-filling stage under no-till direct seeding was more suitable for the growth and development of wheat plants than conventional tillage, and was conducive to maintaining the greenness of wheat plants in the later growth phase and grain-filling, increasing grain weight and, thus, increasing yield.

4.2. Effects of Different Tillage Methods on Wheat Yield

The grain yield of wheat is determined by the product of the number of harvested spikes, the number of grains per spike, and the thousand-grain weight. Compared with conventional tillage, when the water supply during the growing season of winter wheat is less than or equal to 650 mm, no-till combined with straw mulching increases wheat yield by 36.56% under dryland agricultural conditions [33]. The yield of wheat with less tillage or no-till increases by 10.3% compared to conventional tillage, among which the increase of grain number per spike and 1000-grain weight are the main factors contributing to the increased production of wheat [32]. The three-year experimental results of Yang et al. [34] indicate that the average yield of no-till wheat is 5.7% higher than that of conventional tillage, with statistical significance in two of the three years. This is due to the significantly higher thousand-grain weight of no-till wheat over the three years and the significantly higher number of grains per spike in two of the three years compared to conventional tillage. The number of spikes harvested under conventional tillage is significantly higher than that of no-till, while there are no significant differences in grains per spike and thousand-grain weight between the two tillage methods, and the yield of conventional tillage is significantly higher than that of no-till [31]. The results of this study demonstrated that the number of spikes harvested under no-till direct seeding was significantly lower than that under conventional tillage, with a decrease of 10.8%. This was consistent with the conclusions of Han et al. [31]. However, the thousand-grain weight under no-till direct seeding was significantly higher than that under conventional tillage, increasing by 7.9%, and the number of grains per spike was also higher, with an increase of 3.1%. The grain yield of wheat under no-till direct seeding was higher than that under conventional tillage, increasing by 2.7%, which was consistent with the research conclusions of Li et al. [17], Zhang et al. [18], Du et al. [32], Wu et al. [33], and Yang et al. [34], but opposite to the research conclusions of Li et al. [29] and Han et al. [31]. The reasons for this might include significant differences in soil fertility conditions among the experimental plots, substantial variations between different wheat varieties, or considerable discrepancies in the experimental design. In this study, the main reason why the yield of no-till direct seeding is higher than that of conventional tillage was that the combined increase in both grain number per spike and the thousand-grain weight (especially the thousand-grain weight) compensates for the reduction in yield caused by the decrease in the number of harvested spikes. This was basically consistent with the research conclusions of Li et al. [17], Du et al. [32], and Yang et al. [34]. Additionally, this study found that the number of sterile spikelets under no-till direct seeding was significantly lower than that under conventional tillage, with a decrease of 8.6%, which might be one of the main reasons why no-till direct seeding has a higher number of grains per spike compared to conventional tillage.

During the later growth stage of wheat, the dry matter accumulation of vegetative organs, such as stems and leaves, under the straw mulching treatment is significantly higher than that without mulching. The excessive growth of vegetative organs consumes a large amount of nutrients, resulting in a reduced allocation of photosynthetic products to grains [35]. During the harvest period, the dry weight per wheat plant under less tillage and no-till treatments increased by 0.35 g, and the dry weights of spikes, leaves, and stems increased by 13.3%, 10.5%, and 37.0%, respectively, compared to conventional tillage [32]. The results of this study indicated that seven days before the harvest of winter wheat, the dry weight per stem and the dry weight per spike under no-till direct seeding were significantly greater than those under conventional tillage, increasing by 6.1% and 6.4%, respectively. This was basically consistent with the research conclusion of Du et al. [32]. In the late grain-filling stage of winter wheat, no-till direct seeding not only increased the dry matter accumulation in vegetative organs, such as stems and leaves, but also enhanced the dry matter accumulation in the reproductive organ spikes, which was somewhat different from the conclusions of Li et al. [35]. In this study, seven days before the harvest of winter wheat, both the plant moisture content and the spike moisture content under no-till direct seeding were extremely significantly higher than those under conventional tillage, increasing by 38.3% and 38.7%, respectively. In other words, the greenness retention of plants under no-till direct seeding during the late grain-filling stage of winter wheat was significantly better than that of conventional tillage, effectively preventing the early senescence of wheat plants caused by meteorological factors, such as dry hot wind, drought, and high temperatures, which was beneficial for grain-filling and ultimately increased grain weight. This was consistent with the conclusion that soil moisture storage under conventional tillage is lower than that under no-till, leading to earlier senescence of corn plants in the later stages of growth [30]. No-till direct seeding can provide a suitable microclimate for wheat growth compared to conventional tillage. It not only maintains the yield of winter wheat but also reduces the costs associated with the sowing process, improves operational efficiency, and increases soil moisture content. This study provides valuable guidance for the implementation of conservation tillage practices in the NCP, where groundwater overextraction is a severe issue. It can effectively promote the development of sustainable agriculture in this region and beyond.

5. Conclusions

During the wheat grain-filling period, from 10:00 to 18:00, the average air temperature at 20 cm and 100 cm above the ground under no-till direct seeding was significantly lower than that of conventional tillage, while the average relative humidity of the air was significantly higher. At the beginning of grain-filling, the average soil moisture content at 20 cm below the ground under no-till direct seeding was higher than that of conventional tillage. Seven days before the wheat harvest, the moisture content of plants and spike under no-till direct seeding were both extremely significantly greater than those of conventional tillage, and the dry weight per plant and the dry weight per spike were significantly greater than those of conventional tillage. Consequently, the thousand-grain weight of wheat under no-till direct seeding was significantly higher than that of conventional tillage, increasing by 7.9%. The number of wheat sterile spikelets under no-till direct seeding was significantly lower than that under conventional tillage, with a decrease of 8.6%, and the number of grains per spike was higher than that of conventional tillage, with an increase of 3.1%. The yield of no-till direct seeding was 2.7% higher than that of conventional tillage.

Author Contributions: Conceptualization, C.Z. and Z.D.; methodology, Z.D., J.Z., S.Y. and C.Z.; validation, C.Z.; formal analyses, J.Z., Z.D., S.L. and S.Y.; investigation, Z.D., S.Y., S.L., P.F., J.W. and Y.L.; resources, C.Z. and X.W.; software, P.F.; data curation, Z.D., S.L., P.F., Y.L. and X.W.; writing— original draft, Z.D.; writing—review and editing, J.W., C.Z., J.Z. and S.Y.; funding acquisition, C.Z. and X.W.; supervision, C.Z., J.Z. and S.Y. All authors have read and agreed to the published version of the manuscript.

Funding: This research was supported by the Major Project of Yunan Science and Technology (202302AE09002002), the Science and Technology Project of Wuhu City (20231y05), and the Hebei Academy of Agriculture and Forestry Sciences Youth Science and Technology Talent Domestic Training Project (C24R0307).

Data Availability Statement: The original contributions presented in the study are included in the article; further inquiries can be directed to the corresponding authors.

Conflicts of Interest: The authors declare that they have no known competing financial interests or personal relationships that could have appeared to influence the work reported in this paper.

References

1. Li, Q.Q.; Dong, B.D.; Qiao, Y.Z.; Liu, M.Y.; Zhang, J.W. Root growth, available soil moisture, and moisture-use efficiency of winter wheat under different irrigation regimes applied at different growth stages in North China. *Agric. Water Manag.* **2010**, *97*, 1676–1682. [CrossRef]
2. Wang, Q.M.; Hu, F.L.; Chai, Q. Effect of conservation tillage on natural resources utilization efficiency and sustainability of integrated wheat-maize intercropping system. *Chin. J. Eco-Agric.* **2019**, *27*, 1344–1353. (In Chinese with English abstract) [CrossRef]
3. Bottinelli, N.; Angers, D.A.; Hallaire, V.; Michot, D.; Guillou, C.L. Tillage and fertilization practices affect soil aggregate stability in a humic cambisol of Northwest France. *Soil Tillage Res.* **2017**, *170*, 14–17. [CrossRef]
4. Wang, X.B.; Wu, H.J.; Dai, K.; Zhang, D.C.; Feng, Z.H.; Zhao, Q.S.; Wu, X.P.; Jin, K.; Cai, D.X.; Oenema, O.; et al. Tillage and crop residue effects on rainfed wheat and maize production in northern China. *Field Crop Res.* **2012**, *132*, 106–116. [CrossRef]
5. Ren, Z.J.; Han, X.J.; Feng, H.X.; Wang, L.F.; Ma, G.; Li, J.H.; Lv, J.J.; Tian, W.Z.; He, X.H.; Zhao, Y.N.; et al. Long-term conservation tillage improves soil stoichiometry balance and crop productivity based on a 17-year experiment in a semi-arid area of northern China. *Sci. Total Environ.* **2024**, *908*, 168283. [CrossRef]
6. Su, Z.Y.; Zhang, J.S.; Wu, W.L.; Cai, D.X.; Lv, J.J.; Jiang, G.H.; Huang, J.; Gao, J.; Hartmann, R.; Gabriels, D. Effects of conservation tillage practices on winter wheat water-use efficiency and crop yield on the Loess Plateau, China. *Agric. Water Manag.* **2007**, *87*, 307–314. [CrossRef]
7. Pittelkow, C.M.; Liang, X.Q.; Linquist, B.A.; Van Groenigen, K.J.; Lee, J.; Lundy, M.E.; Van, G.N.; Six, J.; Venterea, R.T.; Van, K.C. Productivity limits and potentials of the principles of conservation agriculture. *Nature* **2015**, *517*, 365–368. [CrossRef]
8. Brouder, S.M.; Gomez-Macpherson, H. The impact of conservation agriculture on smallholder agricultural yields: A scoping review of the evidence. *Agric. Ecosyst. Environ.* **2014**, *187*, 11–32. [CrossRef]
9. Giller, K.E.; Corbeels, M.; Nyamangara, J.; Triomphe, B.; Affholder, F.; Scopel, E.; Tittonell, P. A research agenda to explore the role of conservation agriculture in African smallholder farming systems. *Field Crop Res.* **2011**, *124*, 468–472. [CrossRef]
10. Zhang, H.L.; Lal, R.; Zhao, X.; Xue, J.F.; Chen, F. Opportunities and challenges of soil carbon sequestration by conservation agriculture in China. *Adv. Agron.* **2014**, *124*, 1–36. [CrossRef]

11. Lampurlanés, J.; Plaza-Bonilla, D.; Álvaro-Fuentes, J.; Cantero-Martínez, C. Long-term analysis of soil water conservation and crop yield under different tillage systems in Mediterranean rainfed conditions. *Field Crop Res.* **2016**, *189*, 59–67. [CrossRef]

12. Liu, S.; Zhang, X.Y.; Kravchenko, Y.; Iqbal, M.A. Maize (*Zea mays* L.) yield and soil properties as affected by no tillage in the black soils of China. *Acta Agric. Scand. Sect. B-Soil Plant Sci.* **2015**, *65*, 554–565. [CrossRef]

13. Schwartz, R.C.; Baumhardt, R.L.; Evett, S.R. Tillage effects on soil water redistribution and bare soil evaporation throughout a season. *Soil Tillage Res.* **2010**, *110*, 221–229. [CrossRef]

14. Feng, F.X.; Huang, G.B.; Yu, A.Z.; Chai, Q.; Tao, M.; Li, J. Effects of different conservation tillage measures on winter wheat water use in Wuwei oasis irrigated area. *Chin. J. Appl. Ecol.* **2009**, *20*, 1060–1065. (In Chinese with English abstract) [CrossRef]

15. Huang, Y.W.; Tao, B.; Zhu, X.C.; Yang, Y.J.; Liang, L.; Wang, L.X.; Pierre-Andre, J.; Tian, H.Q.; Ren, W. Conservation tillage increases corn and soybean water productivity across the Ohio River Basin. *Agric. Water Manag.* **2021**, *254*, 106962. [CrossRef]

16. Parihar, C.M.; Yadav, M.R.; Jat, S.L.; Singh, A.K.; Kumar, B.; Pooniya, V.; Pradhan, S.; Verma, R.K.; Jat, M.L.; Jat, R.K.; et al. Long-term conservation agriculture and intensified cropping systems: Effects on growth, yield, water, and energy-use efficiency of maize in northwestern India. *Pedosphere* **2018**, *28*, 952–963. [CrossRef]

17. Li, L.L.; Huang, G.B.; Qin, S.H.; Yu, A.Z. Effect of conservation tillage on dry matter accumulating and yield of winter wheat in oasis area. *Acta Agron. Sin.* **2011**, *37*, 514–520. (In Chinese with English abstract) [CrossRef]

18. Zhang, X.C.; Pan, X.; Wang, H.L. Research on contrast test of wheat no tillage seeding and deep plowing rotary tillage seeding. *J. Chin. Agric. Mech.* **2016**, *37*, 28–31. (In Chinese with English abstract) [CrossRef]

19. Teng, J.L.; Hou, R.X.; Dungait, J.A.J.; Zhou, G.Y.; Kuzyakov, Y.; Zhang, J.B.; Tian, J.; Cui, Z.L.; Zhang, F.S.; Delgado-Baquerizo, M. Conservation agriculture improves soil health and sustains crop yields after long-term warming. *Nat. Commun.* **2024**, *15*, 8785. [CrossRef]

20. Tian, S.Z.; Ning, T.Y.; Wang, Y.; Liu, Z.; Li, G.; Li, Z.J.; Lal, R. Crop yield and soil carbon responses to tillage method changes in North China. *Soil Tillage Res.* **2016**, *163*, 207–213. [CrossRef]

21. Cambron, T.W.; Deines, J.M.; Lopez, B.; Patel, R.; Liang, S.Z.; Lobell, D.B. Further adoption of conservation tillage can increase maize yields in the western US corn Belt. *Environ. Res. Lett.* **2024**, *19*, 054040. [CrossRef]

22. Mathers, C.; Heitman, J.; Huse, A.; Locke, A.; Osmond, D.; Woodley, A. No-till imparts yield stability and greater cumulative yield under variable weather conditions in the southeastern USA piedmont. *Field Crop Res.* **2023**, *292*, 108811. [CrossRef]

23. Lv, Y.J.; Zhang, X.L.; Gong, L.; Huang, S.B.; Sun, B.L.; Zheng, J.Y.; Wang, Y.J.; Wang, L.C. Long-term reduced and no tillage increase maize (*Zea mays* L.) yield and yield stability in Northeast China. *Eur. J. Agron.* **2024**, *158*, 127217. [CrossRef]

24. Jiang, F.H.; Xue, X.W.; Zhang, L.Y.; Zuo, Y.Y.; Zhang, H.; Zheng, W.; Bian, L.M.; Hu, L.L.; Hao, C.L.; Du, J.H.; et al. Soil wind erosion, nutrients, and crop yield response to con-servation tillage in North China: A field study in a semi-arid and wind erosion region after 9 years. *Field Crop Res.* **2024**, *316*, 109508. [CrossRef]

25. Guo, X.Y.; Wang, H.; Yu, Q.; Wang, R.; Wang, X.L.; Li, J. Effects of tillage on soil moisture and yield of wheat-maize rotation field in Weibei upland plateau. *Sci. Agric. Sin.* **2021**, *54*, 2977–2990. (In Chinese with English abstract) [CrossRef]

26. Kan, Z.R.; Liu, Q.Y.; He, C.; Jing, Z.H.; Virk, A.L.; Qi, J.Y.; Zhao, X.; Zhang, H.L. Responses of grain yield and water use efficiency of winter wheat to tillage in the North China Plain. *Field Crop Res.* **2020**, *249*, 107760. [CrossRef]

27. Wang, X.; Qi, J.Y.; Liu, B.Y.; Kan, Z.R.; Zhao, X.; Xiao, X.P.; Zhang, H.L. Strategic tillage effects on soil properties and agricultural productivity in the paddies of southern China. *Land Degrad. Dev.* **2020**, *31*, 1277–1286. [CrossRef]

28. Foley, J.A.; Ramankutty, N.; Brauman, K.A.; Cassidy, E.S.; Gerber, J.S.; Johnston, M.; Mueller, N.D.; O'Connell, C.; Ray, D.K.; West, P.C.; et al. Solutions for a cultivated planet. *Nature* **2011**, *478*, 337–342. [CrossRef]

29. Li, Q.Q.; Chen, Y.H.; Liu, M.Y.; Zhou, X.B.; Yu, S.L.; Dong, B.D. Effects of irrigation and straw mulching on microclimate characteristics and water use efficiency of winter wheat in North China. *Plant Prod. Sci.* **2008**, *11*, 161–170. [CrossRef]

30. Irmak, S.; Kukal, M.S.; Mohammed, A.T.; Djaman, K. Disk-till vs. no-till maize evapotranspiration, microclimate, grain yield, production functions and water productivity. *Agric. Water Manag.* **2019**, *216*, 177–195. [CrossRef]

31. Han, H.; Ning, T.; Li, Z. Effects of tillage and weed management on the vertical distribution of microclimate and grain yield in a winter wheat field. *Plant Soil Environ.* **2013**, *59*, 201–207. [CrossRef]

32. Du, J.L.; Hou, H.P.; Lu, D.Q.; Li, B.; Xie, J.; Yang, Y.A.; Tian, H. Effects of less and no tillage on soil physicochemical properties and wheat yield. *China Agric. Technol. Ext.* **2020**, *36*, 69–71. (In Chinese with English abstract)

33. Wu, B.Y.; Ma, D.K.; Shi, Y.; Zuo, G.Q.; Chang, F.; Sun, M.Q.; Yin, L.N.; Wang, S.W. Optimizing tillage practice based on water supply during the growing season in wheat and maize production in northern China. *Agric. Water Manag.* **2024**, *300*, 108923. [CrossRef]

34. Yang, Y.H.; Ding, J.L.; Zhang, Y.H.; Wu, J.C.; Zhang, J.M.; Pan, X.Y.; Gao, C.M.; Wang, Y.; He, F. Effects of tillage and mulching measures on soil moisture and temperature, photosynthetic characteristics and yield of winter wheat. *Agric. Water Manag.* **2018**, *201*, 299–308. [CrossRef]

35. Li, Q.Q.; Chen, Y.H.; Wu, W.; Yu, S.Z.; Zhou, X.B.; Dong, Q.Y.; Yu, S.L. Effects of straw mulching and irrigation on solar energy utilization efficiency of winter wheat farmland. *Chin. J. Appl. Ecol.* **2006**, *17*, 243–246. (In Chinese with English abstract) [CrossRef]

 *agronomy*

Article

Performance and Stability for Grain Yield and Its Components of Some Rice Cultivars under Various Environments

Mohamed S. Abd El-Aty [1], Mahmoud I. Abo-Youssef [2], Fouad A. Sorour [1], Mahmoud Salem [2], Mohamed A. Gomma [2], Omar M. Ibrahim [3], Mohammad Yaghoubi Khanghahi [4], Wahidah H. Al-Qahtani [5], Mostafa A. Abdel-Maksoud [6] and Amira M. El-Tahan [3,*]

[1] Agronomy Department, Faculty of Agriculture, Kafr Elsheikh University, Kafr El-Sheikh 33516, Egypt
[2] Rice Research Department, Field Crops Research Institute, Agriculture Research Center, Giza 12619, Egypt
[3] Plant Production Department, Arid Lands Cultivation Research Institute, The City of Scientific Research and Technological Applications, SRTA City, New Borg El-Arab City 21934, Egypt
[4] Department of Soil, Plant and Food Sciences (DiSSPA), University of Bari Aldo Moro, 70126 Bari, Italy
[5] Department of Food Sciences & Nutrition, College of Food and Agricultural Sciences, King Saud University, P.O. Box 270677, Riyadh 11352, Saudi Arabia
[6] Botany and Microbiology Department, College of Science, King Saud University, P.O. Box. 2460, Riyadh 11451, Saudi Arabia
* Correspondence: aeltahan@srtacity.sci.eg

Abstract: Refine current agricultural practices considering environmental changes are crucial for finding tolerant rice varieties that can meet the demands of human consumption. To this end, stability analysis assesses a crop genotype's ability to adapt to various conditions. Therefore, the objective of this study was to (1) examine the interaction between rice genotypes and environmental conditions; (2) evaluate the stability of twelve rice genotypes using various stability methods; (3) identify representative environments for multi-environment testing; and (4) determine superior genotypes for specific environments. The evaluated rice cultivars were Sakha 101, Sakha 104, Sakha 105, Sakha 106, Sakha 107, Sakha 108, Giza 177, Giza 178, Giza 179, Giza 182, Egyptian Yasmine, and Sakha super 300. The experiment followed a strip-plot design, with three replications. The findings revealed significant differences among the rice varieties across various environments for the majority of the assessed characteristics. The joint regression analysis of variance demonstrated highly significant differences among rice cultivars for all the studied traits in terms of genotype-by-environment interaction (G × E). The statistical significance of the interaction between genetic and environmental factors was evident for all variables demonstrating heritable variation among the rice cultivars, specifically Sakha 108, Sakha 104, Giza 177, and Giza 178, concerning grain yield per feddan. These rice cultivars exhibited stability parameters that were not significantly different from unity for the regression coefficient (b_i) and from zero for the deviations from regression (S^2d_i) for those traits. Overall, stability criteria are essential for ensuring reliable rice production, meeting human consumption, advancing genetic improvement, and promoting environmental sustainability in agriculture.

Keywords: stability; rice varieties; environment; grain yield

Citation: El-Aty, M.S.A.; Abo-Youssef, M.I.; Sorour, F.A.; Salem, M.; Gomma, M.A.; Ibrahim, O.M.; Yaghoubi Khanghahi, M.; Al-Qahtani, W.H.; Abdel-Maksoud, M.A.; El-Tahan, A.M. Performance and Stability for Grain Yield and Its Components of Some Rice Cultivars under Various Environments. *Agronomy* 2024, 14, 2137. https://doi.org/10.3390/agronomy14092137

Academic Editors: Mariola Staniak, Małgorzata Szostek and Ewa Szpunar-Krok

Received: 23 August 2024
Revised: 11 September 2024
Accepted: 17 September 2024
Published: 19 September 2024

1. Introduction

Rice, a vital cereal that is cultivated as a staple food crop, faces challenges in meeting human consumption demands, which are expected to increase by an additional 112 million metric tons by the year 2035 [1]. To address this growing need, it is imperative to optimize existing farming methods, manage water resources effectively, and explore opportunities for developing new rice varieties [2].

Rice cultivation spans various agroecology and cropping systems, including rain-fed highland and lowland areas, irrigated fields, and mangrove ecosystems [3]. While upland rice constitutes a relatively minor portion (approximately 11%) of global rice output, it holds significant importance in local rice production in both arid and semiarid regions [4,5]. The challenges posed by limited water resources for lowland rice agriculture are exacerbated by the impacts of climate change. In response, upland rice emerges as a potentially sustainable solution to tackle food security concerns, as suggested by [3,5,6].

Before commercial release, plant breeders typically evaluate the performance of the new varieties across various environments to assess their stability and consistency [7]. Genotypes that demonstrate consistent performance across diverse environments are preferred, but the complexity of genotype–environment interactions poses challenges in genotype selection [8–10]. Therefore, it is crucial to employ statistical methods to evaluate genotype stability before commercial release [11,12]. To this end, the applied statistical methods for assessing genotype stability can be classified into parametric and nonparametric stability statistics [13,14]. Parametric stability statistics encompass various univariate and multivariate methods. Univariate methods comprise techniques such as Wricke's equivalence (Wi2) [15], Shukla's stability variance (S^2_Shu.) [16], coefficient of variance (CV) [17], Environmental variance (S^2_Env.) [18], Mean-variance component (q) [19], GE variance component (q') [20], and regression coefficient (b_i) [21], among others. On the other hand, multivariate approaches include the additive main effects and multiplicative interaction (AMMI) model [22] and the GGE biplot method [23], which can accurately predict genotype-by-environment interactions and identify optimal genotypes across diverse testing environments [24].

The objectives of this study are to (1) study the interaction of genotypes and the environment; (2) determine the stability of rice genotypes with various stability methods; (3) determine representative environments for multi-environment testing; (4) determine superior genotypes in specific environments; and (5) identify high-yielding rice genotypes with consistent performance across six distinct locations over two consecutive years (2020–2021) (three sowing dates during two seasons). This is achieved by evaluating the effectiveness of various univariate and multivariate stability criteria. The identification of high-yielding rice genotypes is essential for ensuring food security and enabling adaptation to climate change in rice-producing regions worldwide.

2. Materials and Methods

This study, conducted at the rice research farm Sakha Agricultural station in Kafr Elsheikh, Egypt, aimed to assess 12 rice cultivars over the 2020 and 2021 seasons, which were subjected to three different sowing dates.

2.1. Rice Cultivars and Experimental Design

Twelve rice cultivars, namely Sakha 101, Sakha 104, Sakha 105, Sakha 106, Sakha 107, Sakha 108, Giza 177, Giza 178, Giza 179, Giza 182, Egyptian Yasmine, and Sakha super 300, were evaluated. The experiment was set using a split-plot design with three replications. Sowing dates were allocated in the main plots, and rice varieties were applied in the sub-plots. The plot size was 2 m × 5 m (10 m^2).

The chemical and mechanical analyses of the soil and organic matter were conducted following the methods described by [25] at the Agricultural Research Center, Ministry of Agriculture, Egypt. The chemical and physical properties of the soil at the experimental site, measured at a depth of 0–30 cm, are shown in Table 1.

Table 1. Mechanical and chemical soil analyses of the experimental site during the 2020 and 2021 seasons.

Soil Properties	2020	2021
Mechanical:		
Clay %	57.22	56.56
Silt %	28.20	27.31
Sand %	14.58	16.13
Texture	Clayey	Clayey

Table 1. *Cont.*

Soil Properties	2020	2021
Chemical:		
Organic Matter (O.M)%	1.54	1.52
pH (1:2.5 soil suspension)	8.34	8.45
Ec (ds.m^{-1})	3.34	3.20
Total N (ppm)	467.50	464
Available P (ppm)	14.69	14.32
Available K (ppm)	390	381
Soluble anions, meq.L^{-1}:		
HCO_3^-	6.38	6.20
Cl^-	9.25	9.10
Soluble Cations, meq.L^{-1}:		
Ca^{2+}	10.86	10.72
Mg^{2+}	5.15	5.13
Na^+	2.13	2.09
K^+	15.71	15.05
Total carbonate %	14.23	14.14

Weather data for the two seasons were collected from the Agriculture Research Center, Field Crops Research Institute, and Rice Research and Training Center in Sakha, Kafr El-Sheikh, Egypt (Table 2).

Table 2. The monthly minimum and maximum air temperature ($^\circ$C) and relative humidity (%) at Sakha Agricultural research station during the 2020 and 2021 rice growing seasons.

Month	Day	2020					2021				
		Air Temp.		RH %		Rain	Air Temp.		RH %		Rain
		Min	Max	7.30	13.30	(mm/day)	Min	Max	7.30	13.30	(mm/day)
	1–10	20.98	28.89	76.0	46.9	0	25.4	30.7	77.3	47.7	0
May	11–20	24.36	32.93	66.6	41.5	0	26.3	30.3	78.0	44.3	0
	21–31	22.93	29.72	71.5	48.9	0	25.8	30.9	77.9	44.7	0
	1–10	26.7	33.4	67.3	43.5	0	27.7	32.6	78.7	46.7	0
June	11–20	25.5	33.1	76.0	44.3	0	27.8	31.9	81.5	54.2	0
	21–30	26.8	34.2	83.9	52.1	0	28.8	32.9	80.6	53.2	0
	1–10	26.5	33.9	82.5	57.7	0	29.4	34.7	84.6	57.2	0
July	11–20	25.9	33.7	83.0	54.2	0	29.1	35.2	83.4	54.9	0
	21–31	26.0	32.9	81.7	58.6	0	28.4	34.4	85.6	59.1	0
	1–10	26.2	34.3	84.8	57.6	0	28.8	34.2	86.8	57.2	0
August	11–20	26.3	33.0	82.7	56.8	0	29.2	34.9	84.6	58.3	0
	21–31	25.5	36.2	85.0	54.4	0	27.4	35.6	86.2	50.5	0
	1–10	24.3	33.27	85.4	53.7	0	25.9	33.1	86.0	49.8	0
September	11–20	24.93	33.42	82.6	51.3	0	26.0	33.6	87.1	49.5	0
	21–30	23.63	31.09	82.2	50.1	0	26.7	31.5	85.0	51.7	0
	1–10	20.3	31.5	85.6	50.0	0	24.5	30.0	83.3	53.1	0
October	11–20	23.0	29.84	78.4	56.0	0	23.6	27.9	81.5	56.6	0
	21–31	21.90	28.04	83.2	59.8	0	23.9	28.1	78.5	54.6	0

2.2. Field Preparation and Planting

The experiment was carried out in plots with seven rows, each 5 m in length, and a plant spacing of 20 m × 20 cm. Bed nursery and permanent field preparation followed the guidelines provided by the Rice Research and Training Center (RRTC). Phosphorus fertilizer in the form of mono-super phosphate (15%) at the rate of 36 kg P_2O_5/ha and potassium in the form of potassium sulfate (48% K_2O) at the rate of 57 kg K_2O/ha were applied during land preparation. Nitrogen in the form of urea (46.5% N) was applied in two splits, i.e., two-thirds as basal application incorporated into the soil immediately before flooding and one-third after 30 days from the first dose.

2.3. Transplantation and Spacing

Seedlings were transplanted from the nursery to the permanent field after 25 days.

Transplantation was performed into sub-sub-plots using plant spacing (20×20, cm) for distances between hills and rows. Seedlings were transplanted into hills at a density of 3–4 seedlings per hill with the following schedule: E1 = the first sowing date in the first season 20 April 2020; E2 = the second sowing date in the first season 10 May 2020; E3 = the third sowing date in the first season 30 May 2020; E4 = the first sowing date in the second season 20 April 2021; E5 = the second sowing date in the second season 10 May 2021; and E6 = the third sowing date in the second season 30 May 2021.

2.4. Studied Characteristics

This study proposed the evaluation of additional attributes, including the number of filled spikelets per panicle, seed set percentage, 1000-grain weight, and grain yield per plant. Overall, this study utilized a rigorous experimental design and methodology to evaluate the performance of different rice cultivars under varying sowing dates during the two seasons, providing valuable insights into rice production in the region.

2.5. Statistical Analyses

The stability and adaptability of rice genotypes were determined using parametric and nonparametric stability in addition to multivariate analysis. Parametric stability included linear regression coefficient (b_i) and AMMI stability indexes, whereas nonparametric stability included S_i. Multivariate analysis was performed using AMMI and GGE biplot.

Stability studies were conducted using the Eberhart and Russell model [26] with *metan* package in R software 2021, with particular attention to the $G \times E$ factor. A genotype with a regression coefficient of 1 ($b_i = 1$) and a deviation not statistically significant from zero ($S^2d_i = 0$) is deemed stable, exhibiting a unity response.

Mean comparisons among genotypes were conducted using the LSD test, employing a significance threshold of 5%. To calculate variance components, ANOVA was performed using a split-plot design. The GenStat statistical tool (12th edition) facilitated this analysis. A Bayesian network is a probabilistic directed acyclic graph, where nodes represent variables such as traits and edges represent causal or conditional dependencies among these variables. The aim of analysis algorithms in Bayesian networks is to infer the relations of conditional or causal dependence among network variables based on the Bayes theorem. So, Bayesian networks belong to the class of structural network models. Bayesian network was built using the package *bnlearn* in R software.

3. Results

Substantial variations were observed in the mean square values for the number of filled grains per panicle, seed set percentage, 1000-grain weight, and grain yield per panicle among the different Egyptian rice types and sowing dates (Table 3). Variations in the habitats and genotypes had a significant impact on all the tested traits. Moreover, clear interactions were observed between years and environments, as well as between genotypes and years, across all traits except for the 1000-grain weight, where the disparity was not statistically significant. Furthermore, the analysis revealed a substantial and significant mean square of interaction across genotypes, years, and environments for the number of filled grains per panicle and seed set percentage (Table 3). The results demonstrated that the investigated genotypes had varying responses to environmental variables, highlighting the significance of evaluating genotypes across multiple environments to identify the most well-suited genotypes for specific situations. The environmental mean squares were the primary contributor to the overall mean squares for all traits. Furthermore, the variations attributed to different environments were greater than the variations resulting from the interactions between genotypes and environments for all the observed traits. Hence, the primary factor influencing the variations in the performance of rice genotypes in these trials was the environment, rather than the differences arising from genotype-by-environment interaction.

Table 3. Mean square of seed set (%), 1000-grain weight, and grain yield/plant of the 12 rice cultivars.

SOV	Df	MS			
		Number of Filled Grains/Panicle	Seed Set (%)	1000-Grain Weight	Grain Yield/Plant
Years (Y)	1	87.53 **	296.83 **	1.38	343.27 **
Date (D)	2	2420.52 **	590.29 **	5.59 **	204.71 **
Y × D	2	53.78 **	132.24 **	0.00	21.80 **
Rep/D/Y = (Ea)	9	25	1.0	0.50	2.0
Genotypes (G)	11	3310.30 **	71.60 **	71.22 **	18.92 **
G × Y	11	105.81 **	15.30 **	0.05	3.84 **
G × D	22	129.95 **	9.21 **	3.01 **	0.66
G × Y×D	22	37.35 **	8.13 **	0.05	1.36
Error/D/years (Pooled Error) = (Eb)	135	14.38	0.71	0.04	0.51

** indicate $p < 0.01$, respectively.

3.1. Environment Effect

The mean number of filled grains per panicle decreased by 11.6 grains when comparing the sowing dates condition to the third sowing date condition (Table 4). When comparing the second and third sowing dates to the first sowing date, the seed set percentage, 1000-grain weight (in grams), and grain yield per plant were lower. The results demonstrated that the diverse growing conditions led to a significant difference in average yields, ranging from 45.0 g/plant under favorable initial planting date conditions to 42.5 g/plant under the third stress condition (Table 5).

Table 4. Means of the 12 rice genotypes over years and environments.

Item	No. of Filled Grains/Panicle	Seed Set %	1000-Grain Weight (g)	Grain Yield/Plant (g)
First year	137.5	93.9	27.3	45.6
Second year	136.2	91.6	27.4	43.1
Env. 1	142.9	95.8	27.7	45.8
Env. 2	136.2	92.4	27.3	44.8
Env. 3	131.3	90.1	27.1	42.5
Mean overall	136.82	92.76	27.36	44.36
LSD at 0.05	1.54	0.33	0.07	0.43

Table 5. AMMI analysis for the 12 rice genotypes.

Source	Df	Sum Sq	Mean Sq	F Value	Pr (>F)	Proportion	Accumulated
ENV	5	7.96×10^2	1.59×10^2	111.18	1.27×10^{-9}	-	-
REP(ENV)	12	1.72×10^1	1.43×10^0	2.77	2.18×10^{-3}	-	-
GEN	11	2.08×10^2	1.89×10^1	36.56	8.63×10^{-35}	-	-
GEN:ENV	55	8.67×10^1	1.58×10^0	3.05	1.01×10^{-7}	-	-
PC1	15	4.79×10^1	3.20×10^0	6.18	0.00×10^0	55.3	55.3
PC2	13	2.25×10^1	1.73×10^0	3.34	2.00×10^{-4}	25.9	81.2
PC3	11	1.19×10^1	1.08×10^0	2.09	2.52×10^{-2}	13.7	94.9
PC4	9	4.38×10^0	4.87×10^{-1}	0.94	4.93×10^{-1}	5.1	100
PC5	7	4.07×10^{-3}	5.80×10^{-4}	0	1.00×10^0	0	100
Residuals	132	6.83×10^1	5.18×10^{-1}	-	-	-	-
Total	270	1.26×10^3	4.68×10^0	-	-	-	-

The AMMI 1 biplot (Figure 1a) represents the relationship between grain yield and IPCA 1. In this plot, rice genotypes that fall on the same vertical line indicate similarity in yield, while those on the same horizontal line indicate similarity in interaction pattern. According to the AMMI 1 biplot, the rice genotypes Sakha104, Sakha105, Sakha106, Sakha108, Giza177, Giza178, and SR300 exhibited a high level of stability in grain production, and they are considered extensively adapted lines (Figure 1a). The rice genotypes Sakha101, Sakha107, Giza179, Giza182, and Egyptian Yasmine exhibited relatively low stability in grain yield due to their considerable distance from the origin and their specific adaptability to particular environments. The genotype S101 exhibited superior performance in the E1 and E2 environments, while the genotypes G182 and EY were specifically adapted to environments E3 and E5, respectively. Furthermore, the S106, S108, and G178 genotypes were specifically identified as being well suited to environment E4. In the AMMI 2 biplot (Figure 1b), the scores of IPCA1 plotted against IPCA2 show the stability of the genotypes across different environments. Environments E1, E2, and E4 had relatively small radii and exhibited weak interacting forces, whereas environments E3, E5, E6, and E7 had long radii and exerted substantial interaction. Genotypes S104, S105, and G178 were located in close proximity to the origin, indicating their insensitivity to environmental interaction forces. Conversely, the remaining genotypes were situated further from the zero line, signifying their heightened responsiveness. The genotypes S106 and EY were found to be the most well suited for site E6, whereas the genotypes S108 and SR300 showed a close similarity in environment E4.

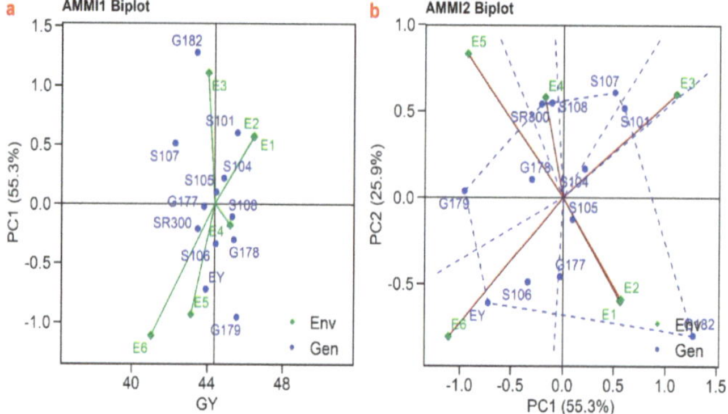

Figure 1. (**a**) AMMI 1 (additive main effects and multiplicative interaction) biplot for grain yield of the 12 rice genotypes and 6 environments using genotypic and environmental scores; (**b**) AMMI 2 biplot for grain yield showing the interaction of IPCA1 (interaction principal component axes) against IPCA2 scores of the 12 rice genotypes in 6 environments.

3.2. Mean Performance

The average number of filled grains per panicle varied between 121.06 for Giza 179 and 168.0 grains for Giza 178, as shown in Table 6. The seed set percentages of the twelve rice genotypes exhibited substantial variations, as indicated by the data presented in Table 6. The seed set percentage was highest for Sakha 101 and lowest for E. Yasmin.

The mean 1000-grain weight was highest for Sakha 106, Sakha 105, and Sakha 108, with values of 29.1, 29.0, and 29.0, respectively. The genotype Giza 178 (22.1) had the lowest value among all genotypes (Table 7). The data presented clearly demonstrate that the genotypes had substantial variations in grain production per plant, which were statistically significant (Table 7). The Sakha 108, Giza 179, and Giza 178 varieties exhibited the highest average weights per plant, measuring 45.6, 45.5, and 45.4 g, respectively. The genotype Sakha 107 had the lowest yield, measuring 42.3 g per plant. The grain production per plant varied between 42.3 and 45.6 g per plant across the different genotypes. The analysis of variations was conducted on the combined data from six different environments, comprising three sowing dates and two seasons. Tables 6 and 7 demonstrate notable variations in features across different contexts. The genotypic variation (s2g) was greater than the error variation (s2e) for the number of filled grains, seed set percentage, 1000-grain weight, and grain yield per plant. This indicates that environmental effects have little influence on these traits. Conversely, the s2g was significantly larger than s2 for all traits, indicating that genetic effects play a crucial role in the expression of these traits (Table 7).

Table 6. The average number of filled grains/panicle and seed sets for the twelve rice cultivars in different environments.

Genotypes	No. of Filled Grains/Panicle							Seed Set %						
	E1	E2	E3	E4	E5	E6	Mean	E1	E2	E3	E4	E5	E6	Mean
Sakha 101	144.3	142.0	134.0	145.3	139.0	135.0	139.94	97.7	96.9	94.8	97.9	93.2	90.5	95.2
Sakha 104	141.7	138.0	130.3	140.4	133.3	129.0	135.46	97.3	96.3	94.4	96.7	93.7	90.4	94.8
Sakha105	130.0	124.3	118.3	129.7	124.0	119.0	'124.22	96.7	96.3	93.7	98.6	92.7	88.8	94.5
Sakha 106	127.3	128.3	123.3	131.0	129.0	126.0	127.50	95.1	94.1	91.5	94.6	94.6	93.8	93.9
Sakha 107	126.0	123.3	120.0	127.0	127.3	123.5	124.51	93.3	92.6	90.5	93.8	89.8	87.6	91.3
Sakha 108	152.7	149.0	143.3	151.0	147.0	140.4	147.23	95.5	94.7	93.1	94.7	90.7	86.7	92.6
Giza177	141.7	136.7	129.3	142.7	133.3	128.1	135.30	97.4	96.6	94.3	99.5	90.6	87.1	94.3
Giza178	188.3	163.3	173.0	189.3	150.0	144.0	168.00	95.8	94.2	92.2	98.9	86.2	81.4	91.4
Giza179	129.7	124.3	120.3	130.7	128.3	125.1	126.41	96.7	93.8	93.7	97.3	94.5	91.9	94.7
Giza 182	123.7	117.7	115.3	124.7	124.0	121.0	121.06	91.5	90.7	88.5	92.8	90.4	88.0	90.3
Egyptian Yasmine	149.3	152.7	138.0	150.3	143.7	135.3	144.89	93.2	91.2	89.8	93.8	88.0	83.5	89.9
Sakha super 300	155.7	150.3	142.7	156.7	141.0	137.3	147.3	94.7	92.2	90.8	95.1	84.6	85.1	90.4
Sakha 101	142.53	137.49	132.32	143.23	134.99	130.31	136.82	95.41	94.13	92.28	96.14	90.75	87.90	92.78
LSD 0.05	1.3	15.1	1.7	1.52	3.64	2.01	0.14	0.14	1.32	0.11	0.93	2.40	2.01	0.55
LSD 0.01	1.7	20.6	2.4	2.06	4.95	2.73	0.19	0.19	1.79	0.15	1.27	3.26	2.74	0.72

Table 7. Average 1000-grain weight and grain yield/plant for the twelve rice cultivars over environments.

Genotypes	1000-Grain Weight							Grain Yield/Plant						
	E1	E2	E3	E4	E5	E6	Mean	E1	E2	E3	E4	E5	E6	Mean
Sakha 101	28.6	28.4	28.3	28.8	28.6	28.5	28.5	47	47	44.9	46.6	44.3	41.0	45.0
Sakha 104	28.3	28.2	27.7	28.5	28.4	27.9	28.2	47.1	47.1	44.2	46.8	43.1	40.9	44.9
Sakha105	28.1	28.1	28.9	28.3	28.2	29.0	28.4	46.8	46.8	43.8	45.2	43.2	40.8	44.4
Sakha 106	28.3	28.5	28.9	28.5	28.7	29.1	28.6	46.9	46.8	43.2	44.7	43.5	41.6	44.4
Sakha 107	27.5	27.4	27.2	27.7	27.5	27.3	27.4	44.0	44.0	43.5	42.7	41.1	38.4	42.3
Sakha 108	29.1	29.0	28.9	29.3	29.2	29.0	29.1	47.8	47.8	46.9	46.6	44.7	41.5	45.6
Giza177	28.3	28.3	27.9	28.5	28.4	28.1	28.2	46.0	46.0	43.7	43.8	42.3	41.2	43.8
Giza178	22.1	22.0	21.9	22.2	22.2	22.1	22.1	47.3	47.3	44.7	46.1	44.7	42.2	45.4
Giza179	27.5	27.5	27.1	27.7	27.6	27.2	27.4	47.0	47.0	44.1	46.9	45.0	43.3	45.5
Giza 182	27.3	24.2	22.6	27.5	24.4	22.8	24.8	46.6	46.6	44.0	43.7	40.2	39.3	43.4
Egyptian Yasmine	27.7	27.4	27.3	27.9	27.6	27.4	27.5	45.8	45.8	42.6	44.6	42.6	42.1	43.9
Sakha super 300	28.1	27.9	27.8	28.2	28.1	28.0	28.0	45.0	45.0	43.5	44.5	42.8	40.2	43.5
Sakha 101	27.58	27.24	27.04	27.76	27.41	27.20	27.35	46.45	46.44	44.02	44.18	43.13	41.04	44.37
LSD 0.05	0.1	0.5	0.1	0.11	0.55	0.11	0.13	1.1	1.1	0.9	1.28	1.51	1.40	0.47
LSD 0.01	0.2	0.7	0.2	0.16	0.74	0.15	0.17	1.4	1.4	1.2	1.74	2.05	1.91	0.61

The acquired results are consistent with those reported by [4]. The phenotypic (PCV) and genotypic (GCV) coefficients of variability were estimated and found to have modest variances for all traits evaluated across the six environments (Table 8). It is noteworthy that certain rice cultivars consistently exhibited similar characteristics across various conditions, including the number of panicles per plant, panicle weight, the number of filled grains per panicle, seed set percentage, 1000-grain weight, and grain yield per plant. However, certain types of rice varied in performance across different environments. For example, the rice cultivar Sakha 108 showed superiority in terms of the number of filled grains per panicle, 1000-grain weight, and grain weight per panicle in environment E1. On the other hand, Giza 178 performed better in terms of the number of filled grains per panicle, while Sakha 101 had a higher seed set percentage.

Table 8. AMMI-based stability 12 indexes (low numbers indicate stability) and superiority index based on stability and yield (high numbers indicate superiority).

GEN	GY	ASTAB	ASI	ASV	AVAMGE	DA	DZ	EV	FA	MASI	MASV	SIPC	ZA	SI
S101	45.6	0.64	0.36	1.37	2.63	1.48	0.44	0.07	2.18	0.36	1.6	1.26	0.26	77.57
S104	44.9	0.85	0.13	0.48	2.85	1.34	0.64	0.14	1.8	0.17	1.04	1.26	0.17	78.82
S105	44.4	0.11	0.06	0.24	1.2	0.49	0.22	0.02	0.24	0.07	0.43	0.51	0.07	82.82
S106	44.4	0.36	0.23	0.88	1.96	1.06	0.34	0.04	1.13	0.23	1.18	0.87	0.18	69.20
S107	42.3	1.18	0.32	1.23	3.35	1.77	0.69	0.16	3.13	0.34	1.74	1.86	0.31	23.47
S108	45.3	0.35	0.15	0.59	2.16	0.97	0.37	0.04	0.95	0.16	1.07	0.87	0.14	89.28
G177	43.8	0.62	0.12	0.47	2.21	1.18	0.53	0.09	1.39	0.15	1.08	1.13	0.14	68.08
G178	45.4	0.11	0.17	0.66	1.46	0.64	0.17	0.01	0.4	0.17	0.68	0.45	0.1	92.19
G179	45.5	0.95	0.53	2.05	4.4	1.93	0.49	0.08	3.73	0.53	2.05	1.15	0.29	71.91
G182	43.4	2.26	0.73	2.83	6.6	2.87	0.8	0.21	8.24	0.73	3.11	2.14	0.48	17.20
E.Y.	43.9	0.92	0.43	1.66	3.1	1.78	0.53	0.09	3.16	0.43	1.93	1.45	0.31	45.92
SR300	43.5	0.4	0.18	0.7	2.17	1.05	0.39	0.05	1.1	0.19	1.14	1	0.17	57.95

AMMI-based stability parameter (ASTAB), AMMI stability index (ASI), AMMI stability value (ASV), sum across environments of the absolute value of GEIs modeled by AMMI (AVAMGE), Annicchiarico's D parameter values (DA), Zhang's D parameter (DZ), sums of the averages of the squared eigenvector values (EV), stability measure based on fitted AMMI model (FA), modified AMMI stability index (MASI), modified AMMI stability value (MASV), sums of the absolute value of the IPC scores (SIPC), absolute value of the relative contribution of IPCs to the interaction (Za), superiority index (SI).

3.3. Stability Parameters

The predictability of genotypes for the number of filled grains per panicle varied from 0.04 for Sakha 106 to 1.37 for Sakha super 300 (Table 9). The examined genotypes were classified into three categories based on the regression coefficient (b_i) values. The initial group consisted of the most genetically stable genotypes, specifically Sakha 107, which exhibited the coefficient of regression b_i values equal to 1. The second group consisted of genotypes that were better suited to unfavorable environments, namely Sakha 104, Sakha 106, and Giza 178 (with the lowest b_i values). On the other hand, genotypes Sakha 101, Sakha 105, Sakha 108, Giza 177, Giza 179, Giza 182, E. Yasmin, and Sakha super 300 were sensitive to input and adapted to high-potential environments. Regarding the grain yield, the rice cultivars Sakha 106, Sakha 107, Sakha 108, and Giza 178 had a bi value close to 1 (Table 9), suggesting that these cultivars were well suited to varied environmental conditions. Genotypes Giza 177, E. Yasmin, and Sakha super 300 exhibited b_i values below 1, showing their suitability for unfavorable conditions. Sakha 101, Sakha 104, Sakha 105, and Giza 182 displayed a b_i value greater than 1.

Table 9. Mean performance and phenotypic stability measurements of the number of filled grains/panicle and seed set (%) for the twelve genotypes in three environments.

Genotypes	Number of Filled/Grains Panicle		Seed Set (%)		1000-Grain Weight (g)		Grain Yield/Plant (g)	
	b_i	S^2d_i	b_i	S^2d_i	b_i	S^2d_i	b_i	S^2d_i
Sakha 101	1.08	1.863	0.94	0.106	0.59	0.002	1.21	0.315
Sakha 104	0.85	1.525	0.80	0.343	1.09	0.027	1.20	0.276
Sakha105	1.23	1.601	1.12	0.273	−0.96	0.127	1.11	0.027
Sakha 106	0.04	4.594 **	0.13	1.908	−0.85	0.046	0.96	0.395
Sakha 107	0.96	5.946 **	0.77	0.045	0.64	0.002	0.98	0.788
Sakha 108	1.13	1.971	1.04	1.085	0.48	0.004	1.00	0.244
Giza177	1.33	0.249	1.47	0.480	0.76	0.014	0.88	0.270
Giza178	0.67	144.307 **	2.06	1.439	0.34	0.006	0.92	0.105
Giza179	1.04	8.322 **	0.57	1.256	0.83	0.016	0.72	0.514
Giza 182	1.05	12.466 **	0.50	1.089	7.77	0.509	1.40	1.166
Egyptian Yasmine	1.27	16.268 **	1.22	0.242	0.84	0.000	0.75	0.484
Sakha super 300	1.37	5.268 **	1.39	3.204 *	0.48	0.003	0.86	0.176

Symbols * and ** indicate $p < 0.05$, $p < 0.01$, respectively.

3.4. Environmental Indices

An environment index is a quantitative metric that assesses the appropriateness of a specific environment in a given area. By analyzing these indicators, we may determine the optimal conditions required to fully use the genetic potential. If an environment has a positive index value, it can be inferred that it is a favorable environment for genotypes. The indices of E1 and E4 were 2.08 and 2.07, respectively. The value for E4 was 0.8, while the values for E3, E5, and E6 were −0.37, −1.25, and −3.34, correspondingly.

Figure 2 displays the results of a GGE biplot analysis, which is widely regarded as the most efficient method for summarizing both genotypes and the interaction between genotypes and the environment. This analysis helps select the most suitable genotype for each environment and evaluate the stability of the tested genotypes. Figure 2a demonstrates that genotypes located closer to the origin exhibit greater stability compared to those located further away from the origin. Furthermore, the environments were grouped into distinct clusters, namely (E1, E2, E3, and E4) and (E5 and E6). Furthermore, the genotype characterized by a long-dashed line has a higher GY (grain yield) compared to the genotype with a short-dashed line. The genotypes positioned on the vertices of the polygon in Figure 2b are the most responsive to different settings, meaning that they show greater variability in their performance. On the other hand, the genotypes inside the polygon are more stable across environments, indicating that they are less influenced by changes in environmental conditions. The lines of equality partition the graph (Figure 2b) into five sectors. A total of six habitats were preserved, divided into two sectors, and further categorized into two mega-environments.

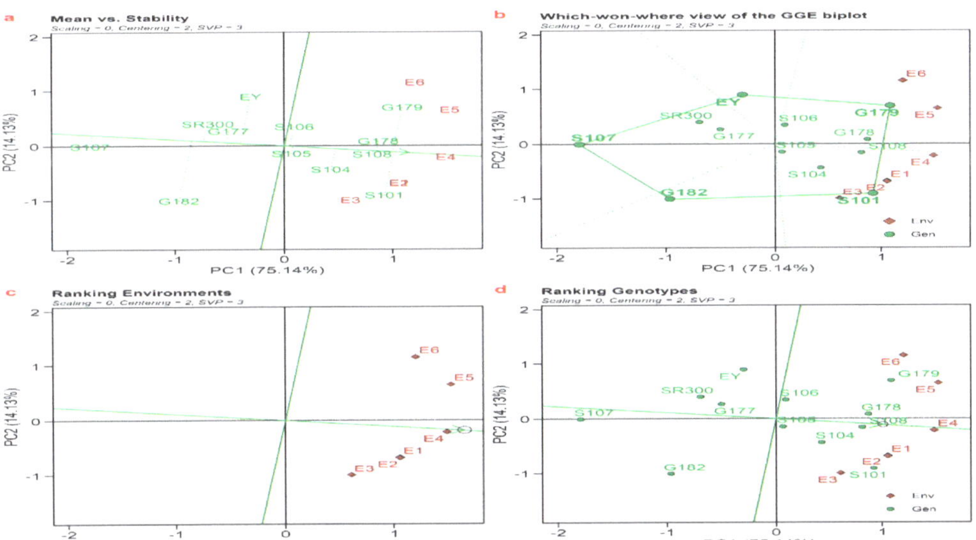

Figure 2. (**a**) Means versus stability; (**b**) polygon view of GGE biplot (which-won-where model); (**c**) ranking environment; (**d**) ranking genotypes.

In the first mega-environment, which included E1, E2, E3, and E4, the top-performing varieties were Sakha 101, Sakha 104, Sakha 105, and Sakha 108. Conversely, in the second mega-environment encompassing E5 and E6, the leading varieties were Sakha 106, Giza 178, and Giza 179. Figure 2c,d depict the respective rankings of environments and genotypes.

The Bayesian network in Figure 3 depicts the impact of treatments on traits, as well as the influence of traits on other traits, using the Bayesian information criterion (BIC). A greater negative BIC value implies a more pronounced effect. The data showed that genotypes had a significant and direct impact on plant height (PH), days to heading

(DTH), grain weight (PW), thousand-kernel weight (TKW), the total grain number per plant (TGNP), and the filled grain number per plant (FGNP). Additionally, the influence of genotypes on FGNP was mediated indirectly through its direct effect on TGNP. Conversely, the environment had a significant direct impact on PL (primary productivity) and PNPP (primary net primary productivity), as well as an indirect impact on PL through PNPP. Additionally, the environment had an indirect impact on FGNP (final gross net primary productivity) and PW (plant water) through PNPP, as well as an indirect impact on GY (growth yield) through PL. PW was influenced by both genotypes and PNPP, but GY was significantly impacted by HI. To summarize, genetics exerted a greater influence on traits compared to the environment.

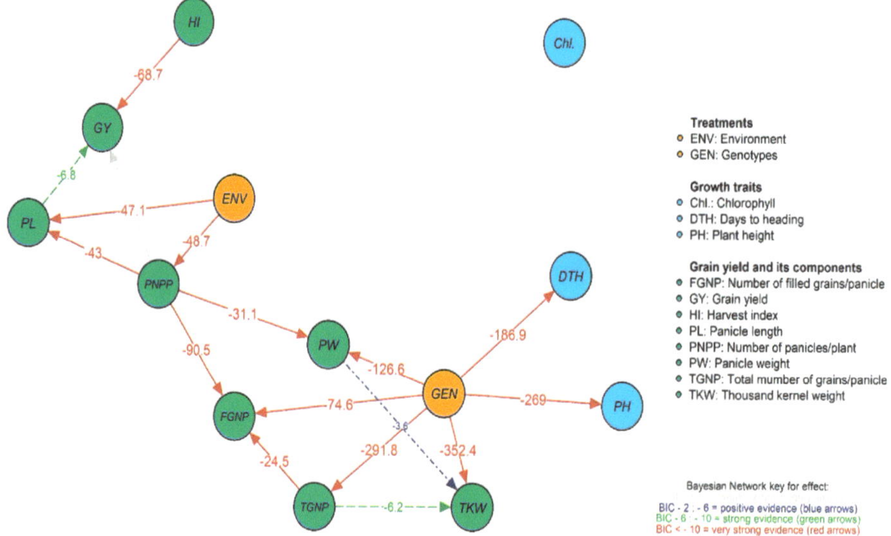

Figure 3. Bayesian network.

3.5. Stability and Superiority Indexes

The total of twelve stability indexes, along with one superiority index, is displayed (Table 9). Based on the stability indexes, Sakha 105 exhibited the highest level of stability among the genotypes, whereas Giza 182 displayed the lowest level of stability. The stability indices primarily consider the stability of genotypes in different conditions while disregarding the yield potential of the genotype. In contrast, the superiority index takes both the stability and prospective yield of the genotype into consideration. According to the superiority index, Giza 178 exhibited the highest level of superiority among the genotypes, whilst Giza 182 displayed the lowest level of superiority.

4. Discussion

Multi-environmental trials are crucial in plant breeding projects since they assess crop yield and adaptability across several settings. To enhance the accuracy of genotype selection, one can assess the genotype-by-environment interaction using mega-environment analysis, genotype evaluation, and multi-environment stability analysis [27]. Genetic diversity can be achieved by breeding inter-subspecific cultivars or accessions having distinct genetic lineages [28–30]. Environmental variability can be obtained by increasing the magnitudes of location factors such as geographic distance, variances in climatic regions (temperate versus tropical), and differences in elevation [31,32].

Our breeding operations primarily prioritize the preservation of the commercial worth of cultivars. Crop enhancements are primarily achieved by the interbreeding of superior cultivars within restricted genetic pools, resulting in limited genetic diversity among the cultivars [33–35]. Furthermore, the test environments were located in geographically proximate agriculturally advantageous flat regions. The stability of a cultivar/genotype pertains to the degree of variation in the phenotypic manifestation of crop performance across diverse environmental conditions. The aim of this study was to assess the idea of "dynamic" stability, which refers to the performance of a stable genotype in different environments. Consequently, not all genotypes exhibited the same response to the environment [36,37].

The results of our study indicate that the genotypes we tested in various locations over two consecutive years are extremely susceptible to climatic zones and environmental factors [38]. The variations seen may be attributed to the disparities in terrain and climatic circumstances among various sites and years where the experiments were carried out. Achieving effective breeding outcomes for yield and related traits in rice necessitates precise measurement and analysis of genotype, environment, years, and their interactions within the breeding protocol [39]. The current discovery of noteworthy sources of variation has been previously documented in rice and other cereal crops [10,27]. Significant effects of both linear and nonlinear components of G × E interaction were seen for all examined traits, as evidenced by highly significant mean squares resulting from G × E (linear) interaction and pooled deviation. The reported results are consistent with those obtained by [40]. The grain production of different rice varieties exhibited significant variations in response to changes in climatic variables, such as planting dates and years [41].

Hence, the cultivar that exhibits consistent and predictable crop output is suitable for implementing risk management strategies to mitigate yield losses in challenging lowland environments [42]. A statistically significant difference has been observed across cultivars and settings in a linear manner, as opposed to a nonlinear manner, in which the deviations are aggregated. This finding highlights the significance of the interplay between genotype and environment in determining grain yield performance and its components across several locations [28,43]. Therefore, the phenotypic stability assessment took into account both the linear (b_i) and nonlinear (S^2d_i) components of G × E interactions [23,44]. Moreover, it has been documented that a high average value of the linear regression coefficient, equivalent to the nonlinear coefficient, is advised for achieving a diverse range. Furthermore, it has been suggested that nonlinear regression can be employed to assess stability, whereas linear regression can be utilized to evaluate the impact of different environmental circumstances on varietal response [45]. Hence, it is crucial to take into account the average and the deviation from the regression of each cultivar when assessing the stability and linear regression for evaluating the responsiveness of the variety. Stresses can limit the growth resources available to plants, resulting in a reduction in the size of plant organs such as leaves, tillers, and spikes [41].

The stability regression coefficient (b_i) and deviation from regression (S^2d_i) for the genotypes under study are shown in Table 7. A stable genotype refers to a genotype that exhibits a consistently high level of performance, with a regression coefficient (b_i) equal to 1 and a deviation from regression of 0. This indicates that the system is well suited for favorable situations, but its performance will be subpar in unfavorable environments. Therefore, these rice cultivars possess significant value for conducting investigations in harsh environments, as the identified types have been suggested for their suitability in unfavorable environmental conditions, as indicated in Table 9 [2].

In the context of selecting and developing superior rice cultivars, genetic studies on yield and yield quality serve as crucial benchmarks for both farmers and industry professionals. As highlighted in previous research, evaluations incorporating various stability statistics offer a more comprehensive understanding of genotype performance across different environments. For instance, a study on sweet potato genotypes in West Java, Indonesia, demonstrated the importance of genotype-by-environment interactions (GEIs) in identifying high-yield and stable genotypes. This research utilized a randomized block design across three environments and found significant effects of genotypes, environments, and GEIs on yield and quality traits. Specifically, GEIs contributed substantially to the variability observed in yield, sweetness, moisture content, tuber diameter, and tuber length. We identified five genotypes (G4, G6, G7, G31, and G32) as both high-yielding and stable across environments. Furthermore, genotype-by-yield ×trait (GYT) analysis revealed seven superior genotypes based on yield and quality, namely G7, G15, G4, G20, G6, G31, and G14. Notably, G4, G6, G7, and G31 were recognized for their high and stable yields combined with favorable quality traits. These findings underscore the value of integrating multiple stability measures and GYT analysis to identify and release superior varieties [46].

The results of pooled ANOVA for stability according to Eberhart and Russell [29], as well as the AMMI model, revealed that the variance of genotypes was significant for rice grain yield, indicating that the performance of rice genotypes was different; the genotypes also had differential responses to the changes in years and locations (environments).

In this study, the genotype Sakha 105 had regression coefficients (b_i) of 1.11 and was observed to be stable. According to the Eberhart and Russell [29] model, a slope of >1.0 with high mean yield and nonsignificant squared deviation (S^2d_i) is suitable for a favorable environment.

The AMMI model defines the genotype–environment interaction. The AMMI model revealed that a major part of the variation in rice grain yield is explained by the environment, which indicates environmental diversity. The AMMI1 biplot analysis showed that the variation was due to the main effect of rice grain yield and the interaction effect. Genotypes with IPCA1 scores close to zero were characterized as having low interaction effects and consequently considered stable.

5. Conclusions

This study highlights that environmental factors are the primary drivers of variations in rice yield and related traits among Egyptian rice genotypes, with genotype–environment interactions playing a secondary role. Certain genotypes, like Sakha 104 and Giza 178, demonstrated stable performance across different environments, while others were better suited to specific conditions. Notably, Sakha 106, Sakha 107, Sakha 108, and Giza 178 showed b_i values close to 1, indicating they are well adapted across diverse environments. In contrast, Giza 177, E. Yasmin, and Sakha Super 300, with b_i values below 1, were more suitable for poor environments, while Sakha 101, Sakha 104, Sakha 105, and Giza 182, with bi values above 1, were better suited to favorable environments but may underperform in less favorable conditions. These findings underscore the importance of multi-environmental trials for selecting stable, high-yielding rice cultivars. The results provide a foundation for targeted breeding programs aimed at optimizing rice productivity by focusing on genotypes that consistently perform well across various growing conditions, considering both stability and adaptability.

Author Contributions: Conceptualization, M.S.A.E.-A., M.I.A.-Y., F.A.S., M.S., M.A.G., O.M.I., M.Y.K., W.H.A.-Q., M.A.A.-M. and A.M.E.-T.; methodology, M.S.A.E.-A., M.I.A.-Y., F.A.S., M.S., M.A.G., O.M.I., M.Y.K., W.H.A.-Q., M.A.A.-M. and A.M.E.-T.; software, M.S.A.E.-A., M.I.A.-Y., F.A.S., M.S., M.A.G., O.M.I., M.Y.K., W.H.A.-Q., M.A.A.-M. and A.M.E.-T.; validation, M.S.A.E.-A., M.I.A.-Y., F.A.S., M.S., M.A.G., O.M.I., M.Y.K., W.H.A.-Q., M.A.A.-M. and A.M.E.-T.; formal analysis, M.S.A.E.-A., M.I.A.-Y., F.A.S., M.S., M.A.G., O.M.I., M.Y.K., W.H.A.-Q., M.A.A.-M. and A.M.E.-T.; investigation, M.S.A.E.-A., M.I.A.-Y., F.A.S., M.S., M.A.G., O.M.I., M.Y.K., W.H.A.-Q., M.A.A.-M. and A.M.E.-T.; resources, M.S.A.E.-A., M.I.A.-Y., F.A.S., M.S., M.A.G., O.M.I., M.Y.K., W.H.A.-Q., M.A.A.-M. and A.M.E.-T.; data curation, M.S.A.E.-A., M.I.A.-Y., F.A.S., M.S., M.A.G., O.M.I., M.Y.K., W.H.A.-Q., M.A.A.-M. and A.M.E.-T.; writing—original draft preparation, M.S.A.E.-A., M.I.A.-Y., F.A.S., M.S., M.A.G., O.M.I., M.Y.K., W.H.A.-Q., M.A.A.-M. and A.M.E.-T.; writing—review and editing, M.S.A.E.-A., M.I.A.-Y., F.A.S., M.S., M.A.G., O.M.I., M.Y.K., W.H.A.-Q., M.A.A.-M. and A.M.E.-T.; visualization, M.S.A.E.-A., M.I.A.-Y., F.A.S., M.S., M.A.G., O.M.I., M.Y.K., W.H.A.-Q., M.A.A.-M. and A.M.E.-T.; supervision, M.S.A.E.-A., M.I.A.-Y., F.A.S., M.S., M.A.G., O.M.I., M.Y.K., W.H.A.-Q., M.A.A.-M. and A.M.E.-T.; project administration, M.S.A.E.-A., M.I.A.-Y., F.A.S., M.S., M.A.G., O.M.I., M.Y.K., W.H.A.-Q., M.A.A.-M. and A.M.E.-T.; funding acquisition, M.S.A.E.-A., M.I.A.-Y., F.A.S., M.S., M.A.G., O.M.I., M.Y.K., W.H.A.-Q., M.A.A.-M. and A.M.E.-T. All authors have read and agreed to the published version of the manuscript.

Funding: The authors extend their appreciation to the researchers supporting Project Number (RSP2024R293), King Saud University, Riyadh, Saud Arabia.

Data Availability Statement: The original contributions presented in the study are included in the article, further inquiries can be directed to the corresponding author.

Conflicts of Interest: The authors declare no conflicts of interest.

References

1. Habib, M.A.; Azam, M.G.; Haque, A.; Hassan, L.; Khatun, M.S.; Nayak, S.; Abdullah, H.M.; Ullah, R.; Ali, E.A.; Hossain, N.; et al. Climate-smart rice (*Oryza sativa* L.) genotypes identification using stability analysis, multi-trait selection index, and genotype-environment interaction at different irrigation regimes with adaptation to universal warming. *Sci. Rep.* **2024**, *14*, 13836. [CrossRef] [PubMed]
2. El-Aty, M.S.A.; Katta, Y.S.; Abd, A.E.M.B.E.; Mahmoud, S.; Ibrahim, O.M.; Wali, A.M.; El-Shehawi, A.M.; Elseehy, M.M.; El-Saadony, M.T.; El-Tahan, A.M. Assessment of grain quality traits in rice under normal and water deficit condition. *Saudi J. Biol. Sci.* **2022**. [CrossRef]
3. Al Azzawi, T.N.I.; Khan, M.; Hussain, A.; Shahid, M.; Imran, Q.M.; Mun, B.-G.; Lee, S.-U.; Yun, B.-W. Evaluation of Iraqi Rice Cultivars for Their Tolerance to Drought Stress. *Agronomy* **2020**, *10*, 1782. [CrossRef]
4. Abd-El-Aty, M.S.; Abo-Youssef, M.I.; Bahgt, M.M.; Ibrahim, O.M.; Faltakh, H.; Nouri, H.; Korany, S.M.; Alsherif, E.A.; AbdElgawad, H.; El-Tahan, A.M. Mode of gene action and heterosis for physiological, biochemical, and agronomic traits in some diverse rice genotypes under normal and drought conditions. *Front. Plant Sci.* **2023**, *14*, 1108977. [CrossRef] [PubMed]
5. Akanksha, A.; Jaiswal, H.K. Combining ability studies for yield and quality parameters in basmati rice (*Oryza sativa* L.) genotypes using diallel approach. *Electron. J. Plant Breed.* **2019**, *10*, 9–17. [CrossRef]
6. Aslam, M.M.; Rashid, M.A.R.; Siddiqui, M.A.; Khan, M.T.; Farhat, F.; Yasmeen, S.; Khan, I.A.; Raja, S.; Rasool, F.; Sial, M.A.; et al. Recent Insights into Signaling Responses to Cope Drought Stress in Rice. *Rice Sci.* **2022**, *29*, 105–117. [CrossRef]
7. Aswidinnoor, H.; Listiyanto, R.; Rahim, S.; Holidin; Setiyowati, H.; Nindita, A.; Ritonga, A.W.; Marwiyah, S.; Suwarno, W.B. Stability analysis, agronomic performance, and grain quality of elite new plant type rice lines (*Oryza sativa* L.) developed for tropical lowland ecosystem. *Front. Sustain. Food Syst.* **2023**, *7*, 1147611. [CrossRef]
8. Awad-Allah, M.M.A. Heterosis and Combining ability Estimates using Line x Tester Analysis to Develop Wide Compatibility and Restorer Lines in Rice. *J. Agric. Chem. Biotechnol.* **2020**, *11*, 383–393. [CrossRef]
9. SC, M.K.; Singh, S.K.; Khaire, A.; Korada, M.; Majhi, P.K.; Singh, D.K. An analytical approach integrating GGE-Biplot and AMMI techniques for assessing genotype-environment interactions and yield stability in rice (*Oryza sativa* L.) genotypes. *Emergent Life Sci. Res.* **2023**, *9*, 168–176. [CrossRef]
10. El-Malky, M.; Al-Daej, M. Studies of Genetic Parameters and Cluster analysis of some Quantitative Characters through Diallel analysis of rice (*Oryza Sativa* L.). *Vegetos* **2018**, *2018*, 1–10.
11. El-Mowafi, H.; Reda, M.; Abdallah, R. Combining Ability Analysis for Agronomic and Yield Attributing Traits in Hybrid Rice. *Egypt. J. Plant Breed.* **2015**, *19*, 2195–2219. [CrossRef]
12. El-Mowafi, H.F.; AlKahtani, M.D.; Abdallah, R.M.; Reda, A.M.; Attia, K.A.; El-Hity, M.A.; El-Dabaawy, H.E.; Husnain, L.A.; Al-Ateeq, T.K.; EL-Esawi, M.A. Combining Ability and Gene Action for Yield Characteristics in Novel Aromatic Cytoplasmic Male Sterile Hybrid Rice under Water-Stress Conditions. *Agriculture* **2021**, *11*, 226. [CrossRef]
13. Gaballah, M.M.; Attia, K.A.; Ghoneim, A.M.; Khan, N.; El-Ezz, A.F.; Yang, B.; Xiao, L.; Ibrahim, E.I.; Al-Doss, A.A. Assessment of Genetic Parameters and Gene Action Associated with Heterosis for Enhancing Yield Characters in Novel Hybrid Rice Parental Lines. *Plants* **2022**, *11*, 266. [CrossRef]
14. Ganapati, R.K.; Rasul, G.; Sarker, U.; Singha, A.; Faruquee, M. Gene action of yield and yield contributing traits of submergence tolerant rice (*Oryza sativa* L.) in Bangladesh. *Bull. Natl. Res. Cent.* **2020**, *44*, 8. [CrossRef]
15. Wricke, G. Uber eine methode zur erfassung der okologischen streubreite in feldversuchen. *Z. Pflanzenzucht.* **1962**, *47*, 92–96.
16. Shukla, G.K. Some statistical aspects of partitioning genotype-environmental components of variability. *Heredity* **1972**, *29*, 237–245. [CrossRef]
17. Francis, T.R.; Kannenberg, L.W. Yield Stability Studies in Short-Season Maize. I. A Descriptive Method for Grouping Genotypes. *Can. J. Plant Sci.* **1978**, *58*, 1029–1034. [CrossRef]
18. Roemer, J. Sinde die ertagdreichen sorten ertagissicherer? *Mitt DLG* **1917**, *32*, 87–89.
19. Plaisted, R.L.; Peterson, L.C. A technique for evaluating the ability of selections to yield consistently in different locations or seasons. *Am. J. Potato Res.* **1959**, *36*, 381–385. [CrossRef]
20. Plaisted, R.I. A shorter methods for evaluating the ability of selections to yield consistently over locations. *Am. J. Potato Res.* **1960**, *37*, 166172. [CrossRef]
21. Finlay, K.W.; Wilkinson, G.N. The analysis of adaptation in a plant-breeding programme. *Aust. J. Agric. Res.* **1963**, *14*, 742–754. [CrossRef]
22. Gauch, H.G.; Zobel, R.W. Predictive and postdictive success of statistical analyses of yield trials. *Theor. Appl. Genet.* **1988**, *76*, 1–10. [CrossRef] [PubMed]

23. Yan, W. GGEbiplot—A Windows Application for Graphical Analysis of Multienvironment Trial Data and Other Types of Two-Way Data. *Agron. J.* **2001**, *93*, 1111–1118. [CrossRef]

24. Gauch, H.G.; Piepho, H.; Annicchiarico, P. Statistical Analysis of Yield Trials by AMMI and GGE: Further Considerations. *Crop. Sci.* **2008**, *48*, 866–889. [CrossRef]

25. Page, A.L.; Miller, R.H.; Keeney, D.R. *Methods of Soil Analysis. Part 2. Chemical and Microbiological Properties*; American Society of Agronomy: Madison, WI, USA, 1982; Volume 1159.

26. Freeman, G.H.; Gomez, K.A.; Gomez, A.A. Statistical Procedures for Agricultural Research with Emphasis on Rice. *Biometrics* **1978**, *34*, 721. [CrossRef]

27. Herwibawa, B.; Sakhidin; Haryanto, T.A.D. Agronomic performances of aromatic and non-aromatic M_1 rice under drought stress. *Open Agric.* **2019**, *4*, 575–584. [CrossRef]

28. Parimala, K.; Raju, C.S.; Kumar, S.S.; Reddy, S.N. Stability analysis over different environments for grain yield and its components in hybrid rice (*Oryza sativa* L.). *Electron. J. Plant Breed.* **2019**, *10*, 389–399. [CrossRef]

29. Kumar, C.P.S.; Sathiyabama, R.; Suji, D.B.; Muraleedharan, A. Estimation of heterosis for earliness and certain growth characters in rice (*Oryza sativa* L.). *Plant Arch.* **2020**, *20*, 1429–1432.

30. Kumari, J.; Mahatman, K.K.; Sharma, S.; Singh, A.K.; Adhikari, S.; Bansal, R.; Kaur, V.; Kumar, S.; Yadav, M.C. Recent Advances in Different Omics Mechanism for Drought Stress Tolerance in Rice. *Russ. J. Plant Physiol.* **2022**, *69*, 18. [CrossRef]

31. Kunnam, J.; Pinta, W.; Ruttanaprasert, R.; Bunphan, D.; Thabthimtho, T.; Aninbon, C. Stability of Phenols, Antioxidant Capacity and Grain Yield of Six Rice Genotypes. *Plants* **2023**, *12*, 2787. [CrossRef]

32. El-Nashart, A.B.; El-Nwehy, S.S.; Rezk, A.E.-H.I.; Ibrahim, O.M. Improving seed and oil yield of sunflower grown in calcareous soil under saline stress conditions. *Asian J. Crop. Sci.* **2017**, *9*, 35–39. [CrossRef]

33. Manjunath, K.; Chandramohan, Y.; Shankar, V.; Balram, M.; Lavuri, K. Genetic and molecular studies on fertility restoration in rice (*Oryza sativa* L.). *Genotypes* **2020**, *10*, 944–950.

34. Ranjith, S.A.; Ram, R.S.; Saravanan, K.R.; Karthikeyan, P.G.; Anbananthan, V.; Sathiyanarayanan, G. Studies on heterosis breeding for qualitative and quantitative traits in rice (*Oryza Sativa* L.). *Plant Arch.* **2020**, *20*, 1349–1353.

35. Rasheed, A.; Hassan, M.; Aamer, M.; Bian, J.; Xu, Z.; He, X.; Yan, G.; Wu, Z. Iron toxicity, tolerance and quantitative trait loci mapping in rice—A review. *Appl. Ecol. Environ. Res.* **2020**, *18*, 7483–7498. [CrossRef]

36. Shrestha, J.; Kushwaha, U.K.S.; Maharjan, B.; Kandel, M.; Gurung, S.B.; Poudel, A.P.; Karna, M.K.L.; Acharya, R. Grain Yield Stability of Rice Genotypes. *Indones. J. Agric. Res.* **2020**, *3*, 116–126. [CrossRef]

37. Suvi, W.T.; Shimelis, H.; Laing, M.; Mathew, I.; Shayanowako, A.I.T. Determining the Combining Ability and Gene Action for Rice Yellow Mottle Virus Disease Resistance and Agronomic Traits in Rice (*Oryza sativa* L.). *Agronomy* **2020**, *11*, 12. [CrossRef]

38. Chandramohan, Y.; Krishna, L.; Srinivas, B.; Rukmini, K.; Sreedhar, S.; Prasad, K.S.; Kishore, N.S.; Rani, C.V.D.; Singh, T.V.J.; Jagadeeshwar, R. Stability analysis of short duration rice genotypes in Telangana using AMMI and GGE Bi-plot models. *Environ. Conserv. J.* **2023**, *24*, 243–252. [CrossRef]

39. Thirumalai, R.; Anbananthan, V.; Azmath, S. Genetic Effects and Combining Ability for Yield and its Component Traits in Rice (*Oryza sativa* L.) Using Line X Tester. *Int. J. Sci. Res.* **2018**, *7*, 1478–1483.

40. Zewdu, Z.; Abebe, T.; Mitiku, T.; Worede, F.; Dessie, A.; Berie, A.; Atnaf, M. Performance evaluation and yield stability of upland rice (*Oryza sativa* L.) varieties in Ethiopia. *Cogent Food Agric.* **2020**, *6*, 1842679. [CrossRef]

41. Singh, A.K.; Kumar, S. Heterosis and combining ability analysis for yield and its components in rice (*Oryza sativa* L.). *J. Pharmacogn. Phytochem.* **2019**, *8*, 3172–3181. [CrossRef]

42. El-Hadi, A.H.A.; Kash, K.S.; El-Mowafi, H.F.; Anis, G.B. The utilization of cytoplasmic male sterile (cms) and restorer lines in the developing of hybrid rice. *J. Agric. Chem. Biotechnol.* **2013**, *4*, 263–274. [CrossRef]

43. Hasan, M.K.; Kulsum, M.U.; Hossain, E.; Hossain, M.M.; Rahman, M.M.; Rahmat, N.M.F. Combining ability analysis for iden-tifying elite parents for heterotic rice hybrids. *Acad. J. Agric. Res.* **2015**, *3*, 70–75.

44. Ali, M.A.; Ghazy, A.I.; Alotaibi, K.D.; Ibrahim, O.M.; Al-Doss, A.A. Nitrogen efficiency indexes association with nitrogen recovery, utilization, and use efficiency in spring barley at various nitrogen application rates. *Agron. J.* **2022**, *114*, 2290–2309. [CrossRef]

45. Wu, C.; Cui, K.; Li, Q.; Li, L.; Wang, W.; Hu, Q.; Ding, Y.; Li, G.; Fahad, S.; Huang, J.; et al. Estimating the yield stability of heat-tolerant rice genotypes under various heat conditions across reproductive stages: A 5-year case study. *Sci. Rep.* **2021**, *11*, 13604. [CrossRef]

46. Maulana, H.; Solihin, E.; Trimo, L.; Hidayat, S.; Wijaya, A.A.; Hariadi, H.; Amien, S.; Ruswandi, D.; Karuniawan, A. Genotype-by-environment interactions (GEIs) and evaluate superior sweet potato (*Ipomoea batatas* [L.] Lam) using combined analysis and GGE biplot. *Heliyon* **2023**, *9*, e20203. [CrossRef]

agronomy

Article

Evaluation of the Potential Use of Wild Relatives of Tomato (*Solanum pennellii*) to Improve Yield and Fruit Quality Under Low-Input and High-Salinity Cultivation Conditions

Maria Gerakari [1], Anastasia Kyriakoudi [2], Dimitris Nokas [3], Ioannis Mourtzinos [2], Evangelia G. Chronopoulou [3], Eleni Tani [1,*] and Ilias Avdikos [4,*]

1   Laboratory of Plant Breeding and Biometry, Agricultural University of Athens, 11855 Athens, Greece; mgerakari@aua.gr
2   Department of Food Science and Technology, School of Agriculture, Forestry and Natural Environment, Aristotle University of Thessaloniki, 54124 Thessaloniki, Greece; ankyria@agro.auth.gr (A.K.); mourtzinos@agro.auth.gr (I.M.)
3   Laboratory of Enzyme Technology, Department of Biotechnology, School of Applied Biology and Biotechnology, Agricultural University of Athens, 11855 Athens, Greece; d.nokas@aua.gr (D.N.); exronop@aua.gr (E.G.C.)
4   Laboratory of Vegetable Crop Science, Department of Agriculture, International Hellenic University, Sindos, 57400 Thessaloniki, Greece
*   Correspondence: etani@aua.gr (E.T.); avdikos.elias@ihu.gr (I.A.)

Citation: Gerakari, M.; Kyriakoudi, A.; Nokas, D.; Mourtzinos, I.; Chronopoulou, E.G.; Tani, E.; Avdikos, I. Evaluation of the Potential Use of Wild Relatives of Tomato (*Solanum pennellii*) to Improve Yield and Fruit Quality Under Low-Input and High-Salinity Cultivation Conditions. *Agronomy* **2024**, *14*, 3042. https://doi.org/10.3390/agronomy14123042

Academic Editors: Mariola Staniak, Ewa Szpunar-Krok, Małgorzata Szostek and Alessandro Miceli

Received: 3 November 2024
Revised: 12 December 2024
Accepted: 17 December 2024
Published: 20 December 2024

check for
updates

Abstract: Salinity stress is a major abiotic factor limiting tomato (*Solanum lycopersicum*) production, particularly in arid and semi-arid regions. Utilizing genetic resources from wild tomato relatives, such as *Solanum pennellii*, through the exploitation of introgression lines (ILs) provides a promising strategy to enhance salt tolerance. This study evaluates the performance of nine tomato genotypes, including one commercial tomato hybrid (Formula F_1) and eight ILs under three different soil salinity levels (1.88, 6.44, and 8.63 mS/cm), trying to identify salt-tolerant lines that maintain yield and fruit quality. Morphological characteristics, gas exchange parameters, yield traits, fruit quality characteristics, and antioxidant activity were assessed. High-performance liquid chromatography (HPLC) was employed to quantify the levels of carotenoids, namely lycopene and β-carotene, of fruits in selected genotypes. Additionally, total antioxidant capacity was measured in leaves, using DPPH, FRAP and FOLIN assays. The results indicate that out of all the evaluated characteristics, four plant-related traits, four fruit-related traits, one gas exchange parameter, and three productivity-related traits presented strong correlations to total yield (g/plant). These 12 traits could be considered as potential indexes for genotype salinity tolerance discrimination and could be utilized as an efficient marker tool for distinguishing tolerant genotypes to salinity stress, allowing breeders to reduce the time-consuming process of developing new salinity-tolerant varieties. Regarding genotypes' ranking based on the relative performance of agronomic traits under a salinity regime of 8.63 mS/cm compared to a salinity regime of 1.88 mS/cm, IL6-6 exhibited significant tolerance to high-salinity conditions compared to the commercial hybrid and other ILs, like IL8-9. This tolerant IL maintained higher plant growth, yield, and fruit quality traits, including elevated levels in its fruits' carotenoids and leaves' antioxidant capacity, under severe salinity conditions, highlighting its potential for breeding programs targeting saline environments. ILs can help maintain productivity and fruit quality under salinity stress, making them a promising solution for sustainable tomato cultivation in salinity-affected regions. These findings, combined with previous results, suggest that tomato introgression lines offer a valuable genetic resource for developing tomato varieties suitable for harsh environments.

Keywords: tomato (*Solanum lycopersicum*); introgression lines (ILs); *Solanum pennellii*; fruit quality; salinity stress; antioxidant capacity; stress tolerance; nutritional value; yield

1. Introduction

Climate change is a result of human activities, especially the combustion of fossil fuels, deforestation, and industrial processes, which cause significant and persistent alterations in global climate patterns. These changes have wide-ranging consequences, affecting areas, such as agriculture, public health, water resources, energy production, and biodiversity [1]. Climate change has aggravated abiotic stressing factors, like increased temperature, drought, atmospheric pollution, and salinity stress, subjecting plants to harsh and extreme climates. These challenging conditions negatively impact the morphological, developmental, cellular, and molecular processes within plants [2].

The detrimental impact of excessive mineral salts, such as Na^+ and Cl^-, on plants is known as salt or salinity stress and composes a significant abiotic stressor that limits plant growth and productivity [3,4]. Soil salinity is a critical issue in modern agriculture, posing significant challenges to crop productivity by disrupting plant water uptake and causing osmotic and ionic stress. This leads to stunted growth, reduced photosynthetic activity, and ultimately lower yields [5,6]. Arid and semi-arid regions, like Greece, characterized by low rainfall, high evapotranspiration rates, elevated temperatures, and limited water availability, are more affected by salinity stress [7]. Soil salinity occurs due to the accumulation of salts, either water-soluble, like sodium chloride ($NaCl$) and sodium carbonate (Na_2CO_3), or partially water-soluble salts, such as calcium chloride ($CaCl_2$) [7]. A soil type is classified as saline when its electrical conductivity (EC) reaches 4 mS/cm, which significantly disrupts plant metabolism, leading to symptoms, like chlorosis, necrosis in plants and, consequently, low yield and fruit quality [8]. To mitigate the effects of salinity stress, plants have developed various adaptive mechanisms, including maintaining ionic balance by regulating their Na^+ concentration and Na^+/K^+ ratio [9,10]. They also accumulate osmoprotectants, like proline, to reduce cytoplasmic osmotic potential, as well as ascorbic acid, glycine betaine and signaling molecules, such as ethylene and hydrogen peroxide (H_2O_2). H_2O_2 belongs to the family of reactive oxygen species (ROS), and they are usually found to increase at the first stages of exposure to a stressing factor, but after the signal induction from the plants' metabolism, they are found to demonstrate reduced accumulation in the plants' tissues [7,11,12]. To this end, plants activate various ROS-scavenging enzymes, including ascorbate peroxidase (APX) and catalase (CAT) [10,13].

Tomato (*Solanum lycopersicum* L.), one of the most important vegetable crops with high nutritional and economic value [14], is particularly sensitive to salinity stress. High levels of salt in the soil impair fruit yield and quality by altering physiological processes, like ion balance, water use efficiency, and nutrient absorption. These changes result in smaller fruits, reduced sugar and acid content, and overall poor market value. Understanding and improving salt tolerance in tomato varieties is a key area of research to sustain productivity in areas affected by salinization [15,16]. Wild tomato species often exhibit higher tolerance to salt due to their evolution in non-optimum environments. Tomato introgression lines (ILs) can provide a valuable solution to addressing salinity stress by introducing salt-tolerance traits from wild relatives into cultivated varieties. These lines are developed through hybridization between cultivated tomatoes and wild species, followed by backcrossing to transfer specific genomic regions, or "introgressions", associated with desirable traits [17]. By introducing these genetic regions into cultivated tomatoes, breeders can enhance the crop's ability to cope with high-salinity without compromising fruit quality or yield [18]. Several research studies examine the impact of soil salinity on root development in tomato ILs of *Solanum lycopersicum* × *Solanum pennellii*, identifying salt-tolerant lines and loci involved in root growth under saline conditions through testing different $NaCl$ concentrations and analyzing introgression lines for potential genetic markers of salt stress tolerance [19]. In view of climate change, it is essential to adopt sustainable agricultural practices and breed tomato cultivars to improve crop adaptability to environmental stressors, like soil salinity, while maintaining crop yield and quality. Soil salinity is a major abiotic stress factor in Greece, leading to land degradation and a decrease in quality and yield for tomato cultivation. Moreover, the narrow genetic base and the absence of resistance traits in the currently cultivated varieties and hybrids of tomato creates the demanding issue of how to breed commercial varieties that are well-adapted to the saline soils of the country. Exploiting native germplasm or wild relatives to introduce novel allelic combinations into existing varieties is an efficient strategy used in tomato breeding programs to enhance resistance to biotic and abiotic challenges, with special attention recently given to introgression lines [20–26]. The present research study contributes to the exploration of agriculturally useful traits among morphological, physiological, and fruit quality characteristics of the ILs studied, related to their tolerance in soils characterized by different levels of salinity content. Thus, this study seeks to address a critical issue for breeders and producers by examining and proposing specific traits that can be utilized as indexes for the selection of salt-tolerant genotypes. Utilizing and exploiting these genetic resources in future breeding programs to create commercial tomato varieties tolerant to abiotic stressors, like salinity, can provide an effective solution to the crop production problems arising due to climate change, helping farmers to maintain tomato yield and quality in low-input farming systems [27].

Towards this end, in the present study, nine tomato genotypes (eight ILs and one commercial variety) were cultivated under three different soil salinity conditions (low, medium, and high salinity). For all the genotypes, phenotypic characterization and descriptive characteristics were measured according to UPOV. Fruit quality was monitored by conducting resistance of the flesh to pressure, total soluble solids (°Brix), and pH measurements. Gas exchange parameters were also determined. Fruit quality characteristics, total carotenoids, lycopene and β-carotene concentrations were determined for four selected genotypes through high-performance liquid chromatography (HPLC). Three out of the four genotypes were selected according to their extreme resistance and susceptibility to salt stress for the determination of their antioxidant capacity, which was determined by the Folin–Ciocalteu reagent method, DPPH radical scavenging activity, and ferric reducing antioxidant power assay (FRAP). Finally, we intended to provide information on whether the ILs previously screened for tolerance to the holoparasitic weed *Orobanche* sp. [17] are well-adapted to the salinity conditions of Greek soils, and, therefore, whether they can be incorporated in the next phases of tomato breeding programs for biotic/abiotic stress tolerance. Additionally, under saline soils, these ILs can be utilized as rootstocks in grafting experiments.

2. Materials and Methods

2.1. Plant Material and Soil Analysis

In the present work, one commercial tomato (*S. lycopersicum*) hybrid "Formula F$_1$", with high adaptability in Greek environmental conditions [28] and eight ILs (*S. lycopersicum x S. pennellii*) developed by Eshed in 1995 [29], were studied regarding their tolerance to soil salinity stress (Figure 1). These specific genotypes have previously been studied in experiments regarding their tolerance against broomrape parasitism, which consists of a severe biotic stressor [17], in the framework of the scientific project «ZeroParasitic» (project ID: 1485). Since these genotypes exhibited increased tolerance to biotic stresses, they were selected to be tested for their resistance to abiotic stresses and specifically to salinity stress. The nine genotypes were cultivated in an organic (low-input) cultivation system at the following three different soil salinity conditions: A: 1.88 mS/cm, B: 6.44 mS/cm, and C: 8.63 mS/cm. For tomato plants, soil salinity values above 2.5 to 3.0 mS/cm are generally considered stressful, with significant impacts on plant growth, fruit yield, and quality [30]; thus, in this research we considered treatment A: 1.88 mS/cm to be our control treatment [31].

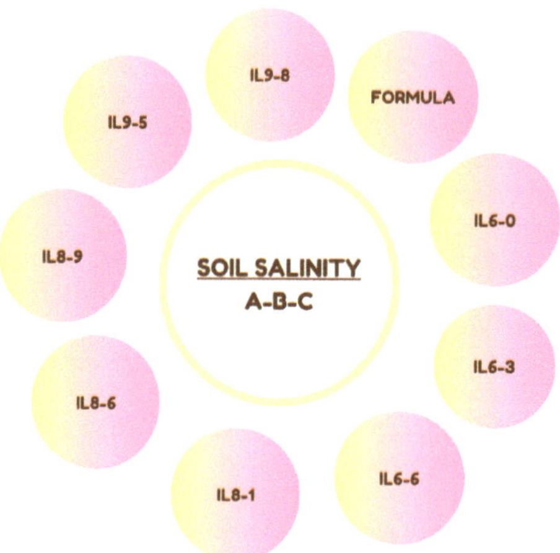

Figure 1. The nine genotypes tested for the three soil salinity conditions: (A) 1.88 mS/cm, (B) 6.44 mS/cm, and (C) 8.63 mS/cm.

Soil analysis was conducted across the three soil salinity conditions, revealing the following characteristics: the soil was classified as sandy loam, comprising 60% sand, 33% silt, and 7% clay, which was consistent across all experimental plots. Under the different soil salinity levels (electrical conductivity, (EC): A = 1.88 mS/cm, B = 6.44 mS/cm, and C = 8.63 mS/cm), the pH exhibited a slight increase, ranging from 7.88 at soil condition A to 7.90 at condition B and 7.98 at condition C. Calcium carbonate ($CaCO_3$) content was absent in all soil salinity conditions. The organic matter content remained consistent across the soil conditions (approximately 2%). The sodium (Na) concentration, however, increased significantly with higher salinity levels, rising from 200 mg/kg at salinity level A to 497 mg/kg at level B and to 970 mg/kg at level C, indicating substantial salt accumulation in the soil. Regarding the freshwater used in the experiment, it exhibited a pH of 7.86 and an EC of 0.483 mS/cm. Continuous irrigation with freshwater was applied during the experimental procedure across all three soil conditions.

2.2. Methodology

The experiment was conducted under organic farming conditions, during spring–summer 2024, in a non-heated greenhouse at the farm of the Alexandria Campus of International Hellenic University, Sindos-Thessaloniki (40°40′ N lat., 22°47′ E long., 4 m alt.). The experimental plant material was prepared conventionally. Tomato seeds were sown in a peat-based substrate and left at 25 °C for 5 days to germinate. After 20 days, the seedlings were transplanted into individual seedling trays filled with peat substrate and grown for another 20 days without any fertilizer application. Irrigation was carried out every two days. Subsequently, the seedlings were transplanted into the field with a planting distance of 50 cm between plants within rows and 1 m between rows. Drip irrigation was applied every two days to ensure an adequate water supply. In each of the three sub-experiments with different soil salinity levels (A, B, and C), a randomized complete block design (RCBD) was used, with three replications, with each experimental unit consisting of ten plants. Organic cropping practices were followed with no chemical applications. All observations were taken on an individual plant basis, and for each entry, descriptive, qualitative, and yield characteristics were determined. Sampling of leaves and fruits for further experimentation was conducted 72 days after transplantation (72 D.A.T.).

2.3. Traits Evaluated

2.3.1. Descriptive and Gas Exchange Characteristics

Phenotypic characterization and descriptive characteristics were measured according to the UPOV system concerning plants, leaves, flowers, and fruits of all the genotypes studied (https://www.upov.int/test_guidelines/en/list.jsp, code 044, accessed on 18 June 2024). More specifically 32 characteristics were measured: plant growth type, number of inflorescences on the main stem, plant height, height of the 4th inflorescence, length of internodes, anthocyanin coloration of the stem, leaf attitude, leaf length, leaf width, leaf's intensity of green color, attitude of the petiole of leaflets in relation to the petiole, type of leaf, blistering of leaves, type of inflorescence, flower color, pedicel abscission layer, pubescence of pedicle, presence of green shoulder in fruits, extent of green shoulder, intensity of green color of shoulder, intensity of green color excluding the shoulder, fruit's shape in the longitudinal section, fruit's ribbing, fruit's depression at the pedicel end, fruit's shape at the blossom end, size of pedicle scar, fruit's polar diameter, fruit's equatorial diameter, fruit's internal color, thickness of pericarp, number of carpels, and peduncle size inside the fruit. Ten out of the thirty-two estimated characteristics were selected to be presented based on their importance and on their response under the three treatments with different soil salinity concentrations.

Gas exchange parameters (net rate of CO_2 assimilation or net photosynthesis rate (A); transpiration rate (E); intercellular CO_2 (ci); and stomatal conductance (gs)) were measured using a portable open-circuit gas-exchange instrument (LCPro, ADC BioScientific Ltd., Hoddesdon, UK), equipped with a broad leaf chamber enclosing 6.25 cm^2 of leaf area. Water use efficiency (WUE) was calculated as the ratio of net photosynthesis rate to transpiration rate (A/E) [32].

Two red ripe stage fruits were harvested (the first and second fruit of the second cluster of each plant) for all nine genotypes and the three salinity treatments applied, and their nutritional value was evaluated. Fully developed tomato leaves from the upper third of each plant were also collected.

2.3.2. Fruit Quality Characteristics

Determination of Total Soluble Solids (TSS), pH, and Resistance of the Flesh to Pressure of the Fruit

The total soluble solids (TSS) were determined with the use of an Atago PR-100 hand refractometer on the juice taken from each fruit sample. The flesh's resistance to pressure (firmness) (kg) was measured using a Chatillon penetrometer with a plunger of 3.4 mm in diameter and 9.5 mm in length. pH measurements were also conducted to determine the fruit quality.

High-Performance Liquid Chromatography (HPLC)

Regarding the nutritional value of the fruits under the three soil salinity conditions, four genotypes were selected for further experimentation: the commercial hybrid, Formula, as well as two potentially tolerant (IL6-6 and IL8-6) ILs and one susceptible (IL8-9) IL, according to the results of the descriptive characteristics. Individual major carotenoids, namely lycopene and β-carotene, were determined using HPLC. The tomato fruit samples were frozen and subsequently lyophilized using a HyperCOOL HC8080 lyophilizer (Gyrozen Co., Ltd., Incheon, Republic of Korea) ($-80\ °C$, 0.1 mbar), so that the dried material could be used for the determination of carotenoids. Specifically, a certain amount of the lyophilized sample (~0.25 g) was used for the analysis. A total of two consecutive extractions were performed to ensure the complete decolorization of the plant material. Carotenoids were extracted using cold acetone at a solid–solvent ratio of 1:40 (w/v). The extraction was carried out using a UltraTurrax homogenizer (IKA, Stanfen, Germany) at 11,000 rpm for 10 min. Then, the mixture was centrifuged (6000 rpm, 10 min), the supernatant was collected, and the solid residue was re-extracted with fresh acetone until it was completely discolored. The carotenoid content of the tomato fruits, expressed in mg per 100 g of fruit dry material, was determined using HPLC. In particular, the HPLC system consisted of a Marathon IV series HPLC pump (Rigas Laboratories, Thessaloniki, Greece), an injection valve with a 20 µL fixed loop (Rheodyne Cotaki, CA, USA), and a UV6000 LP diode array detector (DAD; Thermo Separation Products, San Jose, CA, USA). Separation was carried out isocratically on a Kromasil 100 C18 5 µm (250 × 4.6 mm i.d.) column (MZ Analysentechnik GmbH, Mainz, Germany) as described by Mantzouridou and Tsimidou [33]. Column temperature was set at 30 °C with the aid of a Timberline TL-50 controller and maintained as such in a TL-340 column heater. The elution system consisted of a mixture of acetone–acetonitrile (60:40, *v*/*v*). The flow rate was 1.2 mL/min. The injection volume was 20 µL. The analytical sample was prepared after proper dilution (1:2, *v*/*v*) with methanol and filtration through 0.45 µm PTFE filters (Frisenette, Knebel, Denmark). The identification of the major components of the examined plant extracts was achieved by comparing their retention times and spectral characteristics with the available literature data and by using standard compounds. The quantitative determination of carotenoids was performed using appropriate calibration curves of lycopene and β-carotene. The results were expressed as mg/100 g of dry plant material.

2.3.3. Determination of Total Leaf Antioxidant Capacity

Based on the previously mentioned descriptive characteristics three genotypes were selected according to their tolerance or susceptibility to salt stress for further study, namely Formula (medium tolerant), IL6-6 (tolerant), and IL8-9 (susceptible). The total antioxidant capacity of leaf methanolic extracts (using 80% MeOH) was determined for the lowest and the highest soil salinity treatments (A and C, respectively) and different methods were applied [31]. For the methanolic extracts, 1 g of frozen tomato leaf was weighed and dissolved using a pestle and mortar in 4 mL of 80% MeOH (Sigma-Aldrich, St. Louis, MO, USA).

The Folin–Ciocalteu reagent method was performed according to Singleton et al.'s protocol with several modifications [34] to estimate the total phenolic content of the leaves. Plant extract was mixed with Folin–Ciocalteu reagent. After 5 min, saturated sodium carbonate solution (Na_2CO_3 25% *w*/*v*) was then added to the reaction mixture and incubated for 60 min. The absorbance was recorded at 725 nm and phenolic content was estimated using gallic acid as standard.

DPPH (2,2-Diphenyl-1-picrylhydrazyl) radical scavenging activity was examined according to Brand-Williams et al.'s method with some modifications [35]. DPPH (Sigma-Aldrich, St. Louis, MO, USA) radical solution (0.1 mM) was mixed with methanolic extract (0.4 mg each) in an 8:2 proportion and incubated under dark for 10 min at room temperature (RT). Absorbance was recorded at 517 nm against solvent blank, and percentage inhibition was calculated by using the following formula: $I\% = (Abs_{control} - Abs_{sample}/Abs_{control}) \times 100$.

A ferric reducing antioxidant power assay (FRAP) was also conducted. The FRAP reagent was prepared by adding acetate buffer (300 mM, pH 3.6), TPTZ (10 mM) in HCl (40 mM), and ferric chloride ($FeCl_3$) (20 mM) (Sigma-Aldrich, Sigma-Aldrich, St. Louis, MO, USA). Then, 1.5 µL of methanolic extract was mixed with the FRAP reagent for total volume of 200 µL, and the absorbance was recorded at 593 nm after 10 min of incubation.

2.3.4. Statistic Analysis and Visualizations

The results of the present study were statistically analyzed by conducting Ma ltifactor ANOVA (*p* value < 0.05), while for the homogenous subsets, the Duncan test was performed using the statistical package SPSS (Version 18.0) [36]. Visualizations and cross-validation of the results given by the statistical package SPSS were generated using R packages (version 4.3.2; R Core Team, 2023) [37]. Duncan's test was performed using the agricolae package in R for the analysis of variance (ANOVA) results [38]. For each task, box plots, bar plots, and heatmaps were created using ggplot2 [39] (version 4.3.3), reshape2 [40] (version 4.3.3), and dplyr [41] (version 1.1.2), respectively.

3. Results

3.1. Descriptive and Gas Exchange Characteristics

The descriptive characteristics of the genotypes were assessed in accordance with the UPOV system. Several of these characteristics are presented in the following tables. The nine genotypes did not differ statistically in terms of the number of inflorescences on the main stem, measured 72 days after transplantation (72 D.A.T.). However, the IL 6-6 variety exhibited the highest number of inflorescences (6.44). Soil salinity significantly reduced this characteristic in the highest salinity treatment (C), which is a typical response to severe salinity stress. Additionally, all genotypes presented decreased plant height and leaf size under the higher salinity treatments (B and C). Among the genotypes, 'Formula' displayed the greatest plant height and leaf size, followed by IL8-9, IL6-0, and 'IL6-6' (regarding plant height) and 'IL6-6' (regarding leaf size). In contrast, 'IL8-9' exhibited the lowest values for leaf size (Table 1). There were significant decreases of 5.5% and 16.3% in leaf length and significant decreases of 10% and 20.4% in leaf width under salinity treatments B and C, respectively. For the height of fourth inflorescence, no statistical differences between the three treatments were found. Taking into account the fact that the height of the fourth inflorescence is, in general, preferred to be short, the best performance was observed in the introgression lines IL6-3, IL6-0, IL8-1, and IL6-6 (Table 1).

Table 1. Descriptive characteristics regarding the number of inflorescences on the main stem 72 days after transplantation (D.A.T.), plants' height (cm), and leaves' length and width (cm). Salinity treatment regimes: A = 1.88 mS/cm, B = 6.44 mS/cm, and C = 8.63 mS/cm.

		Number of Inflorescences on Main Stem (72 D.A.T.)	Plant Height (cm)	Leaf Length (cm)	Leaf Width (cm)	Height of the Fourth Inflorescence (cm)
Genotype	Formula	6.85 ± 0.64 a	186.84 ± 9.35 a	38.58 ± 1.96 a	30.33 ± 2.48 a	107.75 ± 4.29 a
	IL6-0	6.94 ± 0.51 a	67.17 ± 5.25 bc	26.12 ± 1.06 b	21.61 ± 1.38 bc	36.81 ± 3.96 c
	IL6-3	5.4 ± 0.28 a	53.53 ± 3.91 c	27.14 ± 0.72 b	22.38 ± 1.54 bc	35.74 ± 3.97 c
	IL6-6	6.44 ± 0.18 a	66.33 ± 3.65 bc	29.13 ± 0.78 b	25.79 ± 1.80 b	40.92 ± 4.30 c
	IL8-1	6.17 ± 0.45 a	54.51 ± 1.83 c	26.54 ± 1.06 b	21.81 ± 1.79 bc	37.44 ± 4.59 c
	IL8-6	5.73 ± 0.51 a	60.33 ± 3.25 bc	27.11 ± 0.77 b	22.70 ± 0.96 bc	42.02 ± 3.96 c
	IL8-9	5.34 ± 0.42 a	70.10 ± 5.80 b	27.26 ± 1.06 b	20.69 ± 1.54 c	59.97 ± 4.29 b
	IL9-5	5.62 ± 0.44 a	59.93 ± 3.74 bc	26.77 ± 1.03 b	23.80 ± 1.42 bc	43.26 ± 4.28 c
	IL9-8	5.36 ± 0.35 a	67.41 ± 4.39 bc	27.00 ± 1.39 b	22.23 ± 1.50 bc	46.95 ± 4.29 bc
Salinity treatment	A	6.33 ± 0.23 a	78.89 ± 7.14 a	30.47 ± 0.55 a	25.99 ± 0.86 a	52.97 ± 2.29 a
	B	6.38 ± 0.29 a	82.20 ± 9.33 a	28.79 ± 0.60 b	23.39 ± 0.93 b	50.58 ± 2.54 a
	C	5.18 ± 0.26 b	60.27 ± 5.14 b	25.50 ± 0.60 c	20.70 ± 1.03 c	46.74 ± 2.48 a

Letters indicate the statistically significant differences as discriminated using Duncan's multiple range test (DMRT) (p value < 0.05).

Measurements of fruit characteristics, including polar and equatorial diameter, pericarp thickness, peduncle size within the fruit, and number of carpels were also conducted. High-salinity conditions impacted most of these traits across all genotypes studied. Specifically, the polar and equatorial diameters exhibited a significant increase under treatment B, while a statistically significant decrease was observed under salinity treatment C. In general, differences among the genotypes regarding polar and equatorial diameters were observed, with 'Formula' exhibiting the highest values (which is normal considering it is a large-fruited tomato variety) followed by 'IL6-6', while the lowest values were exhibited in 'IL8-1' and 'IL8-9' (Table 2). Considering pericarp thickness, no statistical differences were displayed among the three salinity treatments. Regarding the genotypes, IL6-0 did not differ from Formula (they both exhibited the highest values), in contrast to IL8-1 and IL8-9, which exhibited the lowest values (Table 2). Referring to the trait of peduncle size inside the fruit, statistically, the highest values were measured under salinity treatment C, which deteriorates fruit quality. Among the genotypes, IL6-6 and Formula exhibited the most increased values. Considering the trait of the number of carpels, under salinity treatment C, a statistically significant decrease was observed compared to salinity treatments A and B. The introgression lines did not present significant differences; however, a lower number of carpels compared to the large fruited hybrid Formula was observed, as expected (Table 2).

Table 2. Descriptive characteristics measurements regarding the fruits' polar diameter (cm), equatorial diameter (cm), pericarp thickness (mm), peduncle size in the fruit (mm), and number of carpels, according to the UPOV system. Salinity treatment regimes: A = 1.88 mS/cm, B = 6.44 mS/cm, and C = 8.63 mS/cm.

		Polar Diameter (cm)	Equatorial Diameter (cm)	Pericarp Thickness (mm)	Peduncle Size Inside the Fruit (mm)	Number of Carpels
Genotype	Formula	5.19 ± 0.11 a	6.66 ± 0.14 a	6.29 ± 0.12 a	13.44 ± 1.37 ab	4.16 ± 0.14 a
	IL6-0	4.69 ± 0.17 ab	4.26 ± 0.16 bc	5.70 ± 0.24 ab	14.19 ± 1.51 ab	2.23 ± 0.13 b
	IL6-3	4.11 ± 0.09 c	4.44 ± 0.16 b	4.99 ± 0.24 bcd	11.73 ± 0.77 bc	2.62 ± 0.14 b
	IL6-6	4.91 ± 0.15 ab	4.48 ± 0.13 b	5.45 ± 0.17 bc	14.48 ± 0.71 a	2.49 ± 0.14 b
	IL8-1	3.63 ± 0.16 c	3.88 ± 0.18 c	4.34 d ± 0.15 d	10.60 ± 0.62 c	2.63 ± 0.14 b
	IL8-6	4.06 ± 0.18 c	4.10 ± 0.23 bc	5.47 ± 0.25 bc	12.04 ± 0.68 abc	2.58 ± 0.13 b
	IL8-9	3.96 ± 0.31 c	3.93 ± 0.17 c	4.53 ± 0.31 d	12.31 ± 0.62 abc	2.37 ± 0.14 b
	IL9-5	4.61 ± 0.37 b	4.06 ± 0.22 bc	5.08 ± 0.38 bcd	11.93 ± 0.90 abc	2.28 ± 0.14 b
	IL9-8	3.88 ± 0.26 c	4.20 ± 0.18 bc	4.79 ± 0.45 cd	10.71 ± 0.33 c	2.35 ± 0.14 b
Salinity treatment	A	4.31 ± 0.11 b	4.43 ± 0.13 b	5.02 ± 0.15 a	11.64 ± 0.38 b	2.72 ± 0.08 a
	B	4.81 ± 0.16 a	4.96 ± 0.19 a	5.23 ± 0.24 a	11.88 ± 0.40 b	2.72 ± 0.09 a
	C	3.85 ± 0.15 c	3.83 ± 0.17 c	5.34 ± 0.20 a	13.66 ± 0.80 a	2.47 ± 0.08 b

Letters indicate the statistically significant differences as discriminated using Duncan's multiple range test (DMRT) (*p* value < 0.05).

Measurements of the gas exchange parameters did not show statistically significant differences among the genotypes or across the three soil salinity treatments. However, under treatments B and C, the values of internal CO_2 concentration (c_i), transpiration rate (E), and stomatal conductance (g_s) were decreased compared to treatment A. For IL6-6, although there were no statistically significant differences to the rest of the genotypes, it demonstrated some of the most increased c_i, A, and WUE values, while the E and g_s values showed an extreme decline, indicating a potential resistance of that genotype to salinity stress. In contrast, IL8-9, although it demonstrated high c_i, it presented lower values for the majority of the other traits measured, and is considered to be a potentially susceptible genotype to salt stress (Table 3).

Table 3. Gas exchange parameters (c_i: internal CO_2 concentration; E: transpiration rate; g_s: stomatal conductance; A: net photosynthetic rate; WUE: water use efficiency). Salinity treatment regimes: A = 1.88 mS/cm, B = 6.44 mS/cm, and C = 8.63 mS/cm.

		ci (vpm)	E (mmol H$_2$O m^{-2} s^{-1})	gs (mmol m^{-2} s^{-1})	A (μmol CO$_2$ m^{-2} s^{-1})	WUE (A/E)
Genotype	Formula	721.33 ± 96.28 a	2.64 ± 0.26 cd	0.22 ± 0.03 a	32.95 ± 6.30 a	13.85 ± 2.95 a
	IL6-0	698.43 ± 111.47 a	2.35 ± 0.31 d	0.14 ± 0.03 a	19.30 ± 7.29 a	10.07 ± 3.42 a
	IL6-3	533.79 ± 111.48 a	3.30 ± 0.32 abc	0.28 ± 0.03 a	29.31 ± 7.30 a	9.07 ± 3.41 a
	IL6-6	707.33 ± 96.28 a	2.48 ± 0.27 cd	0.21 ± 0.03 a	24.74 ± 6.29 a	12.51 ± 2.95 a
	IL8-1	516.9 ± 96.284 a	3.24 ± 0.26 abcd	0.20 ± 0.03 a	26.69 ± 6.30 a	8.49 ± 2.94 a
	IL8-6	579.50 ± 111.47 a	3.98 ± 0.30 a	0.27 ± 0.03 a	21.59 ± 7.30 a	5.51 ± 3.42 a
	IL8-9	734.44 ± 96.28 a	2.45 ± 0.27 cd	0.14 ± 0.03 a	18.48 ± 6.30 a	8.27 ± 2.96 a
	IL9-5	501.83 ± 96.28 a	2.87 ± 0.26 bcd	0.18 ± 0.03 a	18.42 ± 6.29 a	8.27 ± 2.95 a
	IL9-8	563.94 ± 104.32 a	3.63 ± 0.29 ab	0.27 ± 0.03 a	33.06 ± 6.82 a	9.72 ± 3.12 a
Salinity treatment	A	729.46 ± 58.66 a	3.21 ± 0.16 a	0.24 ± 0.17 a	24.72 ± 3.80 a	9.05 ± 1.80 a
	B	551.38 ± 58.65 b	2.90 ± 0.16 a	0.21 ± 0.17 a	22.71 ± 3.80 a	8.84 ± 1.79 a
	C	575.40 ± 60.25 ab	2.79 ± 0.17 a	0.19 ± 0.18 a	27.57 ± 3.90 a	10.72 ± 1.85 a

Letters indicate the statistically significant differences as discriminated using Duncan's multiple range test (DMRT) (*p* value < 0.05).

Measurements regarding early and total yield characteristics were also determined. Early yield sampling was conducted at 65 DAT, while total yield sampling was carried out at 95 DAT. Fruit number per plant, yield per plant, and weight per fruit were measured to determine both early and total yield. In both early and total yield, the trait of fruit number did not demonstrate any significant differences among the three salinity treatments. However, in comparisons made when comparing the two higher salinity treatments (B and C) to treatment A, increased levels of fruit weight were observed under treatment B (33% in early yield and 19% in total yield measurements) leading to a total yield increase. In contrast, under treatment C, fruit weight declined notably, both in early and total yield (reaching a 29% and 34% reduction, respectively), significantly affecting the total productivity. This change in fruit weight between the two higher salinity treatments led to a corresponding increase in total yield (g/plant) of 33% and 13% under treatment B and a decrease of 42% and 38% under treatment C, in early and total production, respectively (Figure 2).

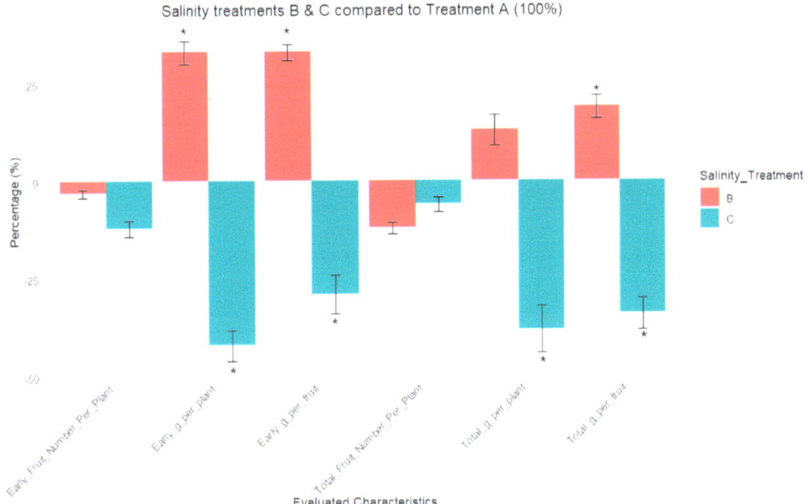

Figure 2. Early and total yield measurements for fruit number per plant, yield per plant, and fruit weight, in early and total production, for salinity treatment values B = 6.44 mS/cm and C = 8.63 mS/cm, given as percentage (%) of salinity treatment A = 1.88 mS/cm. (*) = significant differences, *p* value 0.05.

Similar observations can be noticed in Figures 3–5, where comparisons among the nine genotypes have been conducted for total yield characteristics per plant between the lowest and the highest soil salinity treatments, i.e., A and C, respectively. Regarding the number of fruits per plant, there are no statistically different observations through the different ILs (Figure 3), while yield per plant and weight per fruit demonstrate significant differences between the two salinity conditions among the different genotypes (Figures 4 and 5). More specifically, regarding the number of fruits in total yield, when comparing the lowest (A) and the highest (C) soil salinity, the results indicate that the ILs did not differ statistically (Figure 3). IL6-6 was the only genotype that maintained high values of yield traits under salinity treatment C. The same genotype, in terms of early yield (g/plant), showed an increase of 38% and 2%, while in total yield, it showed an increase of 48% and 18%, for treatments B and C, compared to treatment A, respectively. IL8-9 showed the greatest reduction under treatment C compared to treatment A, in terms of early and total yields (g/plant) of 61% and 64%, respectively. Concerning the different genotypes, IL6-6 and IL8-6 demonstrated some of the highest values in all the evaluated yield characteristics, while IL8-9 represented one of the lowest values regarding the yield per plant trait, both in terms of early and total yield (Table 4).

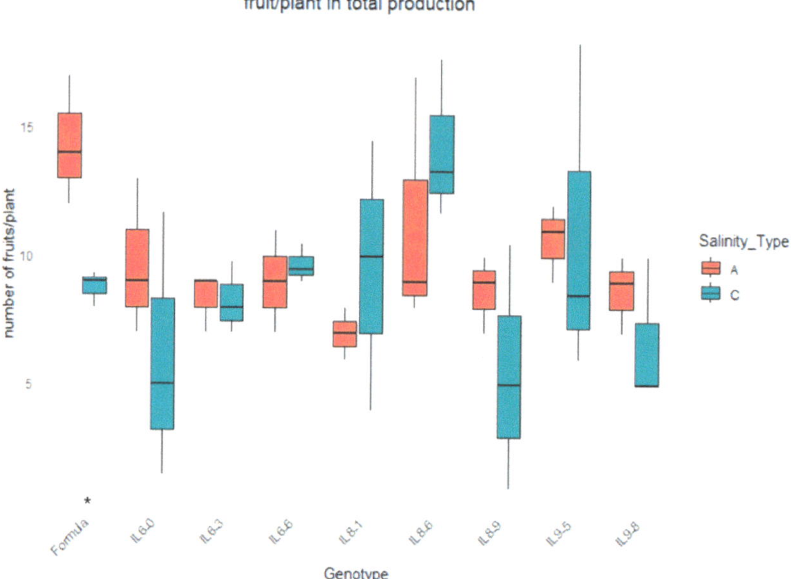

Figure 3. The nine genotypes' number of fruits per plant in total production under the two different salinity conditions (A = 1.88 mS/cm and C = 8.63 mS/cm). (*) = significant differences, *p* value 0.05.

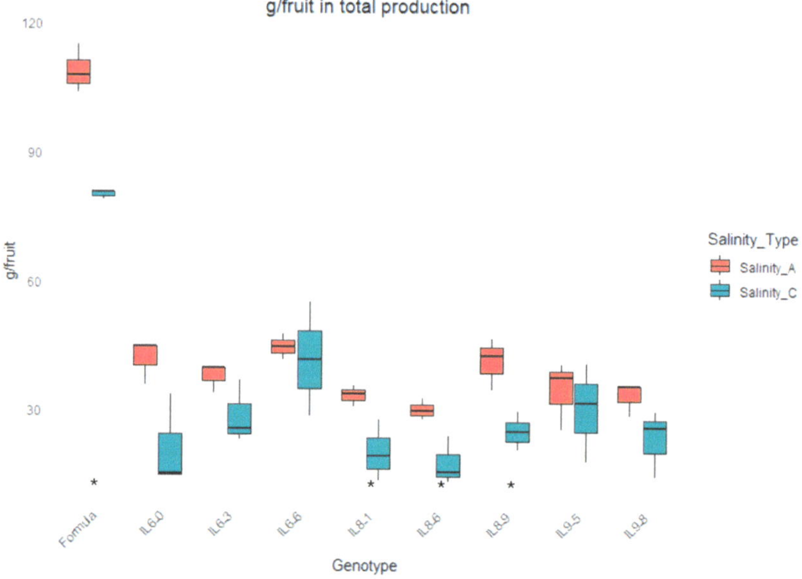

Figure 4. The nine genotypes' fruit weight of total production under the two different salinity conditions (A = 1.88 mS/cm and C = 8.63 mS/cm). (*) = significant differences, *p* value 0.05.

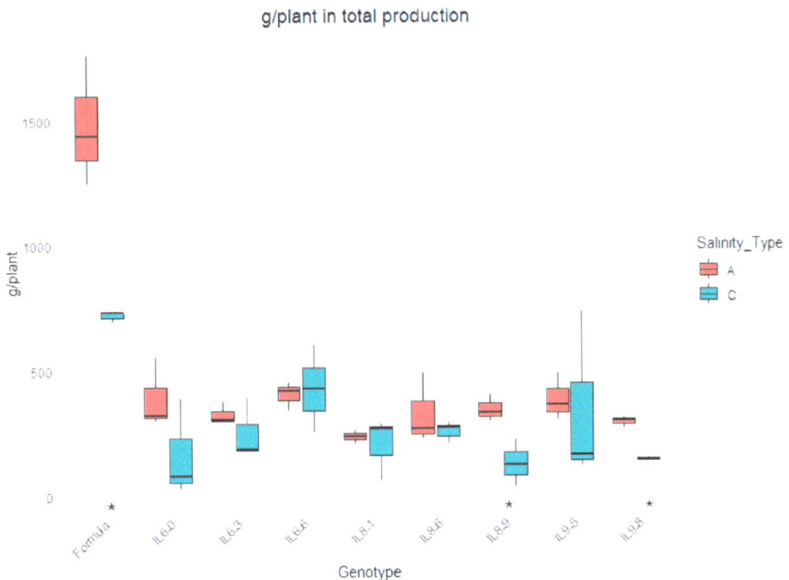

Figure 5. The nine genotypes' yield per plant of total production under the two different salinity conditions (A = 1.88 mS/cm and C = 8.63 mS/cm). (*) = significant differences, *p* value 0.05.

Table 4. Early and total yield measurements for: fruit number per plant; grams per plant; grams/fruit. Salinity treatment values: A = 1.88 mS/cm, B = 6.44 mS/cm, and C = 8.63 mS/cm.

		Early Yield			Total Yield		
		Fruit Number/Plant	g/Plant	g/Fruit	Fruit Number/Plant	g/Plant	g/Fruit
Genotype	Formula	3.64 ± 0.56 d	458.54 ± 79.02 a	124.57 ± 3.96 a	10.49 ± 1.29 ab	1116.23 ± 176.87 a	104.70 ± 5.86 a
	IL6-0	4.94 ± 0.92 cd	209.67 ± 44.84 c	39.79 ± 5.96 bc	7.41 ± 1.19 b	302.33 ± 56.07 b	39.60 ± 5.10 bc
	IL6-3	7.03 ± 0.67 abc	285.36 ± 53.47 bc	37.50 ± 3.82 bc	8.62 ± 0.50 ab	337.41 ± 49.18 b	37.74 ± 3.41 bc
	IL6-6	7.54 ± 0.45 ab	338.56 ± 44.28 b	45.50 ± 4.62 b	10.00 ± 0.62 ab	454.03 ± 68.98 b	44.03 ± 4.64 b
	IL8-1	5.44 ± 0.73 bcd	166.50 ± 9.26 c	32.50 ± 3.71 c	7.31 ± 1.17 b	209.89 ± 23.65 b	31.10 ± 3.70 c
	IL8-6	7.96 ± 0.72 a	242.72 ± 36.41 bc	33.67 ± 5.10 c	11.99 ± 1.26 a	348.56 ± 51.44 b	31.76 ± 4.21 c
	IL8-9	5.13 ± 0.73 bcd	181.13 ± 36.12 c	35.34 ± 3.27 c	6.61 ± 1.09 b	236.44 ± 45.32 b	36.08 ± 2.83 bc
	IL9-5	7.31 ± 0.61 abc	268.73 ± 8.06 bc	35.68 ± 5.10 c	9.31 ± 1.56 ab	341.80 ± 85.18 b	34.27 ± 4.16 bc
	IL9-8	4.99 ± 0.96 cd	172.23 ± 6.64 c	34.48 ± 4.53 c	7.94 ± 0.90 b	277.06 ± 63.47 b	34.27 ± 4.51 bc
Salinity treatment	A	6.38 ± 0.36 a	264.96 ± 20.18 b	44.88 ± 4.30 b	9.37 ± 0.56 a	431.41 ± 68.25 a	45.37 ± 4.48 b
	B	6.18 ± 0.51 a	353.12 ± 39.42 a	59.88 ± 6.52 a	8.28 ± 0.74 a	489.17 ± 78.04 a	54.00 ± 5.26 a
	C	5.60 ± 0.61 a	154.23 ± 21.51 c	31.77 ± 5.39 c	8.85 ± 0.85 a	267.17 ± 42.04 b	30.12 ± 3.54 c

Letters indicate the statistically significant differences as discriminated using Duncan's multiple range test (DMRT) (*p* value < 0.05).

3.2. Fruit Quality Characteristics

The resistance of fruit flesh to pressure and total soluble solids (TSS) were also affected by severe soil salinity stress. A non-significant increase of 5% in environment B and a significant decrease of 14% in environment C were observed for this trait. 'Formula' was identified as the most susceptible genotype to flesh pressure, whereas the ILs displayed varying degrees of tolerance, with 'IL9-5', 'IL8-6', and 'IL6-6' being the most tolerant. TSS levels were generally higher in most ILs and lower in 'Formula'. A progressive and statistically significant decrease for this characteristic was observed among the three treatments, with values of 8% (treatment B) and 31% (treatment C), respectively. IL6-3 demonstrated high TSS values under treatment A, and it maintained its high values under treatment C (Figure 6). pH was not significantly influenced by the different soil salinity conditions, although pH differences were noted among the genotypes, with 'IL6-3', 'Formula', 'IL8-1' and 'IL6-6' presenting the highest values (Table 5). Figure 6 demonstrates the TSS and their significant differences, comparing each genotype in the lowest (A) and in the highest (C) soil salinity condition. Most ILs presented a significant reduction in TSS, while Formula, IL6-0 and IL6-3 reduced the TSS, but not significantly (Figure 6).

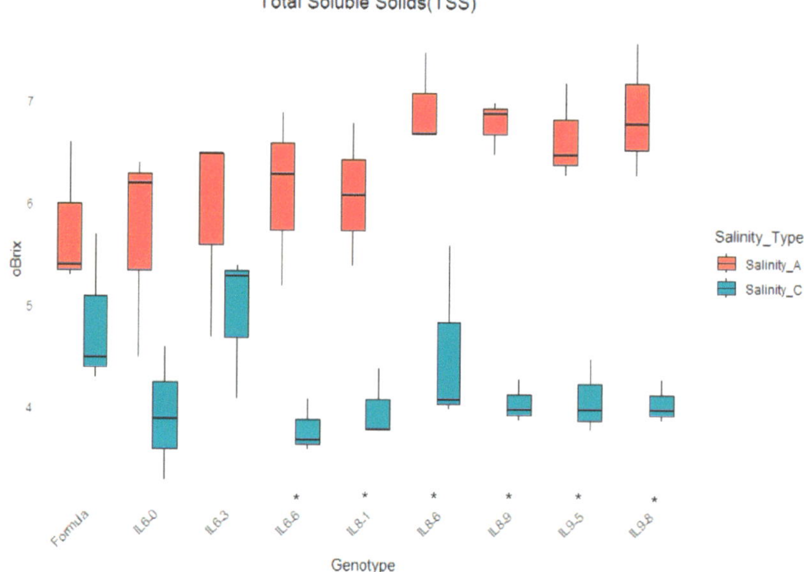

Figure 6. The nine genotypes' total soluble solids (TSS) under the two different salinity conditions (A = 1.88 mS/cm and C = 8.63 mS/cm). (*) symbol indicates the statistically significant differences between the two salinity conditions for each genotype as discriminated using R (version 4.3.2; R Core Team, 2023) (*p* value 0.05).

During the statistical analysis, a multifactorial ANOVA was employed to reveal whether the interaction between the genotypes and the three soil salinity levels had a significant effect on the studied characteristics. The results showed statistically significant interaction only for polar diameter, where the interaction of genotypes and salinity levels had a significant effect on this characteristic with a *p* value of 0.012 (<0.05).

Four genotypes were selected for further investigation of their fruit quality under the three soil salinity conditions, based on their descriptive, gas exchange, and yield characteristics. 'Formula', a commercial tomato hybrid used as a reference for salinity tolerance, was compared with 'IL6-6' and 'IL8-6', identified as resistant lines, and 'IL8-9', discriminated for its susceptibility to salinity stress. The results show that in the normal salinity condition (A), 'Formula' has a total carotenoid (T.C.) content of approximately 172 mg/100 g dry material (D.M.), but this value is reduced notably under salinity stress, reaching 89 mg/100 g D.M. under treatment B (a 48% reduction) and 74 mg/100 g D.M. under treatment C (a 57% reduction). Interestingly, the highest T.C. content under normal salinity was found in 'IL6-6', with 237 mg/100 g D.M. Notably, under severe salinity stress, 'IL6-6' retains 155 mg/100 g D.M. under treatment C (35% reduction), only 17 mg lower than 'Formula' under normal conditions, indicating that 'IL6-6' preserves high nutritional value even under high salinity. While 'IL8-6' and 'IL8-9' initially show lower T.C. accumulation under normal salinity (A) compared to the other two genotypes, under severe salinity stress, the T.C. levels in 'IL8-9' are only slightly reduced, and in 'IL8-6', they remain stable (Figure 7a). A similar trend is observed in lycopene and beta-carotene accumulations, further emphasizing that the ILs, particularly 'IL6-6', maintain higher fruit quality under salinity stress compared to 'Formula' (Figure 7b,c).

Table 5. Measurements of fruit quality characteristics, regarding the fruits' resistance of flesh to pressure (Kg), total soluble solids (°Brix), and pH, according to the UPOV system. Salinity treatment regimes: A = 1.88 mS/cm, B = 6.44 mS/cm, and C = 8.63 mS/cm.

		Resistance of Flesh to Pressure (Kg)	TSS (°Brix)	pH
Genotype	Formula	2.53 ± 0.26 $_e$	5.40 ± 0.24 $_{cd}$	4.58 ± 0.03 $_{ab}$
	IL6-0	3.63 ± 0.10 $_{bcd}$	4.61 ± 0.37 $_e$	4.55 ± 0.02 $_b$
	IL6-3	3.28 ± 0.16 $_d$	5.62 ± 0.28 $_{abc}$	4.67 ± 0.02 $_a$
	IL6-6	3.80 ± 0.14 $_{abc}$	4.76 ± 0.41 $_{de}$	4.56 ± 0.02 $_{ab}$
	IL8-1	3.76 ± 0.20 $_{bcd}$	5.36 ± 0.43 $_{cd}$	4.57 ± 0.02 $_{ab}$
	IL8-6	3.99 ± 0.09 $_{ab}$	5.90 ± 0.38 $_{abc}$	4.52 ± 0.02 $_b$
	IL8-9	3.60 ± 0.30 $_{bcd}$	6.07 ± 0.42 $_{ab}$	4.50 ± 0.05 $_b$
	IL9-5	4.17 ± 0.12 $_a$	5.70 ± 0.37 $_{abc}$	4.51 ± 0.03 $_b$
	IL9-8	3.41 ± 0.14 $_{cd}$	6.14 ± 0.50 $_a$	4.40 ± 0.02 $_c$
Salinity treatment	A	3.70 ± 0.11 $_a$	6.25 ± 0.17 $_a$	4.55 ± 0.02 $_a$
	B	3.87 ± 0.10 $_a$	5.75 ± 0.19 $_b$	4.55 ± 0.02 $_a$
	C	3.20 ± 0.15 $_b$	4.31 ± 0.14 $_c$	4.53 ± 0.02 $_a$

Letters indicate the statistically significant differences as discriminated using Duncan's multiple range test (DMRT) (p value < 0.05).

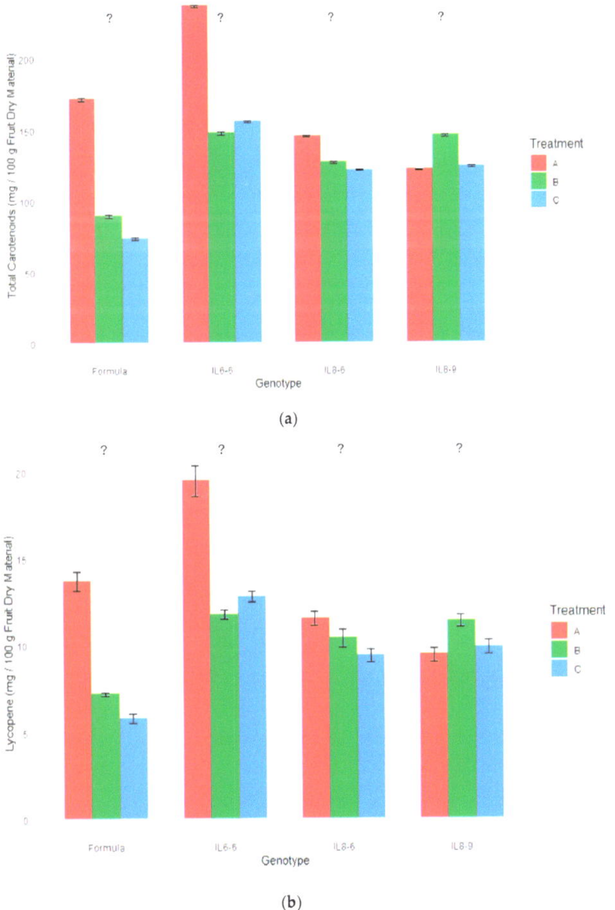

Figure 7. *Cont.*

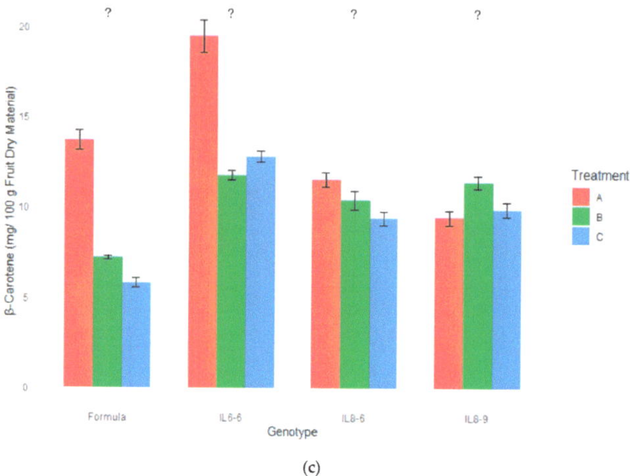

(c)

Figure 7. (**a**) Total carotenoids (T.C.) accumulation in the four tomato genotypes under the three salinity treatments. (**b**) Lycopene accumulation in the four tomato genotypes under the three salinity treatments. (**c**) Beta-carotene accumulation in the four tomato genotypes under the three salinity treatments. (A = 1.88 mS/cm, B = 6.44 mS/cm, and C = 8.63 mS/cm). "?" indicates the statistically significant differences between the three salinity conditions for each genotype as discriminated using R (version 4.3.2; R Core Team, 2023) (*p* value 0.05).

3.3. Leaf Total Antioxidant Capacity

The antioxidant defense system in leaves, including antioxidant compounds (total phenols measured by the FOLIN method) and antioxidant capacity assessed using two different methods (DPPH and FRAP) were also evaluated. Three out of the previous four genotypes were selected for further experimentation considering the descriptive, photosynthetic, yield, and nutritional value results (Formula as control, IL6-6 as the tolerant genotype, and IL8-9 as the susceptible genotype to salinity). In terms of total phenols using the FOLIN method, no statistically significant (ns) differences were observed when comparing each genotype individually across the two salinity treatments. However, when comparing the genotypes within each salinity treatment, it was found that the total phenol content was significantly higher in the two ILs compared to Formula under treatment A (Figure 8).

The determination of antioxidant capacity using the DPPH method showed that only 'IL8-9' exhibited a significant reduction in inhibition percentage (%) under salinity stress (C) compared to normal salinity conditions (A). When comparing the mean inhibition among genotypes, differences were observed across all genotypes under treatment A and between IL6-6 and IL8-9 under treatment C. IL6-6 maintains inhibition (%) under both treatments (Figure 9).

Similar results were observed with the FRAP method, where antioxidant capacity was significantly reduced under treatment (C) for 'IL8-9' compared to treatment (A). Comparisons between the genotypes under treatment (A) revealed that 'IL6-6' had a significantly higher antioxidant capacity compared to Formula. However, under treatment (C), no significant differences were found between the genotypes' responses to high salinity (Figure 10).

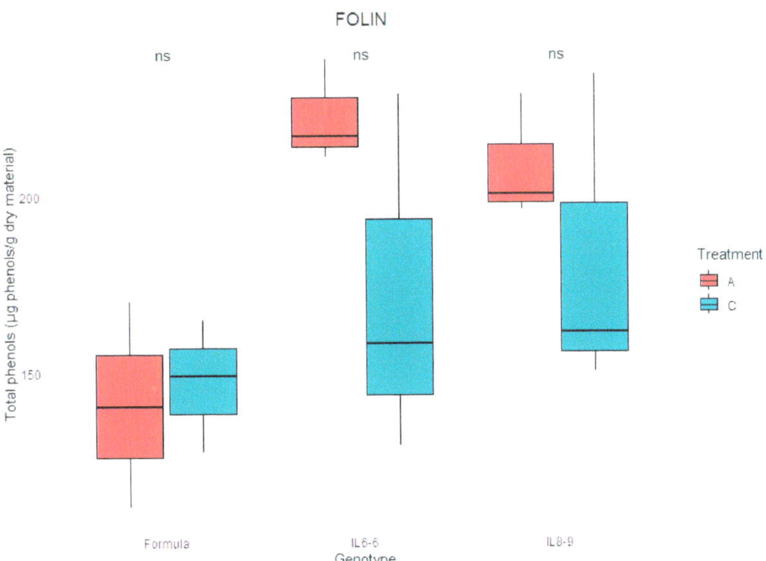

Figure 8. Total phenols content for the three genotypes between the two soil salinity treatments (A = 1.88 mS/cm, C = 8.63 mS/cm). ns = no significant differences among the comparisons, p value < 0.05). "ns" indicates "non-statistically significant" differences between the three salinity conditions for each genotype as discriminated using R (version 4.3.2; R Core Team, 2023) (p value 0.05).

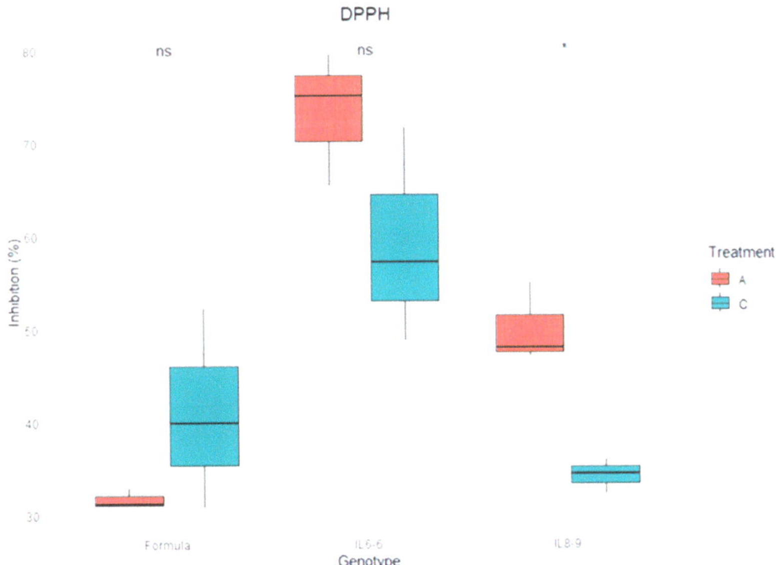

Figure 9. Antioxidant capacity (DPPH) for the three genotypes between the two soil salinity treatments (A = 1.88 mS/cm, C = 8.63 mS/cm). ns = no significant differences among the comparisons, (*) = significant differences, p value 0.05.

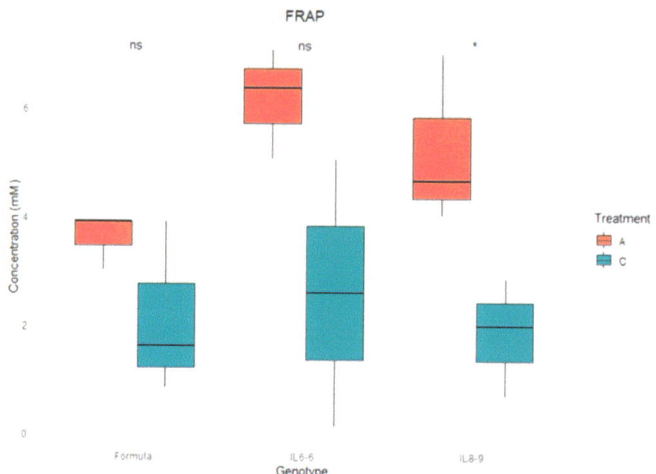

Figure 10. Antioxidant capacity (FRAP) for the three genotypes between the two soil salinity treatments (A = 1.88 mS/cm, C = 8.63 mS/cm). ns = no significant differences among the comparisons, (*) = significant differences, *p* value 0.05.

3.4. Trait Correlations and Genotype Performance Under Salinity Stress

Breeding for crop yield maintenance is the most critical quantitative trait regarding salinity tolerance. In Figure 11, the Pearson correlation index of total yield with other traits highlights key characteristics that could serve as potential indicators of genotype tolerance to salinity stress. In the analysis conducted, considering the three different soil salinity treatments, agronomic traits related to plant height and fruit morphology exhibit strong positive correlations with total yield (g/plant). Specifically, traits, such as plant height (cm), height of the fourth inflorescence, and length of three internodes, display a robust positive correlation with total yield. Similarly, fruit-related parameters, including polar diameter, equatorial diameter, pericarp thickness, and the number of carpels, show a strong positive association with yield, highlighting that larger fruit size and greater pericarp thickness enhance yield potential in varying salinity conditions. Moderate positive correlations are observed with other traits, such as the resistance of flesh to pressure (Kg) and peduncle size inside the fruit (mm). In contrast, biochemical parameters, including FOLIN, FRAP, DPPH, β-carotene, and lycopene, display weak or no correlation with total yield. Physiological traits related to gas exchange parameters, such as WUE, A, gs, and ci, present mixed correlations, which are predominantly weak or neutral, indicating that these traits are not significantly associated with yield variations in response to salinity stress. Overall, these findings underscore the greater importance of agronomic traits, particularly those related to plant height, fruit size, and structural integrity, in association with yield under salinity stress, compared to biochemical and specific physiological parameters (Figure 11). The traits that were found to be strongly related to yield could potentially be utilized as indexes for genotype salinity tolerance discrimination.

The heatmap in Figure 12 illustrates the ranking of the nine studied genotypes based on the specific agronomic traits that correlated significantly (Pearson correlation coefficient higher than 0.5) with yield, when comparing their gain or loss (%), in soil salinity environments (treatment C compared to treatment A). Each cell color represents the score (1–9) for a given trait in each genotype, with darker colors indicating lower scores, which are preferable (i.e., lower scores represent better performance). Genotypes IL6-6, IL6-3, and IL9-5 show darker cells for many traits, indicating more favorable trait performance under treatment C compared to treatment A. Conversely, some genotypes, such as IL9-8, IL8-9, and IL 6-0, display lighter colors for multiple traits, suggesting a comparatively weaker performance in response to treatment C. This heatmap provides a visual assessment of genotype performance for multiple traits, aiming to identify salt-tolerant genotypes suitable for high-salinity environments (Figure 12).

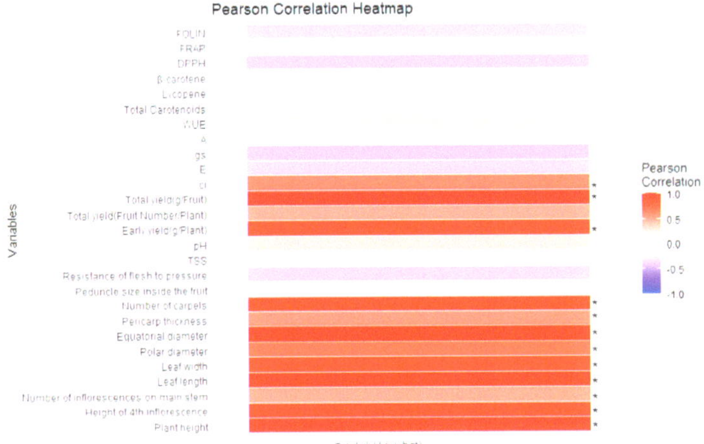

Figure 11. Pearson correlation heatmap displaying the correlation between various agronomic, physiological, and biochemical traits and total yield (g/plant) in tomato genotypes in all soil salinity treatments. Colors represent correlation strength, with intense red indicating strong positive and intense blue strong negative correlations, (*) = significant differences, *p* value 0.05. Abbreviations of the measured traits as presented on the above heatmap (Folin–Ciocalteu reagent method—FOLIN; ferric reducing antioxidant power assay—FRAP; 2,2-Diphenyl-1-picrylhydrazyl—DPPH; water use efficiency—WUE; net photosynthetic rate—A; stomatal conductance—gs; transpiration rate—E; internal CO_2 concentration—ci; total soluble solids—TSS).

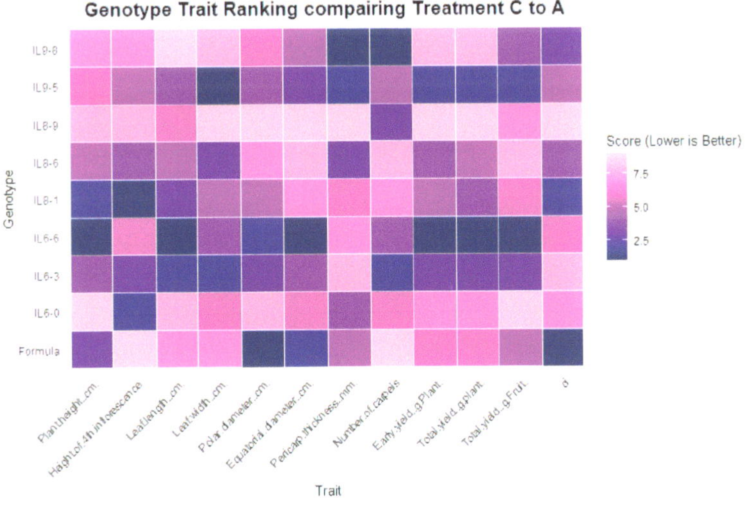

Figure 12. Heatmap showing genotype rankings based on the relative performance of agronomic traits under treatment C compared to treatment A. Darker colors represent lower scores, indicating better performance.

4. Discussion

The present study highlights the potential use of tomato introgression lines (ILs) for cultivation under adverse environmental conditions without compromising fruit quality. The findings demonstrate the nine tomato genotypes' responses in salinity stress through the evaluation of descriptive characteristics, gas exchange parameters, productivity, fruit quality, nutritional value, and antioxidant capacity.

The traits for which notable genotypic variations were not found are probably linked to the genomic segments that are shared by the genotypes rather than the introgressed fragments [24]. Only the features exhibiting genotypic differences will be explicitly discussed, since the differences favored by the introgressed segments are the ones of interest.

Among all the traits measured, no significant differences were observed under the three salinity regimes (soil salinity of 1.88 mS/cm, 6.44 mS/cm, and 8.63 mS/cm) for some of them, like the number of inflorescences, pericarp thickness, E (transpiration rate), gs (stomatal conductance), A (net photosynthetic rate), WUE (water use efficiency), early fruit number/plant, fruit number/plant, and pH. In a 2014 study, Barrios-Masias et al. [42] investigated the ecophysiological traits related to WUE and yield in California processing tomatoes, using introgression lines (ILs) with benefits in leaf gas exchange-related traits that are relevant to crop improvement under dry and hot environments. Although it was previously shown in ILs of other plant species [43] that estimating the net assimilation rate could be a useful method for differentiating between genotypes that are tolerant to salt stress and those that are not, no statistically significant differences were observed in our studied ILs. However, IL6-6 exhibited high water use efficiency (WUE) under saline conditions, a trait that also mediates drought tolerance, as recently described [44,45].

Regarding the polar diameter and equatorial diameter of the fruit, there was an increase in fruit size in salinity treatment B while a significant decrease in fruit size was observed in salinity treatment C. Similarly, for the related trait of fruit weight, a notable increase was observed under treatment B in both early and total yield, whereas a significant decrease occurred under treatment C. In a recent review article, Guo et al. in 2022 [15] addressed the impact of salinity on fruit morphological traits in tomatoes, summarizing findings from various studies on how salt stress affects fruit size, indicating that severe salinity stress can affect fruit size, potentially leading to more irregular or smaller fruit. Similarly, in our study, it was observed that the fruit morphological traits are generally affected by high-salinity regimes.

Leaf size and soluble solids demonstrated a significant and gradual decrease in values as salinity levels increased. Relevant results regarding leaf morphology are also revealed in a study conducted by Rebah et al., where they studied three tomato ILs by measuring morphological parameters, like total leaf area, thirty-two days after transplantation [22]. Similarly, Ali et al., 2021, demonstrated a smaller reduction in growth parameters, such as leaf number and shoot length, under salinity treatments in tomato ILs compared to the commercial varieties [23], which is in accordance with our findings regarding plants' height. Furthermore, Chitwood et al. in 2013 examined leaf morphological characteristics across an IL population. The study identified specific genomic regions linked to traits, such as leaf complexity, curvature, and shape, and the findings suggest that introgressed segments from *S. pennellii* contribute to morphological diversity in leaf traits, revealing complex genetic interactions that underlie leaf development and shape in tomatoes [46]. However, according to several research studies, while morphological size and weight are often reduced, salinity stress can increase certain quality parameters in low soil salinity concentrations, such as soluble solid content, potentially due to reduced water content in tomato fruit, which leads to the increased concentration of sugars and organic acids, resulting in higher TSS levels and enhanced flavor intensity [15]. Nonetheless, our work indicates there are some ILs, like IL6-3, that maintain high levels of total soluble solids under severe salinity stress.

Regarding traits related to the nutritional value of the fruit—total carotenoids, lycopene, and β-carotene—genotypes responded in a different manner in the two saline environments compared to treatment A (and not necessarily in accordance to their general response to increased salinity). In accordance with this, Rebah et al. (2018) compared the responses of three tomato introgression lines to salinity stress and found a decrease in the carotenoid content in both the tolerant and the sensitive genotype [22]. In our study, IL6-6 under severe salinity stress maintained its nutritional value, almost at the same level as Formula under normal salinity conditions, highlighting its value as breeding material.

Concerning traits associated with leaf antioxidant capacity (total phenolics, total antioxidants via DPPH, and total antioxidants via FRAP), similar or lower concentrations were generally observed under treatment C compared to treatment A. The findings of Frary et al., 2010, reported an opposite response, when studying the wild tomato (*S. pennellii*) and other introgression lines that accumulated more antioxidant components (phenolics, flavonoids, SOD, CAT, and APX) than the cultivated tomato under the salt stress (150 mM NaCl) [47]. This can be attributed to the small duration of their experiment (21 days) and the different environmental conditions.

The most important feature, however, whose response appears as the general consequence of the action of all other characteristics and mechanisms of salinity tolerance, is the overall yield performance of genotypes. For this characteristic, therefore, in the comparison between treatments, treatment B did not differ significantly from treatment A, while C lagged significantly behind both other salinity treatments. In fact, the increase in value in terms of fruit weight under treatment B, a characteristic that, in contrast to the number of fruits per plant, seems to be influenced by salinity, implies a corresponding increase in yield in this treatment, whereas the reduction in values under treatment C in fruit weight has an effect on the corresponding reduction in yield. It appears that tomato genotypes react differently to different salinity treatments and, in this case, between 6.44 mS/cm and 8.63 mS/cm, there is a change in the effect of salinity on plant's yield performance.

During research on unraveling characteristics and criteria for tolerance to salinity of tomato genotypes, it may help to focus on those characteristics that are significantly associated with the overall yield. Estimating the absolute values of these characteristics at the different salinity levels is not sufficient without a proportionate assessment of the relative change (%) in the values of these characteristics under the different salinity regimes. After estimating the relative change in the values of these characteristics and classifying the genotypes from 1 to 9 accordingly from salinity level C to salinity level A, genotypes IL6-6 and IL8-9 were observed to be in the first and last position, respectively.

Therefore, it is necessary to assess the behavior of these two genotypes as two extreme classes, namely tolerance (IL6-6) and susceptibility/non-tolerance (IL8-9) to salinity, to better understand the critical characteristics that give or indicate tolerance. Table 6 below presents the classification of genotypes in terms of measured characteristics as general averages in all three salinity treatments (Table 6). It is observed that the tolerant genotype IL6-6 is found to have the vast majority of characteristics in the first place in the ranking, while the sensitive one (IL8-9) has a majority of characteristic in last place in the ranking. Therefore, the salinity-tolerant genotype not only has the best relative behavior in terms of change in the values of its studied characteristics under high salinity (treatment C) compared to treatment A, but it is also generally characterized by high values in the studied characteristics.

In conditions of very high (extreme) salinity, it seems that the yield depends more on the weight of the fruit rather than on the number of fruits per plant, a condition that prevails at normal salinity levels. This is evidenced by the higher correlation between yield and fruit weight (0.85 *p* value 0.01) compared to the correlation of yield with the number of fruits per plant (0.59, *p* value 0.01), in the present study. This is also in accordance with Pessoa et al., 2023, where 10 tomato ILs were subjected to drought stress and one of the characteristics that was correlated to increased tolerance was fruit weight rather than the number of fruits per plant [24].

Table 6. The classification of the two genotypes (IL6-6 as the resistant genotype and IL8-9 as the susceptible genotype to salinity), in terms of measured characteristics as general averages in all three salinity treatments. The classification was conducted from 1 to 9 accordingly from salinity level C to salinity level A. Number 1 indicates the best performance in response to salinity stress while 9 indicates the worst performance in response.

Genotype	IL6-6	IL8-9
Number of inflorescences on main stem (72 D.A.T.)	3	9
Plant height	5	2
Height of the 4th inflorescence	4	8
Leaf length	2	3
Leaf width	2	9
Polar diameter	2	7
Equatorial diemeter	2	8
Pericarp thickness	4	8
Number of carpels	5	6
Peducle size inside the fruit	9	6
ci	3	1
E	3	2
gs	6	3
A	5	8
WUE	2	7
Early fruit number/plant	2	6
Early g/plant	2	7
Early g/fruit	2	6
Total fruit number/plant	3	8
Total g/plant	2	8
Total g/fruit	2	5
Resistance of flesh to pressure	3	6
TSS	8	2
pH	4	8
Total carotenoids	1	3
Lycopene	1	3
β-carotene	1	3
DPPH	1	2
FRAP	1	2
FOLIN	1	2

For total carotenoids, lycopene, β-carotene, the classification scale is 1 to 4, as four out of the total nine genotypes were measured. For DPPH, FRAP, and FOLIN, the classification scale is 1 to 3, as three out of the total nine genotypes were measured.

Some of these ILs have been tested in the field under high *Orobanche* sp. Infestation stress and performed well [17]. For instance, line IL6-3 performed well in both saline soils and soils infected with *Orobanche* sp. It would be of great agronomic importance to test the resilient lines under combinatorial stress and monitor their performance. These results could potentially contribute to increased crop resilience under scenarios of abiotic/biotic stress combinations.

5. Conclusions

In conclusion, this study emphasizes the value of utilizing tomato ILs for breeding programs for the cultivation of tomatoes under saline conditions without sacrificing fruit quality and yield. Genotypes, like IL6-6, which showed resilience in yield and fruit weight under high salinity, emerged as potential candidates for breeding programs aiming to improve tomato's salinity tolerance. Critical traits, such as fruit weight, plant height, and leaf length, under saline conditions proved to be indicators of tolerance. Additionally, high yield correlated more closely to fruit weight than fruit number at elevated salinity levels. These insights offer a framework for breeders: focusing on these key traits and relative changes in trait values under varying salinity levels can guide the selection of tolerant genotypes with stable yields and high quality under saline soils. Leveraging these ILs with possible combined resilience to abiotic/biotic stressors can pave the way for breeding high-quality tomato cultivars suited for a plethora of adverse environments, especially given the increasing challenges posed by climate change.

Author Contributions: Conceptualization, I.A. and E.T.; methodology, I.A.; software, M.G.; validation, I.A. and M.G.; formal analysis, I.A., M.G., A.K., D.N. and E.G.C.; investigation, M.G., D.N. and A.K.; resources, I.A., I.M., E.T. and E.G.C.; data curation, I.A., I.M., E.T. and E.G.C.; writing—original draft preparation, I.A., M.G. and E.T.; writing—review and editing, I.A., E.T. and M.G.; visualization, M.G.; supervision, I.A. and E.T.; project administration, I.A. and E.T.; funding acquisition, I.A., I.M., E.T. and E.G.C. All authors have read and agreed to the published version of the manuscript.

Funding: This research received no external funding.

Data Availability Statement: Further data supporting the reported results can be made available upon reasonable request from the corresponding authors.

Acknowledgments: The research work was partly supported by the Hellenic Foundation for Research and Innovation (HFRI) under the 3rd Call for HFRI PhD Fellowships (Proposal Number: 06001).

Conflicts of Interest: The authors declare no conflicts of interest.

References

1. Rawat, A.; Kumar, D.; Khati, B.S. A Review on Climate Change Impacts, Models, and Its Consequences on Different Sectors: A Systematic Approach. *J. Water Clim. Chang.* **2023**, *15*, 104–126. [CrossRef]
2. Chaudhry, S.; Sidhu, G.P.S. Climate Change Regulated Abiotic Stress Mechanisms in Plants: A Comprehensive Review. *Plant Cell Rep.* **2022**, *41*, 1–31. [CrossRef] [PubMed]
3. Parihar, P.; Singh, S.; Singh, R.; Singh, V.P.; Prasad, S.M. Effect of Salinity Stress on Plants and Its Tolerance Strategies: A Review. *Environ. Sci. Pollut. Res.* **2015**, *22*, 4056–4075. [CrossRef]
4. Mall, M.; Kumar, R.; Akhtar, M.Q. Horticultural Crops and Abiotic Stress Challenges. In *Stress Tolerance in Horticultural Crops*; Elsevier: Amsterdam, The Netherlands, 2021; pp. 1–19.
5. Deinlein, U.; Stephan, A.B.; Horie, T.; Luo, W.; Xu, G.; Schroeder, J.I. Plant Salt-Tolerance Mechanisms. *Trends Plant Sci.* **2014**, *19*, 371–379. [CrossRef]
6. Gruda, N.S.; Dong, J.; Li, X. From Salinity to Nutrient-Rich Vegetables: Strategies for Quality Enhancement in Protected Cultivation. *Crit. Rev. Plant Sci.* **2024**, *43*, 327–347. [CrossRef]
7. Nawaz, K.; Hussain, K.; Majeed, A.; Khan, F.; Afghan, S.; Ali, K. Fatality of Salt Stress to Plants: Morphological, Physiological and Biochemical Aspects. *Afr. J. Biotechnol.* **2010**, *9*, 5475–5480.
8. Munns, R.; Tester, M. Mechanisms of Salinity Tolerance. *Annu. Rev. Plant Biol.* **2008**, *59*, 651–681. [CrossRef] [PubMed]
9. Gharsallah, C.; Fakhfakh, H.; Grubb, D.; Gorsane, F. Effect of Salt Stress on Ion Concentration, Proline Content, Antioxidant Enzyme Activities and Gene Expression in Tomato Cultivars. *AoB Plants* **2016**, *8*, plw055. [CrossRef]
10. Toscano, S.; Romano, D.; Ferrante, A. Molecular Responses of Vegetable, Ornamental Crops, and Model Plants to Salinity Stress. *Int. J. Mol. Sci.* **2023**, *24*, 3190. [CrossRef] [PubMed]
11. Zhu, T.; Deng, X.; Zhou, X.; Zhu, L.; Zou, L.; Li, P.; Zhang, D.; Lin, H. Ethylene and Hydrogen Peroxide Are Involved in Brassinosteroid-Induced Salt Tolerance in Tomato. *Sci. Rep.* **2016**, *6*, 35392. [CrossRef] [PubMed]
12. Kahlaoui, B.; Hachicha, M.; Misle, E.; Fidalgo, F.; Teixeira, J. Physiological and Biochemical Responses to the Exogenous Application of Proline of Tomato Plants Irrigated with Saline Water. *J. Saudi Soc. Agric. Sci.* **2018**, *17*, 17–23. [CrossRef]
13. Sofo, A.; Scopa, A.; Nuzzaci, M.; Vitti, A. Ascorbate Peroxidase and Catalase Activities and Their Genetic Regulation in Plants Subjected to Drought and Salinity Stresses. *Int. J. Mol. Sci.* **2015**, *16*, 13561–13578. [CrossRef] [PubMed]
14. Ouattara, S.S.S.; Konate, M. The Tomato: A Nutritious and Profitable Vegetable to Promote in Burkina Faso. *Alex. Sci. Exch. J.* **2024**, *45*, 11–20. [CrossRef]
15. Guo, M.; Wang, X.-S.; Guo, H.-D.; Bai, S.-Y.; Khan, A.; Wang, X.-M.; Gao, Y.-M.; Li, J.-S. Tomato Salt Tolerance Mechanisms and Their Potential Applications for Fighting Salinity: A Review. *Front. Plant Sci.* **2022**, *13*, 949541. [CrossRef]
16. Roșca, M.; Mihalache, G.; Stoleru, V. Tomato Responses to Salinity Stress: From Morphological Traits to Genetic Changes. *Front. Plant Sci.* **2023**, *14*, 1118383. [CrossRef]
17. Gerakari, M.; Kotsira, V.; Kapazoglou, A.; Tastsoglou, S.; Katsileros, A.; Chachalis, D.; Hatzigeorgiou, A.G.; Tani, E. Transcriptomic Approach for Investigation of *Solanum* Spp. Resistance upon Early-Stage Broomrape Parasitism. *Curr. Issues Mol. Biol.* **2024**, *46*, 9047. [CrossRef]
18. Gur, A.; Zamir, D. Unused Natural Variation Can Lift Yield Barriers in Plant Breeding. *PLoS Biol.* **2004**, *2*, e245. [CrossRef]
19. Ferrari, G.; Americo, S.; Prazzoli, M.L.; Beretta, M. Investigating Genetic Control of Salt Stress Tolerance in Tomato Commercial Hybrid Cultivars and *Solanum pennellii* Introgression Lines. In Proceedings of the XVI International Symposium on Processing Tomato 1351, San Juan, Argentina, 21 March–1 April 2022; pp. 165–174.

20. Kapazoglou, A.; Gerakari, M.; Lazaridi, E.; Kleftogianni, K.; Sarri, E.; Tani, E.; Bebeli, P.J. Crop Wild Relatives: A Valuable Source of Tolerance to Various Abiotic Stresses. *Plants* **2023**, *12*, 328. [CrossRef]
21. Kissoudis, C.; Sunarti, S.; Van De Wiel, C.; Visser, R.G.; van der Linden, C.G.; Bai, Y. Responses to Combined Abiotic and Biotic Stress in Tomato Are Governed by Stress Intensity and Resistance Mechanism. *J. Exp. Bot.* **2016**, *67*, 5119–5132. [CrossRef] [PubMed]
22. Rebah, F.; Ouhibi, C.; Alamer, K.H.; Msilini, N.; Nasri, M.B.; Stevens, R.; Attia, H. Comparison of the Responses to NaCl Stress of Three Tomato Introgression Lines. *Acta Biol. Hung.* **2018**, *69*, 464–480. [CrossRef] [PubMed]
23. Ali, A.A.M.; Romdhane, W.B.; Tarroum, M.; Al-Dakhil, M.; Al-Doss, A.; Alsadon, A.A.; Hassairi, A. Analysis of Salinity Tolerance in Tomato Introgression Lines Based on Morpho-Physiological and Molecular Traits. *Plants* **2021**, *10*, 2594. [CrossRef]
24. Pessoa, H.P.; Dariva, F.D.; Copati, M.G.F.; de Paula, R.G.; de Oliveira Dias, F.; Gomes, C.N. Uncovering Tomato Candidate Genes Associated with Drought Tolerance Using *Solanum pennellii* Introgression Lines. *PLoS ONE* **2023**, *18*, e0287178. [CrossRef] [PubMed]
25. Dharbaranyam, B.; Sakthivel, K.; Venkataraman, G. De Novo Domestication of Wild and Halophilic Plants for Designing Salt-Tolerant Crops: Acceleration through CRISPR. In *Genetics of Salt Tolerance in Plants*; Ganie, S.A., Wani, S.H., Eds.; CABI: Wallingford, UK, 2024; pp. 144–168, ISBN 978-1-80062-301-9.
26. Pompeiano, A.; Moles, T.M.; Viscomi, V.; Scartazza, A.; Huarancca Reyes, T.; Guglielminetti, L. Behind the Loss of Salinity Resistance during Domestication: Alternative Eco-Physiological Strategies Are Revealed in Tomato Clade. *Horticulturae* **2024**, *10*, 644. [CrossRef]
27. Avdikos, I.D.; Tagiakas, R.; Mylonas, I.; Xynias, I.N.; Mavromatis, A.G. Assessment of Tomato Recombinant Lines in Conventional and Organic Farming Systems for Productivity and Fruit Quality Traits. *Agronomy* **2021**, *11*, 129. [CrossRef]
28. Avdikos, I.D.; Tagiakas, R.; Tsouvaltzis, P.; Mylonas, I.; Xynias, I.N.; Mavromatis, A.G. Comparative Evaluation of Tomato Hybrids and Inbred Lines for Fruit Quality Traits. *Agronomy* **2021**, *11*, 609. [CrossRef]
29. Eshed, Y.; Zamir, D. An Introgression Line Population of *Lycopersicon pennellii* in the Cultivated Tomato Enables the Identification and Fine Mapping of Yield-Associated QTL. *Genetics* **1995**, *141*, 1147–1162. [CrossRef] [PubMed]
30. Acosta-Motos, J.R.; Ortuño, M.F.; Bernal-Vicente, A.; Diaz-Vivancos, P.; Sanchez-Blanco, M.J.; Hernandez, J.A. Plant Responses to Salt Stress: Adaptive Mechanisms. *Agronomy* **2017**, *7*, 18. [CrossRef]
31. Azeem, M.; Pirjan, K.; Qasim, M.; Mahmood, A.; Javed, T.; Muhammad, H.; Yang, S.; Dong, R.; Ali, B.; Rahimi, M. Salinity Stress Improves Antioxidant Potential by Modulating Physio-Biochemical Responses in Moringa Oleifera Lam. *Sci. Rep.* **2023**, *13*, 2895. [CrossRef]
32. Beer, C.; Ciais, P.; Reichstein, M.; Baldocchi, D.; Law, B.E.; Papale, D.; Soussana, J.-F.; Ammann, C.; Buchmann, N.; Frank, D.; et al. Temporal and Among-site Variability of Inherent Water Use Efficiency at the Ecosystem Level. *Glob. Biogeochem. Cycles* **2009**, *23*, 2008GB003233. [CrossRef]
33. Mantzouridou, F.; Tsimidou, M.Z. On the Monitoring of Carotenogenesis by Blakeslea Trispora Using HPLC. *Food Chem.* **2007**, *104*, 439–444. [CrossRef]
34. Singleton, V.L.; Rossi, J.A. Colorimetry of Total Phenolics with Phosphomolybdic-Phosphotungstic Acid Reagents. *Am. J. Enol. Vitic.* **1965**, *16*, 144–158. [CrossRef]
35. Brand-Williams, W.; Cuvelier, M.-E.; Berset, C. Use of a Free Radical Method to Evaluate Antioxidant Activity. *LWT Food Sci. Technol.* **1995**, *28*, 25–30. [CrossRef]
36. SPSS Inc. *PASW Statistics 18*; SPSS Inc.: Chicago, IL, USA, 2009.
37. RStudio Team. *RStudio: Integrated Development Environment for R*, Version 4.3.2.; Computer Software: Boston, MA, USA, 2023.
38. Felipe de Mendiburu. Agricolae: Statistical Procedures for Agricultural Research. 2019. Available online: https://scholar.google. com/scholar?hl=el&as_sdt=0,5&q=Felipe+de+Mendiburu+(2023).+agricolae:+Statistical+Procedures+for+Agricultural+Research. +R+package+version+1.3-6.+https://CRAN.R-project.org/package=agricolae&btnG= (accessed on 13 November 2024).
39. Wickham, H. *Ggplot2: Elegant Graphics for Data Analysis*; Springer: New York, NY, USA, 2016.
40. Wickham, H. Reshaping Data with the Reshape Package. *J. Stat. Softw.* **2007**, *21*, 1–20. [CrossRef]
41. Wickham, H.; François, R.; Henry, L.; Müller, K.; Vaughan, D. *Dplyr: A Grammar of Data Manipulation*, R Package, Version 1.1.2.; Computer Software: Boston, MA, USA, 2023.
42. Barrios-Masias, F.H.; Chetelat, R.T.; Grulke, N.E.; Jackson, L.E. Use of Introgression Lines to Determine the Ecophysiological Basis for Changes in Water Use Efficiency and Yield in California Processing Tomatoes. *Funct. Plant Biol.* **2013**, *41*, 119–132. [CrossRef] [PubMed]
43. Dulai, S.; Molnár, I.; Molnár-Láng, M. Changes of Photosynthetic Parameters in Wheat/Barley Introgression Lines during Salt Stress. *Acta Biol. Szeged.* **2011**, *55*, 73–75.
44. Machado, J.; Heuvelink, E.; Vasconcelos, M.W.; Cunha, L.M.; Finkers, R.; Carvalho, S.M.P. Exploring Tomato Phenotypic Variability under Combined Nitrogen and Water Deficit. *Plant Soil* **2024**, *496*, 123–138. [CrossRef]
45. Cirillo, V.; Ruggiero, A.; Caullireau, E.; Di Covella, F.S.; Francesca, S.; Grillo, S.; Batelli, G.; Maggio, A. Isolating Water and Nitrogen Stress Tolerant Genotypes Among Tomato Landraces and *Solanum pennellii* Backcross Inbred Lines. *J. Plant Growth Regul.* **2024**, 1–12. [CrossRef]

46. Chitwood, D.H.; Kumar, R.; Headland, L.R.; Ranjan, A.; Covington, M.F.; Ichihashi, Y.; Fulop, D.; Jiménez-Gómez, J.M.; Peng, J.; Maloof, J.N. A Quantitative Genetic Basis for Leaf Morphology in a Set of Precisely Defined Tomato Introgression Lines. *Plant Cell* **2013**, *25*, 2465–2481. [CrossRef] [PubMed]
47. Frary, A.; Göl, D.; Keleş, D.; Ökmen, B.; Pınar, H.; Şığva, H.Ö.; Yemenicioğlu, A.; Doğanlar, S. Salt Tolerance in *Solanum pennellii*: Antioxidant Response and Related QTL. *BMC Plant Biol.* **2010**, *10*, 58. [CrossRef] [PubMed]

Article

Inconsistent Yield Response of Forage Sorghum to Tillage and Row Arrangement

Christine C. Nieman [1,*], Jose G. Franco [2] and Randy L. Raper [3]

1 USDA-ARS Dale Bumpers Small Farms Research Center, 6883 South Hwy 23, Booneville, AR 72927, USA
2 USDA-ARS Dairy Forage Research Center, 1925 Linden Dr., Madison, WI 53706, USA; jose.franco@usda.gov
3 Oklahoma Agricultural Experiment Station, Field and Research Service Units, Oklahoma State University, 139 Agriculture Hall, Stillwater, OK 74078, USA
* Correspondence: christine.c.nieman@usda.gov; Tel.: +1-479-675-3834

Abstract: Forage sorghum is an alternative source for biofuel feedstock production and may also provide forage for livestock operations. Introducing biofuel feedstock as a dual-use forage to livestock operations has the potential to increase the adoption of biofuel feedstock production. However, additional technical agronomic information focusing on tillage, row arrangement, and harvest date for forage sorghum planted into pasturelands intended for dual use is needed. Three tillage treatments, disking and rototilling (RT), chisel plow (CP), and no tillage (NT), and two row arrangement treatments, single-row planting with 76.2 cm rows and twin rows of 17.8 cm on 76.2 cm centers, were tested for effects on forage sorghum yield in a 3-cut system. This study tested two sites in Booneville, AR, from 2010 to 2012. Several interactions with year were detected, likely due to large precipitation differences within and among years. The year greatly affected the yield, with greater ($p < 0.05$) yields in year 1 compared to years 2 and 3 in both locations. No till resulted in lower yields in some years and harvest dates, though no clear trend was detected among tillage treatments over years. Twin rows generally did not improve yield, except for the third harvest date at one location. No strong trends for tillage or row arrangement effects were observed in this study. Inconsistencies may have resulted from the strong influence of year or interactions of multiple factors, which may challenge producers interested in utilizing forage sorghum for biofuels and livestock feed.

Keywords: forage sorghum; tillage; twin row; biofuels; biomass

Citation: Nieman, C.C.; Franco, J.G.; Raper, R.L. Inconsistent Yield Response of Forage Sorghum to Tillage and Row Arrangement. *Agronomy* **2024**, *14*, 1510. https://doi.org/10.3390/agronomy14071510

Academic Editors: Ewa Szpunar-Krok, Mariola Staniak and Małgorzata Szostek

Received: 13 June 2024
Revised: 8 July 2024
Accepted: 10 July 2024
Published: 12 July 2024

1. Introduction

Cellulosic feedstocks from grain and forage sorghum (*Sorghum bicolor* L.) are acceptable sources for bioethanol production [1,2]. Sorghums are tolerant to drought stress [3], with greater growth and yield in arid environments compared to corn [4], also commonly used for bioethanol. Forage sorghums can also be grazed by or harvested for livestock, providing feed of adequate quality for cattle production [5,6]. With the potential for dual use, both biofuel and livestock feed, integrated crop–livestock operations may be significant in expanding the use of forage sorghum for biofuels. Forage sorghum could be planted with the intention to harvest for biofuel, but in the event of feed shortage, it could be redirected to livestock feed. However, additional agronomic evaluations of reduced tillage systems, seed bed preparation for transitioning from pasture to forage sorghum production, and planting strategies that optimize yield, such as single or twin row, are still needed.

Reduced tillage systems may be a priority for integrated farms. Marginal lands, or areas otherwise prone to soil erosion under tillage, are often relegated to grazing or forage acreage [7]. Reduced tillage results in greater residues on the soil surface, providing several soil health benefits, including improved soil structure, reduced soil erosion, and increased soil organic matter, while tillage may exacerbate these issues on marginal lands [8].

Sorghum yield responses to tillage vary widely by variety, previous field management, soil type, and climatic conditions. A 5 yr, no-till study in the southern high plains, with a wheat–sorghum–graze rotation, observed reduced sorghum yields, likely due to increased soil water storage, and recommended tillage prior to sorghum planting [9], while an 18 yr study conducted in Nebraska determined that no-till plots had greater sorghum yields compared to tilled after wheat harvest in a wheat–sorghum–fallow rotation [10]. In short, in a two-year study in Alabama, sorghum grain yields were greater when no till planted into crimson clover or wheat than tilled systems [11]. Though yield responses may vary, economic evaluations generally agree that tillage systems require greater costs than no-till systems that are not always recouped by greater crop production [12].

Crop row arrangement impacts yield potential [13–16]. Increased corn yields of 2.7–10% have been observed with row spacing narrower than 76.2 cm under favorable growing conditions, while yields similar to those of wider rows ($\geq$76.2 cm) have been observed under unfavorable growing conditions [16]. Greater yields in narrower rows ($<$76.2 cm) may result from greater light interception [17] and greater space between plants in a row that reduce competition among plants [18]. Though, yield advantages are inconsistent, with some studies observing a yield reduction or no yield advantage [15,19,20].

The twin-row arrangement generally involves two rows planted close together, usually within 15–20-cm, with the twin rows 76.2 cm apart (76.2 cm centers). Twin rows have been shown to increase profitability and water conservation for soybean and cotton crops in areas with irrigation agriculture, such as the Mississippi River Valley Alluvial Aquifer [21]. Less information is available for dry land forage sorghum, but interactions may exist between row arrangement and arid environments, with advantages to wider rows when water is limited [22]. In Texas, sorghum yield under irrigation was greater with twin rows (30 cm) to single rows (96.5 cm) at the same population (161,000–198,000 plants ha^{-1}; [23]). Novacek et al. [15] observed similar yields among twin rows (20 cm on 76 cm centers) under irrigation versus single rows (76 cm) in dry land conditions.

Before dual-use livestock feed and biofuel feedstock systems can be recommended for integrated farms, research is needed on successful forage sorghum establishment methods on pasture lands and row-planting configurations that may be beneficial in dry land environments. Our overall objective was to evaluate tillage type and single- and twin-row arrangements on forage sorghum yield in a 3-cut system for biofuel feedstock, considering its alternative use as livestock feed for integrated farms.

2. Materials and Methods

2.1. Description of Field Site and Experimental Layout

Two sites of 0.12 ha each at the USDA-ARS Dale Bumpers Small Farms Research Center near Booneville, AR, USA (35.1401° N, 93.9216° W), were selected for the study area, the "north location" (NL) and "south location" (SL). The previous management of the research site was common bermudagrass, continuously grazed by cow–calf pairs. The soil type on both sites was Leadvale silt loam (fine-silty, siliceous, semiactive, thermic Typic Fragiudult). Pre-study soil fertility analyses for NL indicated 20.3 and 97.0 ppm for P and K, respectively, and a pH of 6.25, and 39.3 and 73.3 ppm for P and K, respectively, and a pH of 6.25 for SL (University of Arkansas Diagnostic Laboratory, Fayetteville, AR, USA). The experimental design was a randomized complete block, 3 $\times$ 2 factorial design, with 3 tillage treatments and single- or twin-row arrangement and 4 blocks. Each site was divided into four 9.1 m $\times$ 18.3 m blocks and contained 6 experimental units of 9.14 m $\times$ 3.05 m. Treatments were allocated randomly to one of six plots within each block in year one. Plot assignments remained the same for all three years, 2010–2012.

2.2. Tillage and Planting Treatments

Three tillage treatments of different intensities were implemented prior to planting. The most intense tillage (RT), designed to leave the least amount of residue, was created by disking, followed by rototilling (Howard Rotovator 41–90, Napa, CA, USA). The medium-intensity tillage (CP), designed to leave partial residue, was accomplished with a chisel plow (W.T. Graham, Amarillo, TX, USA). The final treatment was no till (NT), designed to leave the most amount of residue. Tillage was completed annually 1 d prior to planting for RT and CP.

Single-row planting with 76.2 cm rows was compared to twin-row arrangement of 17.8 cm on 76.2 cm centers. Single rows were planted with a no-till drill (John Deere 1590; Deere & Company, Moline, IL, USA). Twin rows were planted with a twin-row planter prototype (Hesston Manufacturing Co., Hesston, KS, USA). Each year, all plots received glyphosate [N-(phosphonomethyl) glycine] at a rate of 1.0 kg ae ha^{-1} one week prior to planting and received 670 kg ha^{-1} of 12-12-33 (N-P-K) fertilizer at planting and 67 kg ha^{-1} of N when plants reached 30.5 cm or 30 d after planting. Forage sorghum cv. 'Enorma' was planted on NL for all three years. Forage sorghum cv. 'SCG BMR 105' was planted in SL in 2010, but the variety was unavailable thereafter and cv. 'Bundle King BMR' was planted in 2011 and 2012. The seeding rate for all forage sorghum varieties was 7.8 kg ha^{-1}. Planting dates on both sites were 5 June 2010, 14 June 2011, and 15 May 2012. Saturated soils prevented early planting in 2010 and 2011.

2.3. Data Collection

Forage sorghum was harvested three times per year in both locations. The first harvest date (H1) was determined based on seed supplier recommendations for estimated number of days from planting to maximum biomass production, as would be recommended for production for biofuel feedstock. The second date (H2) was approximately 30 d after the first harvest, and the final harvest (H3) was 30 d after the second harvest or just prior to a forecasted freeze. Each year, harvest dates were the same for NL and SL. Forage sorghum was harvested on 10 September, 21 October, and 4 November in year 1; 15 September, 13 October, and 20 October, in year two; and 16 August, 12 September, and 19 October, in year three. All plants within 1.5 m of two adjacent rows were harvested at 15.2 cm stubble height and weighed. A subsample of five plants was taken from the bulk sample, weighed, and dried at 55 °C until no further weight loss to determine dry matter for the bulk sample.

2.4. Statistical Analysis

Although the treatments and experimental design were replicated on both sites, NL and SL, two different forage sorghum varieties were used, and, therefore, sites were analyzed separately. Response variables were tested for normality with the UNIVARIATE procedure of SAS and met the assumptions of normality. The GLIMMIX procedure in SAS (Version 9.4; SAS Institute) was used to test the effects of year, tillage, row arrangement, and their interactions, with block as a random variable and harvest dates treated as repeated measures, on yield per harvest. For total yield, yields from each harvest per year were summed and analyzed with GLIMMIX procedure in SAS for effects of year, tillage, row arrangement, and their interactions with block as a random variable. For all variables, the LSMEANS option was used to generate individual treatment means. Significance was declared at $p \leq 0.05$.

2.5. Environmental Data

Temperature and precipitation for 2010–2012 and the 30 yr average (Figure 1) show slightly above-normal temperatures for all three years and highly variable precipitation. In 2010, precipitation was low in early spring and at planting, but above-average precipitation occurred in July and September, providing excellent growing conditions for forage sorghum. Drought occurred in both 2011 and 2012. In 2011, precipitation was abundant in April and May, but June, July, and September experienced lower-than-average rainfall. Historic drought occurred in 2012 beginning in April and extended throughout the growing season through September. Precipitation varied greatly over the three-year study.

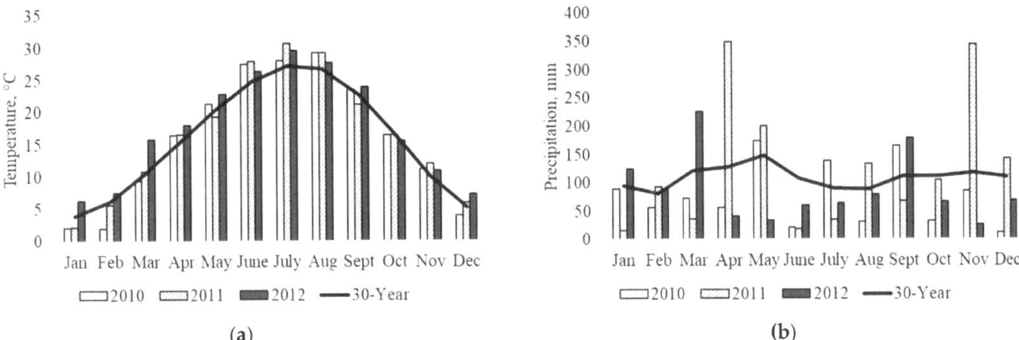

Figure 1. (**a**) Average monthly temperature and (**b**) total monthly precipitation near Booneville, AR, for 2010, 2011, 2012, and the 30-year average (1990–2020). Weather obtained from the National Climatic Data Center (NOAA; https://www.ncdc.noaa.gov/cdo-web accessed on 9 May 2024).

3. Results

3.1. North Location

Yield per harvest was affected by a year × harvest date × row arrangement interaction (p = 0.003; Figure 2a). In the three-way interaction, the greatest differences appear to be among years, with the greatest yields in 2010 when precipitation was greater. Yields from H1 were greater (p < 0.05) than all other harvest dates in 2010, 2011, and for 2012 except for yields from single rows for H3.

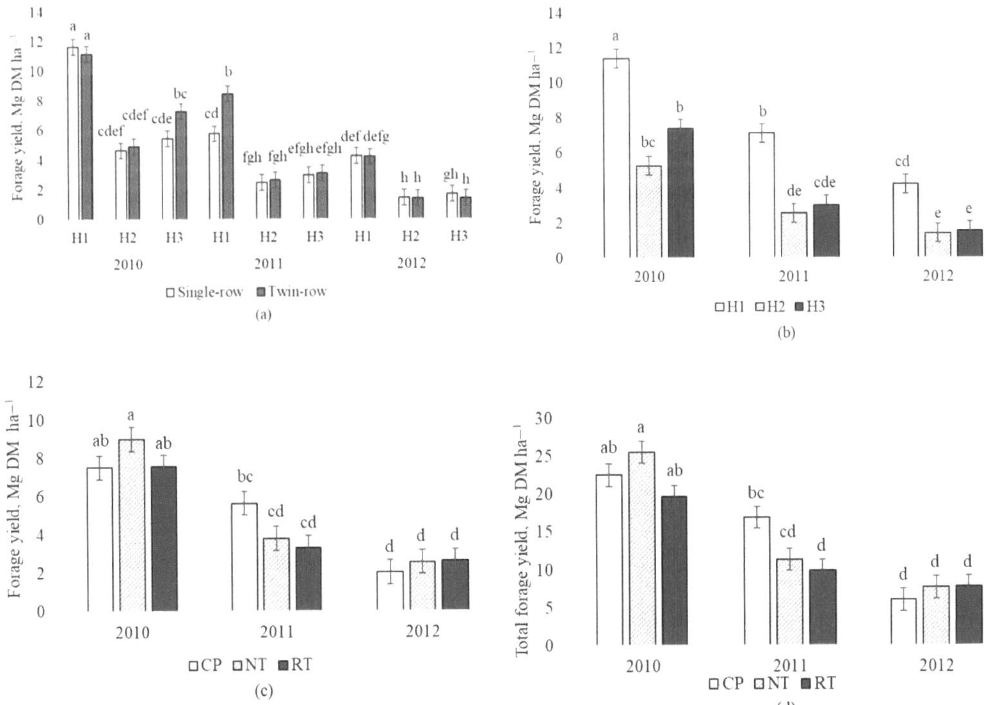

Figure 2. North Location: (**a**) Dry matter (DM) yields for forage sorghum planted in twin rows and single rows at harvest date 1 (H1), harvest date 2 (H2), and harvest date 3 (H3) for 2010, 2011, and 2012.

Effects of row arrangement $\times$ harvest date $\times$ year. $p < 0.01$. SEM = 0.53 Mg DM ha^{-1}. (**b**) Dry matter yields for forage sorghum from H1, H2, and H3 in 2010, 2011, and 2021. Effects of year $\times$ harvest date. $p < 0.01$. SEM = 0.54. (**c**) Dry matter yields for forage sorghum after chisel plow (CP), no-till (NT), or rototilling in 2010, 2011, and 2012. Tillage $\times$ year. $p < 0.01$. SEM = 0.63. (**d**) Total forage sorghum DM yields for CP, NT, and RT for 2010, 2011, and 2012. Tillage $\times$ Year $p < 0.01$. SEM = 1.47. Means without a common letter differ ($p < 0.05$).

Multiple two-way interactions were detected for yield per harvest, year $\times$ harvest date ($p < 0.01$; Figure 2b) and year $\times$ tillage ($p < 0.01$; Figure 2c). The greatest ($p < 0.05$) yields were consistently observed in H1 across years (Figure 2b), while H2 and H3 did not differ within years. Yield per harvest (Figure 2c) was greatest ($p < 0.05$) in 2010 and for CP in 2011, likely due to a drought that occurred in 2011 and 2012.

Total yield, the sum of all harvests per year, was affected by a year $\times$ tillage interaction ($p < 0.01$; Figure 2d). The total yield differed greatly among years and generally decreased per year from 2010 to 2012 due to increasingly lower precipitation through the experiment. In 2011, total yields were greater ($p < 0.05$) for CP compared to RT, though NT did not differ from either tillage treatment; total yield did not differ by tillage treatments within the year for 2010 or 2012.

3.2. South Location

No three-way interactions were observed for SL, but two-way interactions were observed for harvest date $\times$ row arrangement ($p = 0.03$; Figure 3a), and several year interactions were observed, including year $\times$ tillage ($p < 0.01$; Figure 3b), year $\times$ row arrangement ($p < 0.01$; Figure 3c), and year $\times$ harvest date ($p < 0.01$; Figure 3d). Yield per harvest date was greater ($p < 0.05$) with the twin-row arrangement for H3 across years; other row arrangements and harvest dates did not differ. In 2010, no differences in yield per harvest were observed for tillage treatments (Figure 3b), but yields for NT were lower ($p < 0.05$) in 2011 and 2012, while CP and RT did not differ. The yield per harvest was not affected by row arrangement in 2010 or 2012 (Figure 3c), but the yield from twin rows was greater ($p > 0.05$) than single rows in 2011. However, the yields per harvest for single rows in 2011 were considerably lower ($p < 0.05$) than all years, including 2012 when drought was more severe. The harvest date and year affected the yield per harvest, yields for each harvest differed ($p < 0.05$) in 2010, with the greatest yields ranked from H2, H1, to H3, while yields in 2011 did not differ among harvest dates, and in 2012, yields were greater ($p < 0.05$) for H3, while H1 and H2 did not differ from each other.

The total yield was affected by year $\times$ row arrangement ($p < 0.01$) and year $\times$ tillage interactions ($p < 0.01$). The total yield was greatest in 2010 and for CP in 2012 (Figure 4). Tillage treatments did not affect the yields in 2010, though lower yields were observed for NT in 2011 and 2012. The CP and RT tillage treatments did not differ from one another in 2011 or 2012. No differences in row arrangement were observed for 2010 or 2012, but the yield for twin rows was greater ($p < 0.01$) than single rows in 2011.

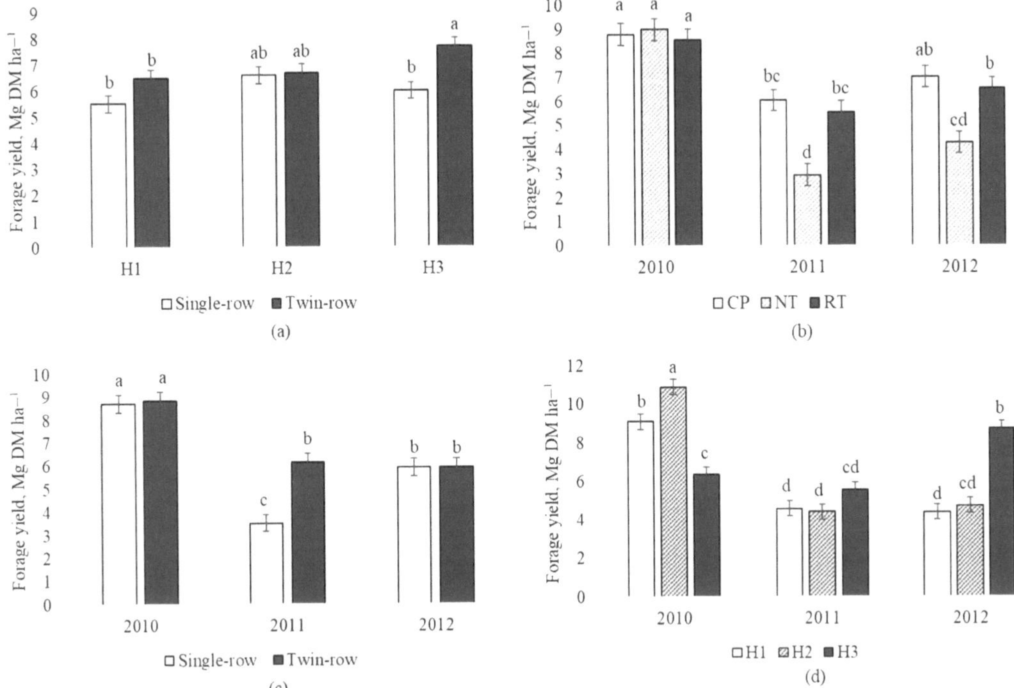

Figure 3. South Location: (**a**) Dry matter (DM) yields across years for forage sorghum planted in twin-row and single row at harvest date 1 (H1), harvest date 2 (H2), and harvest date 3 (H3). Effects of row arrangement $\times$ harvest date. $p = 0.03$. SEM = 0.32 Mg DM ha^{-1}. (**b**) Dry matter yields for forage sorghum after chisel plow (CP), no-till (NT), or rototilling in 2010, 2011, and 2012. Tillage $\times$ year. $p < 0.01$. SEM = 0.45. (**c**) Dry matter yields for forage sorghum planted in twin-row and single row in 2010, 2011, and 2012. Tillage $\times$ year. $p < 0.01$. SEM = 0.37. (**d**) Dry matter yields for forage sorghum from H1, H2, and H3 in 2010, 2011, and 2021. Effects of year $\times$ harvest date. $p < 0.01$. SEM = 0.40. Means without a common letter differ ($p < 0.05$).

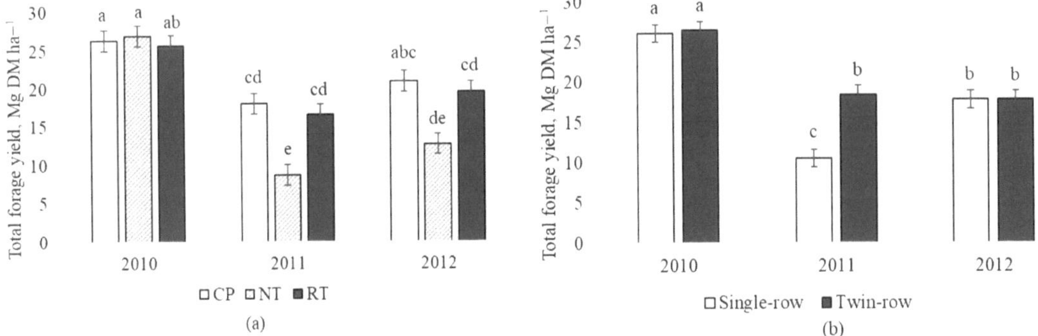

Figure 4. South Location: (**a**) Total dry matter (DM) yields for forage sorghum planted after chisel plow (CP), no-till (NT), or rototilling (RT) in 2010, 2011, and 2012. Tillage $\times$ year. $p < 0.01$. SEM = 1.35 Mg DM ha^{-1}. (**b**) Total dry matter (DM) yields for forage sorghum planted in twin-row and single in 2010, 2011, and 2012. Tillage $\times$ year. $p < 0.01$. SEM = 1.12 Mg DM ha^{-1}. Means without a common letter differ ($p < 0.05$).

4. Discussion

Due to the use of different forage sorghum varieties in NL and SL, locations were analyzed separately. However, the questions and results were similar, and, therefore, results from both locations will be discussed in this section together. Additionally, it is important that caution be used when interpreting data from SL. A different forage sorghum variety was used in 2010 than in 2011 and 2012, and though strongly diverse environmental conditions appear to have driven yield differences among years, the change in variety is a confounding factor with year, and it is unknown which factor most contributed to differences among treatments. Further, the variability in precipitation within and among years was a substantial factor in this study and may potentially mask impacts from other implemented factors.

No trends in tillage effects on forage sorghum yield were observed for NL. Forage sorghum yields were reduced with NT in 2011 and 2012 in SL. Differences in SL may be caused by the different forage sorghum variety used, and without soil moisture and temperature data, interpretations of the effects of tillage on forage sorghum yield are limited. Several studies have observed greater or equal yields for grain sorghum in no-till systems [24,25], while others observed greater productivity in tilled systems [26–28]. Soil moisture and temperature affect the timing of germination and plant emergence [29,30], and both soil moisture and temperature are influenced by tillage and crop residue. Surface residue cover in no-till systems can insulate the soil surface and prevent soil drying, thereby preserving soil moisture for seed germination and crop growth [31]. Generally, areas with low annual rainfall benefit from the moisture conservation provided in no-till systems [32,33]. However, tillage in clay-based or poorly drained soils may improve yield [27,34,35] by creating a seedbed with improved drainage and higher soil temperatures because of low surface residue [36]. Surface residues also have the potential to interfere with adequate seed to soil contact in no-till planting because of inadequate cutting by double-disk openers that cause "hair pinning" [37]. When hair pinning occurs, residues enter the planting channel with the seed, preventing adequate seed to soil contact [37].

Tillage is an important consideration for introducing dual-purpose biofuel production on integrated farms, especially those that have significant acreage in pasture. Pasture-land is generally relegated to non-productive or highly erodible landscapes [7] better suited to conservation tillage or NT establishment. Although NT may have resulted in lower yields in some years, producers will need to weigh the environmental benefits (reduced soil erosion, reduced soil compaction, increased microbial activity [38]) to the potential for more successful establishment and greater yields with tillage in some years. Tillage also has important economic considerations, with greater costs associated with tillage that are not always recouped with increased yields [12]. Short-term mixed results from no-till establishment may discourage some integrated farms from adoption, but analysis of long-term studies indicates no statistical difference in long-term yields from no till compared to conventional tillage for corn, soybean, and wheat [39]. The long-term data also indicate that profits per area tend to be greater for no till compared to conventional tillage due to lower farm operation costs with no till [39].

For NL, no clear trends were observed for the effects of row arrangement on yield, though twin rows produced greater yields in 2011 for H1. For SL, twin rows increased the yield in 2011, and for H3, across years. When planted at the same population density, twin rows allow for more intra-row spacing, reducing interspecies competition [18], potentially resulting in greater yields. Twin rows also allow for greater capture of solar radiation [15,17,40] and weed suppression [41]. Snider et al. [42] observed greater forage sorghum yields with 19.05 cm spacing compared to 38.1 cm and 76.2 cm at two different sites, one each in Alabama and Arkansas, over three years. Snider et al. [42] concluded that the greater yield was a result of increased stem density. Wider row spacing may improve the yield in crops in dry lands, while twin rows have been shown to have greater yields than single rows with high-quality soils and under optimal growing conditions [15,22,43,44]. Fernandez et al. [23] observed greater sorghum grain yields in the twin-row configuration compared to single rows under irrigation, but no difference in non-irrigated plots, indicating that narrow row spacing may be advantageous under optimal conditions but not different from single rows under conditions of limited soil moisture. It is possible that the increased yields for twin rows in H1 in 2011 for NL were a result of twin rows, taking advantage of early adequate moisture provided by heavy April and May precipitation, and single rows were unable to compensate when drought began in July and extended through the fall, resulting in similar yields in H2 and H3.

In SL, twin rows produced greater forage sorghum yields in 2011 but not in 2010 or 2012. Although forage sorghum varieties can vary greatly in response to planting density [45], potentially, the twin-row arrangement was ideal for the variety planted in 2011, and conditions were such that the yield advantage could be expressed, but not in 2012 because of severe drought conditions. Additionally, for SL, twin rows from H3 outyielded single rows at H3 across years. Generally, in studies conducted throughout the Midwest, single-row and twin-row arrangements have not resulted in a yield advantage in corn grain systems [40,46]. However, greater solar radiation reception has been observed in corn planted in twin rows up to the V9 growth stage but not at earlier or later growth stages [47]. Similarly, greater solar radiation capture was observed in early canopy closure in twin rows, though the radiation interception advantage declined at later vegetative stages and with increasing plant density [40]. Potentially, forage sorghum in the twin-row configuration experienced accelerated regrowth due to increased solar radiation reception, which may have been particularly important for H3 with decreasing day length. However, this is speculative as the response is likely very variety dependent and NL did not follow a similar trend.

The total yield was exceptional for both experiments in 2010, averaging 22.6 Mg DM ha^{-1} for NL and 26.3 Mg DM ha^{-1} for SL. However, the yield at each harvest may be a more important factor to consider for dual-use biofuel feedstock and livestock feed systems. When considering a variety for dual use, the regrowth potential may be the greater benefit as it provides more management flexibility. Forage sorghum yield can be spread across multiple cuttings and, considering weather and feed inventory, the producer can keep forage for livestock feed or sell for biofuel feedstock without assuming major yield losses in following cuttings. For example, in NL, greater forage sorghum yields at H1 indicate the correct estimation for peak biomass production, the first cutting could be sold as a biofuel feedstock, and regrowth could be reserved for livestock feed. For SL, yields at harvest differed by year and are also likely confounded by the change in variety. However, it appears that peak biomass may have occurred later for the variety in 2010, based on the greater yield in H2 in SL. If that is consistently true, a producer could graze early in the season and sell the regrowth for biofuel feed stock. Further, if historic drought occurs, such as in 2012, the producer may also choose to keep all forage for livestock, buffering the impacts of drought on the livestock operation.

5. Conclusions

Our findings, while variable, support the literature that suggests no-till establishment can result in reduced yields in some instances, but not consistently, and twin rows, overall, do not result in a greater yield advantage, as compared to single-row arrangements. Differences in forage sorghum yield at different harvest dates also indicate the potential of dual-use forage sorghum. Forage sorghum in a multi-harvest system provides flexibility to sell biofuel feedstock when growth is optimal or redirect to livestock feed if environmental conditions result in low forage inventory. Overall, despite the potential of forage sorghum, the lack of clear trends across both studies indicates the influence of multiple factors and interactions that cause inconsistencies in the forage sorghum yield, which may challenge producers interested in dual-use forage sorghum systems.

Author Contributions: Conceptualization, R.L.R.; methodology, R.L.R.; validation, R.L.R., C.C.N. and J.G.F.; formal analysis, C.C.N. and J.G.F.; investigation, R.L.R.; resources, R.L.R.; data curation, R.L.R. and C.C.N.; writing—original draft preparation, C.C.N.; writing—review and editing, C.C.N., J.G.F. and R.L.R.; visualization, C.C.N. and J.G.F.; supervision, R.L.R.; project administration, R.L.R. All authors have read and agreed to the published version of the manuscript.

Funding: This research received no external funding.

Data Availability Statement: Data will be made available in the USDA National Agricultural Library.

Conflicts of Interest: The authors declare no conflicts of interest. Any opinions, findings, conclusions, or recommendations expressed in this publication are those of the authors and do not necessarily reflect the view of the USDA.

References

1. Szambelan, K.; Nowak, J.; Frankowski, J.; Szwengiel, A.; Jeleń, H.; Burczyk, H. The comprehensive analysis of sorghum cultivated in Poland for energy purposes: Separate hydrolysis and fermentation and simultaneous saccharification and fermentation methods and their impact on bioethanol effectiveness and volatile by-products from the grain and the energy potential of sorghum straw. *Bioresour. Technol.* **2018**, *250*, 750–757. [PubMed]
2. Batog, J.; Frankowski, J.; Wawro, A.; Lacka, A. Bioethanol production from biomass of selected sorghum varieties cultivated as main and second crop. *Energies* **2020**, *13*, 6291. [CrossRef]
3. Safian, N.; Naderi, M.R.; Torabi, M.; Soleymani, A.; Salemi, H.R. Corn (*Zea mays* L.) and sorghum (*Sorghum bicolor* (L.) Moench) yield and nutritional quality affected by drought stress. *Biocatal. Agric. Biotechnol.* **2022**, *45*, 102486. [CrossRef]
4. Bhattarai, B.; Singh, S.; West, C.P.; Ritchie, G.L.; Trostle, C.L. Effect of deficit irrigation on physiology and forage yield of forage sorghum, pearl millet, and corn. *Crop Sci.* **2020**, *60*, 2167–2179. [CrossRef]
5. Dann, H.M.; Grant, R.J.; Cotanch, K.W.; Thomas, E.D.; Ballard, C.S.; Rice, R. Comparison of brown midrib sorghum-sudangrass with corn silage on lactational performance nutrient digestibility in Holstein dairy cows. *J. Dairy Sci.* **2008**, *91*, 663–672. [CrossRef] [PubMed]
6. McCuistion, K.C.; McCollum, F.T., III; Greene, L.W.; MacDonald, J.; Bean, B. Performance of stocker cattle grazing 2 sorghum-sudangrass hybrids under various stocking rates. *Prof. Anim. Sci.* **2011**, *27*, 92–100. [CrossRef]
7. Wu, L.; Wu, L.; Bingham, I.J.; Misselbrook, T.H. Projected climate effects on soil workability and trafficability determine the feasibility of converting permanent grassland to arable land. *Agric. Syst.* **2022**, *203*, 103500. [CrossRef]
8. Kumar, K.; Goh, K. Crop residues and management practices: Effects on soil quality. Soil nitrogen dynamics, crop yield, and nitrogen recovery. *Adv. Agron.* **2000**, *68*, 197–319.
9. Baumhardt, R.L.; Schwartz, R.C.; MacDonald, J.C.; Tolk, J.A. Tillage and cattle grazing effects on soil properties and grain yields in a dryland wheat–sorghum–fallow rotation. *Agron. J.* **2011**, *103*, 914–922. [CrossRef]
10. Wicks, G.A.; Smika, D.E.; Hergert, G.W. Long-term effects of no-tillage in a winter wheat (*Triticum aestivum*)-sorghum (*Sorghum bicolor*)-fallow rotation. *Weed Sci.* **1988**, *36*, 384–393. [CrossRef]
11. Bishnoi, U.R.; Mays, D.A.; Fabasso, M.T. Response of no-till and conventionally planted grain sorghum to weed control method and row spacing. *Plant Soil* **1990**, *129*, 117–120. [CrossRef]
12. Varner, B.T.; Epplin, F.M.; Strickland, G.L. Economics of no-till versus tilled dryland cotton, grain sorghum, and wheat. *Agron. J.* **2011**, *103*, 1329–1338. [CrossRef]
13. Porter, K.B.; Jensen, M.E.; Sletten, W.H. The effect of row spacing, fertilizer and planting rate on the yield and water use of irrigated grain sorghum 1. *Agron. J.* **1960**, *52*, 431–434. [CrossRef]
14. Grichar, W.J. Row spacing, plant populations, and cultivar effects on soybean production along the Texas Gulf Coast. *Crop Manag.* **2007**, *6*, 1–6. [CrossRef]
15. Novacek, M.J.; Mason, S.C.; Galusha, T.D.; Yaseen, M. Twin rows minimally impact irrigated maize yield, morphology, and lodging. *Agron. J.* **2013**, *105*, 268–276. [CrossRef]
16. Licht, M.A.; Parvej, M.R.; Wright, E.E. Corn yield response to row spacing and plant population in Iowa. *Crop Forage Turfgrass Manag.* **2019**, *5*, 1–7. [CrossRef]
17. Andrade, F.H.; Calviño, P.; Cirilo, A.; Barbieri, P. Yield responses to narrow rows depend on increased radiation interception. *Agron. J.* **2002**, *94*, 975–980. [CrossRef]
18. Bullock, D.G.; Nielsen, R.L.; Nyquist, W.E. A growth analysis comparison of corn grown in convention and equidistant plant spacing. *Crop Sci.* **1988**, *28*, 254–258. [CrossRef]
19. Farnham, D.E. Row spacing, plant density, and hybrid effects on corn grain yield and moisture. *Agron. J.* **2001**, *93*, 1049–1053. [CrossRef]
20. Lee, C.D. Reducing row widths to increase yield: Why it does not always work. *Crop Manag.* **2006**, *5*, 1–7. [CrossRef]
21. Quintana-Ashwell, N.; Anapalli, S.S.; Pinnamaneni, S.R.; Kaur, G.; Reddy, K.N.; Fisher, D. Profitability of twin-row planting and skip-row irrigation in a humid climate. *Agron. J.* **2022**, *114*, 1209–1219. [CrossRef]

22. Lyon, D.J.; Pavlista, A.D.; Hergert, G.W.; Klein, R.N.; Shapiro, C.A.; Knezevic, S.; Mason, S.C.; Nelson, L.A.; Baltensperger, D.D.; Elmore, R.W.; et al. Skip-row planting patterns stabilize corn grain yields in the central Great Plains. *Crop Manag.* **2009**, *8*, 1–8. [CrossRef]

23. Fernandez, C.J.; Fromme, D.D.; Grichar, W.J. Grain sorghum response to row spacing and plant populations in the Texas Coastal Bend Region. *Int. J. Agron.* **2012**, *2012*, 238634. [CrossRef]

24. Sainju, U.M.; Whitehead, W.F.; Singh, B.P.; Wang, S. Tillage, cover crops, and nitrogen fertilization effects on soil nitrogen and cotton and sorghum yields. *Eur. J. Agron.* **2006**, *25*, 372–382. [CrossRef]

25. Tarkalson, D.D.; Hergert, G.W.; Cassman, K.G. Long-term effects of tillage on soil chemical properties and grain yields of dryland winter wheat-sorghum/corn-fallow rotation in the Great Plains. *Agron. J.* **2006**, *98*, 26–33. [CrossRef]

26. Bordovsky, D.G.; Choudhary, M.; Gerard, C.J. Tillage effects on grain sorghum and wheat yields in the Texas Rolling Plains. *Agron. J.* **1998**, *90*, 638–643. [CrossRef]

27. Sweeney, D.W. Twenty years of grain sorghum and soybean yield response to tillage and N fertilization of a claypan soil. *Crop Forage Turfgrass Manag.* **2017**, *3*, 1–7. [CrossRef]

28. Matowo, P.R.; Pierzynski, G.M.; Whitney, D.A.; Lamond, R.E. Long term effects of tillage and nitrogen source, rate, and placement on grain sorghum production. *J. Prod. Agric.* **1997**, *10*, 141–146. [CrossRef]

29. Kaspar, T.C.; Erbach, D.C.; Cruse, R.M. Corn response to seed-row residue removal. *Soil Sci. Soc. Am. J.* **1990**, *54*, 1112–1117. [CrossRef]

30. Schneider, E.C.; Gupta, S.C. Corn emergence as influenced by soil temperature, matric potential, and aggregate size distribution. *Soil Sci. Soc. Am. J.* **1985**, *49*, 415–422. [CrossRef]

31. Fortin, M.C. Soil temperature, soil water, and no-till corn development following in-row residue removal. *Agron. J.* **1993**, *85*, 571–576. [CrossRef]

32. Wang, X.B.; Cai, D.X.; Hoogmoed, W.B.; Oenema, O.; Perdok, U.D. Developments in conservation tillage in rainfed regions of North China. *Soil Tillage Res.* **2007**, *93*, 239–250. [CrossRef]

33. Hansen, N.C.; Allen, B.L.; Baumhardt, R.L.; Lyon, D.J. Research achievements and adoption of no-till, dryland cropping in the semi-arid U.S. Great Plains. *Field Crops Res.* **2012**, *132*, 196–203. [CrossRef]

34. DeFelice, M.S.; Carter, P.R.; Mitchell, S.B. Influence of tillage on corn and soybean yield in the United States and Canada. *Crop Manag.* **2006**, *5*, 1–17. [CrossRef]

35. Al-Kaisi, M.M.; Archontoulis, S.V.; Kwaw-Mensah, D.; Miguez, F. Tillage and crop rotation effects on corn agronomic response and economic return at seven Iowa locations. *Agron. J.* **2015**, *107*, 1411–1424. [CrossRef]

36. Mahboubi, A.A.; Lal, R. Long-term tillage effects on changes in structural properties of two soils in central Ohio. *Soil Tillage Res.* **1998**, *45*, 107–118. [CrossRef]

37. Creamer, N.G.; Dabney, S.M. Killing cover crops mechanically: Review of recent literature and assessment of new research results. *Am. J. Altern. Agric.* **2022**, *17*, 32–40.

38. Lal, R. Soil carbon sequestration and aggregation by cover cropping. *J. Soil Water Conserv.* **2015**, *70*, 329–339. [CrossRef]

39. Che, Y.; Rejesus, R.M.; Cavigelli, M.A.; White, K.E.; Aglasan, S.; Knight, L.G.; Dell, C.; Hollinger, D.; Lane, E.D. Long-term economic impacts of no-till adoption. *Soil Secur.* **2023**, *13*, 100103. [CrossRef]

40. Robles, M.; Ciampitti, I.A.; Vyn, T.J. Dynamics of maize plant responses to a twin-row spatial arrangement at multiple plant densities. *Agron. J.* **2012**, *104*, 1747–1756. [CrossRef]

41. Begna, S.H.; Hamilton, R.I.; Dwyer, L.M.; Stewart, D.W.; Smith, D.L. Effects of population density and planting pattern on the yield and yield components of leafy reduced-stature maize in a short-season area. *J. Agron. Crop Sci.* **1997**, *179*, 9–17. [CrossRef]

42. Snider, J.L.; Raper, R.L.; Schwab, E.B. The effect of row spacing and seeding rate on biomass production and plant stand characteristics of non-irrigated photoperiod-sensitive sorghum (*Sorghum bicolor* (L.) Moench). *Ind. Crops Prod.* **2012**, *37*, 527–535. [CrossRef]

43. Karlen, D.L.; Camp, C.R. Row spacing, plant population, and water management effects on corn in the Atlantic Coastal Plain. *Agron. J.* **1985**, *77*, 393–398. [CrossRef]

44. Thapa, S.; Xue, Q.; Stewart, B.A. Alternative planting geometries reduce production risk in corn and sorghum in water-limited environments. *Agron. J.* **2020**, *112*, 3322–3334. [CrossRef]

45. Caravetta, G.J.; Cherney, J.H.; Johnson, K.D. Within-row spacing influences on diverse sorghum genotypes: I. Morphology. *Agron. J.* **1990**, *82*, 206–210. [CrossRef]

46. Rusk, R.; Sievers, J.L. Comparison of Twin Row and 30-in. Row Corn. *Iowa State Univ. Res. Demonstr. Farms Prog. Rep.* **2010**, *268*. Available online: http://lib.dr.iastate.edu/farms_reports/268 (accessed on 7 July 2024).

47. Barr, R.L.; Mason, S.C.; Novacek, M.J.; Wortmann, C.S.; Rees, J.M. Row Spacing and Seeding Rate Recommendations for Corn in Nebraska G2216. University of Nebraska-Lincoln Extension. 2013. Available online: http://extensionpublications.unl.edu/assets/pdf/g2216.pdf (accessed on 7 July 2024).

Article

Short Crop Rotation under No-Till Improves Crop Productivity and Soil Quality in Salt Affected Areas

Aziz Nurbekov [1,2,*], Muhammadjon Kosimov [1], Makhmud Shaumarov [1], Botir Khaitov [3,*] , Dilrabo Qodirova [2], Husniddin Mardonov [2] and Zulfiya Yuldasheva [2]

[1] Food and Agriculture Organization (FAO), Regional Office in Uzbekistan, 2, University Str., Qibray District, Tashkent 100140, Uzbekistan; muhammadjon.kosimov@fao.org (M.K.); makhmud.shaumarov@fao.org (M.S.)
[2] Faculty of Agrobiology, Tashkent State Agrarian University, Tashkent 100140, Uzbekistan
[3] International Center for Biosaline Agriculture (ICBA), Regional Office for Central Asia and the South Caucasus, Tashkent 100084, Uzbekistan
* Correspondence: aziz.nurbekov@fao.org (A.N.); b.khaitov@biosaline.org.ae (B.K.)

Abstract: Soil productivity and crop yield were examined in response to legume-based short crop rotation under conventional (CT) and no-till (NT) tillage practices in saline meadow-alluvial soils of the arid region in Bukhara, Uzbekistan. Compared with the CT treatment, crop yield was consistently higher under NT, i.e., winter wheat 9.63%, millet 9.9%, chickpea 3.8%, and maize 10.7% at the first experiment cycle during 2019–2021. A further crop productivity increase was observed at the second experiment cycle during 2021–2023 under NT when compared to CT, i.e., winter wheat 17.7%, millet 31.2%, chickpea 19.6%, and maize 19.1%. An increase in total phyto residue by 20.9% and root residue by 25% under NT compared to CT contributed to the improvement in soil structure and played a vital role in the sustained improvement of crop yields. In turn, the increased residue retention under NT facilitated soil porosity, structural stability, and water retention, thereby improving soil quality and organic matter content. Soil salinity more significantly decreased under NT than in CT, reducing salinity buildup by 18.9% at the 0–25 cm and 32.9% at the 75–100 cm soil profiles compared to CT. The total forms N and P were significantly increased under NT when compared to CT, while the efficiency of the applied crop rotation was essential. This study showed the essential role of the NT method with legume-based intensive cropping in the maintenance of soil health and crop yield, thereby touching on recent advances in agro-biotechnology and the sustainable land management of drylands.

Keywords: conventional tillage; no-till; crop yield; bulk density; structural stability; saline soil; arid region

Citation: Nurbekov, A.; Kosimov, M.; Shaumarov, M.; Khaitov, B.; Qodirova, D.; Mardonov, H.; Yuldasheva, Z. Short Crop Rotation under No-Till Improves Crop Productivity and Soil Quality in Salt Affected Areas. *Agronomy* **2023**, *13*, 2974. https://doi.org/10.3390/agronomy13122974

Academic Editors: Mariola Staniak, Ewa Szpunar-Krok and Małgorzata Szostek

Received: 27 September 2023
Revised: 12 October 2023
Accepted: 19 October 2023
Published: 1 December 2023

1. Introduction

Cotton was the main agricultural crop for Uzbekistan during the Soviet empire (1924–1991). As a result, this monoculture practice caused extreme soil and environmental degradation and the pollution of water resources and the environment. After the country's independence in 1991, the cotton area gradually decreased while the share of wheat production increased substantially to ensure food security in the region [1]. Population growth puts pressure on agricultural production, leading to intensified land usage along with extensive application of chemicals, i.e., fertilizers and pesticides, which has led to more environmental contamination [2].

Soil salinity and drought are the two most significant constraints that adversely affect the productivity of crops in dry areas like Uzbekistan [3]. The detrimental effects of soil salinity on plant growth and development are followed by reduced crop productivity, dysfunctional cropping systems, an imbalance in mineral nutrition, and specific ionic toxicity [4]. Therefore, it is essential to transition to climate-resilient food and feed crops and adopt innovative cropping systems and sustainable technologies that efficiently use available resources.

Proper land management strategies as a main driver of sustainable agriculture can be implemented to diversify cropping systems, enhance crop production, improve food security, and raise nutritional well-being in the long term [5,6]. In this regard, farmers' knowledge and qualifications are essential in managing efficient cropping systems and improving soil organic matter. Soil health should be perceived as a guarantee of sustainable agriculture and food security in changing climate conditions and natural resource depletion [7]. Expansion of agricultural lands is not an option in this region when water resource scarcity is high [8]. In this situation, crop diversification must be used sparingly, focusing on developing climate resilience in farming without harming the environment.

Tillage might have various impacts on the production system in the short and long terms since soil quality directly impacts plants' capacity to grow and build an appropriate root system to exploit water and nutrients. Therefore, the increase in soil physical structure and subsequent mineralization of N under conventional tillage increase crop production in the short term [9]. However, extended continuous cultivation tends to lower soil porosity through the mechanical breakdown of the structure and compaction of the topsoil [10]. On degraded soils, conventional tillage may lead to slight improvements in crop output in the short term; however, over the long term, it may be neutral or lead to yield reductions due to deterioration of soil structure. As explained in recent studies, continuous tillage lowers the amount of soil organic matter, causes soil erosion and subsoil compaction, and also releases more greenhouse gases [11].

In contrast, the no-till practice as a main component of conservation agriculture (CA) is essential in managing dry arid soils, which facilitates the buildup of SOM and soil nutrients. Long-term application of continuous no-till (NT) increases the stability and microporosity of degraded soils while increasing crop yields [12]. The actions of the soil biota, which flourish under NT, enhance nutrient cycling and soil microbial activity substantially. Significant tillage-induced structural improvements in degraded soils can start 3 to 5 years after NT set-up, and increased crop yields usually accompany it [13]. Despite the benefits of NT, this agrotechnology was not widely adopted by crop producers in arid zones, yet the effect of NT on crop yield and soil characteristics depends on climatic variables, crop rotation, and soil management strategies, and is thereby defined as site-specific [14]. In NT systems, the diversification and intensification of crop rotations is a new climate-smart strategy that attempts to enhance crop yield while lowering environmental impact [15].

Our hypothesis was that intensive legume-based short-term crop rotation under NT will improve soil quality in dryland agriculture, thus increasing crop yield in the short term. A key question is whether the short crop rotation system under the NT practice could offer a potential solution as a climate-resilient agriculture measure to achieve food security in the long term in degraded marginal regions. This study aimed to improve crop productivity and soil health using a short crop rotation system under a climate change impact scenario in harsh and saline environments.

2. Materials and Methods

2.1. Research Initiative

Since 2018, the Food and Agriculture Organization of the United Nations (FAO) has been implementing the regional project "Integrated natural resources management in drought-prone and salt-affected agricultural production landscapes in Central Asia and Turkey" (CACILM-2) funded by the Global Environment Facility and other donors. The project aims to scale up integrated natural resources management (INRM) practices that minimize pressures and negative impacts on natural resources, reduce risks and vulnerability, and enhance capacity of rural communities to cope with and adapt to drought and salinity.

One of the project's target countries is Uzbekistan, where land degradation, soil salinization, water scarcity, and climate change pose serious challenges for agricultural production and food security. The project has been working with the Ministry of Agriculture and other national partners to integrate resilience into policy, legal and institutional frameworks for INRM, as well as to provide incentives for climate-smart agriculture at national and sub-national levels.

The project's main objective is to help rural communities cope with or adapt to the effects of land degradation, soil salinization, water scarcity, and climate change, which threaten their livelihoods and food security. The project does this by promoting sustainable land management practices that minimize pressures and negative impacts on natural resources, reduce risks and vulnerability, and enhance resilience. The project also supports the implementation of the United Nations Convention to Combat Desertification (UNCCD) and its National Action Plans in the country.

This research work, conducted from 2018 to 2023 in the demonstration project territories of Bukhara Region in Uzbekistan, aims to enhance soil productivity in saline arable lands. By applying short-rotation schemes and conservation agriculture practices such as zero tillage, mulching and integrated pest management, this work provides scientific evidence and practical solutions for improving soil quality and crop yield in saline conditions.

2.2. Climate and Soil Characteristics

This research was conducted in saline meadow-alluvial soils of the Bukhara region in the southwest of Uzbekistan in the 2019–2022 growing seasons. This site lies in the central Kyzylkum desert, 206 m above sea level, with $39.46°$ N and $64.25°$ E geographical coordinates.

The harsh climate of this arid region is characterized by sharp climate variability with recurring periods of severe drought. This region has a total of 270–280 frost-free days, with an average annual evapotranspiration of up to 2000 mm. Annual rainfall ranges between 90 and 150 mm; however, the main part of this precipitation falls between January and April, before the start of the vegetation period. The year 2021 was drier during the study period with only 93.0 mm precipitation. The warmest period was July with $39 °C$ and the coldest temperature was $0 °C$, recorded in January. Accordingly, the highest relative moisture was about 70% in January, while it reached the lowest point in July, ranging between 25 and 30%.

The texture of soils is heavy-textured sandy loam. The low precipitation and excessive evaporation caused a buildup of soluble salts, contributing to an increased salinity level in the soil. The soil salinity varied between 4.8 and 7.4 dS/m, and the pH index was 7.6, which is slightly alkaline. The humus content was 1.1%, relatively low in the soil surface and further decreased in the deepest horizons. The soil chemical structure consisted of total N 0.07%, P 0.14%, and K 0.8%, and in available forms NO_3 44.3 mg kg^{-1}, P_2O_5 13.5 mg kg^{-1}, and K_2O 120 mg kg^{-1} at the 0–30 cm depth of the soil (Figure 1).

2.3. Experiment Design

A field experiment was set up in Autumn, 2019 by planting winter wheat (*Triticum aestivum* L.)—Alekseevich variety. The next June millet (*Panicum miliaceum* L.) Saratovskaya-853 variety was planted, which was harvested in October 2020 and forage pea (*Pisum sativum* L.)—Vostok-84 variety—was the next crop in this short crop rotation system. The planted forage pea was harvested in May 2021 and maize (*Zea mays* L.)—NS-205 hybrid—occupied the experiment plot. After harvesting maize in September 2021, winter wheat was planted for the second cycle. Therefore, each short crop rotation cycle continued for two years and the experiment was conducted during the 2019–2021 and 2021–2023 growing seasons. A strip-plot design was used in an area of 1200 m^2 with a total of two experimental units (CT and NT) installed in three replications, each with an area of 100 m^2.

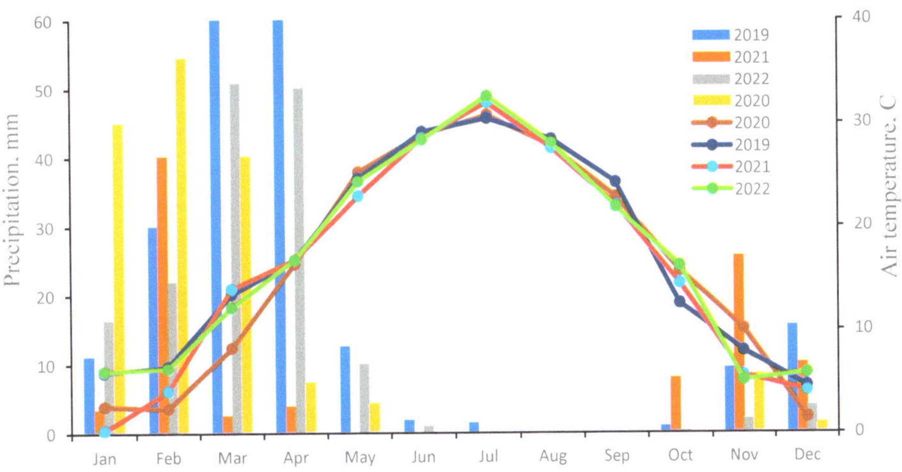

Figure 1. Weather conditions during the experiment years 2019–2023.

Salt leaching was conducted two times during the January–March period before the experiment was set up. During the vegetation period, surface furrow irrigation was applied 4 times with a norm of 1100–1200 m^3 hectare^{-1} based on crop water requirement. The following rates of mineral fertilizers were applied: $N_{220}P_{180}K_{60}$ kg ha^{-1} for winter wheat, $N_{150}P_{70}K_{30}$ kg ha^{-1} for millet, $N_{50}P_{25}K_{20}$ kg ha^{-1} for forage pea, and $N_{300}P_{180}K_{120}$ kg ha^{-1} for maize. Traditional furrow irrigation was conducted at a norm of 900–1000 m^3 for each hectare and 4–5 times during the vegetation period depending on crop water requirements, but the same norm was strictly applied to each plot of the experiment.

A digital recorder (L93-4, Hangzhou Logger Technology Co., Ltd., Hangzhou, China) was used to measure the soil's temperature. This device automatically records soil temperature at one-hour intervals during the whole growth period and was installed in the surface 20 cm soil profile of the planting row in each plot.

A Vence tudo SA 17600 seed planter was employed for seed planting in the NT plots. On the other hand, several agromashinaries were employed for CT, i.e., mouldboard plower, tiller, discing, rotavator, and leveling operations, which consumed a lot more resources and time. All other agronomic operations in vegetation periods such as weeding, cultivation, plant protection, and other measures were conducted similarly for all experiment plots according to local agronomic practices.

2.4. Chemical Content of Soil and Plant Samples

Soil samples were taken in sealable plastic bags at 4 suggested depths of 0–25, 25–50, 50–75, and 75–100 cm based on the envelop method from the experimental plots. The collected soil samples were air-dried for two weeks at room temperature, followed by grounding and sieving through a 2 mm mesh before chemical analysis. The standard methods developed by Ryan, Estefan, and Rashid [16] were used to determine soil physical (soil texture, bulk density, pH, EC parameters) and chemical characteristics (NPK, humus content). An amount of 10 mL of 50% perchloric acid was poured on 0.5 g soil samples placed in a tube, and, after shaking for a minute, 1 mL of concentrated sulfuric acid was added and the process was followed by decomposition by heating on a hot plate. Kjeldahl distillation, the vanadate method, and an inductively coupled plasma spectrophotometer were employed to analyze total N, P_2O_5, and K_2O, respectively. The Tyurin method was used to determine soil organic matter [17]. The standard core method was used to calculate soil bulk density (g cm^{-3}) in each soil layer.

2.5. Statistical Analyses

Using the CropSTAT software (Version 7.2.2007), analysis of variance (ANOVA) was performed on the collected data, which included crop yield, residue retention, soil nutrient content, and soil physical and chemical characteristics. The Tukey test was used to determine whether there was a trait difference at $p < 0.05$.

3. Results and Discussion

3.1. Crop Yield

A consistent increase in crop yield with significant differences ($p \leq 0.05$) under NT is associated with the proper land management of a short crop rotation system. At the first experiment cycle during 2019–2011, crop yield was consistently higher under NT as compared to CT, i.e., winter wheat 9.63%, millet 9.9%, chickpea 3.8%, and maize 10.7% (Table 1). A further crop productivity increase was observed at the second experiment cycle during 2021–2023 under NT when compared to CT, i.e., winter wheat 17.7%, millet 31.2%, chickpea 19.6%, and maize 19.1%.

Table 1. Crop yield (t ha^{-1}).

Experiment Cycles	Tillage	Winter Wheat	Millet	Forage Pea	Maize
2019–2021	CT	6.54 c	2.51 c	1.85 b	12.59 b
	NT	7.17 b	2.76 c	1.92 b	13.94 a
2021–2023	CT	6.76 c	3.11 b	1.68 d	11.87 c
	NT	7.94 a	4.08 a	2.01 a	14.14 a
LSD$_{0.05}$		0.53	0.45	0.09	1.23

Means separated by identical lower-case letters (a, b, c, d) in each column are significantly different at $p \leq 5\%$.

The interactive effect of the experiment cycles/years and the tillage practices was not significant; it was likely that longer than this two-year crop rotation cycle is needed to produce significant changes in crop yield. Although a slight yield increase was observed in winter wheat (3.4%), it did not reach a significant level under CT when the first experiment cycle was compared against the second one. Under similar conditions, the grain yield of millet significantly increased by 23.9%. On the contrary, forage pea and maize yields decreased significantly under CT when the first experiment cycle was compared against the second one.

NT with an intensive cropping system composed of totally different families and grown in sequence showed the greatest benefit on crop productivity. Integrating legume such as forage pea into the cropping system might have an essential role in increasing crop yield. It is worth mentioning that legumes enrich soil with nitrogen fixed from air up to 300 kg per ha depending on environmental factors [18]. The combined positive effects of NT and legume-based short crop rotation systems increase water use efficiency, reduce soil erosion, improve soil properties, and enhance crop productivity [19,20].

A slow and steady impact of the used practices on crop yield is also expected in the long term [21]. As the positive effect of NT is well documented, a consistent increase in crop yield was achieved due to a positive synergism between NT and the legume-based short crop rotation system.

3.2. Residue Retention

Total crop residue retention was enhanced significantly ($p < 0.005$) by the NT practice as compared to the CT (Table 2), with a substantial increase exhibited at the second experiment cycle. The main part of root residues was found at the 0–30 cm soil layer in both tillage treatments, i.e., 86.2% under NT and 87.5% under CT. However, root residue retentions were 16.7% and 33.9% higher under NT as compared to CT in the first and second experiment cycles, respectively. Similarly, phyto residues were 17.4% and 24.4% higher under NT than in CT, respectively, for the above-mentioned experiment periods.

Table 2. Phyto and root residues (t ha^{-1}).

Cycle	Soil Treatments	Phyto Residues	Root Residues		Total Residues
			0–30 cm	30–50 cm	
2019–2021	CT	19.5 b	15.9 b	2.1 b	37.5 b
	NT	22.9 ab	18.2 ab	2.8 a	43.9 a
2021–2023	CT	18.8 b	14.7 c	2.1 b	35.6 b
	NT	23.4 a	19.4 a	3.1 a	45.9 a
LSD$_{0.05}$		2.15	1.23	1.04	3.65

Means separated by identical lower-case letters (a, b, c) in each column are significantly different at $p \leq 5\%$.

The integration of organic matter enhances or rebalances the soil's structure, enriches soil organic matter (SOM) and nutrients, and enhances cation exchange and water retention capacities. Soil health also depends on the quality and quantity of SOM, which regulates many soil functions, i.e., soil organic carbon (SOC), soil biodiversity, availability and cycling of plant nutrients, soil porosity, aeration, water-holding capacity and hydraulic conductivity, thermal properties, and mechanical strength [22]. As mentioned in the previous studies, the amount, quality, and decomposition rate of crop residues returned to the soil are directly impacted by agricultural management strategies including reduced tillage and crop rotations [23]. Furthermore, covering soil surface with residues protects from water and wind erosion and salt accumulation, and reduces soil temperatures in arid regions. The beneficial effect of the legume-based short crop rotation used in this study is associated with nutritious and high-quality stubble generated by legumes.

3.3. Soil Chemical Characteristics

The tested soil quality indicators were significantly ($p < 0.005$) affected by the practiced tillage treatments (Table 3). Soil salinity was positively affected by the short crop rotation system under the NT treatment, reducing salinity buildup by 18.9% at the 0–25 cm and 32.9% at the 75–100 cm soil depths compared to CT.

Table 3. Soil chemical analysis.

Soil Depth (cm)	Salinity (ds m^{-1})	Soil pH	Humus (%)	Total Forms (%)			Available Forms (mg per kg)		
				N	P	K	NO$_3$	P$_2$O$_5$	K$_2$O
				CT					
0–25	8.8 b	7.72 a	1.055 bc	0.072 b	0.136 d	0.84 a	44.3 c	13.5 b	120.4 b
25–50	8.4 c	7.70 a	1.033 c	0.070 b	0.150 b	0.79 a	47.2 b	9.0 c	106.0 c
50–75	9.1 b	7.70 a	0.801 d	0.058 c	0.145 c	0.76 a	50.3 a	5.0 d	65.1 e
75–100	10.5 a	7.63 a	0.611 e	0.046 d	0.141 c	0.66 b	49.2 a	2.0 e	55.4 e
				NT					
0–25	7.4 d	7.69 a	1.160 a	0.081 a	0.155 b	0.87 a	48.3 b	14.0 b	207.1 a
25–50	7.5 d	7.67 a	1.097 b	0.072 b	0.165 a	0.87 a	54.2 a	24.0 a	130.0 b
50–75	7.6 d	7.63 a	0.886 d	0.063 c	0.155 b	0.81 a	52.1 a	8.0 c	106.0 c
75–100	7.9 c	7.58 a	0.759 de	0.050 d	0.136 d	0.69 b	49.9 b	3.0 e	96.3 d
LSD$_{0.05}$	0.6	0.45	0.57	0.008	0.009	0.014	4.12	8.65	50.68

Means separated by identical lower-case letters (a, b, c, d, e) in each column are significantly different at $p \leq 5\%$.

The results showed that pH level was lower under NT despite no significant difference being observed. It turned out that short crop rotation under the NT and CT treatments may not affect soil pH level in the short term.

Soil humus content at the 0–25 soil depth was 1.16% under NT as compared to 1.055% under CT, exhibiting a 9.95% increase in soil organic matter. Similar results were found in all the studied soil profiles. As expected, humus content decreased with an increase in soil depth.

The total amounts of N and P significantly increased due to the application of NT compared to CT. Total N and P at the 0–25 cm soil profile were 12.5% and 13.9% higher under NT compared to CT treatment. However, the highest p values were found at the 25–50 cm soil profile, showing a 10% increase under the NT practice compared to CT. The tested tillage systems influenced total K content, but a significant difference was not observed in this index.

The practiced NT treatment coupled with a short crop rotation system significantly ($p < 0.05$) influenced available forms of nutrients, i.e., NO_3, P_2O_5, and K_2O. Positive effects in these nutrients were detected in all the studied soil profiles. For example, averaged across the soil profiles, NO_3 was increased by 7.1% under NT compared to CT during the two experiment cycles. Similarly, P_2O_5 and K_2O were increased significantly, suggesting a difference in the effectiveness of the studied NT and CT treatments on soil nutrient balance.

As stated by Boselli et al. [24], organic matter accumulation under NT was considerably higher due to crop residue retention, increasing microbial biomass, diversity, and forming a more suitable environment for the coexistence of soil microorganisms. It is well known that soil salinity also has a negative impact on several morphological, physiological, and biochemical properties in crops [25]. Following four years of NT experiments, Nurbekov et al. [26] reported that NT had the lowest soil salinity level of all evaluated practices due to the reduced evaporation and upward salt movement in the soil profiles. The used short crop rotation under NT positively influenced nutrient recycling and balancing in the soil, which might be related to high crop residue levels, improved soil microbial activity, and rejuvenated natural processes. The application of NT combined with legume-based short crop rotation resulted in increased soil nutrients such as C, N, and P compared to CT-based rotation. More notably, the applied land management practice facilitated the retention of maximum crop residues and decreased soil salinity in this arid region.

3.4. Soil Physical Characteristics

Table 4 indicates that NT application coupled with short crop rotation showed some improving trends in the studied soil physical characteristics at the end of the four-year experiment. Soil bulk density was significantly reduced under NT than in CT in saline meadow-alluvial soils. The significant effect of this land management practice on the improvement in soil bulk density reached 8–13.4%, affecting in-depth soil horizons to a greater extent.

Table 4. Soil physical properties.

Soil Depth (cm)	Bulk Density, (g/cm³)	Total Porosity (%)	Soil Moisture Content (%)	Water Infiltration Rate (cm s⁻¹)
		CT		
0–25	1.42 c	46.21 b	14.16 c	24.5 a
25–50	1.52 b	43.07 b	15.18 c	23.4 a
50–75	1.59 a	40.67 c	19.54 b	22.8 a
75–100	1.60 a	40.30 c	21.67 a	21.9 a
		NT		
0–25	1.31 d	49.42 a	17.39 b	22.9 a
25–50	1.34 d	48.46 a	18.47 b	21.8 a
50–75	1.38 c	46.77 b	20.11 b	21.0 a
75–100	1.41 c	45.65 b	23.48 a	21.4 a
$LSD_{0.05}$	0.57	3.23	3.04	3.65

Means separated by identical lower-case letters (a, b, c, d) in each column are significantly different at $p \leq 5\%$.

Similarly, soil porosity was significantly influenced by the implemented tillage management ($p < 0.05$). Continuous tillage may instantly improve soils' overall porosity by producing a few uneven macropores. In this study, after four years NT resulted in

6.94–13.3% higher soil porosity compared to the CT treatment. Higher soil porosity significantly impacted both the root system development and the nutrient and water availability for crops.

Soil moisture values also showed a positive effect of the NT practice, increasing by 22.8% at the 0–25 cm and 8.4% at the 75–100 cm soil horizons. Likewise, the water infiltration rate slightly improved under NT, but did not reach a significant level in this short-term study.

NT had significantly lowered soil temperature compared to the CT treatment. The soil temperature was lower by up to 9.2% in June, 10.7% in July, 13.7% in August, and 17.5% in September under NT than that in the CT treatment (Figure 2). The reduced soil temperatures during summer likely had some positive effects on crop performance at this experiment, which is in line with previous reports [27]. This intensive short crop rotation system under NT turned out to be effective to moderate soil temperature by intercepting incoming solar radiation with vegetative biomass and phyto residues cover on the soil surface [28]. Phyto residues play a crucial role for reducing the maximum soil temperature by as much as 5 °C in summer and increasing the minimum soil temperature by about 1–2 °C in winter [29]. This lowered soil temperature in summer can contribute to increase soil water storage by reducing inefficient evaporation, especially during drought period. The application of this practice also affects soil microbial activity, SOM composition, soil biochemical processes and structural properties, thereby systematically restores soil health [30]. In this case, the effectiveness primarily depends on the amount of residues left on the soil surface and their quality.

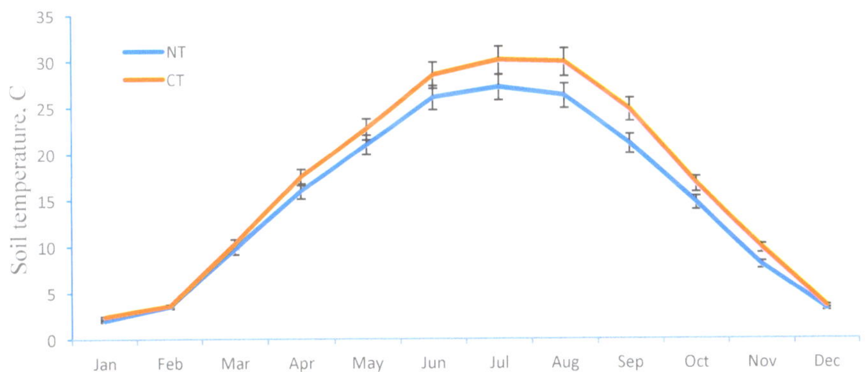

Figure 2. Effect of tillage treatment (NT and CT) on soil temperature (averaged across the experiment years).

Several recent reports highlighted the multifunctional effects of NT management scenarios on soil physical characteristics, i.e., increasing soil moisture retention, while decreasing soil bulk density and temperature [31,32]. Regardless of their composition, gained crop residues under NT improved the physical and chemical characteristics of the soil, which in turn significantly increased soil fertility [33]. The abundant literature indicates that NT lowers wind and water erosion, enhances water penetration and storage, reduces loss of nutrients, improves soil biological activity, and boosts soil organic matter which are indicators of soil health settings [34,35]. Steady improvement of soil quality under NT contributes to the productivity of subsequent row crops, suggesting a potential solution for food security and overall agricultural sustainability in water-limited regions [36]. More recently, the beneficial effects of no-tillage and crop residue management on the characteristics of irrigated silty loam soil in Uzbekistan were documented for a rotation of winter wheat and maize for two years, followed by cotton for another two years [37].

This permanent soil covering maintained via residue retention can result in equal or even better agricultural outcome while drastically decreasing the requirement for and expense of production resources, including fuel, seeds, agrochemicals, water, and labor [38,39]. However, the expanding body of data on productivity, economic, and environmental advantages is still based on a few studies, necessitating site-specific investigations [40,41]. Despite many positive showcases on land degradation and successful practices for their recovery/rehabilitation, NT is still not widely practised in arid regions. This is mostly because of a need for modern NT machinery, knowledge, and experience [26].

4. Conclusions

The results of this four-year study revealed that soil productivity and crop yield in response to conventional (CT) and no-till (NT) tillage practices in saline meadow-alluvial soils of arid regions has the potential to increase the sustainability of farming systems and their resilience to devasting environmental and climatic problems. Residue retention was considerably higher under NT than in CT, which was one of the main components of the soil-structure-maintenance processes, thereby improving ecosystem services in this short-term study. The effect of NT along with the legume-based crop rotation system was more pronounced on increasing crop yield and soil humus and NPK contents, and on decreasing soil salinity and temperature; thus, it may be a more resource-saving, cost-effective, and environmentally friendly method.

Our results suggest that the shift toward NT coupled with a short crop rotation system and involvement of legume can enhance crop yield and combat on-going land degradation in salt-affected arid lands.

Author Contributions: Conceptualization, A.N.; methodology, M.K.; software, M.S.; validation, M.S. and Z.Y.; formal analysis, B.K.; investigation, Z.Y.; resources, H.M.; data curation, A.N. and D.Q.; writing—original draft, B.K.; writing—review and editing, B.K. and A.N.; supervision, A.N. All authors have read and agreed to the published version of the manuscript.

Funding: This study was carried out within the GEF-funded project entitled "Integrated natural resources management in drought-prone and salt-affected agricultural production landscapes in Central Asia and Turkey" (CACILM-2).

Data Availability Statement: Data are contained within the article.

Acknowledgments: The authors thank Shavkat Shodiev and Zarafshon farm workers in Romitan district, Bukhara province, Uzbekistan, for providing technical support and collaboration.

Conflicts of Interest: The authors declare no conflict of interest.

References

1. Rustamova, I.; Primov, A.; Karimov, A.; Khaitov, B.; Karimov, A. Crop Diversification in the Aral Sea Region: Long-Term Situation Analysis. *Sustainability* **2023**, *15*, 10221. [CrossRef]
2. Conrad, C.; Lamers, J.P.A.; Ibragimov, N.; Löw, F.; Martius, C. Analysing irrigated crop rotation patterns in arid Uzbekistan by the means of remote sensing: A case study on post-Soviet agricultural land use. *J. Arid Environ.* **2016**, *124*, 150–159. [CrossRef]
3. Allanov, K.; Sheraliev, K.; Ulugov, C.; Ahmurzayev, S.; Sottorov, O.; Khaitov, B.; Park, K.W. Integrated effects of mulching treatment and nitrogen fertilization on cotton performance under dryland agriculture. *Commun. Soil Sci. Plant Anal.* **2019**, *50*, 1907–1918. [CrossRef]
4. Wang, J.; Hussain, S.; Sun, X.; Zhang, P.; Javed, T.; Dessoky, E.S.; Ren, X.; Chen, X. Effects of Nitrogen Application Rate Under Straw Incorporation on Photosynthesis, Productivity and Nitrogen Use Efficiency in Winter Wheat. *Front. Plant Sci.* **2022**, *13*, 862088. [CrossRef] [PubMed]
5. Sithole, N.J.; Magwaza, L.S.; Mafongoya, P.L. Conservation agriculture and its impact on soil quality and maize yield: A South African perspective. *Soil Tillage Res.* **2016**, *162*, 55–67. [CrossRef]
6. Li, G.D.; Schwenke, G.D.; Hayes, R.C.; Lowrie, A.J.; Lowrie, R.J.; Poile, G.J.; Oates, A.A.; Xu, B.; Rohan, M. Can legume species, crop residue management or no-till mitigate nitrous oxide emissions from a legume-wheat crop rotation in a semi-arid environment? *Soil Tillage Res.* **2021**, *209*, 104910. [CrossRef]
7. Fu, B.; Chen, L.; Huang, H.; Qu, P.; Wei, Z. Impacts of crop residues on soil health: A review. *Environ. Pollut. Bioavail.* **2021**, *33*, 164–173. [CrossRef]

8.	Ibragimov, N.; Djumaniyazova, Y.; Khaitbaeva, J.; Babadjanova, S.; Ruzimov, J.; Akramkhanov, A.; Lamers, J. Simulating crop productivity in a triple rotation in the semi-arid area of the Aral Sea Basin. *Int. J. Plant Prod.* **2020**, *14*, 273–285. [CrossRef]

9.	Duval, M.E.; Galantini, J.A.; Iglesias, J.O.; Canelo, S.; Martinez, J.M.; Wall, L. Analysis of organic fractions as indicators of soil quality under natural and cultivated systems. *Soil Tillage Res.* **2013**, *131*, 11–19. [CrossRef]

10.	Amoakwah, E.; Arthur, E.; Frimpong, K.A.; Parikh, S.J.; Islam, R. Soil organic carbon storage and quality are impacted by corn cob biochar application on a tropical sandy loam. *J. Soils Sediments* **2020**, *20*, 1960–1969. [CrossRef]

11.	Ghorbi, S.; Ebadi, A.; Parmoon, G.; Siller, A.; Hashemi, M. The Use of Faba Bean Cover Crop to Enhance the Sustainability and Resiliency of No-Till Corn Silage Production and Soil Characteristics. *Agronomy* **2023**, *13*, 2082. [CrossRef]

12.	Jahangir, M.M.R.; Islam, S.; Nitu, T.T.; Uddin, S.; Kabir, A.K.M.A.; Meah, M.B.; Islam, R. Bio-Compost-Based Integrated Soil Fertility Management Improves Post-Harvest Soil Structural and Elemental Quality in a Two-Year Conservation Agriculture Practice. *Agronomy* **2021**, *11*, 2101. [CrossRef]

13.	Anastasi, U.; Scavo, A. Cropping Systems and Agronomic Management Practices of Field Crops. *Agronomy* **2023**, *13*, 2328. [CrossRef]

14.	So, H.B.; Grabski, A.; Desborough, P. The impact of 14 years of conventional and no-till cultivation on the physical properties and crop yields of a loam soil at Grafton NSW, Australia. *Soil Tillage Res.* **2009**, *104*, 180–184. [CrossRef]

15.	Nurbekov, A. Conservation agriculture in Uzbekistan—Status and perspectives. *Agro Inform* **2022**, 27–32. Available online: https://cyberleninka.ru/article/n/conservation-agriculture-in-uzbekistan-status-and-perspectives (accessed on 26 September 2023).

16.	Ryan, J.; Estefan, G.; Rashid, A. *Soil and Plant Analysis Laboratory Manual*, 2nd ed.; International Center for Agricultural Research in the Dry Areas (ICARDA): Allepo, Syria, 2001.

17.	NIAST. *Method of Soil and Plant Analysis*; National Institute of Agricultural Science and Technology (NIAST): Suwon, Republic of Korea, 2000.

18.	Zimmer, S.; Messmer, M.; Haase, T.; Piepho, H.-P.; Mindermann, A.; Schulz, H.; Habekuß, A.; Ordon, F.; Wilbois, K.-P.; Heß, J. Effects of soybean variety and Bradyrhizobium strains on yield, protein content and biological nitrogen fixation under cool growing conditions in Germany. *Eur. J. Agron.* **2016**, *72*, 38–46. [CrossRef]

19.	Lal, R. Climate change and soil degradation mitigation by sustainable management of soils and other natural resources. *Agric. Res.* **2012**, *1*, 199–212. [CrossRef]

20.	Malobane, M.E.; Nciizah, A.D.; Mudau, F.N.; Wakindiki, I.I. Tillage, crop rotation and crop residue management effects on nutrient availability in a sweet sorghum-based cropping system in marginal soils of South Africa. *Agronomy* **2020**, *10*, 776. [CrossRef]

21.	Schröder, P.; Beckers, B.; Daniels, S.; Gnädinger, F.; Maestri, E.; Marmiroli, N.; Mench, M.; Millan, R.; Obermeier, M.M.; Oustriere, N.; et al. Intensify production, transform biomass to energy and novel goods and protect soils in Europe-A vision how to mobilize marginal lands. *Sci. Total Environ.* **2018**, *616*, 1101–1123. [CrossRef]

22.	Zuber, S.M.; Behnke, G.D.; Nafziger, E.D.; Villamil, M.B. Crop rotation and tillage effects on soil physical and chemical properties in Illinois. *Agron. J.* **2015**, *107*, 971–978. [CrossRef]

23.	Li, Z.; Yang, X.; Cui, S.; Yang, Q.; Yang, X.; Li, J.; Shen, Y. Developing sustainable cropping systems by integrating crop rotation with conservation tillage practices on the Loess Plateau, a long-term imperative. *Field Crops Res.* **2018**, *222*, 164–179. [CrossRef]

24.	Boselli, R.; Fiorini, A.; Santelli, S.; Ardenti, F.; Capra, F.; Maris, S.C.; Tabaglio, V. Cover crops during transition to no-till maintain yield and enhance soil fertility in intensive agro-ecosystems. *Field Crops Res.* **2020**, *255*, 107871. [CrossRef]

25.	Haruna, S.I.; Nkongolo, N.V. Tillage, Cover Crop and Crop Rotation Effects on Selected Soil Chemical Properties. *Sustainability* **2019**, *11*, 2770. [CrossRef]

26.	Nurbekov, A.; Akramkhanov, A.; Kassam, A.; Sydyk, D.; Ziyadaullaev, Z.; Lamers, J.P.A. Conservation Agriculture for combating land degradation in Central Asia: A synthesis. *AIMS Agric. Food* **2016**, *1*, 144–156. [CrossRef]

27.	Piva, J.T.; Dieckow, J.; Bayer, C.; Zanatta, J.A.; de Moraes, A.; Pauletti, V.; Tomazi, M.; Pergher, M. No-till reduces global warming potential in a subtropical Ferralsol. *Plant Soil* **2012**, *361*, 359–373. [CrossRef]

28.	Ogle, S.M.; Alsaker, C.; Baldock, J.; Bernoux, M.; Breidt, F.J.; McConkey, B.; Vazquez-Amabile, G.G. Climate and soil characteristics determine where no-till management can store carbon in soils and mitigate greenhouse gas emissions. *Sci. Rep.* **2019**, *9*, 11665. [CrossRef] [PubMed]

29.	Blanco-Canqui, H.; Shaver, T.M.; Lindquist, J.L.; Shapiro, C.A.; Elmore, R.W.; Francis, C.A.; Hergert, G.W. Cover crops and ecosystem services: Insights from studies in temperate soils. *Agron. J.* **2015**, *107*, 2449–2474. [CrossRef]

30.	Nyambo, P.; Thengeni, B.; Chiduza, C.; Araya, T. Tillage, crop rotation, residue management and biochar influence on soil chemical and biological properties. *S. Afr. J. Plant Soil* **2021**, *38*, 390–397. [CrossRef]

31.	Namozov, F.; Islamov, S.; Atabaev, M.; Allanov, K.; Karimov, A.; Khaitov, B.; Park, K.W. Agronomic Performance of Soybean with Bradyrhizobium Inoculation in Double-Cropped Farming. *Agriculture* **2022**, *12*, 855. [CrossRef]

32.	Page, K.L.; Dang, Y.P.; Dalal, R.C. The ability of conservation agriculture to conserve soil organic carbon and the subsequent impact on soil physical, chemical and biological properties and yield. *Front. Sustain. Food Syst.* **2020**, *4*, 31. [CrossRef]

33.	Sun, W.; Canadell, J.G.; Yu, L.; Yu, L.; Zhang, W.; Smith, P.; Huang, Y. Climate drives global soil carbon sequestration and crop yield changes under conservation agriculture. *Glob. Change Biol.* **2020**, *26*, 3325–3335. [CrossRef] [PubMed]

34. Rodríguez, M.P.; Domínguez, A.; Moreira Ferroni, M.; Wall, L.G.; Bedano, J.C. The diversification and intensification of crop rotations under no-till promote earthworm abundance and biomass. *Agronomy* **2020**, *10*, 919. [CrossRef]
35. Etemadi, F.; Hashemi, M.; Barker, A.V.; Zandvakili, O.R.; Liu, X. Agronomy, nutritional value, and medicinal application of faba bean (*Vicia faba* L.). *Hortic. Plant J.* **2019**, *5*, 170–182. [CrossRef]
36. Islam, R.; Glenney, D.C.; Lazarovits, G. No-till strip row farming using yearly maize-soybean rotation increases yield of maize by 75%. *Agron. Sustain. Dev.* **2015**, *35*, 837–846. [CrossRef]
37. Pulatov, A.; Egamberdiev, O.; Karimov, A.; Tursunov, M.; Kienzler, S.; Sayre, K.; Tursunov, L.; Lamers, J.P.A.; Martius, C. Introducing conservation agriculture on irrigated meadow alluvial soils (Arenosols) in Khorezm. Uzbekistan. In *Cotton, Water, Salts and Soums—Economic and Ecological Restructuring in Khorezm, Uzbekistan*; Martius, C., Rudenko, I., Lamers, J.P.A., Vlek, P.L.G., Eds.; Springer: Dordrecht, The Netherlands, 2012; pp. 195–217.
38. Xin, X.; Yang, W.; Zhang, J.; Ding, S. Tillage and residue management for long-term wheat-maize cropping in the North China Plain: I. Crop yield and integrated soil fertility index. *Field Crops Res.* **2018**, *221*, 157–165.
39. Büchi, L.; Wendling, M.; Amossé, C.; Jeangros, B.; Sinaj, S.; Charles, R. Long and short term changes in cropyield and soil properties induced by the reduction of soil tillage in a long term experiment in Switzerland. *Soil Tillage Res.* **2017**, *174*, 120–129. [CrossRef]
40. Gura, I.; Mnkeni, P.N.S. Crop rotation and residue management effects under no till on the soil quality of a Haplic Cambisol in Alice, Eastern Cape, South Africa. *Geoderma* **2019**, *337*, 927–934. [CrossRef]
41. Yang, R.; Su, Y.; Kong, J. Effect of tillage, cropping, and mulching pattern on crop yield, soil C and N accumulation, and carbon footprint in a desert oasis farmland. *Soil Sci. Plant Nutr.* **2017**, *63*, 599–606. [CrossRef]

Article

Finnish Farmers Feel They Have Succeeded in Adopting Cover Crops but Need Down-to-Earth Support from Research

Pirjo Peltonen-Sainio [1,*], Lauri Jauhiainen [2] and Hannu Känkänen [2]

1   Natural Resources Institute Finland (Luke), Latokartanonkaari 9, FI-00790 Helsinki, Finland
2   Natural Resources Institute Finland (Luke), Tietotie 2, FI-31600 Jokioinen, Finland;
    lauri.jauhiainen@luke.fi (L.J.); hannu.kankanen@luke.fi (H.K.)
*   Correspondence: pirjo.peltonen-sainio@luke.fi

Abstract: In Finland, there is an ongoing adoption and learning process considering the cultivation of cover crops (CCs). The primary aim is to claim the benefits of CCs for agricultural production and ecosystems, which are both appreciated by Finnish farmers. A farmer survey with 1130 respondents was carried out to build an up-to-date understanding of how farmers have succeeded with CCs and whether they intend to continue with the use of CCs and to collect farmers' views on knowledge gaps that should be filled by research or better knowledge sharing. The studied groups were farmers who had selected CCs as a registered measure in 2020 to receive agricultural payments. Data came from the Finnish Food Authority. Organic farmers were slightly more positive: they have had longer experience with CCs, but organic production is also more dependent on the ecosystem services provided by CCs. A high share of respondents agreed that their experiences with CCs have improved over time and were confident that CCs had become a permanent element of their production systems. Most of the farmers also agreed that the area under CCs would expand significantly in Finland and considered the cultivation of CCs as an effective measure to improve soil conditions. They often considered that challenges in adopting CCs were exaggerated and disagreed that bad experiences prevented them from expanding or continuing the use of CCs. The agricultural payment available for Finnish farmers to support the cultivation of CCs is quite reasonable (EUR 97 + EUR 50 per hectare) to compensate for any economic risks of CCs. Free word answers from the farmers highlighted research needs (in descending order) in the following areas: crop protection, sowing practices, the use of diverse CCs and their mixtures, and impacts on yield and profitability. Many of these are universal, i.e., have been reported elsewhere. Younger farmers ($\leq$50 years) highlighted profitability, which is, in many European countries, a key barrier to the deployment of CCs. Farmers from the east and north regions, where the growing season is short, highlighted alternative CC choices as a knowledge gap.

Keywords: adoption; cover crop; management; experience; farmer survey; knowledge gap

Citation: Peltonen-Sainio, P.; Jauhiainen, L.; Känkänen, H. Finnish Farmers Feel They Have Succeeded in Adopting Cover Crops but Need Down-to-Earth Support from Research. *Agronomy* **2023**, *13*, 2326. https://doi.org/10.3390/agronomy13092326

Academic Editors: Mariola Staniak, Ewa Szpunar-Krok and Małgorzata Szostek

Received: 14 August 2023
Revised: 1 September 2023
Accepted: 4 September 2023
Published: 5 September 2023

1. Introduction

Interest and need for cultivation of multifunctional cover crops (CCs) has increased markedly around the world. In the future, large-scale implementation is needed to support the transition towards green, year-round soil cover and achieve the numerous ecosystem services that CCs may provide [1]. The potential advantages for crop production include higher yields [2,3], nutrient cycling [4,5], weed control [6,7], improved soil health [8–10], and adaptation to climate change [11]. On the other hand, ecosystem services provided by CCs include reduced nutrient leaching and erosion [12,13], pesticide acquisition [14], and carbon sequestration [15,16]. Monitoring the impacts on farms may be challenging because the required follow-up period to manifest changes might be long and complicated by variable weather. Furthermore, the performance and impacts of CCs may vary greatly depending on regional and local conditions, farming systems, land use in the past, and many other farm characteristics [17,18]. Farmers can control these with context-specific, tailored management options [19,20]. Due to the multidimensional implications of integrating CCs into crop production systems [21], successful adoption calls for systems thinking [22]. In Finland, there is an ongoing learning and implementation period concerning the cultivation of CCs with the primary aim of gaining various benefits for both agricultural production and the ecosystem that Finnish farmers often appreciate [23].

It is important to understand whether farmers feel that they have succeeded or failed in implementing CCs as part of their production system and to identify the key knowledge gaps that might paralyze the adoption intensity of CCs. For example, according to researchers and organic farmers, the key knowledge gaps that may limit the adoption of CCs-based no-tillage technique in Europe were selection of CCs choices, sowing methods, measures to increase biomass but prevent competition, and termination methods of CCs stands [24,25]. The farmers' community, per se, represents a large-scale living lab for the use of CCs. Research should fill in the current, locally relevant knowledge gaps with new, verified experimental results, implement the existing public but missed data and understanding, and thereby support a transition towards rewarding and encouraging the use of multifunctional CCs [19]. Furthermore, discussion forums and close dialogue between researchers, advisors, and farmers are needed in addition to the development of decision support systems to address farmers' key knowledge gaps in a region and to give additional certainty for the deployment of CCs [26–29].

According to the farm surveys carried out in EU regions, policy has been so far the strongest determinant of the adoption rates and intensity in the use of CCs, and the adoption measures have been largely shaped by the Nitrates Directive and the Common Agricultural Policy's greening requirements [30]. Finnish farmers can get an agricultural payment of EUR 97 per hectare for the cultivation of CCs, and if the cover stand is left to overwinter, an additional EUR 50 per hectare, at most, on 30% of the field area of a farm. The primary target is to support the expansion of the field area allocated for CCs and the provision of services provided by CCs for the ecosystem and simultaneously reduce the risk that the economic costs may slow down adoption [31]. Thereby, the agricultural payment can be considered as a security measure for a farmer to cover costs when purchasing CC seeds, making changes to crop protection and management, and investing in machinery to facilitate the successful sowing and establishment of CCs [32]. Such agricultural payments are often crucial at the early stages of a green transition to push operators to become familiar with novel practices, learn by doing, and find the best practices that promote success to avoid early failure. The need for financial support is highlighted in many farmer surveys [33–35] because CCs affect whole farm profitability in various ways, not only through establishment and termination costs [21].

This study was carried out in the form of a farmer survey with the aim of gathering up-to-date knowledge on the success of Finnish farmers with the cultivation of CCs, their intentions to continue the use of CCs, and whether and how this varies depending on the farm conditions. To support the permanent transition towards the use of CCs in Finland, we collected farmers' views on knowledge gaps that should be filled either by research or knowledge sharing based on already published literature on comparable growing conditions.

2. Materials and Methods

2.1. Implementation of the Farmer Survey and Utilized Background Data

The farmer survey was carried out in Finland in the spring of 2021. In total, the details of 7025 farms (16% of Finnish farms in 2021) were requested and received from the registry of the Finnish Food Authority (FFA). These included the farm identification number, farm type, location, parcel identification number with CCs, and the farmer's email address. The farmers whose data was sought from the FFA were organic and conventional farms that applied for agricultural payments for CCs in 2020. Agricultural payments were registered on the field parcel scale. With these definitions, the FFA provided requested data on 5593 conventional farms and 1432 organic farms. As the farmers were contacted by email, only those whose email addresses were available in the registry of the FAA were surveyed. The total number of invitees was 6493.

The survey started on 16 March 2021 and ended on 11 April 2021. One reminder message was sent on 30 March 2021. In addition to 13 statements under the title "Your success with CCs?" analyzed in this paper, the survey included other main themes with 51 statements, all dealing with CCs [23,32,36,37]. The farmers had five answer choices for each of the 13 statements: 1 = fully disagree, 2 = agree, 3 = neither disagree nor agree, 4 = agree, and 5 = fully agree. Statements of the whole survey were formulated based on findings reported in recent peer-reviewed papers [11,12,15,38,39]. The questionnaire was test run with four farmer-researchers who had long-term experience of CCs in Finland. Based on dialogue and comments, the questionnaire was finalized, and clarity was improved when needed. In addition to the statements, farmers were given an opportunity to freely list the future research needs that they considered important to support the cultivation, success, and further expansion of CCs.

The median of the respondent's answering time was some 15 min and 42 s (lower quartile 11 min and 31 s; upper quartile 24 min and 10 s). The farmers could save the answers and continue answering later. In total, 1130 farmers answered the survey, which corresponded to a 17.4% response rate of the farmers contacted. An additional 362 viewed or started to fill in the survey without completing and returning it by the deadline. Thereby, there were $N = 1130$ for 13 statements on success with CCs, while 244 farmers freely listed from one to ten future research needs. The most frequently used keywords were selected from the farmers' free word sentences, harmonized, and presented as a word cloud with the share of respondents who used each of the most common words or compound words.

Only the farmer's age ($\leq$50 and >50 years) and education (basic, vocational, college level, and university) were given by the farmers as background information used in the survey. None of these were found to markedly contribute to the farmers' views on their success with CCs. The rest of the background information was available for 2020 in the registry of the FFA by using the farm identity number. After merging the datasets, the respondents were grouped for statistical analyses according to (1) the farming system (organic and conventional) they operated, (2) the farm type they operated (cereal, special crop, horticulture, cattle, pig, poultry and horse/sheep farm), (3) the farm size (<40, 40–79, 80–119 and $\geq$120 ha), and (4) the geographical region their farms were located in (merging 16 Centers for Economic Development, Transport and the Environment (ELY Centers) to form four main regions: South-, West-, East/North Finland and the inland region). Furthermore, for each responding farm, the share of land devoted to cereals, grassland, special crops, peas (*Pisum sativum* L.), faba beans (*Vicia faba* L.), spring and winter oilseed rape (*Brassica napa* L.), and turnip rape (*B. rapa* L.), caraway (*Carum carvi* L.) and other crops, e.g., potatoes (*Solanum tuberosum* L.) and sugar beets (*Beta vulgaris* var. *altissima*), were assembled from the registry of the FFA and grouped as <25%, 25–50%, and $\geq$50% for cereals, grasslands and other crops, and as 0%, <10%, and $\geq$10% for special crops (lower cultivation areas). In addition to the land use for cash crops, the number of CCs (1–5, 5–10, and >10) that a farmer had cultivation experience with was used as background data [32]. Furthermore, the responding farmers were asked to select species they had used as CCs and to indicate the degree of experience with each species: 1 = none, 2 = very little, 3 = somewhat, 4 = a lot, and 5 = very much.

2.2. Statistical Analyses

The non-response bias was assessed by comparing the characteristics of the respondents who returned the survey to those of non-respondents (Table S1). The compared characteristics were the region, farming system, farm type, farm size, farm cereal area, grassland area, special crop area, and other crop area. In such cases, weights were used to bring the sample and the population more in line. This non-response weighting was considered in the statistical analysis. However, such weights were needed only for a few respondents, and hence, the total effect remained marginal, if any, and therefore this procedure was not used. No significant distortions of representativeness were found. The response rate was close to that of the contacted farmers (i.e., 17.4% ± ~2%) for the regions, farm sizes, the shares of land devoted to different crop groups on a farm, or for organic and conventional farms (232 and 898 responses, respectively). Considering the farm types, cattle and pig farms were slightly, but not significantly, under-represented [23]. Based on a preliminary examination of the data, the answers of all the 1130 respondents were considered acceptable and were used for statistical analyses.

The relationship between row and column variables was tested using the Cochran–Mantel–Haenszel (CMH) test. The row variables were formed from ten characteristics of the respondents (e.g., the region, farming system, farm type, farm size, and farmer's education), and the column variables were the results of the farmers' views on their success with CCs based on 13 survey statements (Figure 1) indicated with a 5-point Likert scale. This scale was used because it offers a range of options, provides valid data, enables cross-tabulation of respondents' answers to statements in different sections of the questionnaire, and helps to reduce bias and ambiguity. Typically, both variables were ordinal scales, and the correlation statistic of the CMH with 1 degree of freedom was used. If a row variable was not on an ordinal scale, as in the case of the region and farm type, ANOVA (Row Mean Scores, RMS) statistics for the CMH were used. ANOVA (RMS) tests were used for all pairwise comparisons. All analyses were performed using the SAS software, FREQ, and GLM procedures. The statistical analyses on the number of CCs that farmers had experience with (grouped as 1–5, 6–10, and >10) were also based on the CMH and ANOVA tests.

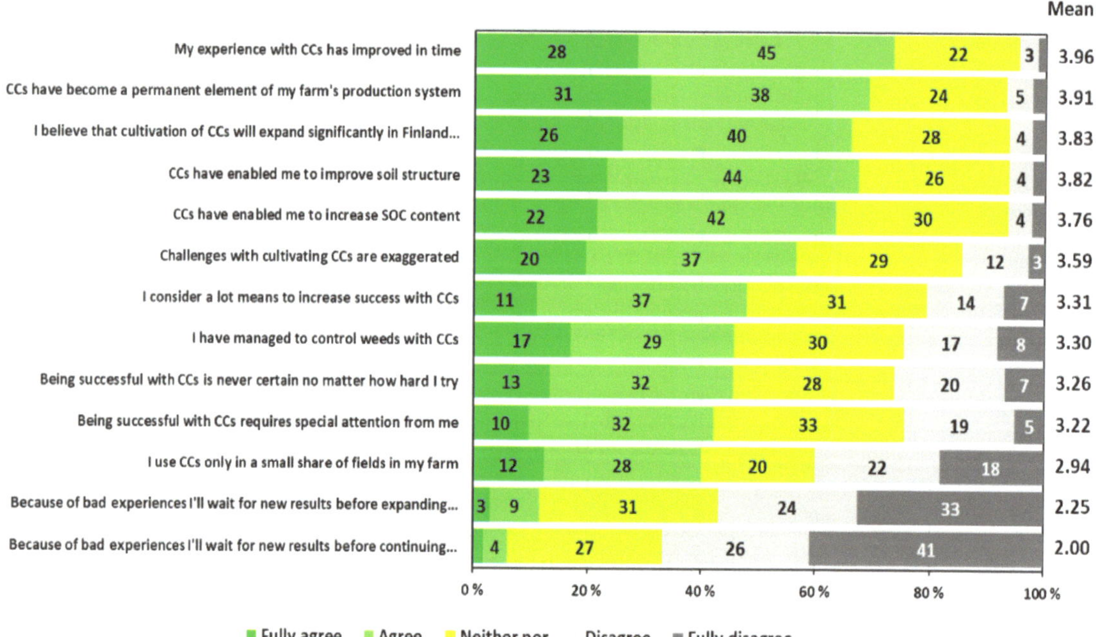

Figure 1. The distribution and means (in order of decreasing value) of the farmers' answers (*N* = 1130) to the statements concerning success with the cultivation of cover crops (CCs). The answer choices were: 1 = fully disagree, 2 = disagree, 3 = neither agree nor disagree, 4 = agree, and 5 = fully agree. The share of each answer choice is shown within each bar except in the case of being <3%.

The relationship between farmers' experiences with different groups of CCs was demonstrated by a graphical method. Groups were grasses, legumes, and special CCs. The experience with different groups did not vary greatly from each other when considering responses to statements on success with CCs, and this was presented by smoothing techniques, cubic spline, using the SAS/GRAPH. In addition, the Spearman rank-correlation coefficients were calculated for all groups.

3. Results

3.1. Farmers' Success with Cover Crops Based on 13 Survey Statements

The farmers were most positive on the statement that their experiences with CCs had improved over time (Figure 1). They also largely considered that CCs had become a permanent element of their farming systems and believed that the cultivation of CCs would expand significantly in Finland in the future, along with gained experiences. In total, 67% of farmers agreed that they had already been successful in improving the soil structure. In total, 64% considered that the SOC content had increased with the use of CCs, while 46% also had good experiences of controlling weeds with CCs.

The majority of farmers agreed that challenges with the cultivation of CCs were exaggerated, while 45% felt that success with CCs was never certain, no matter how much they tried (Figure 1). In total, 48% of the respondents had considered a variety of means to increase their success with CCs, and 42% responded that CCs required special attention to manage them successfully. In total, 40% of farmers have, so far, used CCs only on a small share of fields. In total, 12% of respondents considered that due to their bad experiences, they would wait for new results before expanding the cultivation of CCs on their farms, while only 6% said that they would wait for further understanding before continuing the use of CCs.

3.2. Success Depending on Number and Variety of Cover Crops

Based on the survey statements, the farmers were found to be more successful with CCs when they cultivated a higher number of CCs on a farm (Figure 2). The differences were significant depending on the numbers of CCs used by farmers for statements such as the experience with CCs has improved over time, CCs have become a permanent element of the farming system, CCs have improved the soil conditions, challenges with CCs are exaggerated, I consider means to increase success with CCs, and the cultivation of CCs will expand along with gained experience. On the other hand, farmers with experience with a number of CCs were those who most frequently agreed that they used CCs only on a small share of fields on the farm and would wait before expanding or continuing the cultivation of CCs on their farms (Figure 2).

When CCs were assessed in more detail by grouping them into categories of grasses, legumes, and special CCs, it was found that, in general, the experience with different CC species groups did not vary greatly from each other when considering the response to statements on success with CCs (Figure 3). However, the degree of experience with winter types of CCs did not correlate with the farmer's responses to any of the statements. The correlation between the farmer's experiences with grasses and legumes as CCs and their response was significant ($p < 0.0001$) for nine statements but quite identical. The correlations were positive for all three CC species groups for statements on the persistent and expanding role of CCs in the farming system, improved soil conditions, and weed control, while the correlations were significant only for grasses and legumes as CCs for the statement on exaggerated challenges. On the other hand, such correlations were always negative ($p < 0.0001$) for statements about the use of CCs on small shares of fields only and the expansion and continuation of cultivation due to bad experiences.

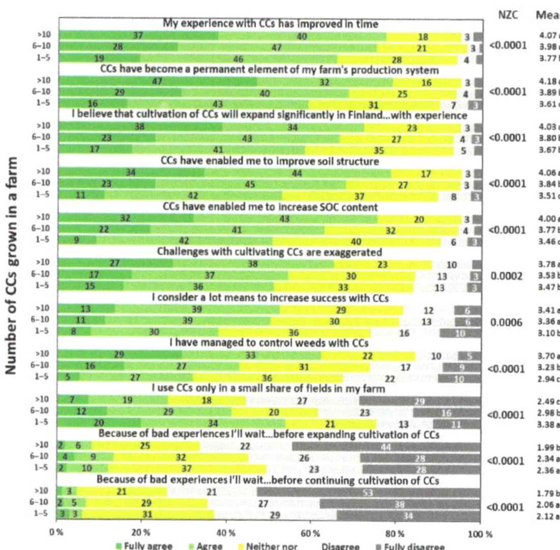

Figure 2. The shares and mean values of the farmers' answers (1 = fully disagree, 2 = disagree, 3 = neither agree nor disagree, 4 = agree, and 5 = fully agree) to statements on success with cover crops (CCs) depending on their number (1–5, 6–10, and >10) on a farm (NZC, nonzero correlation). Pairwise comparisons are shown next to the means: means with the same letter for each statement do not differ significantly. The shares of answers are shown within each bar except in the case of being <3.

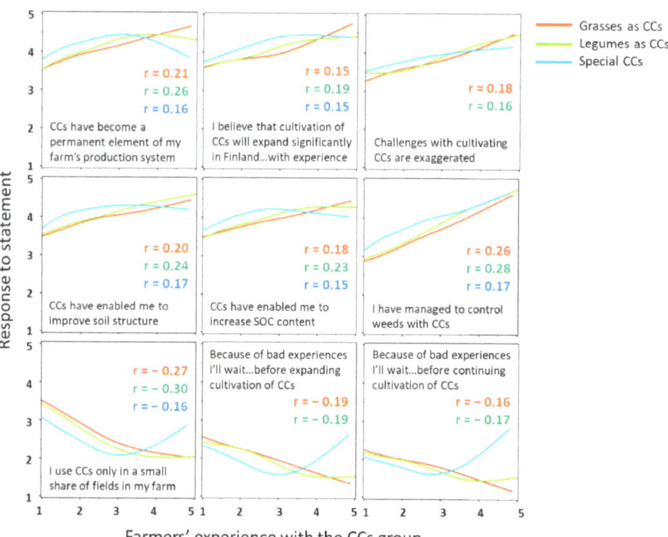

Figure 3. Correlations (*p* < 0.0001) between farmers experiences (1 = none, 2 = very little, 3 = somewhat, 4 = a lot, and 5 = very much) with different groups of cover crops (CCs) and responses to statements on success with CCs (1 = fully disagree, 2 = disagree, 3 = neither agree nor disagree, 4 = agree, and 5 = fully agree). Correlation coefficients (r) are shown for each CC group.

3.3. Success Depending on Farm and Farmer Characteristics

Organic farmers were more positive concerning their success with CCs compared to conventional farmers (Figure 4). For example, only 1–3% of organic producers disagreed that their experiences had improved in time, CCs had become a permanent element of the production system, the cultivation of CCs would expand significantly in Finland, and the use of CCs had improved the soil conditions. The share of conventional farmers who disagreed was 6–8%. While 84% of organic farmers agreed that they had managed to control weeds with CCs, the corresponding share of conventional farmers was only 36%. In total, 9% of organic and 48% of conventional farmers agreed that they used CCs only on a small share of field parcels. The differences between farming systems were smallest for statements that one can never be certain about the success of CCs and that CCs require special attention when incorporated into the cropping system. Only 3% of organic farmers agreed that due to bad experiences, they would wait for new results before expanding or continuing the cultivation of CCs, while the shares were 14% for expanding and 7% for continuing in the case of conventional farmers (Figure 4).

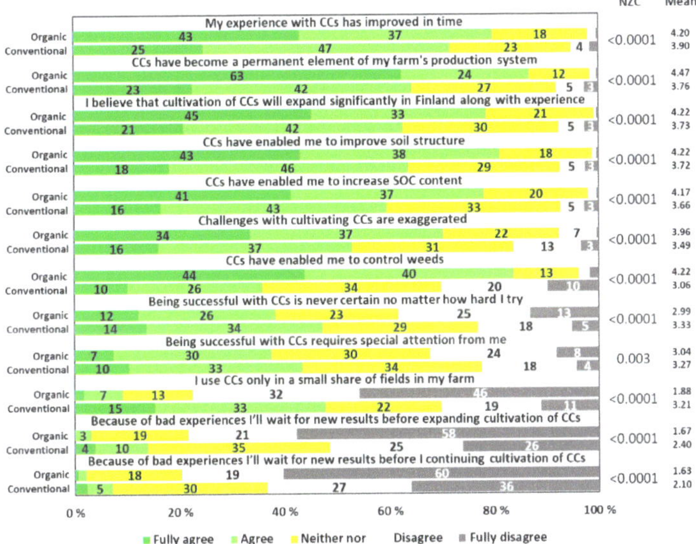

Figure 4. The shares and mean values of the farmers' answers (1 = fully disagree, 2 = disagree, 3 = neither agree nor disagree, 4 = agree, and 5 = fully agree) to statements on success with cover crops (CCs) depending on the farming system (NZC, nonzero correlation). The shares of answers are shown within each bar except in the case of being <3.

Farmers with larger farms tended to more frequently consider means to increase success with CCs, while farmers with small farms used CCs more often only on a small share of fields (Figure 5). On large farms, the farmers had been less successful in controlling weeds with CCs, and when the farm size exceeded 120 hectares, farmers were less certain about their success with CCs no matter how hard they tried.

When considering differences between farm types, farmers with cereal, special crop, and pig farms tended to differ, especially from those with cattle farms (Figure 6). Farmers on cattle farms agreed less frequently that being successful with CCs required special attention and was never certain, as well as often considering multiple means to increase success with CCs. On the other hand, they were more positive than farmers with cereal, special crops, and pig farms about the success of controlling weeds with CCs. Farmers were more positive about their success as well as the impacts, role, and future of CCs on their farms when they had a lower share of cereal area on the farm when compared to >50% land area (Figure 7). The case was opposite to grassland areas. Farmers who cultivated special crops on their farms paid special attention to CCs and more often considered various means to improve their success with CCs but were also more uncertain about their success (Figure S1). Farmers who had higher shares of arable land (>50%) for other types of crops (potatoes and sugar beet) on their farms tended to be more positive on the role, impacts, and future of CCs on their farms than those with a <25% share (Figure S2).

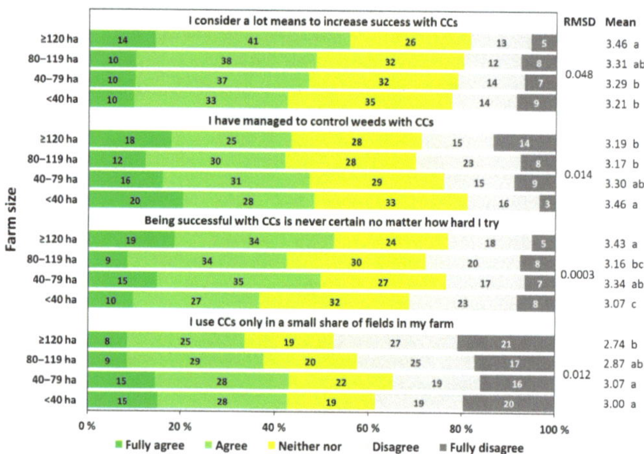

Figure 5. The shares and mean values of the farmers' answers (1 = fully disagree, 2 = disagree, 3 = neither agree nor disagree, 4 = agree, and 5 = fully agree) to statements on success with cover crops (CCs) depending on the farm size (RMSD, row mean scores difference). Means with the same letter for each statement do not differ significantly.

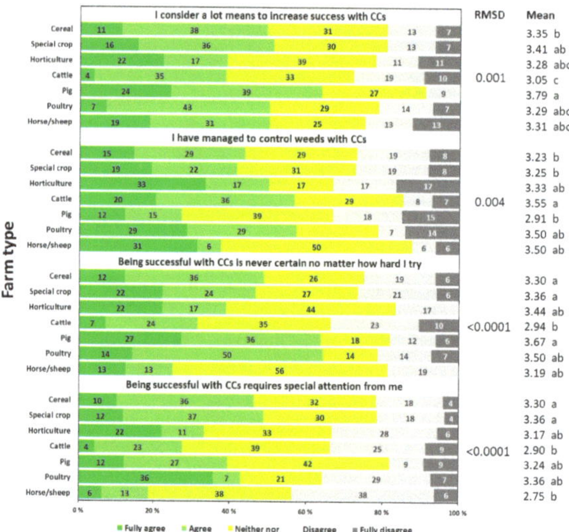

Figure 6. The shares and mean values of the farmers' answers (1 = fully disagree, 2 = disagree, 3 = neither agree nor disagree, 4 = agree, and 5 = fully agree) to statements on success with cover crops (CCs) depending on the farm type (RMSD, row mean scores difference). Means with the same letter for each statement do not differ significantly.

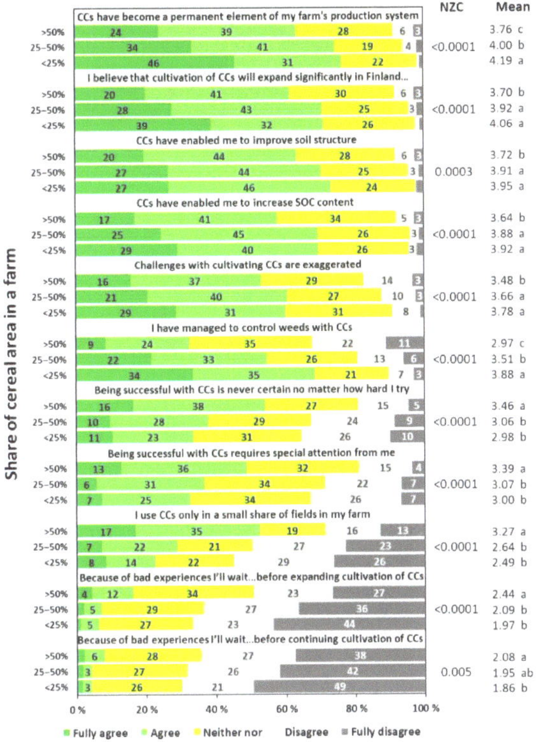

Figure 7. The shares and mean values of the farmers' answers (1 = fully disagree, 2 = disagree, 3 = neither agree nor disagree, 4 = agree, and 5 = fully agree) to statements on success with cover crops (CCs) depending on the cereal area on a farm (NZC, nonzero correlation). Means with the same letter for each statement do not differ significantly. The shares of the answers are shown within each bar except in the case of being <3.

Despite the many differences found in the farmers' views on their success with CCs depending on the farm characteristics, the farmer's education and age did not seem to have any major impacts on how successful they considered to be with CCs. The only exception was that university-educated farmers most frequently considered that CCs had become a permanent element of their production system.

3.4. Farmers' Wishes for Research to Support Success in the Cultivation of Cover Crops

In total, 22% of the respondents brought up topics on CCs that require additional research in Finnish growing conditions. The number of topics varied from one to ten depending on the respondent. Most frequently (17%), the farmers indicated that crop protection was a topic deserving more research attention (Figure 8). In some cases, they expressed this more precisely to concern weed control, the use of pesticides, the herbicide tolerance of CCs, as well as the impacts on disease pressure and weed infestation. Topics such as the timing of sowing (13%) as well as sowing methods, the use of CC mixtures, and different CC species were all highlighted by 11% of the respondents. In total, 10% of farmers mentioned the impacts of CCs on yields, and 9% mentioned the impact on profitability, while 7% highlighted themes such as differences in the growth performance of CCs, impacts on weed control and nutrients, harvesting methods, support for implementation, and subsidies. The rest of the topics shown in Figure 8 were brought up by less than 7% of the respondents.

Figure 8. Word cloud showing the research topics that were most frequently expressed by the farmers (*N* = 244) in their free word answers. The share of farmers who mentioned the topic (if ≥7%) is shown next to the expressed research need (two in the case it crosses both topics).

Only some differences were found between farm and farmer characteristics concerning their views on supportive research needs. Organic farmers more frequently emphasized than conventional farmers the need to better share gained experiences and research outcomes, as did those who had low cereal areas (<25%) on their farms and/or experience with special crops (Table 1). Younger farmers (≤50 years) were especially interested in results on the profitability of CC cultivation. Furthermore, farmers in East- and North Finland needed more information about the use of different CC species and sowing methods than those in South Finland. Farmers having very large farms (>120 ha) tended to highlight knowledge gaps on the impacts of CCs on weeds more frequently than those with smaller farms (Table 1).

Table 1. Odds ratios (OR) with 95% confidence limits (CL) showing the significant impacts of the farm characteristics on the probability that a farmer indicated the specific research need in his/her free word answer. When the odds ratio is <1.00, a farmer proposed this research topic significantly less frequently, and when it is >1.00, significantly more frequently than the reference group (following "vs."), provided that the CL did not include the value 1.00.

Characteristics	*p*-Value	OR	95% CL	
Sowing methods:				
West Finland vs. South Finland		2.74	0.75	10.01
Inland vs. South Finland	0.06	0.43	0.04	4.27
East/North Finland vs. South Finland		5.51	1.11	27.29
CC species and mixtures:				
West Finland vs. South Finland		0.73	0.26	2.07
Inland vs. South Finland	0.02	0.35	0.07	1.78
East/North Finland vs. South Finland		3.83	1.10	13.29
Impacts of CCs on weeds:				
Farm size <40 ha vs. >120 ha		0.32	0.10	1.06
Farm size 40–79 ha vs. >120 ha	0.05	0.34	0.11	1.03
Farm size 80–119 ha vs. >120 ha		0.20	0.04	0.94
Profitability of CC use:				
Respondent's age ≤50 years vs. >50 years	0.05	2.53	1.00	6.38
Knowledge sharing between farmers:				
Conventional vs. organic	0.04	0.38	0.15	0.98
Cereal area <25% vs. >50%		2.99	1.02	8.77
Cereal area 25–50% vs. >50%	0.09	1.09	0.33	3.57
Special crop area 0% vs. >10%		0.32	0.12	0.85
Special crop area <10% vs. >10%	0.07	0.55	0.11	2.69

4. Discussion

The cultivation area of CCs has substantially increased in Finland during the last ten years, from about one percent of the agricultural land area (23,000 hectares) [12] to 138,000 hectares in 2020 when the farmer survey was carried out. The area has only slightly increased for the next two years. When considering the cultivation area of field crops by excluding grasslands and environmental fallows, the current area under CCs could be increased tenfold. Ongoing and potential future expansion is likely to face challenges. Hence, the transition requires targeted support provided by research and extension services to fill in the knowledge gaps by involving the farmers in the dialogue [27,29,33,40,41]. The exchange of experiences and knowledge sharing within the farmer community is also very valuable, as was highlighted by the respondents of this survey (Figure 8) and especially by organic producers, those with low areas on cereals while high in special crops (Table 1).

4.1. Successes and Positive Prospects for Cover Crops Coupled with Uncertainties and Feeling of Demandingness

Farmer's perceptions about the likely pros and cons of CCs impact whether they implement CCs into their cropping systems [21], i.e., perceptions roughly divide farmers into adopters and non-adopters. Depending on the survey, adopters were characterized to be, e.g., environmentally more conscious, inquisitive, and receptive, acknowledge potential benefits of CCs, and have available machinery, while non-adopters had, e.g., perceptions about barriers (economic, management), and more rented land [29,40,42,43]. This survey was focused on the experiences of CC adopters only. This likely explains why responses to various statements on success with the cultivation of CCs were, in general, positive (Figure 1), while they differed systematically depending on the farming system (Figure 4). Statements highlighting the success and positive prospects for CCs were agreed with more frequently by organic than conventional farmers, while the situation was the opposite for statements that indicated uncertainties and the demands of CC use. Organic farmers indicated "fully agreed" choices, while conventional farmers "agreed" and made "neither nor" choices more frequently on the positively worded statements. Organic and conventional farmers differed in terms of their rationales for the cultivation of CCs: organic farmers especially valued multifunctionality and various ecosystem services from CCs [44]. Farmers use CCs to increase soil health, decrease soil disturbance, control weeds, and improve nutrient cycling [5,6,8,9,20,45] and, thereby, promote crop yields [2]. Organic farmers usually had more experience with CCs than conventional farmers. Failures may, however, limit the further adoption of CCs [19,24], especially in conventional production. Only a very low share of respondents agreed that they waited for new results before expanding (12%) or continuing (6%) the use of CCs on their farms (Figure 1). In this light, the positiveness of conventional farmers was a surprise, knowing the shorter history of CCs as part of the production systems and that the yield benefits gained with CCs vary greatly depending on climate, management, and soil properties [2]. In Finland, agricultural payments for the cultivation of CCs (EUR 97 per hectare) balance the risks related to the benefits and profitability of their use. This is important during the transition towards regenerative crop production systems [1,30]. Not least, because in recent surveys lack of financial support was found to be a key barrier for deployment of CCs [33–35]. Interestingly, in another study, the majority of subsidized farmers said that they continue to grow CCs even if the incentive is ceased [46].

Farmers seemed to do better with CCs in the case that their land use was not cereal-dominated (Figure 7), which may be attributable to limited experience in cultivating special, diversifying crops [32]. On the other hand, CCs were used especially on cereal farms with the likely aim of restoring the soil structure and condition [23]. Respondents were altogether more positive toward CCs when they were familiar with a high number of CC species (Figure 2), although their experiences did not vary greatly depending on CC groups: grasses, legumes, and special CCs (Figure 3). A higher diversity of CC species and the use of CC mixtures may reinforce success when adopting CCs on Finnish farms, not least due to the variable weather, which is typical for high-latitude conditions [47]. Acknowledging differences in the functional traits of CCs supports the provision of different services for food production and the ecosystem and enables win-win solutions [48–50]. Organic production leans on various ecosystem services provided by CCs [17,38], and hence, farmers often have more experience with the cultivation of CCs. These may have further supported the shift toward the use of more diverse CC species and their mixtures compared to conventional farming [51]. The difference between farming systems was most striking in how successful farmers were in controlling weeds with CCs (Figure 4). In organic production, CCs act as an important biological tool, but their weed suppression capacity varies depending on the properties of the CCs, as well as the conditions and used crop management [6,7,52–54]. Successes with CCs did not greatly differ depending on the farm size—nor in terms of controlling weeds—but farmers with larger farms agreed more frequently that they considered the means to increase success with CCs a lot (Figure 5).

4.2. Farmers' To-Do List for Research to Support Success with Cover Crops

The farmers indicated a high number of diverse, down-to-earth topics around CCs that require near-future research activities. The knowledge gaps did not strongly vary according to the farm characteristics (Table 1). Crop protection was the most frequently mentioned management practice that would require further know-how (Figure 8), with more precise expressions on growth, suppression, and control of weeds, herbicide tolerance of CCs, impacts of CCs on disease pressure, and pesticide use. Cover crops may either ease, strengthen, or have no effect on biotic risks [55–58]. Concern about crop protection risks was found to be one of the major barriers to the large-scale adoption of CCs in many regions of Europe [33]. In particular, farmers with larger field areas tended to highlight the need for a better understanding of crop protection in the presence of CCs (Table 1).

Many of the knowledge gaps highlighted by Finnish farmers to be relevant in high-latitude conditions were also indicated by farmers from more southern regions of Europe [24,25]. For example, farmers indicated the need for more research on the sowing time and methods. The successful establishment of both crop and CC stands is critical, especially in the case of under-sowing. Early summer drought is common in Finland [59] and may interfere with the growth performance of both crops and CCs and alter the competition between them [37] depending on the sowing method and CC species [60]. Farmers in East- and North Finland needed more information about sowing methods than those in South Finland, which is likely attributable to the lower share of field parcels with CCs and, hence, less experience with cultivation methods for both CCs and special crops due to limitations caused by the northern climate [61,62]. Likely for the same reasons, farmers in East- and North Finland (Table 1) more frequently expressed the need for locally relevant results on the use of different CC species and their mixtures to support the adoption of diverse CCs. Some 10% of the farmers mentioned the impacts of CCs on yield and specified these as yield losses caused by competition with CCs as well as elevation of crop yields in the short term (e.g., nitrogen effects of leguminous CCs) and long term (e.g., through impacts on soil quality). In total, 9% of farmers highlighted profitability. Younger farmers ($\leq$50 years) were especially interested in having more results on the profitability of the cultivation of CCs, i.e., costs compared to benefits and consideration of the monetary value of the short- and long-term impacts on yields and ecosystem services [3,7]. The benefits of CCs tend to accumulate over time; therefore, many surveys have indicated that economic barriers prevent deployment [33–35]. Localized economic information is essential to promote adoption [29]. Even though many farmers are, in general, very interested in using satellite and drone images to aid decision-making [63–65], they did not highlight remote sensing in their free word answers despite the potential.

5. Conclusions

A high share of survey respondents felt that their experiences with CCs had improved in time and were confident that CCs have become a permanent element on their farms and that the area under CCs will expand significantly in Finland. They also felt that CCs had been an effective measure to improve the soil structure and increase the SOC content, though they felt they had managed to use CCs to control weeds to a lesser extent. Weed control has been perceived as a challenge for the deployment of CCs, as well as in many other regions. The Finnish farmers also felt that the challenges to adopting CCs were exaggerated. The majority of the respondents disagreed that because of bad experiences, they would wait for new results before expanding or continuing the use of CCs. Overall, organic farmers were more positive, which is attributable to more experience with CCs but also their dependence on various ecosystem services that CCs provide as a regenerative practice. Nonetheless, conventional farmers were also confident in their success with CCs. Financial compensation available for Finnish farmers might explain the positive views on adopting CCs when compared to surveys conducted elsewhere. Such a payment compensates for the economic costs and risks of the introduction of CCs. Therefore, the current subsidies offer a good opportunity for a powerful learning-by-doing method to support the adoption of CCs. Therefore, subsidies sound justified as long as the transition towards regenerative agriculture proceeds. The Finnish farmers' free word answers highlighted research needs on topics that were also highlighted in surveys from very different production systems and conditions. They especially listed crop protection, sowing practices, the use of CC mixtures, and a wider choice of CC species, as well as the potential impacts on yield and profitability. The research needs were not widely divided. Younger farmers ($\leq$50 years) who are often more educated highlighted studies on profitability, while farmers from East- and North Finland indicated they would be interested in experiments on diverse CC choices. Dialogue among farmers' community and researchers is needed 1) to share good management practices, as encouraged by organic farmers, and 2) to focus on the highlighted knowledge gaps, e.g., on crop protection and CCs establishment. The response rate was, in general, high (17.4%), but it is possible that positive experiences with CCs encouraged farmers to answer the survey.

Supplementary Materials: The following supporting information can be downloaded at https://www.mdpi.com/article/10.3390/agronomy13092326/s1, Table S1: Results of sampling bias analysis comparing respondents to non-respondents of the survey with total of 6493 invited farmers (17.4% response rate) and in parenthesis respondents to non-respondents of 1354 invited organic farmers (17.1% response rate); Figure S1: The shares and mean values of the farmers' answers (1 = fully disagree, 2 = disagree, 3 = neither agree nor disagree, 4 = agree, and 5 = fully agree) to statements on success with cover crops (CCs) depending on the area under special crops (e.g., rapeseed, grain legumes) on a farm. Pairwise comparisons are shown next to the mean.; Figure S2: The shares and mean values of the farmers' answers (1 = fully disagree, 2 = disagree, 3 = neither agree nor disagree, 4 = agree, and 5 = fully agree) to statements on success with cover crops (CCs) depending on the area under other types of crops (potatoes and sugar beet) on a farm. Pairwise comparisons are shown next to the mean. The shares of the answers are shown within each bar except in the case of being <3. Reference [23] is cited in the Supplementary Materials.

Author Contributions: Conceptualization, P.P.-S.; methodology, P.P.-S., L.J. and H.K.; validation, L.J. and P.P.-S.; formal analysis, L.J.; investigation, P.P.-S., L.J. and H.K.; resources, P.P.-S. and L.J.; data curation, L.J. and P.P.-S.; writing—original draft preparation, P.P.-S.; writing—review and editing, P.P.-S., L.J. and H.K.; visualization, P.P.-S.; project administration, P.P.-S.; funding acquisition, P.P.-S. All authors have read and agreed to the published version of the manuscript.

Funding: This work was financed by the Ministry of Agriculture and Forestry in Finland, project Evergreen Revolution with Cover Crops—Best Practices to Enhance C Sequestration (IKIVIHREÄ), grant no. VN/5082/2021-MMM-2 (Catch the Carbon program).

Data Availability Statement: Not applicable.

Acknowledgments: We thank all the 1130 pioneer farmers who answered the survey.

Conflicts of Interest: The authors declare no conflict of interest.

References

1. EASAC. *Regenerative Agriculture in Europe. A Critical Analysis of Contributions to European Union Farm to Fork and Biodiversity Strategies*; EASAC Policy Report 44; EASAC: Brussels, Belgium, 2022; p. 58, ISBN 978-3-8047-4372-4. Available online: www.easac.eu (accessed on 26 August 2023).
2. Fan, F.; van der Werf, W.; Makowski, D.; Ram Lamichhane, J.; Huang, W.; Li, C.; Zhang, C.; Cong, W.-F.; Zhang, F. Cover crops promote primary crop yield in China: A meta-regression of factors affecting yield gain. *Field Crops Res.* **2021**, *271*, 108237. [CrossRef]
3. Chahal, I.; Van Eerd, L.L. Do Cover Crops Increase Subsequent Crop Yield in Temperate Climates? A Meta-Analysis. *Sustainability* **2023**, *15*, 6517. [CrossRef]
4. Hansen, V.; Müller-Stöver, D.; Gómez-Muñoz, B.; Oberson, A.; Magid, J. Differences in cover crop contributions to phosphorus uptake by ryegrass in two soils with low and moderate P status. *Geoderma* **2022**, *426*, 116075. [CrossRef]
5. Hansen, V.; Meilvang, L.V.; Magid, J.; Thorup-Kristensen, K.; Jensen, L.S. Effect of soil fertility level on growth of cover crop mixtures and residual fertilizing value for spring barley. *Eur. J. Agron.* **2023**, *145*, 126796. [CrossRef]
6. Alonso-Ayuso, M.; Gabriel, J.L.; García-González, I.; Del Monte, J.P.; Quemada, M. Weed density and diversity in a long-term cover crop experiment background. *Crop Prot.* **2018**, *112*, 103–111. [CrossRef]
7. Rouge, A.; Adeux, G.; Busset, H.; Hugard, R.; Martin, J.; Matejicek, A.; Moreau, D.; Guillemin, J.-P.; Cordeau, S. Carry-over effects of cover crops on weeds and crop productivity in no-till systems. *Field Crops Res.* **2023**, *295*, 108899. [CrossRef]
8. Adetunji, A.T.; Ncube, B.; Mulidzi, R.; Lewu, F.B. Management impact and benefit of cover crops on soil quality: A review. *Soil Tillage Res.* **2020**, *204*, 104717. [CrossRef]
9. Kim, N.; Zabaloy, M.C.; Guan, K.; Villamil, M.B. Do cover crops benefit soil microbiome? A meta-analysis of current research. *Soil Biol. Biochem.* **2020**, *142*, 107701. [CrossRef]
10. Koudahe, K.; Allen, S.C.; Djaman, K. Critical review of the impact of cover crops on soil properties. *Int. Soil Water Conserv. Res.* **2022**, *10*, 343–354. [CrossRef]
11. Kaye, J.; Quemada, M. Using cover crops to mitigate and adapt to climate change. A review. *Agron. Sustain. Dev.* **2017**, *37*, 4. [CrossRef]
12. Aronsson, H.; Hansen, E.M.; Thomsen, I.K.; Liu, J.; Øgaard, A.F.; Känkänen, H.; Ulén, B. The ability of cover crops to reduce nitrogen and phosphorus losses from arable land in southern Scandinavia and Finland. *J. Soil Water Conserv.* **2016**, *71*, 41–55. [CrossRef]
13. Nouri, A.; Lukas, S.; Singh, S.; Singh, S.; Machado, S. When do cover crops reduce nitrate leaching? A global meta-analysis. *Glob. Chang. Biol.* **2022**, *28*, 4736–4749. [CrossRef] [PubMed]
14. Morrison, B.A.; Xia, K.; Stewart, R.D. Evaluating neonicotinoid insecticide uptake by plants used as buffers and cover crops. *Chemosphere* **2023**, *322*, 138154. [CrossRef] [PubMed]
15. Poeplau, C.; Don, A. Carbon sequestration in agricultural soils via cultivation of cover crops—A meta-analysis. *Agric. Ecosyst. Environ.* **2015**, *200*, 33–41. [CrossRef]
16. Jian, J.; Du, X.; Reiter, M.S.; Stewart, R.D. A meta-analysis of global cropland soil carbon changes due to cover cropping. *Soil Biol. Biochem.* **2020**, *143*, 107735. [CrossRef]
17. Blanco-Canqui, H.; Shaver, T.M.; Lindquist, J.L.; Shapiro, C.A.; Elmore, R.W.; Francis, C.A.; Hergert, G.W. Cover crops and ecosystem services: Insights from studies in temperate soils. *Agron. J.* **2015**, *107*, 2449–2474. [CrossRef]
18. Plumhoff, M.; Connell, R.K.; Bressler, A.; Blesh, J. Management history and mixture evenness affect the ecosystem services from a crimson clover-rye cover crop. *Agric. Ecosyst. Environ.* **2022**, *339*, 108155. [CrossRef]
19. Lamichhane, J.R.; Alletto, L. Ecosystem services of cover crops: A research roadmap. *Trends Plant Sci.* **2022**, *27*, 758–768. [CrossRef]
20. Scavo, A.; Fontanazza, S.; Restuccia, A.; Pesce, G.R.; Abbate, C.; Mauromicale, G. The role of cover crops in improving soil fertility and plant nutritional status in temperate climates. A review. *Agron. Sustain. Dev.* **2022**, *42*, 93. [CrossRef]
21. Plastina, A.; Liu, F.; Miguez, F.; Carlson, S. Cover crops use in Midwestern US agriculture: Perceived benefits and net returns. *Renew. Agric. Food Syst.* **2020**, *35*, 38–48. [CrossRef]
22. Church, S.P.; Lu, J.; Ranjan, P.; Reimer, A.P.; Prokopy, L.S. The role of systems thinking in cover crop adoption: Implications for conservation communication. *Land Use Policy* **2020**, *94*, 104508. [CrossRef]
23. Peltonen-Sainio, P.; Jauhiainen, L.; Mattila, T.; Joona, J.; Hydén, T.; Känkänen, H. Pioneering farmers value agronomic performance of cover crops and their impacts on soil and environment. *Sustainability* **2022**, *14*, 8067. [CrossRef]
24. Vincent-Caboud, L.; Peigné, J.; Casagrande, M.; Silva, E.M. Overview of organic cover crop-based no-tillage technique in Europe: Farmer's practices and research challenges. *Agriculture* **2017**, *7*, 42. [CrossRef]
25. Casagrande, M.; Peigné, J.; Payet, V.; Mäder, P.; Xavier Sans, F.; Blanco-Moreno, J.M.; Antichi, D.; Beeckman, A.; Bigongiali, F.; Cooper, J.; et al. Organic farmers' motivations and challenges for adopting conservation agriculture in Europe. *Org. Agric.* **2016**, *6*, 281–295. [CrossRef]
26. Crossland, M.; Fradgley, N.; Creissen, H.; Howlett, S.; Baresel, J.P.; Finckh, M.R.; Girling, R. An online toolbox for cover crops and living mulches. *Asp. Appl. Biol.* **2015**, *129*, 1–6.
27. Beach, H.M.; Laing, K.W.; Walle, M.V.D.; Martin, R.C. The Current State and Future Directions of Organic No-Till Farming with Cover Crops in Canada, with Case Study Support. *Sustainability* **2018**, *10*, 373. [CrossRef]

28. Jian, J.; Lester, B.J.; Du, X.; Reiter, M.S.; Stewart, R.D. A calculator to quantify cover crop effects on soil health and productivity. *Soil Tillage Res.* **2020**, *199*, 104575. [CrossRef]

29. McCollum, C.; Bergtold, J.S.; Williams, J.; Al-Sudani, A.; Canales, E. Perceived Benefit and Cost Perception Gaps between Adopters and Non-Adopters of In-Field Conservation Practices of Agricultural Producers. *Sustainability* **2022**, *14*, 11803. [CrossRef]

30. Kathage, J.; Smit, B.; Janssens, B.; Haagsma, W.; Adrados, J.L. How much is policy driving the adoption of cover crops? Evidence from four EU regions. *Land Use Policy* **2022**, *116*, 106016. [CrossRef]

31. Bergtold, J.S.; Ramsey, S.; Maddy, L.; Williams, J.R. A review of economic considerations for cover crops as a conservation practice. *Renew. Agric. Food Syst.* **2019**, *34*, 62–76. [CrossRef]

32. Peltonen-Sainio, P.; Jauhiainen, L.; Joona, J.; Mattila, T.; Hydén, T.; Känkänen, H. Sowing and Harvesting Measures to Cope with Challenges of Cover Crops Experienced by Finnish Farmers. *Agronomy* **2023**, *13*, 499. [CrossRef]

33. Hijbeek, R.; Pronk, A.A.; van Ittersum, M.K.; Verhagen, A.; Ruysschaert, G.; Bijttebier, J.; Zavattaro, L.; Bechini, L.; Schlatter, N.; ten Berge, H.F.M. Use of organic inputs by arable farmers in six agro-ecological zones across Europe: Drivers and barriers. *Agric. Ecosyst. Environ.* **2019**, *275*, 42–53. [CrossRef]

34. Strauss, V.; Paul, C.; Dönmez, C.; Löbmann, M.; Helming, K. Sustainable soil management measures: A synthesis of stakeholder recommendations. *Agron. Sustain. Dev.* **2023**, *43*, 17. [CrossRef]

35. Chang, S.-H.-E.; Yi, X.; Sauer, J.; Yin, C.; Li, F. Explaining farmers' reluctance to adopt green manure cover crops planting for sustainable agriculture in Northwest China. *J. Integr. Agric.* **2022**, *21*, 3382–3394. [CrossRef]

36. Peltonen-Sainio, P.; Jauhiainen, L. Come out of a hiding place: How are cover crops allocated on Finnish farms? *Sustainability* **2022**, *14*, 3103. [CrossRef]

37. Peltonen-Sainio, P.; Jauhiainen, L.; Känkänen, H.; Joona, J.; Hydén, T.; Mattila, T.J. Farmers' Experiences of How Under-Sown Clovers, Ryegrasses, and Timothy Perform in Northern European Crop Production Systems. *Agronomy* **2022**, *12*, 401. [CrossRef]

38. Daryanto, S.; Fu, B.; Wang, L.; Jacinthe, P.-A.; Zhao, W. Quantitative synthesis on the ecosystem services of cover crops. *Earth-Sci. Rev.* **2018**, *185*, 357–373. [CrossRef]

39. Meyer, N.; Bergez, J.-E.; Constantin, J.; Belleville, P.; Justes, E. Cover crops reduce drainage but not always soil water content due to interactions between rainfall distribution and management. *Agric. Water Manag.* **2020**, *231*, 105998. [CrossRef]

40. Nowatzke, L.W.; Arbuckle, J.G. Measuring and Predicting Iowa Farmers' Current and Potential Future Use of Cover Crops. *Soc. Nat. Resour.* **2023**, *36*, 755–775. [CrossRef]

41. Lang, Z.; Rabotyagov, S. Socio-psychological factors influencing intent to adopt conservation practices in the Minnesota River Basin. *J. Environ. Manag.* **2022**, *307*, 114466. [CrossRef]

42. Byerly, H.; Kross, S.M.; Niles, M.T.; Fisher, B. Applications of behavioral science to biodiversity management in agricultural landscapes: Conceptual mapping and a California case study. *Environ. Monit. Assess.* **2021**, *193*, 270. [CrossRef] [PubMed]

43. Lee, S.; McCann, L. Adoption of Cover Crops by U.S. Soybean Producers. *J. Agric. Appl. Econ.* **2019**, *51*, 527–544. [CrossRef]

44. Wayman, S.; Kissing Kucek, L.; Mirsky, S.B.; Ackroyd, V.; Cordeau, S.; Ryan, M.R. Organic and conventional farmers differ in their perspectives on cover crop use and breeding. *Renew. Agric. Food Syst.* **2017**, *32*, 376–385. [CrossRef]

45. Zhou, R.; Liu, Y.; Dungait, J.A.J.; Kumar, A.; Wang, J.; Tiemann, L.K.; Zhang, F.; Kuzyakov, Y.; Tian, J. Microbial necromass in cropland soils: A global meta-analysis of management effects. *Glob. Chang. Biol.* **2023**, *29*, 1998–2014. [CrossRef]

46. Cottney, P.; Williams, P.N.; White, E.; Black, L. The perception and use of cover crops within the island of Ireland. *Ann. Appl. Biol.* **2021**, *179*, 34–47. [CrossRef]

47. Peltonen-Sainio, P.; Venäläinen, A.; Mäkelä, H.M.; Pirinen, P.; Laapas, M.; Jauhiainen, L.; Kaseva, J.; Ojanen, H.; Korhonen, P.; Huusela-Veistola, E.; et al. Harmfulness of weather events and the adaptive capacity of farmers at high latitudes of Europe. *Clim. Res.* **2016**, *67*, 221–240. [CrossRef]

48. Elhakeem, A.; van der Werf, W.; Ajal, J.; Lucà, D.; Claus, S.; Vico, R.A.; Bastiaans, L. Cover crop mixtures result in a positive net biodiversity effect irrespective of seeding configuration. *Agric. Ecosyst. Environ.* **2019**, *285*, 106627. [CrossRef]

49. Drost, S.M.; Rutgers, M.; Wouterse, M.; de Boer, W.; Bodelier, P.L.E. Decomposition of mixtures of cover crop residues increases microbial functional diversity. *Geoderma* **2020**, *361*, 114060. [CrossRef]

50. Zhang, C.; Xue, W.; Xue, J.; Zhang, J.; Qiu, L.; Chen, X.; Hu, F.; Kardol, P.; Liu, M. Leveraging functional traits of cover crops to coordinate crop productivity and soil health. *J. Appl. Ecol.* **2022**, *59*, 2627–2641. [CrossRef]

51. Peltonen-Sainio, P.; Jauhiainen, L.; Joona, J.; Mattila, T.J.; Hydén, T.; Känkänen, H. Farm characteristics shape farmers' common and special cover crop choices in Finland. *Int. J. Agric. Sustain.* **2023**, *manuscript revised*.

52. Almoussawi, A.; Lenoir, J.; Spicher, F.; Dupont, F.; Chabrerie, O.; Closset-Kopp, D.; Brasseur, B.; Kobaissi, A.; Dubois, F.; Decocq, G. Direct seeding associated with a mixture of winter cover crops decreases weed abundance while increasing cash-crop yields. *Soil Tillage Res.* **2020**, *200*, 104622. [CrossRef]

53. Smith, R.; Warren, N.; Cordeau, S. Are cover crop mixtures better at suppressing weeds than cover crop monocultures? *Weed Sci* **2020**, *68*, 186–194. [CrossRef]

54. Rouge, A.; Adeux, G.; Busset, H.; Hugard, R.; Martin, J.; Matejicek, A.; Moreau, D.; Guillemin, J.-P.; Cordeau, S. Weed suppression in cover crop mixtures under contrasted levels of resource availability. *Eur. J. Agron.* **2022**, *136*, 126499. [CrossRef]

55. Hossain, S.; Bergkvist, G.; Berglund, K.; Mårtensson, A.; Persson, P. Aphanomyces pea root rot disease and control with special reference to impact of Brassicaceae cover crops. *Acta Agric. Scand. Sect. B Soil Plant Sci.* **2012**, *62*, 477.

56. Lemessa, F.; Wakjira, M. Cover crops as a means of ecological weed management in agroecosystems. *J. Crop Sci. Biotechnol.* **2015**, *18*, 123. [CrossRef]
57. Ait Kaci Ahmed, N.; Galaup, B.; Desplanques, J.; Dechamp-Guillaume, G.; Seassau, C. Ecosystem Services Provided by Cover Crops and Biofumigation in Sunflower Cultivation. *Agronomy* **2022**, *12*, 120. [CrossRef]
58. McNee, M.E.; Rose, T.J.; Minkey, D.M.; Flower, K.C. Effects of dryland summer cover crops and a weedy fallow on soil water, disease levels, wheat growth and grain yield in a Mediterranean-type environment. *Field Crops Res.* **2022**, *280*, 108472. [CrossRef]
59. Peltonen-Sainio, P.; Juvonen, J.; Korhonen, N.; Parkkila, P.; Sorvali, J.; Gregow, H. Climate change, precipitation shifts and early summer drought: An irrigation tipping point for Finnish farmers? *Clim. Risk Manag.* **2021**, *33*, 100334. [CrossRef]
60. St Aime, R.; Noh, E.; Bridges, W.C., Jr.; Narayanan, S.A. Comparison of drill and broadcast planting methods for biomass production of two legume cover crops. *Agronomy* **2022**, *12*, 79. [CrossRef]
61. Peltonen-Sainio, P.; Jauhiainen, L. Large zonal and temporal shifts in crops and cultivars coincide with warmer growing seasons in Finland. *Reg. Environ. Chang.* **2020**, *20*, 89. [CrossRef]
62. Peltonen-Sainio, P.; Jauhiainen, L. Lessons from the past in weather variability: Sowing to ripening dynamics and yield penalties for northern agriculture from 1970 to 2012. *Reg. Environ. Chang.* **2014**, *14*, 1505–1516. [CrossRef]
63. Kumar, V.; Singh, V.; Flessner, M.L.; Haymaker, J.; Reiter, M.S.; Mirsky, S.B. Cover crop termination options and application of remote sensing for evaluating termination efficiency. *PLoS ONE* **2023**, *18*, e0284529. [CrossRef] [PubMed]
64. Kümmerer, R.; Noack, P.O.; Bauer, B. Using High-Resolution UAV Imaging to Measure Canopy Height of Diverse Cover Crops and Predict Biomass. *Remote Sens.* **2023**, *15*, 1520. [CrossRef]
65. Näsi, R.; Mikkola, H.; Honkavaara, E.; Koivumäki, N.; Oliveira, R.A.; Peltonen-Sainio, P.; Keijälä, N.-S.; Änäkkälä, M.; Arkkola, L.; Alakukku, L. Can Basic Soil Quality Indicators and Topography Explain the Spatial Variability in Agricultural Fields Observed from Drone Orthomosaics? *Agronomy* **2023**, *13*, 669. [CrossRef]

Review

Shaping Soil Properties and Yield of Cereals Using Cover Crops under Conservation Soil Tillage

Edward Wilczewski [1,*], Irena Jug [2], Ewa Szpunar-Krok [3], Mariola Staniak [4] and Danijel Jug [2]

1   Department of Agronomy, Bydgoszcz University of Science and Technology, Kaliskiego 7, 85-796 Bydgoszcz, Poland

2   Faculty of Agrobiotechnical Sciences Osijek, Josip Juraj Strossmayer University of Osijek, Vladimira Preloga 1, HR-31000 Osijek, Croatia; ijug@fazos.hr (I.J.); djug@fazos.hr (D.J.)

3   Department of Crop Production, University of Rzeszow, Zelwerowicza 4., 35-601 Rzeszów, Poland; eszpunar@ur.edu.pl

4   Department of Forage Crop Production, Institute of Soil Science and Plant Cultivation-State Research Institute, Czartoryskich 8, 24-100 Puławy, Poland; staniakm@iung.pulawy.pl

*   Correspondence: edward.wilczewski@pbs.edu.pl; Tel.: +48-52-3749-443

Abstract: The aim of this review was to collect current results on the effect of different plants grown as winter and summer cover crops (CC) on the physical, chemical, and biological properties of soil and on the yield of cereal crops grown in a site with CC, using conservation soil tillage. The analyzed studies indicate that CC usually have a positive impact on the physical and biological properties of the soil. Regardless of the plant species used as CC, we can expect an increase in the number of soil microorganisms and an improvement in the activity of soil enzymes. This effect is particularly beneficial in the case of reduced tillage systems. Mixing CC biomass with the topsoil loosens compacted soils and, in the case of light, sandy soils, increasing the capacity of the sorption complex. The size and composition of CC biomass and weather conditions during the vegetation period and during the covering of the soil with plant biomass are of great importance for improving the chemical properties of the soil. A beneficial effect of CC, especially legumes, on the content of the mineral nitrogen in the topsoil is usually observed. Sometimes, an increase in the content of available forms of potassium (K) and/or phosphorus (P) is also achieved. The effect of CC on the content of soil organic carbon (C), total nitrogen (N), or soil pH is less common. CC used in reduced tillage systems can significantly improve the yield and quality of cereal grain, especially when legumes are used as CC in low-fertility soil conditions and at low fertilization levels. However, non-legumes can also play a very positive role in shaping soil properties and improving cereal yield.

Keywords: biological properties; cereal yield; cover crops; chemical properties; physical properties; reduced tillage

Citation: Wilczewski, E.; Jug, I.; Szpunar-Krok, E.; Staniak, M.; Jug, D. Shaping Soil Properties and Yield of Cereals Using Cover Crops under Conservation Soil Tillage. *Agronomy* **2024**, *14*, 2104. https://doi.org/10.3390/agronomy14092104

Academic Editor: Tie Cai

Received: 2 August 2024
Revised: 11 September 2024
Accepted: 13 September 2024
Published: 15 September 2024

1. Introduction

The growing world population results in an increasing demand for food production [1,2]. On the other hand, the area of agricultural land is gradually decreasing. In just 20 years (between 2000 and 2020), the area of agricultural land decreased by 2% [3]. In this period, the world population increased by 27.8% [4]. As a result, the area of agricultural land per capita decreased by 23.3%. Taking these changes into account, we should expect increasing difficulties in meeting the nutritional needs of the growing population. One of the solutions of this problem is increasing crop yield per unit area to meet human food needs. Moreover, it is widely known that meeting human nutritional needs will be possible after changing the way we eat by increasing the consumption of plant products instead of animal products [5,6]. As a result, a decrease in the number of farm animals and the amount of farmyard manure production may be expected. Taking into account the influence of farmyard manure fertilization on the physical, chemical, and biological properties of soil [7,8], especially a decrease in bulk density, penetration resistance, and an increase in hydraulic conductivity, soil aggregation and aggregate-associated carbon, an increase of soil organic carbon as well as an increase in the number of soil microorganisms, we could predict deterioration of the soil properties as a result of reducing of farmyard manure production and usage [9–11]. To prevent unfavorable changes, it is increasingly necessary to supply the soil with another type of organic matter, substituting farmyard manure. To maintain a high level of soil productivity and sustainable agrotechnology on the other hand, it is necessary to look for alternative methods to improve soil biological activity. Green manuring may be an essential agrotechnology practice in conditions without farmyard manure. Winter and summer CC used as green manure may be an effective source of organic matter and may have a very positive effect on the physical properties of soil and its biological activity [12–14]. CC biomass is not as effective as farmyard manure in terms of its impact on soil properties, but a systematic supply of plant biomass from green manures to the soil (every year or every other year) contributes to increasing the resources of soil organic matter (SOM), which has a significant impact on plant production [15,16]. It is an important source of N and other nutrients necessary for plant growth [17]. It also plays an important role in retaining nutrients and water in the soil, which is particularly important in highly weathered soils with a low cation exchange capacity [18]. That reason is why it is so important to cultivate summer CC, which, in addition to providing organic matter to the soil, may reduce the losses of nutrients unused by the main crop. Soil covered with both mulch from summer CC, as well as by live plants from winter CC, is protected against erosion during the winter, and during the sowing period of spring crops, it usually has more favorable moisture and content of available forms of fertilizer components [19–21]. If organic matter is not supplied to the soil, the soil is gradually impoverished. According to estimates by Sanderman et al. [22], agricultural land use has resulted in the loss of 133 Pg of C from the top 2 m of soil over the last 200 years. Alternative production systems and conventional systems can play a fundamental role in soil carbon sequestration, such as conservation tillage, which is characterized by reduced tillage (including no-tillage systems), more diverse crop succession, and permanent vegetation cover, including increased cover cropping frequency. Any simplifications in soil cultivation, limiting interference with the topsoil, and the use of organic fertilizers, contribute to improving soil properties and increasing biological activity, which helps maintain its productivity and fertility [23]. In temperate climate conditions, in soils covered by a reduced tillage system, Szostek et al. [23] obtained higher contents of soil organic carbon (SOC), total N, and SOM fractions than in the conventional tillage system and organic farming.

Different plants used as CC are characterized by a different efficiency and ability to take up and accumulate nutrients. In addition, the rate of mineralization of their biomass is different [24–26]. Many studies have shown the special value of legume crops [21,27,28]. They increase the activity of soil microorganisms, improve the physical and chemical properties of the soil, and suppress weeds [28]. They can also interrupt the development cycles of diseases, insects, and other pests [29]. Legume crop biomass undergo fast mineralization in the soil, which can support the growth of plants grown after them from the very early stages of their development [21]. The beneficial effect of these plants on the soil is particularly important on the conditions of their long-term use as CC [30]. Biomass mineralization of non-legume crops is usually slower, and their subsequent impact mainly concerns traits formed in the later stages of development of plants grown after them [31–33]. Taking this effect into account, the aim of this review was to collect current results on the effect of different plants grown as winter and summer CC on the physical, chemical, and biological properties of soil and on the yield of cereal crops grown in a site with CC, using conservation soil tillage.

2. Cover Crop Biomass

CC, besides its positive impact on the biological, physical, and chemical properties of soil, are an important source of nutrients for plants grown after them [17,18]. Therefore, it is very important to produce significant biomass, which can play an important role as green manure. Therefore, the right choice of plant species used for this purpose is very important. Other factors (sowing parameters, fertilization, and weather conditions) also play an important role. In conservation tillage, the planting of CC is usually carried out into undisturbed soil by opening a narrow trench [30]. For this purpose, seed drills equipped with disk coulters are used (Figure 1).

Figure 1. Sowing of CC seeds in conservation tillage conditions. (Photo: E. Wilczewski).

To obtain a positive effect of CC, it is necessary for plants to produce significant biomass that can substantially affect the growth conditions of the main crop. The amount of biomass produced by CC depends on many factors. The most important are total rainfall, soil fertility, the type of plant grown, mineral fertilization, sowing time, the length of the growing season, frosts, and in the case of winter CC, also the conditions of plant wintering. The total dry matter yield of CC ranges from 1.6 Mg ha^{-1}, in unfavorable conditions [34], to even 24.1 Mg ha^{-1}, in very good soil and water conditions [35]. In average conditions, 2.5–6.5 Mg ha^{-1} dry matter of summer CC may usually be obtained [21,24–26,36]. The disadvantage of summer CC is the high dependence of the biomass yield on weather conditions, especially precipitation during the sowing period and during the vegetative growth of plants [37–39]. This results in high variability over the years, both in terms of the biomass produced and the nutrients accumulated in it [39–42]. Non-legume plants (Figures 2 and 3), especially white mustard, oilseed radish, lacy phacelia, and ryegrass, are characterized by high yield fidelity when grown both as single-species crops and as mixtures [26,34,43–45]. Tribouillois et al. [38], based on studies including 36 species of plants used as CC, stated that some Fabaceae species are sensitive to high temperatures, whereas species from Poaceae and Brassicaceae families were more resistant to water deficit and germinate under a low base water potential.

Figure 2. White mustard (**A**), oilseed radish (**B**), and tansy phacelia (**A–C**) as components of non-legume mixtures of plants grown as summer CC (Photo: M. Staniak).

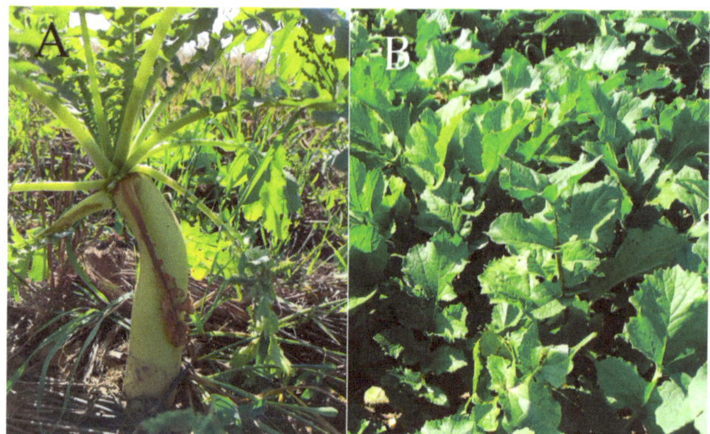

Figure 3. Different varieties of oilseed radish as winter (**A**) or summer (**B**) CC (Photo: D. Jug and E. Wilczewski).

According to Wendling et al. [45], mixtures with high species diversity ensure a stable and high biomass production with a low risk of failure. The dry biomass yield of winter CC usually ranges from 3.5 to 9.5 Mg ha^{-1} [46–51]. Particularly useful as winter CC are hairy vetch, winter rye, ryegrass, crimson clover, winter oilseed rape, and mixtures of these crops. In regions with mild winters, black oat, blue lupin, oilseed radish, and turnip are also useful [46,49]. The share of post-harvest residues in the total biomass yield of winter CC is usually from 11 to 26% [46]. In the total biomass, winter CC can accumulate from 33 to 185 kg N ha^{-1}, depending on the plant species and harvest date [40,47,52,53].

In conservation tillage, the biomass of CC is usually retained on the soil surface [28]. In the second half of autumn or in winter, the plant biomass is usually damaged by frost and covers the soil surface (Figure 4). In the spring, the biomass can be mixed with the soil, or the following crop may be sown directly into the mulch of CC. Plants with stiff stems, such as white mustard, may remain upright until spring (Figure 5). In this case, it is necessary to mow or mechanically break the stems before sowing the main crop.

Figure 4. Field pea as CC in autumn (**A**), winter (**B**), and early spring (**C**) (Photo: E. Wilczewski).

Biomass from CC is used as green manure for various plants (cereals, root crops, potatoes, and vegetables) grown in the following year. Using the entire biomass produced by CC as green manure usually produces better results than using only post-harvest residues for this purpose. The share of it in the entire biomass is usually 20–40%, depending on the plant species and soil and weather conditions [21,24–26]. In a study by Redin et al. [46] conducted in very good soil and weather conditions with the total biomass yield ranging from 6.4 to 24.1 Mg ha^{-1}, this share ranged from 6.3% for Jack bean to 10.5% for Dwarf pigeonpea and Gray mucuna. The highest share of post-harvest residues (72%) was found for Italian ryegrass [34] and a mixture of phacelia with radish and turnip (46.1%) [24]. The uptake of nitrogen by CC and thus the reduction in its leaching from the upper soil layer depended on the plant species, weather conditions, and agrotechnics used [40,41]. A high N uptake potential (90–120 kg ha^{-1}) had legumes [42], white mustard, and fodder radish [31,54,55]. Smaller amounts of N (60–80 kg ha^{-1}) were taken up by rye and oats [40,52] and turnip rape [25,42].

Figure 5. White mustard as CC in autumn (**A**) and late winter (**B**) (Photo: E. Wilczewski).

3. The Influence of Cover Crops on Soil Properties and Reducing the Loss of Nutrients Depending on Tillage System

3.1. Biological Properties

Soil microorganisms play an important role in the processes of decomposition of organic residues and enrichment of the soil with nutrients, biological activity, growth, antibiotic substances, participation in the formation of soil structure, detoxification of agrochemicals, and organic pollutants [56,57]. In this way, they contribute to maintaining the balance in the soil environment. According to Marinari et al. [58], more than 80% of all soil processes are closely linked to the activity of microorganisms.

Soil fertility, and thus its biological activity, can be increased using various agrotechnical treatments, including appropriate crop rotation with catch crops, organic fertilization, regulating water relations in the soil, and appropriate soil cultivation [59,60]. The method of cultivation affects the quantity and quality of organic matter, soil structure and moisture, pH, oxygen content, as well as the number and diversity of soil organisms. Intensive cultivation leads to the degradation of the soil environment, therefore limiting the plowing system and using tillage simplifications to help preserve the natural environment and support natural biological processes in the soil [61]. No-tillage systems improve soil properties, mainly by increasing and maintaining higher levels of organic C, which is crucial for increasing production stability. Significantly higher levels of organic C and total N in simplified systems were observed especially after a long period of soil use in this system: after 10 years [62], after 22 years [63], and even after more than 30 years [64,65], which indicates a high durability and stability of such an agricultural ecosystem.

The method of tillage significantly affects the populations of soil microorganisms. Both the biomass and the diversity of soil microorganisms are much higher in the conservation system compared to plow tillage. This is confirmed by Gajda's research [60], which showed that in the soil under winter wheat cultivated in a reduced tillage system, the number of soil microorganisms was higher by an average of 20% compared to the conventional system, both at a depth of 0–15 cm and 15–30 cm. In reduced tillage, higher dehydrogenase activity and ammonification and nitrification strength of soils were shown, which is also confirmed by other authors [66,67]. Growing rough oats in a no-tillage system has also shown many benefits related to soil microbiology and nutrient cycling [66]. It promoted greater diversity and activity of microorganisms in the soil thanks to the preservation of the natural soil structure, and it also contributed to an increase in the amount of organic C in the soil. Microorganisms, such as bacteria and fungi, processing crop residues, create stable organic material that have a beneficial effect on the soil structure, its ability to retain water, and the availability of nutrients for plants. This in turn contributes to improved soil health, better nutrient cycling, and a more balanced approach to the cultivation system. The method of cultivation also has a significant impact on the enzymatic activity of the soil, which is manifested by greater activity of soil enzymes in simplified systems compared to conventional ones. In a study by Harasim et al. [68], conservation tillage increased the activity of dehydrogenases and urease (on average, by 21.3% and 24.0%, respectively), regardless of the depth, compared to traditional plow tillage. Similarly, in a study by Lupwayi et al. [69], the conservation system showed higher (on average by 50%) activity of β-glucosidase, an enzyme involved in the carbon cycle, compared to the conventional system. The tillage system also affects the populations of earthworms and other mesofauna. Reducing the intensity of tillage reduces soil disturbance. The lack of soil turning protects its natural structure and helps save moisture. In turn, the increased amount of crop residues and organic biomass that characterize these systems provides more food and improves the living conditions of soil organisms [70]. In studies by Lenart and Sławiński [65], under the conditions of long-term (33 years) direct sowing, the number and mass of earthworms were 2–3 times higher, both in spring and autumn, compared to plow cultivation. Also, studies by Pelosi et al. [71] showed that the total biomass of earthworms in the no-tillage system with the participation of CC (alfalfa) was 4.2 times higher than in the conventional system. A greater functional diversity of earthworms was also observed at a lower cultivation intensity [72]. Earthworms help mitigate the effects of soil compaction that accompanies no-till systems. By digging tunnels in the soil, they create macroaggregates that loosen the soil, increase its water stability and permeability, and improve ventilation and air access. This allows plant roots to develop freely, which is especially important in the case of compact soils [73,74]. The method of tillage also affects the number and activity of other soil mesofauna organisms. The Collembola and Acari families are widely represented in the soil. They have a beneficial effect on soil fertility and respond relatively quickly to any changes caused by agricultural practices [75]. For this reason, they are considered a good indicator of the quality of the soil environment [76]. Studies show that greater numbers and diversity of Collembola and Acari are found in soil cultivated using the no-tillage system (direct sowing) compared to the reduced system and traditional tillage [77].

In reduced tillage systems CC play a special role, as they have a beneficial effect on the biological properties of the soil and also fulfill many different ecological functions. Keeping CC on the soil surface reduces the soil temperature range, retains moisture, and stabilizes the gradient of organic matter transformations [78]. In addition, they protect the soil in the winter from erosion caused by the direct impact of rainfall and sun. The biomass of CC plants is also an important source of organic C for soil microorganisms, which help temporarily immobilize and reduce losses of nutrients, thus engaging in C and nutrient cycling [63]. However, it depends largely on the CC species. Productive CC with low lignin content (e.g., clover, common vetch, field pea, white mustard, and blue phacelia) decompose relatively quickly and stimulate organic matter turnover and activate soil microorganisms. Thanks to this, the processes of C and nutrient circulation are balanced, which is the basis of soil fertility. In the case of plants containing a lot of lignin (e.g., winter rye, triticale, sheep, and alfalfa), the decomposition process is hindered by calcium compounds and at the same time stimulated by high N concentration. Organic N can undergo mineralization or be immobilized, which changes the availability of mineral N for subsequent plants and the functioning of the entire ecosystem [79].

Long-term no-tillage farming practices combined with the use of CC have a particularly positive effect on soil microbiome activity. Barel et al. [80] showed that monocultures of perennial ryegrass, white clover, common vetch, common radish, and their mixtures had a beneficial effect on the microbiological properties of the soil, in particular on the biomass of bacteria and fungi, their quantitative ratio, and the concentration of ergosterol. However, some of these properties differed depending on the CC species and the previous main crop. Other authors have shown beneficial effects of hairy vetch [81], common vetch [64], yellow lupine, and pea [82] on the population size and diversity of soil microorganisms. Cultivation of these CC under no-tillage conditions significantly increased microbial biomass (especially Gram-positive bacteria), N mineralization, and promoted the sequestration of more C and N in stable SOM fractions, which had a beneficial effect on the yield of the main crop. In turn, Nevins et al. [67] showed that the rye CC promoted the development of soil bacteria such as Acidobacteria, Proteobacteria, and Actinobacteria, while the white mustard catch crop increased the share of bacteria associated with the plant rhizosphere, such as *Rhizobiales* and *Bacillales*.

The beneficial effect of CC on soil microbiological activity was also demonstrated by Chavarria et al. [83]. The inclusion of mixtures consisting of oats, common vetch, and common radish in the crop rotation contributed to an increase in the activity of soil enzymes, especially dehydrogenase activity and fluorescein diacetate hydrolysis by 38.1% and 35.3%, respectively, compared to the control treatment, and no negative effect of CC on the yield of the main crop was demonstrated. Also, Kwiatkowski et al. [84] showed an increase in soil enzymatic activity (dehydrogenase by 10.4%, urease by 16.3%) in no-tillage system conditions using various CC: white mustard, blue phacelia, and a mixture of field bean and field vetch. Dehydrogenase is one of the indicators of biological soil functionality, which reflects redox processes and a respiration of soil microorganisms, while urease is an enzyme responsible for converting urea into ammonia. Other authors also report an increase in soil enzymatic activity under the influence of CC [12,64,67,82,85–87].

CC may increase the availability of phosphorus (P) for main crops, improving their nutrition with this element and consequently having a positive effect on yields. The key process is the decomposition of CC biomass, but P availability also increases due to an increase in the biomass of soil microorganisms and the abundance of mycorrhiza or phosphatase activity, which affects the P cycle in the soil [13,88]. CC have been shown to be the most effective in systems with low P availability, but the species of plant used as CC, the level of fertilization, and the cultivation system are also important [88].

CC protects the soil against erosion and moisture loss, improves its fertility and structure, and inhibits the growth of weeds through physical and chemical intervention, which is confirmed by numerous authors. In a study by St Aime et al. [49], the use of a mixture of five CCs (peas, rye, incarnate clover, hairy vetch, and oats) grown in the autumn–winter season effectively reduced a weed infestation of the main crop (weed reduction by over 90%) and did not deteriorate water conditions of the habitat. In addition, the biological activity of the soil increased (by 43%) compared to chemical fallow, which contributed to the improvement of soil health indicators, such as soil respirometry, and consequently had a positive effect on the yield of the main crop (soybean). In other studies [89], CC (incarnate clover and rye) reduced the number and biomass of weeds but did not allow for the complete elimination of pre-emergence herbicides in soybean cultivation. However, they had a positive effect on the content of organic matter, the number of bacterial and fungal populations, and the activity of esterase.

A key factor influencing soil microbiology is the water condition. Soil microorganisms need water to carry out their life processes. Optimal moisture levels enable enzyme activity and microbial metabolism. Excessive moisture or soil compaction restricts water flow, which can lead to soil hypoxia, inhibiting microbial activity. In turn, water deficiency restricts the vital functions and activity of soil microorganisms. Some may even enter a dormant state or spore form in response to drought. This is confirmed by the results of Calderon et al. [90], who showed that CC cultivation (mixture of oat, pea, flax, rapeseed, lentil, vetch, white clover, barley, safflower, and phacelia) had a short-term effect on soil microbial activity in semi-arid wheat-based rotations, while irrigation significantly increased soil enzymatic activity. In a semi-arid environment, a longer period was needed to observe the beneficial effects of CC on the microbial community structure, enzymatic activity, and C sequestration in the soil. Water is one of the main factors shaping the structure and activity of microorganisms. It promotes the decomposition of organic biomass, root proliferation, and microbial development. Giving up plow cultivation reduces water losses from the soil by limiting evaporation and increasing water retention, which has a beneficial effect on soil microorganisms [91].

3.2. Physical Properties

Improving soil physical properties is one important step in soil conservation. Cover cropping (different types of crops and time for cropping) can improve soil physical properties, which can lead to the reduction of soil loss and thereby improve soil productivity and environmental quality [92].

CC can modify soil structure indirectly through their effects on SOM content, aggregate stability, and microbial activity. Well-aggregated soil has large, stable pores that promote water infiltration, aeration, and root penetration, while poor soil structure, often caused by compaction or erosion, restricts root growth [93] and water movement, leading to decreased productivity [94,95]. Haruna and Nkongolo [96] observed a 3.5% decrease in soil bulk density as an effect of rye use as CC. It may have profound effects on soil structure through their root growth, exudation of organic compounds, and residue decomposition. Roots of CC play a crucial role in soil structure formation by exerting physical forces that aggregate soil particles and create pore spaces [97]. According to Abdollahi et al. [98], the use of fodder radish as CC increased air-filled porosity at -10 kPa, air permeability, and pore organization reduced the value of blocked air porosity both on reduced and conventional tillage treatments. CC created continuous macropores and in this way improved the conditions for water and gas transport and root growth. CC thus alleviated the effect of tillage pan (e.g., plough pan) compaction in all tillage treatments. Extensive root systems of CC (forage radish, rapeseed, and rye), particularly deep-rooted species, penetrate deep into the soil profile, loosening compacted layers and improving soil aeration and drainage [99–101]. Root exudates, such as polysaccharides and mucilages, act as binding agents, promoting soil aggregation and stability [102]. Additionally, CC roots facilitate the development of macro- and micro-aggregates by providing habitats for soil organisms that contribute to organic matter decomposition and aggregation.

Upon termination, CC contribute organic residues to the soil, which undergo decomposition and incorporation into the soil matrix. The decomposition of CC residues releases carbon compounds and nutrients, stimulating microbial activity [103] and enhancing soil aggregation. Residues from leguminous CC, rich in nitrogen, enhance microbial biomass and promote the formation of stable soil aggregates. The incorporation of CC residues also improves soil water retention by increasing SOM content and enhancing soil structure [104].

Different forms of soil water (gravitational, capillary, hygroscopic, and chemically bound) have different roles, impacts, and importance in complex crop production systems [105]. Soil water dynamics are influenced by factors such as texture, structure, and organic matter content, impacting plant water uptake and drought resilience [106,107]. CC enhance soil aggregate stability, which influences water infiltration and retention. Haruna et al. [92] states that CC can reduce surface seal formation, which increases water infiltration and reduced runoff. Well-aggregated soil has a higher macroporosity, facilitating rapid water infiltration and reducing surface runoff [108–110]. CC contributes to SOM accumulation, which enhances soil water retention capacity and prevents excessive drainage. The incorporation of CC residues increases soil water-holding capacity, particularly in sandy soils with low inherent organic matter content [105]. Furthermore, CC roots facilitate the formation of macropores, allowing water to infiltrate deeper into the soil profile and reducing the risk of waterlogging.

SOM plays a crucial role in soil physics by improving soil structure, water retention, [111,112] and nutrient cycling. It enhances aggregation, increases porosity, and enhances soil's cation exchange capacity, aiding in nutrient retention and availability [105]. CC play a crucial role in enhancing SOM content, thereby contributing to soil fertility, structure, and overall health [113]. The incorporation of CC (perennial ryegrass, rapeseed, and rye) into agricultural systems promotes the accumulation of organic residues, which decompose and enrich the soil with C and nutrients [114]. CC influence SOM through various mechanisms, including the following: (I) aboveground biomass production: producing aboveground biomass, including leaves, stems, and roots, which contribute to SOM when incorporated into the soil; (II) belowground root exudation: the extensive root systems of CC release organic compounds, known as root exudates, into the soil.

SOM plays a crucial role in preventing soil compaction and maintaining soil structure [111], and it contributes to compaction prevention through various mechanisms, including the following: (a) Improved soil structure: SOM promotes the formation and stabilization of soil aggregates, which enhance soil structure and resistance to compaction [115] (the binding agents present in SOM, such as humic substances and polysaccharides, facilitate the aggregation of soil particles, creating stable soil aggregates that resist compression). (b) Increased soil porosity: SOM creates macropores and micropores within the soil matrix [116]. These pores provide pathways for water infiltration, root penetration, and gas exchange, reducing the likelihood of soil compaction. Additionally, SOM improves soil water retention capacity, increases soil moisture content, and alleviates the effects of compaction [111]; CC associated with conservation soil tillage plays a crucial role in soil aggregation, organic matter decomposition, and nutrient cycling, thereby contributing to soil structure and compaction prevention [89].

CC exert profound effects on the physical properties of soil, influencing soil structure, texture, water dynamics, and compaction. Through their root growth, residue decomposition, and organic matter incorporation, CC enhance soil aggregation, water retention, and aeration, thereby improving soil health and productivity. Understanding the interactions between CC and soil physical properties is essential for implementing sustainable agricultural practices that promote soil conservation, crop resilience, and ecosystem stability [117]. According to Abdollahi and Munkholm [118], five years of using fodder radish as a CC-alleviated plow pan compaction led to a 20–40 cm depth by reducing penetration resistance. A significant interaction between tillage and cover crop treatments indicated the potential benefit of using a combination of cover crops and direct drilling to produce better soil friability.

Studies reporting on CC combined with reduced or conservation tillage are rare. Many studies have focused on investigating partial elements of reduced or conservation tillage or CC, but studies of their combinations are rare [118]. Changes in soil physics as a result of a combination of different approaches to conservation soil tillage and CC [114] are not universal or uniformly accepted, and conducted studies provided different results [119]. Many factors directly or indirectly influence the results of the strongest factors: soil type, meteorological elements (precipitation, temperature, wind, etc.), and soil and crop management. For some parameters of soil physics, when considered the most important (soil compaction, bulk density, soil structure, and porosity), the authors report both positive and negative reactions to combinations of conservation tillage and CC. Soil compaction, bulk density, soil structure, and porosity can be alleviated using CC (mostly forage radish, rapeseed, and rye) [118–122], which implies the possibility for the application of reduced or conservation tillage [100] instead of intensive conventional tillage.

3.3. Chemical Properties

Many studies have shown that the presence of CC affects the chemical composition of the soil. CC can increase the level of SOC or reduce the rate of its depletion as a result of additional C input from aboveground and belowground biomass [109,123]. The accumulation of SOC in the topsoil depends on many factors, including the amount of biomass and CC species, soil type, cultivation system, initial soil C level, and climatic conditions [124–128]. In maize cultivation, the combination of no-tillage and CC systems has a stronger effect on SOC accumulation in the topsoil than the separate use of these practices [129]. The inclusion of CC in no-tillage systems leads to a higher SOC accumulation from 0.1 to 1.0 Mg ha^{-1} yr^{-1} compared to no-tillage and no CC systems [124]. Poeplau and Don [125] estimated that CC can sequester about 0.32 ± 0.08 Mg ha^{-1} yr^{-1} carbon to a depth of 22 cm of soil. In the first few years after CC introduction in different cropping systems, no increase in SOC is usually observed, but the effects of their presence are noticeable in the long term [130]. The lack of accumulation of SOC and soil total N during the year may be due to the rapid degradation of plant residues after their introduction into the soil. [123]. In the experiment conducted in temperate climate conditions (Poland) on two types of soil (Luvisol and Phaeozem), CC biomass (field pea) introduced in subsequent years, compared to the control without CC, had no effect on the soil pH, SOC, or total N content [131]. Also, the incorporation time of CC (in the autumn or mulched and incorporated in the spring vs. a control—without CC) did not significantly affect the chemical properties of typical Alfisol formed of a sandy loam. Other studies have reported contradictory results on this issue. In the CC/rice/CC/rice crop rotation in a no-tillage system, an increase in SOC content at a depth of 0–5 cm was observed from the first year of CC presence. The presence of CC contributed more to soil fertility than fallow [132]. Based on a 2-year study conducted in Missouri, USA, it was shown that the presence of CC in a corn–soybean rotation in a no-tillage system had a positive effect on the SOC content in the topsoil [133]. In a long-term experiment conducted in Switzerland, the presence of well-developed CC, even for only two months of the year, increased the content of SOC and total N in the soil, especially in the case of reduced tillage, and allowed maintaining the wheat yield at the level observed in the plow and minimum tillage treatments [134]. In a 16-year experiment (northern France), the SOC amount remained stable over time in conventional and low-input systems (+3% and +1%, respectively), increased slightly in the organic system (+12%) and increased significantly in the topsoil (0–10 cm) in the conservation agriculture system (+24%). The SOC increase observed in the conservation agriculture system was attributed to one or more of the following practices: zero tillage, permanent soil cover, and crop diversification [135]. Most of the changes in soil chemical properties under the influence of CC are limited to the upper soil layer. Olson and Al-Kaisi [136], after 20 years in the no-tillage system, observed a greater accumulation of SOC in the soil to a depth of 20 cm, but the accumulation of SOC in the 20–35 cm layer was lower than under the influence of moldboard plow cultivation. The accumulation of SOC in deeper no-tillage soil depths may be facilitated by the use of deep-rooting CC [127].

It has been shown in many studies [30,132,134,137,138], that the presence of CC in crop rotations, including legumes, is important not only because of their effect on increasing the level of SOC in the soil or reducing the rate of its depletion but also retaining plant-available N in organic matter. In the case of legumes, the increase in soil fertility is associated not only with the supply of biomass, but also with N from symbiotic fixation of atmospheric N (BNF) [139,140]. Legumes in crop rotations also contribute to solubilizing unsolvable P in the soil [26].

As a result of the decomposition of organic matter, many organic compounds are released into the soil [141]. The use of CC in no-tillage systems can significantly improve soil fertility [132]. In addition to the large amount of roots present in the soil (from 20 to 50% of biomass), plants release root exudates into the soil, containing more than 200 compounds of Ca (organic acids, phenolic acids, amino acids, etc.), which have a major impact on the soil microbial community and biological processes. Root lysis and root exudates contribute significant quantities of carbon deposited in sub-surface soil. Amounts from 5 to 21% of photosynthetically bound C is released into the rhizosphere as root exudates [142]. In addition, soil microorganisms associated with CC root systems have the potential to release nutrients and transform them into forms more available to plants [134], e.g., arbuscular mycorrhizal fungi and rhizobia in the rhizosphere of legumes can increase the availability of N and P for subsequent crops [143,144].

In conventional cereal production systems, nitrate losses due to leaching range from 10% to 30% of applied N [145]. In the long term, CC cultivation is an effective way to reduce N leaching from soil (from 36 to 62%) and maintain long-term nitrate concentrations below 50 mg L^{-1}, and this positive effect decreases with decreasing CC frequency [34]. Based on research conducted in the mid-Atlantic, USA, Hirsh et al. [146] recommend incorporating CC into minimum tillage systems to capture residual N. This allows for a reduction in inorganic N content in the soil. Deep-rooting CC sown in autumn can potentially take up residual N and recycle some of it for following crops. The use of CC promotes a reduction in nitrate content in the upper 60–120 cm soil depths in autumn. Since in temperate climate conditions, an increase in the share of legumes in crop rotations reduces C input to the soil and leads to greater N leaching, the use of CC in cereal–legume rotations is recommended to increase N_2 fixation and reduce nitrate leaching [147]. Based on the analysis of many studies covering different countries, climate zones, and management practices, Abdalla et al. [148] emphasized that CC cultivation significantly reduces N leaching and increases SOC sequestration without significantly affecting direct N_2O emissions. It can also indirectly increase N_2O emissions (NO_3 leaching) due to the introduction of organic substrates, the release of N during decomposition, or the increase in soil water content in the summer [149]. In a study conducted in sandy loam soils in Georgia, USA, a non-legume cover crop (rye) was shown to be superior to a legume (hairy vetch and crimson clover) in its influence to increase SOC and N content [150]. Under these conditions, CC and N fertilization can increase C input and storage in arable and untilled soils, and growing a mixture of hairy vetch and rye results in greater SOC accumulation than monocultures [151]. CC and fertilization with 80–180 kg N ha^{-1} yr^{-1} can increase soil organic C and N concentrations by up to 4–12% compared with no CC or N fertilization. The combination of conservation tillage, a mix of legume and non-legume crops, and reduced N fertilization rates may show greater potential to maintain yields and improve soil and water quality compared with the use of any of these practices alone [140]. In the Canadian prairies, the inclusion of legumes (field pea, lentil, and faba bean) in no-tillage soils increased soil fertility and improved soil N cycling in canola and spring wheat cropping systems. A total of 40–65% in crop N uptake came from in-crop N mineralization, particularly by legumes, compared to 35–60% from N fertilizers, and the magnitude and timing of these benefits depended largely on the legume species, crop purpose (greater in green manure than seed crops), and specific site conditions [152]. In temperate climate conditions (Poland), the cultivation of field pea as a CC increased the mineral N content in the arable layer of Luvisol soil, which was observed in the early spring before sowing and during the barley tillering period (successive plant). It was particularly high in the soil where CC was left for winter as mulch, which can be associated with delayed mineralization of pea biomass [21].

As a result of the decomposition of organic residues, many organic compounds are released into the soil and/or synthesized by the microflora that decompose them [141]. The rate of nutrient release depends on the biochemical and chemical composition of the residue, as well as the microbial activity in the soil [33,153]. N release from crop residues varies with residue quality [154], and the best single predictors for net N mineralization are the C/N ratio and the N content of CC residues [155], while lignin concentration is a good predictor of soil C mineralization. Net N mineralization is significantly and negatively related to both shoot and root C/N ratios. At a soil temperature of 10 °C, the C/N ratio affects N mineralization from CC residues mainly in the first 2 weeks, while net C mineralization is significantly limited by lignin content [156]. Biomass decomposition is faster in crop residues with low C/N ratios, such as legumes [157], where the ratio is <20 [158], which in turn results in a rapid increase in microbial activity leading to increased SOC decomposition [159]. Based on many years of experiments in various regions of France, it has been shown that in soils with a high C/N ratio (>11), N availability may limit

SOC decomposition and the rate of N net mineralization from SOM may decrease due to greater N immobilization by microorganisms [160]. In another study [123] conducted in the temperate, humid region of Washington state, silt loam soil showed increased organic C content in soil with CC that produced residues with C:N ratios >30. The effect of CC on SOC building was related to the size of the C input. Due to the higher CC biomass yield (>4 Mg ha^{-1}), cereal rye and annual ryegrass were better suited for building SOC compared to Austrian winter pea, hairy vetch, and canola.

The rate of decomposition of plant residues introduced into the soil also depends on the CC species. Legumes decompose faster, enabling faster delivery of nutrients to the soil, while non-legumes tend to accumulate and decompose slower. Hence, the combined supply of legume and non-legume residues promotes the differential release of nutrients such as N, K, P, Ca, and Mg into the soil [33]. Also in studies conducted in the no-tillage system in the Brazilian Cerrado on acidic clayey loam soil, CC did not cause changes in soil chemical properties compared to the fallow, but in the CC/cash crops/CC/cash crops rotation, a decrease in soil pH, Al, and H+Al was observed, as well as an increase in Ca, Mg, K, and Fe contents [161]. In temperate climate conditions (Poland), in a reduced tillage system, the use of field pea as CC significantly increased the available K content in Luvisol and Phaeozem soils [131]. Studies by Damon et al. [162] indicate a beneficial effect of CC on P content in the soil. The effect of CC on P uptake by the main crop depends on many factors, and its transfer may occur via CC residues, organic anion exudates, root-exuded enzymes, and microbial interactions. As organic residues decompose, P is released into the soil, which increases its availability to subsequent plants [141]. The effect of CC on soil P content also depends on the soil type. The use of field pea as CC in the reduced tillage system increased the concentration of available P in the Phaeozem soil, but no such effect was found in the Luvisol soil [131]. Villamil et al. [121], using winter CC in a no-tillage corn–soybean rotation for at least 5 years on silt loam soils, observed a decrease in soil P content due to its absorption and transformation of available P into organic forms. From an environmental protection point of view, this is beneficial because it reduces the potential loss of P due to leaching. Long-term use of CC (ruzigrass) caused a decrease in P availability in a no-tillage production system of corn [163] and soybeans [164]. CC cultivation for 4 years, in both conventional and no-tillage cultivation systems, increased SOC content, but nutrient recycling by CC was not sufficient to maintain P, K$^+$, Fe^{3+}, and Mn^{2+} contents in the soil [165].

The literature emphasizes the role of CC in limiting the migration of nutrients into deeper soil layers, in particular N [166], P [131,166], and K [131]. However, there are ambiguous opinions regarding the effect of CC in reduced tillage systems on the chemical composition of the soil at its different levels. Sharma et al. [167] indicate that the use of CC can change the chemical properties of the soil, especially in its top layer. In an experiment conducted in Nebraska, in large-scale production fields, CC was used into a no-tillage seed maize or soybean cropping system. They found a positive effect of CC on the content of exchangeable K, Mg, Na, cation exchange capacity, and soil micronutrients (Zn, B, Fe, and Mn). In the soil, CC increased the exchangeable K concentration below the 5 cm soil layer, the cation exchange capacity, and the exchangeable Mg concentration in the 20–40 cm soil layer. CC also reduced the Na concentration at a 20 to 40 cm soil depth but had no effect on the exchangeable Ca concentrations. CC showed the potential to maintain optimal levels of Zn, B, Fe, and Mn in the 0–5 cm layer. In other studies conducted on Red Latosol in Brazil, in no-tillage and conventional cultivation systems, CC (sun hemp and millet) did not change the chemical properties of the soil at depths of 0–10 cm and 10–20 cm [168].

The effect of CC on soil pH is ambiguous. It can affect soil pH as a result of acidic secretions from plant roots [168], causing a decrease in soil pH [132,168,169]. As a result of many years of conservation practices with no-till, CC, and limited inputs, soils contained more SOC and were more acidic than soils under conventional practices with tillage and bare soil between the two cash crops. The positive effects of conservation practices may be limited by carbon saturation. If the soil reaches its carbon potential, it will not be able to sequester more organic C [163]. In turn, Rankoth et al. [133], growing cereal mixture as CC in a no-till system, observed an increase in the amount of SOC and an increase in soil pH. A temporary increase in soil pH often occurs during the decomposition of residues, and the causes include the following: oxidation of organic acid anions present in decomposing residues, ammonification of residue N, specific adsorption of organic molecules formed during decomposition, and reduction reactions induced by anaerobes [141].

3.4. Reducing the Loss of Nutrients from the Soil

Nutrient loss from soil is a critical challenge facing modern agriculture, with significant implications for agricultural productivity, environmental sustainability, and food security. As it is known, the magnitude of nutrient loss from the soil (such as leaching, volatilization, and reutilization) is under the influence of many factors, external and internal, and anthropogenic and/or natural-driven causes. Excessive nutrient runoff and leaching increase the cost of production for farmers and contribute to water quality deterioration [170], which can lead to soil degradation and reduced crop yields [171]. Therefore, implementing effective strategies to reduce soil nutrient loss is essential to maintain soil fertility [172], protect water quality, and promote sustainable agriculture. There are several strategies (techniques and practices) aimed at reducing soil nutrient loss: catch crops [173], cover crops, crop rotation, mulching, conservation tillage, nutrient management planning, contour farming, buffer strips, the use of nutrient rich amendments, and other soil conservation practices.

Crop rotation is a widely recognized strategy for reducing nutrient loss from soil while promoting soil fertility and crop productivity. By rotating different crops in a planned sequence, it is possible to optimize nutrient intake, disrupt pest and disease cycles, and improve soil health. For example, leguminous crops in rotation can enhance soil nitrogen levels through BNF, reducing the need for synthetic nitrogen fertilizers [27]. Additionally, crop rotation can help balance nutrient inputs and outputs, minimizing the risk of nutrient accumulation or depletion in the soil. Diversified crop rotations lead to improved soil quality, reduced nutrient loss, and enhanced ecosystem services compared to monoculture systems [105]. Mulching is a simple yet effective practice for reducing nutrient loss from soil by conserving soil water, reducing evaporation, regulating soil temperature, minimizing weed growth, minimizing soil compaction, improving soil health, and increasing nutrient status [174]. Organic mulches, such as straw (or other crop/plant residues), hay, or compost, provide a protective layer on the soil surface, reducing the impact of rainfall and minimizing erosion [175]. In addition to erosion control, mulches contribute organic matter to the soil as they decompose, improving soil fertility and nutrient retention [176]. Some studies have shown that mulching can significantly reduce nutrient runoff and leaching, particularly in areas prone to erosion [177].

Reduced/conservation tillage offers promising opportunities for reducing nutrient loss from soil while conserving soil health and structure. Unlike conventional tillage, which involves intensive soil disturbance, conservation tillage practices minimize soil disruption, leaving crop residues on the soil surface. This protective cover reduces erosion, increases water infiltration, and enhances soil organic matter content, thereby improving nutrient retention [94]. Numerous studies have demonstrated the benefits of reduced tillage in reducing nutrient runoff and leaching, particularly in regions with vulnerable soils and high rainfall intensity [178,179].

Developing comprehensive nutrient management plans is essential for optimizing nutrient use efficiency [94] and minimizing nutrient loss from soil [180]. Soil testing plays a crucial role in nutrient management planning, providing valuable information about soil nutrient levels, pH, and texture. By analyzing soil test results, it is possible to make informed decisions regarding fertilizer types, application rates, and timing, thereby reducing the risk of nutrient overapplication and leaching [181]. Nutrient management plans also emphasize the use of best management practices, such as appropriate soil tillage nutrient split applications [182], timing fertilizer applications to coincide with crop nutrients uptake and incorporating organic amendments to improve soil fertility and structure. Numerous studies have demonstrated the effectiveness of nutrient management planning in reducing nutrient loss and improving crop yields across diverse agricultural systems [180,183].

One of the most effective strategies for reducing nutrient loss from soil is the use of CC (mostly a mixture of legume and non-legume species) [184–186]. CC are planted during fallow periods or alongside cash crops to protect the soil from erosion, retain nutrients, and improve soil health. These crops, such as legumes, grasses, and clovers, help prevent soil erosion by covering the ground surface and reducing the negative impact of rainfall. CC, such as legumes and grasses, have extensive root systems that help hold soil in place and prevent erosion. These roots also absorb excess nutrients like nitrogen, phosphorus, and potassium, which otherwise could leach into groundwater or runoff into nearby water bodies [187]. CC also contribute to overall soil health by promoting beneficial microbial activity and enhancing soil biodiversity. Healthy soils are better equipped to retain nutrients and make them available to plants, thus reducing nutrient loss [188]. CC are often used as part of crop rotation systems. Rotating cash crops with cover crops helps break pest and disease cycles, improves soil structure, and reduces nutrient imbalances, ultimately leading to better nutrient retention. In summary, CC contribute to nutrient retention in agricultural systems through various mechanisms, including nutrient uptake, erosion control, organic matter addition, weed suppression, and the overall improvement of soil health [189]. Incorporating CC into farming practices is a sustainable approach to minimize nutrient loss and promote long-term soil fertility and productivity. Several studies have demonstrated the effectiveness of CC in reducing nutrient runoff and leaching, particularly in vulnerable agricultural landscapes [190–192]. By including CC in crop rotations, it is possible to improve soil fertility, reduce nutrient loss, and promote sustainable agricultural practices.

Investigating nitrogen as one of the most important nutrients and the importance of nitrate leaching and soil tillage practices relations, Pereira et al. [193] analyze different studies (for over two decades) and conclude that the no-tillage system is in most cases dominant as a strategy in the reduction in nitrate leaching, compared to conventional tillage. They also stated that the scale of nitrate leaching depends on several factors, including soil compounds (physical, chemical, and biological) and weather conditions. According to Issaka et al. [194], the reduction in nitrogen and phosphorus losses and the improvement of soil nutrient status [41,176] can be achieved by conservation tillage and the inclusion of catch crops [19,24,173,195] and its mixture (usually crops with developed and high potential roots) for a high potential of nutrient capture [196] and effective management, especially in crop rotations [197]. CC in combination with conservation soil tillage, in most cases, has a positive influence on arable crop yields [190,193,198], without significant influences [196] or even decreased crop yields in some cases [199]. In line with the above findings, and for a better understanding of the relationships between different tillage practices and cover crops, in ways of nutrient losses from the soil, further investigations on a global level need to be performed.

4. The Influence of Cover Crops on Growth Conditions and Yield of Cereals

As was shown in the data presented in Section 3, CC used both in conservation and conventional tillage conditions may affect the biological, physical, and chemical conditions of the soil and can protect nutrients from being leached out of the reach of the cereal root system. As a result, the conditions for growth and development of cereals are improved, which often affect improved plant health status and increased grain yield (Table 1). However, this effect is very complex, and some studies indicate the opposite effect or no effect of CC on cereal yield. The results of studies concerning the effect of CC on the occurrence of cereal infection by fungal pathogens are ambiguous. Many authors have shown that under the influence of these crops, there is a significant reduction in the disease index of the succeeding cereals [44,200–202]. In some cases, no significant effect of CC on the health of cereals is observed [203], or there may be even a negative effect [204–206]. Usually, a significant reduction in the infection index, ranging from 15 to 60%, occurs as a result of the cultivation and use of white mustard and/or blue phacelia as CC [44,200–202]. In a study by Mielniczuk et al. [201], the number of colony-forming units decreased by 26% and 34% when using lacy phacelia and white mustard as CC, respectively. A weaker, although still positive effect on the cereal infection index (−4 to −10%), can be expected when using Westerwolds ryegrass [200]. In a study by Lemańczyk and Wilczewski [205], an unfavorable effect of fodder pea on the infection of roots and stem base of spring barley grown with the use of plow tillage was demonstrated. The disease index increased by as much as 60% compared to the control, without CC. However, in a study by Wojciechowski et al. [202], CC in the form of a mixture of fodder pea and field bean caused a reduction in the infection index by approx. 15%. In studies by Kwiatkowski et al. [44], a very significant (by 44.8%) reduction in the root and stem base infection index of spring wheat was demonstrated as a result of growing the legumes with an oat mixture as CC. Therefore, the effect of legumes as a CC on the health of cereals is unpredictable. More clear results regarding white mustard and lacy phacelia may be due to their beneficial effect on the development of bacteria of the genus Trichoderma, which are considered antagonistic to cereal pathogens [201]. White mustard had a particularly strong effect on the development of Trichoderma bacteria, which increased their numbers by 211%. In the case of lacy phacelia, this effect was weaker but still very significant (+98.5%). CC also contributed to a very significant increase in the numbers of other antagonists of pathogenic bacteria for cereals, e.g., Clonostachys rosea (an increase of 339% and 414% for lacy phacelia and white mustard, respectively). Taking into account the published research results, it can be stated that a reduction in the index of infection of roots and the stem base of cereals occurs both in the plowing system and in simplified systems and in direct sowing with no tillage conditions [44,200,202]. The highest average reduction in the infection index occurred under the influence of CC in reduced tillage conditions (−27.6%) and was significantly lower in plow tillage and no tillage conditions (−17.5%).

The influence of CC on weed infestation of cereals grown after them varies depending on agricultural technology, soil properties, and weather conditions during the vegetation period [44,207–210]. Kraska et al. [207] found a significant reduction in weed biomass in spring wheat cultivation. Depending on the plant species used as CC, it ranged from 10.7 to 35.7% in plow tillage conditions and from 9.2 to 30.3% in reduced tillage conditions. The greatest reduction in weed dry mass was achieved after using red clover, lacy phacelia, and white mustard as CC. Weaker effects, although significant, were provided by Westerwolds ryegrass. An even greater reduction in weed biomass was achieved in a study by Kwiatkowski et al. [44] and Darby et al. [210]. In studies by Kwiatkowski et al. [44], regardless of the tillage system, the reduction in weed biomass was about 68% in the case of lacy phacelia and 77–78% for white mustard. A slightly weaker weed control effect was found in the case of CC from the legumes with an oat mixture. These plants used as CC allowed for a reduction in weed dry mass by 40.6% in plow tillage conditions and 47.1% in conservation tillage conditions. Darby et al. [210] achieved a reduction in weed biomass in a field of spring wheat grown in the no-tillage system, from 77% in the case of millet to even 86% in the case with oats as CC. Carrera et al. [209] assumed that the lower weed biomass in CC sites was due to the delay in their germination by the biomass of plants covering the soil.

CC most often had a positive effect on the yield of cereals grown after them [48,97,209–213]. However, some studies did not show a significant effect of CC [36,214], or it was unfavorable [215,216]. The yield-forming effect of CC is also dependent on the tillage system or the plant species used as CC (Table 1). In studies on the effect of fodder radish on spring wheat yield [97], a very strong positive effect of CC was demonstrated in treatments where reduced tillage was applied (+22.6%) and a much smaller effect of this factor on grain yield occurred in plow tillage conditions (+7.2%). A very strong positive effect of oats, millet, and a mixture of oats, clover, and radish on the grain yield of spring wheat was demonstrated in no tillage conditions [210]. Also in Majchrzak's [211] studies, a significant positive effect of CC (white mustard) on the grain yield of spring wheat was found in Albic luvisol soil conditions, especially in no tillage and plow tillage treatments. The author obtained a slightly smaller, although still very positive, effect of white mustard after the use of reduced tillage. Studies on the effect of field pea, as CC, on the grain yield of spring barley also indicate the dependence of this effect on soil conditions (Table 1). In studies by Wilczewski et al. [21], a significant positive effect of this factor was demonstrated in the luvisol soil conditions, whereas in the very fertile black Earth conditions [36], with an average grain yield that was higher by 24%, this factor did not have a significant effect on the yield. These studies did not show the effect of tillage technology on the yield-forming effect of CC. In studies by Toom et al. [217], in the plow tillage system, a varied effect of individual plants used as CC was demonstrated (Table 1). A significantly positive effect on the yield of spring barley grain was demonstrated for forage radish and hairy vetch, while winter turnip rape, winter rye, and bersem clover did not significantly affect the yield of this plant. A very strong influence of the tillage system on the effect of CC in shaping the yield of winter wheat was demonstrated in a study by Wittwer et al. [212], in which white mustard and common vetch contributed to increasing the grain yield of winter wheat by 35%, while in the conditions of plow tillage, no significant effect was found, and in the conditions of no tillage it was found only after the application of common vetch. Regardless of the tillage system, the mixture of vetch and mustard did not significantly affect the yield of winter wheat. The yield-forming effect of CC, especially for winter cereals, depends on water conditions during their growth and development. According to Nielsen and Vigil [216], water use by legumes may reduce subsequent wheat yield. As a result, these authors found a reduction in grain yield of winter wheat after winter and spring pea used as CC by as much as 32.7% in comparison to the yield obtained after fallow (Table 1). Similar results were presented by Lyon et al. [215], who assessed the effect of a mixture of peas and oats on the yield of winter wheat grain.

Numerous studies on maize have shown a very positive effect of CC on grain yield, especially under reduced tillage or no tillage conditions [48,209,212,213]. In plow tillage conditions, the grain yield increase under CC was usually from 2 to 10%; however, in a study by Clark et al. [214], the increase was as much as 43.2% (Table 1). The use of CC under reduced tillage conditions resulted in an increase in the yield of maize grain by 6% and 11%, after oilseed radish and white mustard, respectively, and by 10–61% after using legumes, legume mixtures, or their mixtures with rye or white mustard for this purpose [48,209,212,213]. A positive effect of CC on maize yield was also observed under the no tillage technology [212,213].

Table 1. The influence of CC * on cereal grain yield, depending on tillage system—increase (+) or decrease (−) compared to control, without CC [%].

Cover Crops Species	Soil	Increase or Decrease [%], Depending on Tillage System			References
		Plough Tillage	**Reduced**	**No Tillage**	
		Spring wheat			
Fodder radish	Luvic chernozem	+7.2	+22.6	no data	[97]
White mustard	Albic luvisol	+18.1 to +27.6	+12.0 to +21.5	+13.9 to +39.4	[211]
Mixture: Hairy vetch + rye	Silty Loam	+1.2 n.s. ^	no data	no data	[214]
Millet		no data	no data	+71	
Oat	Silty Loam	no data	no data	+63	[210]
Oat + clover + radish		no data	no data	+51	
		Spring barley			
Field pea	Black earth	+2.6 n.s.	+1.3 n.s.	no data	[36]

Table 1. *Cont.*

Cover Crops Species	Soil	Increase or Decrease [%], Depending on Tillage System			References
		Plough Tillage	Reduced	No Tillage	
Field pea	Luvisol	+6.3	+8.4	no data	[21]
Fodder radish	Cambic Phaeozem (Loamic)	+11.0	no data	no data	[217]
Hairy vetch		+9.0	no data	no data	
Winter turnip rape		+3 n.s.	no data	no data	
Winter rye		+3 n.s.	no data	no data	
Winter wheat					
White mustard	Cambisol	−2.3 n.s.	+30	+2.3 n.s.	[212]
Common vetch		+2.3 n.s.	+35	+7.0	
Common vetch + white mustard		+2.3 n.s.	+4.3 n.s.	+2.3 n.s.	
Field pea	Weld silt loam	−32.7	no data	no data	[216]
Oat + field pea		−29	no data	no data	[215]
Straw mulch	Claypan soil	+6.4 to +9.3	no data	n.s.	[218]
Maize (grain)					
White mustard	Cambisol	−2.0 n.s.	+11	+2.0 n.s.	[212]
Hairy vetch		+8	+61	+17	
Common vetch + white mustard		+4.0 n.s.	+45	+12	
Oilseed radish	Sandy Loam	+2.0 n.s.	+6.0 n.s.	+12	[213]
Hairy vetch	Sandy Loam	+10	+36	+19	
Subterranean clover	Sandy Loam	+9	+10.0	+14	
Hairy vetch	Sandy Acrisol	no data	+3.3 to + 15.3	no data	[48]
Hairy vetch	no information	no data	+43	no data	[209]
Mixture: Hairy vetch + rye		no data	+30	no data	
Mixture: Hairy vetch + rye	Silty Loam	+43.2	no data	no data	[214]

* CC—cover crops; ^ n.s.—not significant.

The average effect of CC on grain yield for all analyzed cereals (winter and spring wheat, spring barley, and corn) was the highest in treatments where reduced tillage was used (+23.1%), significantly lower in the no tillage conditions (+10.7%), and the lowest for plow tillage (Figure 6). Taking into account the above-mentioned negative effect of leguminous CC on the health of cereals, this positive effect on grain yield is probably caused by the fertilizing effect of the leguminous biomass. The significant amounts of N contained in their biomass and the rapid mineralization of post-harvest residues in the soil ensure good availability of N during the tillering and stem shooting period [21]. In conditions of good water supply, this leads to rapid plant growth and, consequently, greater susceptibility to fungal pathogens. Legumes are more effective than plants from other botanical groups in increasing grain yield of spring cereals. This is particularly visible in reduced tillage conditions, where mixtures with legumes showed an equally positive effect on yield as legumes in single-species cultivation. However, some studies indicate their negative effect on the grain yield of winter wheat, especially in conditions of rainfall deficiency.

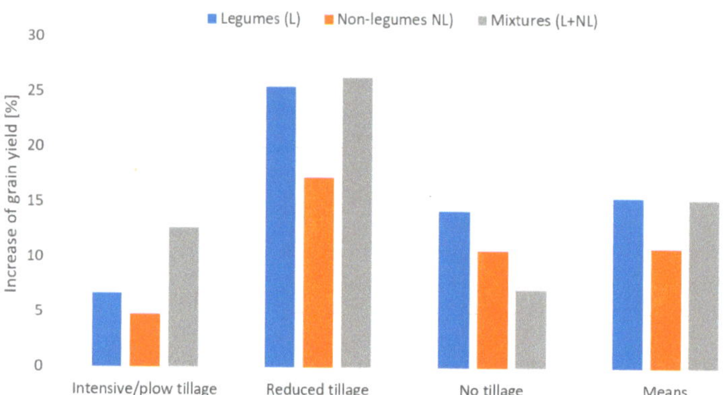

Figure 6. Average impact of plants grown as cover crops on cereals (winter and spring wheat, spring barley, and corn) grain yield, depending on tillage system [21,36,48,97,209–218].

The benefits of using CC are particularly high in conditions of low-intensity cultivation, good water supply to the main plant, and average soil conditions [21,49,212]. In a study by Acosta et al. [48], the yield-forming effect of hairy vetch as CC was 15.3% under the conditions of nitrogen fertilization at a dose of 60 kg ha^{-1} and only 3.3% for fertilization at a dose of 120 kg ha^{-1} N. In a study by Wittwer et al., [212], the inclusion of nitrogen-fixing CC in the organic production systems led to increased yields and substantially contributed to decreasing the yield gaps compared to the conventional systems. Yield differences between conventionally managed plots were reduced from −37% without CC to −19% with the use of CC mixtures.

5. Conclusions

The literature studies concerning the possibilities of shaping soil properties and cereal yields through the use of CC have shown that in the conditions of reduced tillage, they can be an important element allowing maintenance of favorable soil parameters and high plant productivity. However, it is not possible to obtain a beneficial effect of CC in all conditions. Individual plant species used as CC have a varied value in shaping soil properties as well as in terms of their influence on the growth, development, and yield of the main plant. The most studies concerned summer CC, using plants characterized by a fast growth rate (white mustard, oilseed radish, and oat), and plants with lower growth dynamics but other beneficial properties (field pea, common vetch, lupin, and serradella). Among winter CC, the most frequently used were winter rye, hairy vetch, winter turnip rape, clover, and grasses. These plants were grown both in mixtures and as a single crop.

The results presented in the current literature indicate that the influence of CC on the biological properties of soil is relatively certain. Regardless of the plant species used as CC, we can expect an increase in the number of soil microorganisms and an enrichment of their species diversity. There is also an improvement in the activity of soil enzymes (dehydrogenase, urease, and beta-glucosidase). Usually, this influence is particularly beneficial under reduced tillage systems, which are particularly conducive to the development of earthworms and other organisms belonging to the soil mesofauna, especially those from the Collembola and Acari families. CC also usually have a beneficial effect on the physical properties of the soil. However, due to the large amounts of water taken up in late summer and early autumn, they can lead to a temporary reduction in moisture and an increase in soil penetration resistance in the autumn. In early spring, they usually provide favorable physical properties for spring crops. In tillage systems that involve mixing CC biomass with the topsoil, it ensures the loosening of compacted soils and, in the case of light, sandy soils, allowing for an increase in the capacity of the sorption complex.

The influence of CC on the chemical properties of soil is less clear than its influence on the biological and physical properties. The size and composition of CC biomass and weather conditions during the vegetation period and during the covering of soil with plant biomass are of great importance for improving soil chemical properties. A beneficial influence of CC, especially legumes, on the content of the mineral nitrogen in the topsoil is usually observed. Sometimes an increase in the content of available forms of K and/or P is also achieved. The influence of CC on the content of organic C, total N, or soil pH is less common.

CC used in reduced soil tillage systems can significantly improve the yield and quality of cereal grain. The strength of this effect depends on soil conditions and the intensity of agricultural technology. The greatest positive effect on cereal yield is obtained as a result of using legumes as CC, in conditions of less fertile soils, and in conditions of low fertilization levels. In these conditions, we can expect an increase in the grain yield by 15–25% and sometimes even by over 40%. The improvement in yield is mainly related to the more dynamic development of plants, increased numbers of spikes, and the number of grains in the cereal spike. In average soil conditions, the increase in cereal yield is in the range of 5–15%, and in the case of fertile soils and/or the use of full N fertilization, no direct effect of CC on grain yield is observed.

Author Contributions: Conceptualization, E.W., M.S. and E.S.-K.; methodology, E.W., E.S.-K. and M.S.; investigation, E.W., I.J., E.S.-K., M.S. and D.J.; data curation—results compiled and analyzed, E.W., I.J., E.S.-K., M.S. and D.J.; writing—original draft preparation, E.W., I.J., E.S.-K., M.S. and D.J.; review and editing, E.W. All authors have read and agreed to the published version of the manuscript.

Funding: This study has been partly supported by the Croatian Science Foundation under the project "Assessment of conservation soil tillage as advanced methods for crop production and prevention of soil degradation—ACTIVEsoil" (IP-2020-02-2647).

Institutional Review Board Statement: Not applicable.

Data Availability Statement: Not applicable.

Conflicts of Interest: The authors declare no conflicts of interest.

References

1. Holman, F.H.; Riche, A.B.; Michalski, A.; Castle, M.; Wooster, M.J.; Hawkesford, M.J. High throughput field phenotyping of wheat plant height and growth rate in field plot trials using UAV based remote sensing. *Remote Sens.* **2016**, *8*, 1031. [CrossRef]
2. Van Dijk, M.; Morley, T.; Rau, M.L.; Saghai, Y. A meta-analysis of projected global food demand and population at risk of hunger for the period 2010–2050. *Nat. Food* **2021**, *2*, 494–501. [CrossRef] [PubMed]
3. Food and Agriculture Organization of the United Nations. *World Food and Agriculture-Statistical Yearbook*; FAO Documents; FAO: Rome, Italy, 2023. [CrossRef]
4. UN Population Division Data Portal. Available online: https://population.un.org/dataportal/ (accessed on 27 June 2024).
5. Pingali, P.; Boiteau, J.; Choudhry, A.; Hall, A. Making meat and milk from plants: A review of plant-based food for human and planetary health. *World Dev.* **2023**, *170*, 106316. [CrossRef]
6. Ritchie, H. If the World Adopted a Plant-Based Diet, We Would Reduce Global Agricultural Land use From 4 to 1 Billion Hectares. 2021. Available online: https://ourworldindata.org/land-use-diets (accessed on 27 March 2024).
7. Kunj, B.M.; Alam, M.S.; Singh, H.; Bhat, M.A.; Singh, A.K.; Mishra, A.K.; Thomas, T. Influence of farmyard manure and fertilizers on soil properties and yield and nutrient uptake of wheat. *Int. J. Chem. Stud.* **2018**, *6*, 386–390. Available online: https://www.chemijournal.com/archives/?year=2018&vol=6&issue=3&ArticleId=2480&si=false (accessed on 12 April 2024).
8. Dhaliwal, S.S.; Sharma, V.; Verma, V.; Kaur, M.; Singh, P.; Gaber, A.; Laing, A.M.; Hossain, A. Impact of manures and fertilizers on yield and soil properties in a rice-wheat cropping system. *PLoS ONE* **2023**, *18*, e0292602. [CrossRef]
9. Rayne, N.; Aula, L. Livestock manure and the impacts on soil health: A review. *Soil Syst.* **2020**, *4*, 64. [CrossRef]
10. Koninger, J.; Lugato, E.; Panagos, P.; Kochupillai, M.; Orgiazzi, A.; Briones, M.J.I. Manure management and soil biodiversity: Towards more sustainable food systems in the EU. *Agric. Syst.* **2021**, *194*, 103251. [CrossRef]
11. Koninger, J.; Panagos, P.; Jones, A.; Briones, M.J.I.; Orgiazzi, A. In defence of soil biodiversity: Towards an inclusive protection in the European Union. *Biol. Conserv.* **2022**, *268*, 109475. [CrossRef]
12. Piotrowska-Długosz, A.; Wilczewski, E. Influences of catch crop and its incorporation time on soil carbon and carbon-related enzymes. *Pedosphere* **2015**, *25*, 569–579. [CrossRef]

13. Piotrowska-Długosz, A.; Wilczewski, E. Influence of field pea (*Pisum sativum* L.) as catch crops cultivated for green manure on soil phosphorus and P-cycling enzyme activity. *Arch. Agron. Soil Sci.* **2020**, *66*, 1570–1582. [CrossRef]

14. Zhang, P.; Chen, X.; Wei, T.; Yang, Z.; Jia, Z.; Yang, B.; Han, Q.; Ren, X. Effects of straw incorporation on the soil nutrient contents, enzyme activities, and crop yield in a semiarid region of China. *Soil Tillage Res.* **2016**, *160*, 65–72. [CrossRef]

15. Verma, N.S.; Yadav, D.; Chouhan, M.; Bhagat, C.; Kochale, P. Understanding Potential Impact of Green Manuring on Crop and Soil: A Comprehensive Review. *Biol. Forum-Int. J.* **2023**, *15*, 832–839.

16. Joshna, A.; Bokado, K. Barkha, Green Manure for Sustainable Crop Production: A Review. *Int. J. Environ. Clim. Change* **2024**, *14*, 147–156. [CrossRef]

17. Smith, P.; Cotrufo, M.F.; Rumpel, C.; Paustian, K.; Kuikman, P.J.; Elliott, J.A.; McDowell, R.; Griffiths, R.I.; Asakawa, S.; Bustamante, M.; et al. Biogeochemical cycles and biodiversity as key drivers of ecosystem services provided by soils. *Soil* **2015**, *1*, 665–685. [CrossRef]

18. Don, A.; Schumacher, J.; Freibauer, A. Impact of tropical land-use change on soil organic carbon stocks—A meta-analysis. *Glob. Change Biol.* **2011**, *17*, 1658–1670. [CrossRef]

19. Aronsson, H.; Hansen, E.M.; Thomsen, I.K.; Liu, J.; Øgaard, A.F.; Känkänen, H.; Ulén, B.J.J.O. The ability of cover crops to reduce nitrogen and phosphorus losses from arable land in southern Scandinavia and Finland. *J. Soil Water Conserv.* **2016**, *71*, 41–55. [CrossRef]

20. Koudahe, K.; Allen, S.C.; Djaman, K. Critical review of the impact of cover crops on soil properties. *Int. Soil Water Conserv. Res.* **2022**, *10*, 343–354. [CrossRef]

21. Wilczewski, E.; Piotrowska-Długosz, A.; Lemańczyk, G. Properties of Alfisol and yield of spring barley as affected by catch crop. *Zemdirb. Agric.* **2015**, *102*, 23–30. [CrossRef]

22. Sanderman, J.; Hengl, T.; Fiske, G.J. Soil carbon debt of 12,000 years of human land use. *Proc. Natl. Acad. Sci. USA* **2017**, *114*, 9575–9580. Available online: http://www.pnas.org/cgi/doi/10.1073/pnas.1706103114 (accessed on 12 April 2024). [CrossRef]

23. Szostek, M.; Szpunar-Krok, E.; Pawlak, R.; Stanek-Tarkowska, J.; Ilek, A. Effect of different tillage systems on soil organic carbon and enzymatic activity. *Agronomy* **2022**, *12*, 208. [CrossRef]

24. Rinnofner, T.; Friedel, J.K.; Pietsch, G.; Freyer, B.; Kruijff, R. Effect of catch crops on N dynamics and following crops in organic farming. *Agron. Sustain. Dev.* **2008**, *28*, 551–558. [CrossRef]

25. Tuulos, A.; Yli-Halla, M.; Stoddard, F.; Mäkelä, P. Winter turnip rape as a soil N scavenging catch crop in a cool humid climate. *Agron. Sustain. Dev.* **2015**, *35*, 359–366. [CrossRef]

26. Wilczewski, E.; Lemańczyk, G.; Skinder, Z.; Sadowski, C. Effect of nitrogen fertilization on the yielding and health status of selected non-papilionaceous plant species grown in stubble intercrop. *EJPAU Agron.* **2006**, *9*, 4. Available online: http://www.ejpau.media.pl/volume9/issue2/art-04.html (accessed on 12 April 2024).

27. Schipanski, M.; MacDonald, G.; Rosenzweig, S.; Chappell, J.; Bennett, E.; Kerr, R.B.; Blesh, J.; Crews, T.; Drinkwater, L.; Lundgren, J.G.; et al. Realizing Resilient Food Systems. *BioScience* **2016**, *66*, 600–610. [CrossRef]

28. Stagnari, F.; Maggio, A.; Galieni, A.; Pisante, M. Multiple benefits of legumes for agriculture sustainability: An overview. *Chem. Biol. Technol. Agric.* **2017**, *4*, 1–13. [CrossRef]

29. Wani, S.P.; Repula, O.P.; Lee, K.K. Sustainable agriculture in the semi-arid tropics through biological nitrogen fixation in grain legumes. *Plant Soil* **1995**, *174*, 42–45. [CrossRef]

30. Balota, E.L.; Calegari, A.; Nakatani, A.S.; Coyne, M.S. Benefits of winter cover crops and no-tillage for microbial parameters in a Brazilian Oxisol: A long-term study. *Agric. Ecosyst. Environ.* **2014**, *197*, 31–40. [CrossRef]

31. Thorup-Kristensen, K. The effect of nitrogen catch crop species on the nitrogen nutrition of succeeding crops. *Fert. Res.* **1994**, *37*, 227–234. [CrossRef]

32. Skinder, Z.; Wilczewski, E. Forecrop value of non-papilionaceous plants cultivated in stubble intercrop for spring barley under various fertilization conditions. *EJPAU Ser. Agron.* **2004**, *7*, 3.

33. da Silva, J.P.; da Silva Teixeira, R.; Ivo, R.; da Silva Soares, E.M.B.; Lima, A.M.N. Decomposition and nutrient release from legume and non-legume residues in a tropical soil. *Eur. J. Soil Sci.* **2022**, *73*, e13151. [CrossRef]

34. Constantin, J.; Mary, B.; Laurent, F.; Aubrion, G.; Fontaine, A.; Kerveillant, P.; Beaudoin, N. Effects of catch crops, no till and reduced nitrogen fertilization on nitrogen leaching and balance in three long-term experiments. *Agric. Ecosyst. Environ.* **2010**, *135*, 268–278. [CrossRef]

35. Askegaard, M.; Olesen, J.E.; Rasmussen, I.A.; Kristensen, K. Nitrate leaching from organic arable crop rotations is mostly determined by autumn field management. *Agric. Ecosyst. Environ.* **2011**, *142*, 149–160. [CrossRef]

36. Wilczewski, E.; Piotrowska-Długosz, A.; Lemańczyk, G. Influence of catch crop on soil properties and yield of spring barley. *Int. J. Plant Prod.* **2014**, *8*, 391–408. [CrossRef]

37. Brant, V.; Pivec, J.; Fuksa, P.; Neckář, K.; Kocourková, D.; Venclová, V. Biomass and energy production of catch crops in areas with deficiency of precipitation during summer period in central Bohemia. *Biomass Bioenergy* **2011**, *35*, 1286–1294. [CrossRef]

38. Tribouillois, H.; Dürr, C.; Demilly, D.; Wagner, M.-H.; Justes, E. Determination of Germination Response to Temperature and Water Potential for a Wide Range of Cover Crop Species and Related Functional Groups. *PLoS ONE* **2016**, *11*, e0161185. [CrossRef]

39. Wilczewski, E.; Skinder, Z.; Szczepanek, M. Effects of weather conditions on yield of tansy phacelia and common sunflower grown as stubble catch crop. *Pol. J. Environ. Stud.* 2012, 21, pp. 1053–1060. Available online: https://www.pjoes.com/pdf-88839-22698?filename=Effects%20of%20Weather.pdf (accessed on 12 April 2024).

40. Hashemi, M.; Farsad, A.; Sadeghpour, A.; Weis, S.A.; Herbert, S.J. Cover-crop seeding-date influence on fall nitrogen recovery. *J. Plant Nutr. Soil Sci.* **2013**, *176*, 69–75. [CrossRef]

41. Sieling, K. Improved N transfer by growing catch crops—A challenge. *J. Kulturpflanz* **2019**, *71*, 145–160. [CrossRef]

42. Tribouillois, H.; Cohan, J.-P.; Justes, E. Cover crop mixtures including legume produce ecosystem services of nitrate capture and green manuring: Assessment combining experimentation and modelling. *Plant Soil* **2016**, *401*, 347–364. [CrossRef]

43. Kraska, P. Effect of tillage system and catch crop on weed infestation and spring wheat stands (*Triticum aestivum* L.). *Acta Sci. Pol. Agric.* **2012**, *11*, 27–43.

44. Kwiatkowski, C.A.; Harasim, E.; Wesołowski, M. Effects of catch crops and tillage system on weed infestation and health of spring wheat. *J. Agric. Sci. Technol.* **2016**, *18*, 999–1012.

45. Wendling, M.; Charles, R.; Herrera, J.; Amossé, C.; Jeangros, B.; Walter, A.; Büchi, L. Effect of species identity and diversity on biomass production and its stability in cover crop mixtures. *Agric. Ecosyst. Environ.* **2019**, *281*, 81–91. [CrossRef]

46. Redin, M.; Recous, S.; Aita, C.; Chaves, B.; Pfeifer, I.C.; Bastos, L.M.; Pilecco, G.E.; Giacomini, S.J. Root and Shoot Contribution to Carbon and Nitrogen Inputs in the Topsoil Layer in No-Tillage Crop Systems under Subtropical Conditions. *Rev. Bras. Ciênc. Solo* **2018**, *42*, e0170355. [CrossRef]

47. Stein, S.; Hartung, J.; Perkons, U.; Möller, K.; Zikeli, S. Plant and soil N of different winter cover crops as green manure for subsequent organic white cabbage. *Nutr. Cycl. Agroecosyst.* **2023**, *127*, 285–298. [CrossRef]

48. Acosta, J.A.A.; Amado, T.J.C.; de Neergaard, A.; Vinther, M.; da Silva, L.S.; da Silveira Nicoloso, A. Effect of 15n-labeled hairy vetch and nitrogen fertilization on maize nutrition and yield under no-tillage. *Rev. Bras. Ciênc. Solo* **2011**, *35*, 1337–1345. [CrossRef]

49. St Aime, R.; Bridges, W.C., Jr.; Narayanan, S. Fall-winter cover crops promote soil health and weed control in the southeastern clayey soils. *Agron. J.* **2023**, *115*, 242260. [CrossRef]

50. Teasdale, J.R.; Mirsky, S.B.; John, T.; Spargo, J.T.; Cavigelli, M.A.; Maul, J.E. Reduced-tillage organic corn production in a hairy vetch cover crop. *Agron. J.* **2012**, *104*, 621–628. [CrossRef]

51. Mirsky, S.B.; Curran, W.S.; Mortenseny, D.M.; Ryany, M.R.; Shumway, D.L. Timing of cover-crop management effects on weed suppression in no-till planted soybean using a roller-crimper. *Weed Sci.* **2011**, *59*, 380–389. [CrossRef]

52. Schröder, J.J.; Van Dijk, W.; De Groot, W.J.M. Effects of cover crops on the nitrogen fluxes in a silage maize production system. *Neth. J. Agric. Sci.* **1996**, *44*, 293–315. [CrossRef]

53. Komainda, M.; Taube, F.; Kluß, C.; Herrmann, A. Above- and belowground nitrogen uptake of winter catch crops sown after silage maize as affected by sowing date. *Eur. J. Agron.* **2016**, *79*, 31–42. [CrossRef]

54. Arlauskienė, A.; Cesevičienė, J.; Velykis, A. Improving mineral nitrogen control by combining catch crops, fertilisation, and straw management in a clay loam soil. *Acta Agric. Scand. Sec. B—Soil Plant Sci.* **2019**, *69*, 422–431. [CrossRef]

55. Allison, M.F.; Armstrong, M.J.; Jaggard, K.W.; Todd, A.D. Integration of nitrate cover crops into sugarbeet (*Beta vulgaris*) rotations. II. Effect of cover crops on growth, yield and N equirement of sugarbeet. *J. Agric. Sci.* **1998**, *130*, 61–67. [CrossRef]

56. Brooks, P.C.; Cayuela, M.L.; Contin, M.; De Nobili, M.; Kemmitt, S.J.; Mondini, C. The mineralization of fresh and humified soil organic matter by soil microbial biomass. *Waste Manag.* **2008**, *28*, 716–722. [CrossRef]

57. Doran, J.W.; Elliott, E.T.; Paustian, K. Soil microbial activity, nitrogen cycling, and long-term changes in organic carbon pools as related to fallow tillage management. *Soil Tillage Res.* **1998**, *49*, 3–18. [CrossRef]

58. Marinari, S.; Mancinelli, R.; Campiglia, E.; Grego, S. Chemical and biological indicators of soil quality in organic and conventional farming systems in Italy. *Ecol. Indic.* **2006**, *6*, 701–711. [CrossRef]

59. Acosta-Martinez, V.; Acosta-Mercado, D.; Sotomayor-Ramirez, D.; Cruz-Rodriguez, L. Microbial communities and enzymatic activities under different management in semiarid soils. *Appl. Soil Ecol.* **2008**, *38*, 249–260. [CrossRef]

60. Gajda, A.M. Effect of different tillage systems on some microbiological properties of soils under winter wheat. *Int. Agroph.* **2008**, *22*, 201–208.

61. Jackson, L.E.; Calderon, F.J.; Steenwerth, K.L.; Scow, K.M.; Rolston, D.E. Responses of soil microbial processes and community structure to tillage events and implications for soil quality. *Geoderma* **2003**, *114*, 305–317. [CrossRef]

62. Mazzoncini, M.; Sapkota, T.B.; Bàrberi, P.; Antichi, D.; Risaliti, R. Long-term effect of tillage, nitrogen fertilization and cover crops on soil organic carbon and total nitrogen content. *Soil Tillage Res.* **2011**, *114*, 165–174. [CrossRef]

63. Balota, E.L.; Colozzi-Filho, A.; Andrade, D.S.; Dick, R.P. Microbial biomass in soils under different tillage and crop rotation systems. *Biol. Fertil. Soils* **2003**, *38*, 15–20. [CrossRef]

64. Mbuthia, L.W.; Acosta-Martinez, V.; DeBruyn, J.; Schaeffer, S.; Tyler, D.; Odoi, E.; Mpheshea, M.; Walker, F.; Eash, N. Long term tillage, cover crop, and fertilization effects on microbial community structure, activity: Implications for soil quality. *Soil Biol. Biochem.* **2015**, *89*, 24–34. [CrossRef]

65. Lenart, S.; Sławiński, P. Selected soil properties and the occurrence of earthworms under the conditions of direct sowing and mouldboard ploughing. *Fragm. Agron.* **2010**, *27*, 86–93. (In Polish)

66. Bini, D.; Santos, C.A.D.; Bernal, L.P.T.; Andrade, G.; Nogueira, M.A. Identifying indicators of C and N cycling in a clayey Ultisol under different tillage and uses in winter. *App. Soil Ecol.* **2014**, *76*, 95–101. [CrossRef]

67. Nevins, C.J.; Nakatsu, C.; Armstrong, S. Characterization of microbial community response to cover crop residue decomposition. *Soil Biol. Biochem.* **2018**, *127*, 39–49. [CrossRef]

68. Harasim, E.; Antonkiewicz, J.; Kwiatkowski, C.A. The effects of catch crops and tillage systems on selected physical properties and enzymatic activity of loess soil in a spring wheat monoculture. *Agronomy* **2020**, *10*, 334. [CrossRef]

69. Lupwayi, N.Z.; Larney, F.J.; Blackshaw, R.E.; Kanashiro, D.A.; Pearson, D.C. Phospholipid fatty acid biomarkers show positive soil microbial community responses to conservation soil management of irrigated crop rotations. *Soil Tillage Res.* **2017**, *168*, 1–10. [CrossRef]
70. Baldivieso-Freitasa, P.; Blanco-Morenoa, J.M.; Gutiérrez-Lópezd, M.; Peignéc, J.; Pérez-Ferrera, A.; Trigo-Azad, D.; Sansa, F.X. Earthworm abundance response to conservation agriculture practices in organic arable farming under Mediterranean climate. *Pedobiol. J Soil Sci.* **2018**, *66*, 58–84. [CrossRef]
71. Pelosi, C.; Grandeau, G.; Capowiez, Y. Temporal dynamics of earthworm-related macroporosity in tilled and non-tilled cropping systems. *Geoderma* **2017**, *289*, 169–177. [CrossRef]
72. Pelosi, C.; Pey, B.; Hedde, M.; Caro, G.; Capowiez, Y.; Guernion, M.; Peigné, J.; Piron, D.; Bertrand, M.; Cluzeaud, D. Reducing tillage in cultivated fields increases earthworm functional diversity. *Appl. Soil Ecol.* **2014**, *83*, 79–87. [CrossRef]
73. Arai, M.; Miura, T.; Tsuzura, H.; Minamiya, Y.; Kaneko, N. Two-year responses of earthworm abundance, soil aggregates, and soil carbon to no-tillage and fertilization. *Geoderma* **2018**, *332*, 135–141. [CrossRef]
74. Yvan, C.; Stéphane, S.; Stéphane, C.; Pierre, B.; Guy, R.; Hubert, B. Role of earthworms in regenerating soil structure after compaction in reduced tillage system. *Soil Biol. Biochem.* **2012**, *55*, 93–103. [CrossRef]
75. Neave, P.; Fox, C.A. Response of soil invertebrates to reduced tillage systems established on a clay loam soil. *Appl. Soil Ecol.* **1998**, *9*, 423–428. [CrossRef]
76. Hansen, R.A. Diversity in the decomposing landscape. In *Invertebrates as Webmasters in Ecosystems*; Coleman, D.C., Hendrix, P.F., Eds.; CABI Press: Wallingford, UK, 2000; pp. 203–219.
77. Twardowski, J.; Smolis, A.; Kordas, L. The effect of different tillage systems on soil mesofauna. Preliminary studies. *Ann. UMCS Sec. E* **2004**, *59*, 817–824. (In Polish)
78. Hungria, M.; Franchini, J.C.; Brandão-Junior, O.; Kaschuk, G.; Souza, R.A. Soil microbial activity and crop sustainability in a long-term experiment with three soil-tillage and two crop-rotation systems. *Appl. Soil Ecol.* **2009**, *42*, 288–296. [CrossRef]
79. Hobbie, S.E. Plant species effects on nutrient cycling: Revisiting litter feedbacks. *Trends Ecol. Evol.* **2015**, *30*, 357–363. [CrossRef] [PubMed]
80. Barel, J.M.; Kuyper, T.W.; Paul, J.; de Boer, W.; Cornelissen, J.H.C.; De Deyn, G.B. Winter cover crop legacy effects on litter decomposition act through litter quality and microbial community changes. *J. Appl. Ecol.* **2019**, *56*, 132–143. [CrossRef]
81. Frasier, I.; Noellemeyer, E.; Figuerola, E.; Erijman, L.; Permingeat, H.; Quiroga, A. High quality residues from cover crops favor changes in microbial community and enhance C and N sequestration. *Glob. Ecol. Conserv.* **2016**, *6*, 242–256. [CrossRef]
82. Niewiadomska, A.; Majchrzak, L.; Borowiak, K.; Wolna-Maruwka, A.; Waraczewska, Z.; Budka, A.; Gaj, R. The influence of tillage and cover cropping on soil microbial parameters and spring wheat physiology. *Agronomy* **2020**, *10*, 200. [CrossRef]
83. Chavarría, D.N.; Verdenelli, R.A.; Muñoz, E.J.; Conforto, C.; Restovich, S.B.; Andriulo, A.E.; Meriles, J.M.; Vargas-Gil, S. Soil microbial functionality in response to the inclusion of cover crop mixtures in agricultural systems. *Span. J. Agric. Res.* **2016**, *14*, e0304. [CrossRef]
84. Kwiatkowski, C.A.; Harasim, E.; Klikocka-Wiśniewska, O. Effect of catch crops and tillage systems on the content of selected nutrients in spring wheat grain. *Agronomy* **2022**, *12*, 1054. [CrossRef]
85. Majchrzak, L.; Sawinska, Z.; Natywa, M.; Skrzypczak, G.; Głowicka-Wołoszyn, R. Impact of different tillage systems on soil dehydrogenase activity and spring wheat infection. *J. Agric. Sci. Technol.* **2016**, *18*, 1871–1881.
86. Qi, Y.; Zhou, R.; Nie, L.; Sun, M.; Wu, X.; Jiang, F. The Effects of catch crops on properties of continuous cropping soil and growth of vegetables in greenhouse. *Agronomy* **2022**, *12*, 1179. [CrossRef]
87. Sürücü, A.; Özyazici, M.A.; Bayrakli, B.; Kizilkaya, R. Effects of manuring on soil enzyme activity. *Fresenius Environ. Bull.* **2014**, *23*, 2126–2132. Available online: https://avesis.omu.edu.tr/yayin/3c68c793-b228-4e66-9b0e-265da7885ca4/effects-of-green-manuring-on-soil-enzyme-activity (accessed on 13 May 2024).
88. Hallama, M.; Pekrun, C.; Lambers, H.; Kandeler, E. Hidden miners-the roles of cover crops and soil microorganisms in phosphorus cycling through agroecosystems. *Plant Soil* **2019**, *434*, 7–45. [CrossRef]
89. Reddy, K.N.; Zablotowicz, R.M.; Locke, M.A.; Koger, C.H. Cover crop, tillage, and herbicide effects on weeds, soil properties, microbial populations, and soybean yield. *Weed Sci.* **2003**, *51*, 987–994. [CrossRef]
90. Calderon, F.J.; Nielsen, D.; Acosta-Martinez, V.; Vigil, M.F.; Lyons, D. Cover crop and irrigation effects on soil microbial communities and enzymes in semiarid agroecosystems of the central great plains of north America. *Pedosphere* **2016**, *26*, 192–205. [CrossRef]
91. Smagacz, J.; Madej, A. Perspektywiczne kierunki rozwoju różnych systemów uprawy roli. *Stud. Rap. IUNG-PIB* **2012**, *30*, 9–22. (In Polish)
92. Haruna, S.I.; Anderson, S.H.; Udawatta, R.P.; Gantzer, C.J.; Phillips, N.C.; Cui, S.; Gao, Y. Improving soil physical properties through the use of cover crops: A review. *Agrosyst. Geosci. Environ.* **2020**, *3*, e20105. [CrossRef]
93. Reintam, E.; Trükmann, K.; Kuht, J.; Nugis, E.; Edesi, L.; Astover, A.; Noormets, M.; Karin Kauer, K.; Krebstein, K.; Rannik, K. Soil compaction effects on soil bulk density and penetration resistance and growth of spring barley (*Hordeum vulgare* L.). *Acta Agric. Scand. B Soil Plant Sci.* **2009**, *59*, 265–272. [CrossRef]
94. Jug, D.; Đurđević, B.; Birkás, M.; Brozović, B.; Lipiec, J.; Vukadinović, V.; Jug, I. Effect of conservation tillage on crop productivity and nitrogen use efficiency. *Soil Tillage Res.* **2019**, *194*, 104327. [CrossRef]

95. Lipiec, J.; Hatano, R. Quantification of compaction effects on soil physical properties and crop growth. *Geoderma* **2003**, *116*, 107–136. [CrossRef]
96. Haruna, S.I.; Nkongolo, N.V. Cover crop management effects on soil physical and biological properties. *Procedia Environ. Sci.* **2015**, *29*, 13–14. [CrossRef]
97. Głąb, T.; Kulig, B. Effect of mulch and tillage system on soil porosity under wheat (*Triticum aestivum*). *Soil Tillage Res.* **2008**, *99*, 169–178. [CrossRef]
98. Abdollahi, L.; Munkholm, L.J.; Garbout, A. Tillage system and cover crop effects on soil quality: II. *Pore Characteristics. Soil Sci. Soc. Am. J.* **2014**, *78*, 271–279. [CrossRef]
99. Khan, S.R.; Abbasi, M.K.; Hussan, A. Effect of induced soil compaction on changes in soil properties and wheat productivity under sandy loam and sandy clay loam soils: A greenhouse experiment. *Commun. Soil Sci. Plant Anal.* **2012**, *43*, 2550–2563. [CrossRef]
100. Chen, G.; Weil, R. Penetration of cover crop roots through compacted soils. *Plant Soil* **2010**, *331*, 31–43. [CrossRef]
101. Batey, T. Soil compaction and soil management—A review. *Soil Use Manag.* **2009**, *25*, 335–345. [CrossRef]
102. Morel, J.L.; Habib, L.; Plantureux, S.; Guckert, A. Influence of maize root mucilage on soil aggregate stability. *Plant Soil* **1991**, *136*, 111–119. [CrossRef]
103. Drost, S.; Rutgers, M.; Wouterse, M.; de Boer, W.; Bodelier, P.L.E. Decomposition of mixtures of cover crop residues increases microbial functional diversity. *Geoderma* **2020**, *361*, 114060. [CrossRef]
104. Angers, D.A.; Caron, J. Plant-induced changes in soil structure: Processes and feedbacks. *Biogeochemistry* **1998**, *42*, 55–72. [CrossRef]
105. Jug, I.; Jug, D.; Brozović, B.; Vukadinović, V.; Đurđević, B. *Basic of Soil Science and Plant Production*; University Textbook; Faculty of Agrobiotechnical Sciences, University of Josip Juraj Strossmayer in Osijek: Osijek, Croatia, 2022; p. 527, ISBN 978-953-8421-00-6. (In Croatian)
106. O' Geen, A.T. Soil Water Dynamics. *Nature. Educ. Knowl.* **2013**, *4*, 9.
107. Wilczewski, E.; Jug, I.; Lipiec, J.; Gałęzewski, L.; Đurđević, B.; Kocira, A.; Brozović, B.; Marković, M.; Jug, D. Tillage system regulates the soil moisture tension, penetration resistance and temperature responses to the temporal variability of precipitation during the growing season. *Int. Agrophys.* **2023**, *37*, 391–399. [CrossRef] [PubMed]
108. Blanco-Canqui, H.; Holman, J.D.; Schlegel, A.; Tatarko, J.S.; Shaver, T.M. Replacing Fallow with Cover Crops in a Semiarid Soil: Effects on Soil Properties. *Soil Sci. Soc. Am. J.* **2013**, *77*, 1026–1034. [CrossRef]
109. Blanco-Canqui, H.; Shapiro, C.A.; Wortmann, C.S.; Drijber, R.A.; Mamo, M.; Shaver, T.M.; Ferguson, R.B. Soil organic carbon: The value to soil properties. *J. Soil Water Conserv.* **2013**, *68*, 129A–134A. [CrossRef]
110. Calonego, J.; Raphael, J.; Rigon, J.; de Oliveira Neto, L.; Rosolem, C.A. Soil compaction management and soybean yields with cover crops under no-till and occasional chiseling. *Eur. J. Agron.* **2017**, *85*, 31–33. [CrossRef]
111. Six, J.; Conant, R.T.; Paul, E.; Paustian, K. Stabilization mechanisms of soil organic matter: Implications for C-saturation of soils. *Plant Soil* **2002**, *241*, 155–176. [CrossRef]
112. Brady, N.C.; Weil, R.R. *The Nature and Properties of Soils*, 14th ed.; Pearson Prentice Hall: Hoboken, NJ, USA, 2008; ISBN 9780135133873.
113. Shepherd, T.G.; Saggar Newman, R.H.; Ross, C.W.; Dando, J.L. Tillage induced changes in soil structure and soil organic matter fractions. *Soil Res.* **2001**, *39*, 465–489. [CrossRef]
114. De Cima, D.S.; Luik, A.; Reintam, E. Organic farming and cover crops as an alternative to mineral fertilizers to improve soil physical properties. *Int. Agrophys.* **2015**, *29*, 405–412. [CrossRef]
115. Franzluebbers, A.J. Soil organic carbon sequestration and agricultural greenhouse gas emissions in the southeastern USA. *Soil Tillage Res.* **2005**, *83*, 120–147. [CrossRef]
116. Bronick, C.J.; Lal, R. Soil structure and management: A review. *Geoderma* **2005**, *124*, 3–22. [CrossRef]
117. Jug, D.; Jug, I.; Brozović, B.; Vukadinović, V.; Stipešević, B.; Đurđević, B. The role of conservation agriculture in mitigation and adaptation to climate change. *Poljoprivreda* **2018**, *24*, 35–44. [CrossRef]
118. Abdollahi, L.; Munkholm, L.J. Tillage system and cover crop effects on soil quality: I. Chemical, mechanical, and biological properties. *Soil Sci. Soc. Am. J.* **2014**, *78*, 262–270. [CrossRef]
119. Głąb, T.; Ścigalska, B.; Łabuz, B. Effect of crop rotations with triticale (× *Triticosecale* Wittm.) on soil pore characteristics. *Geoderma* **2013**, *202–203*, 1–7. [CrossRef]
120. Chen, G.; Wail, R.R.; Hill, L.R. Effects of compaction and cover crops on soil least limiting water range and air permeability. *Soil Tillage Res.* **2013**, *136*, 61–69. [CrossRef]
121. Villamil, M.B.; Bollero, G.A.; Darmody, R.G.; Simmons, F.W.; Bullock, D.G. No-till corn/soybean systems including winter cover crops. *Soil Sci. Soc. Am. J.* **2006**, *70*, 1936–1944. [CrossRef]
122. Treonis, A.M.; Austin, E.E.; Buyer, J.S.; Maul, J.E.; Spicer, L.; Zasada, I.A. Effects of organic amendment and tillage on soil microorganisms and microfauna. *Appl. Soil Ecol.* **2010**, *46*, 103–110. [CrossRef]
123. Kuo, S.; Sainju, U.M.; Jellum, E.J. Winter cover crop effects on soil organic carbon and carbohydrate in soil. *Soil Sci. Soc. Am. J.* **1997**, *61*, 145–152. [CrossRef]
124. Blanco-Canqui, H.; Shaver, T.M.; Lindquist, J.L.; Shapiro, C.A.; Elmore, R.W.; Francis, C.A.; Hergert, G.W. Cover crops and ecosystem services: Insights from studies in temperate soils. *Agron. J.* **2015**, *107*, 6. [CrossRef]

125. Poeplau, C.; Don, A. Carbon sequestration in agricultural soils via cultivation of cover crops: A meta-analysis. *Agric. Ecosyst. Environ.* **2015**, *200*, 33–41. [CrossRef]
126. Berhe, A.A.; Harte, J.; Harden, J.; Torn, M.S. Significance of the erosion-induced terrestrial carbon sink. *BioScience* **2007**, *57*, 337–346. [CrossRef]
127. Stavi, I.; Lal, R.; Jones, S.; Reeder, R.C. Implications of cover crops for soil quality and geodiversity in a humid-temperate region in the midwestern USA. *Land Degrad. Dev.* **2012**, *23*, 322–330. [CrossRef]
128. Olson, K.; Ebelhar, S.A.; Lang, J.M. Long-term effects of cover crops on crop yields, soil organic carbon stocks and sequestration. *Open J. Soil Sci.* **2014**, *4*, 284–292. [CrossRef]
129. Huang, Y.; Ren, W.; Grove, J.; Poffenbarger, H.; Jacobsen, K.; Tao, B.; Zhu, X.; McNear, D. Assessing synergistic effects of no-tillage and cover crops on soil carbon dynamics in a long-term maize cropping system under climate change. *Agric. For. Meteorol.* **2020**, *291*, 108090. [CrossRef]
130. Acuna, J.C.M.; Villamil, M.B. Short-term effects of cover crops and compaction on soil properties and soybean production in Illinois. *Agron. J.* **2014**, *106*, 860–870. [CrossRef]
131. Wilczewski, E.; Sadkiewicz, J.; Piotrowska-Długosz, A.; Gałęzewski, L. Change of plant nutrients in soil and spring barley depending on the field pea management as a catch crop. *Agriculture* **2021**, *11*, 394. [CrossRef]
132. Nascente, A.S. Soil chemical properties affected by cover crops under no-tillage system. *Soil Sci. Plant Nutr. Rev. Ceres* **2015**, *62*, 4. [CrossRef]
133. Rankoth, L.M.; Udawatta, R.P.; Veum, K.S.; Jose, S.; Alagele, S. Cover crop influence on soil enzymes and selected chemical parameters for a claypan corn–soybean rotation. *Agriculture* **2019**, *9*, 125. [CrossRef]
134. Büchi, L.; Wendling, M.; Amossé, C.; Necpalova, M.; Charles, R. Importance of cover crops in alleviating negative effects of reduced soil tillage and promoting soil fertility in a winter wheat cropping system. *Agric. Ecosyst. Environ.* **2018**, *256*, 92–104. [CrossRef]
135. Autret, B.; Mary, B.; Chenu, C.; Balabane, M.; Girardin, C.; Bertrand, M.; Grandeau, G.; Beaudoin, N. Alternative arable cropping systems: A key to increase soil organic carbon storage?: Results from a 16 year field experiment Agric. *Ecosyst. Environ.* **2016**, *232*, 150–164. [CrossRef]
136. Olson, K.R.; Al-Kaisi, M.M. The importance of soil sampling depth for accurate account of soil organic carbon sequestration, storage, retention and loss. *Catena* **2015**, *125*, 33–37. [CrossRef]
137. Hubbard, R.K.; Strickland, T.C.; Phatak, S. Effects of cover crop systems on soil physical properties and carbon/nitrogen relationships in the coastal plain of southeastern USA. *Soil Tillage Res.* **2013**, *126*, 276–283. [CrossRef]
138. Farmaha, B.S.; Sekaran, U.; Franzluebbers, A.J. Cover cropping and conservation tillage improve soil health in the southeastern United States. *Agron. J.* **2021**, *114*, 296–316. [CrossRef]
139. Creamer, N.G.; Baldwin, K.R. An evaluation of summer cover crops for use in vegetable production systems in North Carolina. *HortScience* **2000**, *35*, 600–603. [CrossRef]
140. Sainju, U.M.; Whitehead, W.F.; Singh, B.P. Agricultural management practices to sustain crop yields and improve soil and environmental qualities. *Sci. World* **2003**, *3*, 768–789. [CrossRef] [PubMed]
141. Haynes, R.; Mokolobate, M. Amelioration of Al toxicity and P deficiency in acid soils by additions of organic residues: A critical review of the phenomenon and the mechanisms involved. *Nutr. Cycl. Agroecosyst.* **2001**, *59*, 47–63. [CrossRef]
142. Kumar, R.; Pandey, S.; Pandey, A. Plant roots and carbon sequestration. *Curr. Sci.* **2006**, *91*, 885–890. Available online: https://www.jstor.org/stable/24094284 (accessed on 13 May 2024).
143. Meng, L.; Zhang, A.; Wang, F.; Han, X.; Wang, D.; Li, S. Arbuscular mycorrhizal fungi and rhizobium facilitate nitrogen uptake and transfer in soybean/maize intercropping system. *Front. Plant Sci.* **2015**, *6*, 339. [CrossRef]
144. Gao, X.; He, Y.; Chen, Y.; Wang, M. Leguminous green manure amendments improve maize yield by increasing N and P fertilizer use efficiency in yellow soil of the Yunnan-Guizhou Plateau. *Front. Sustain. Food Syst.* **2024**, *8*, 1369571. [CrossRef]
145. Meisinger, J.; Delgado, J. Principles for Managing Nitrogen Leaching. *J. Soil Water Conserv.* **2002**, *57*, 485–498. Available online: https://www.jswconline.org/content/57/6/485 (accessed on 13 May 2024).
146. Hirsh, S.M.; Duiker, S.W.; Graybill, J.; Nichols, K.; Weil, R.R. Scavenging and recycling deep soil nitrogen using cover crops on mid-Atlantic, USA farms. *Agric. Ecosyst. Environ* **2021**, *309*, 107274. [CrossRef]
147. Plaza-Bonilla, D.; Nolot, J.-M.; Raffaillac, D.; Justes, E. Cover crops mitigate nitrate leaching in cropping systems including grain legumes: Field evidence and model simulations. *Agric. Ecosyst. Environ.* **2015**, *212*, 1–12. [CrossRef]
148. Abdalla, M.; Hastings, A.; Cheng, K.; Yue, Q.; Chadwick, D.; Espenberg, M.; Truu, J.; Rees, R.M.; Smith, P. A critical review of the impacts of cover crops on nitrogen leaching, net greenhouse gas balance and crop productivity. *Glob. Change Biol.* **2019**, *25*, 2530–2543. [CrossRef] [PubMed]
149. Sedghi, N.; Cavigelli, M.; Weil, R.R. Soil texture, fertilization, cover crop species and management affect nitrous oxide emissions from no-till cropland. *Sci. Total Environ.* **2024**, *1*, 169991. [CrossRef] [PubMed]
150. Sainju, U.M.; Singh, B.P.; Whitehead, W.F. Long-term effects of tillage, cover crops, and nitrogen fertilization on organic carbon and nitrogen concentrations in sandy loam soils in Georgia, USA. *Soil Tillage Res.* **2002**, *63*, 167–179. [CrossRef]
151. Sainju, U.M.; Singh, B.P.; Whitehead, W.F.; Wang, S. Carbon supply and storage in tilled and nontilled soils as influenced by cover crops and nitrogen fertilization. *J. Environ. Qual.* **2006**, *35*, 1507–1517. [CrossRef]

152. St. Luce, M.; Grant, C.A.; Ziadi, N.; Zebarth, B.J.; O'Donovan, J.T.; Blackshaw, R.E.; Harker, K.N.; Johnson, E.N.; Gan, Y.; Lafond, G.P.; et al. Preceding crops and nitrogen fertilization influence soil nitrogen cycling in no-till canola and wheat cropping systems. *Field Crops Res.* **2016**, *191*, 20–32. [CrossRef]
153. Zhou, G.; Cao, W.; Bai, J.; Xu, C.; Zeng, N.; Gao, S.; Rees, R.M.; Dou, F. Co-incorporation of rice straw and leguminous green manure can increase soil available nitrogen (N) and reduce carbon and N losses: An incubation study. *Pedosphere* **2020**, *30*, 661–670. [CrossRef]
154. Chivenge, P.; Vanlauwe, B.; Gentile, R.; Six, J. Organic resource quality influences short-term aggregate dynamics and soil organic carbon and nitrogen accumulation. *Soil Biol. Biochem.* **2011**, *43*, 657–666. [CrossRef]
155. Thomsen, I.K.; Elsgaard, L.; Olesen, J.E.; Christensen, B.T. Nitrogen release from differently aged *Raphanus sativus* L. nitrate catch crops during mineralization at autumn temperatures. *Soil Use Manag.* **2016**, *32*, 183–191. [CrossRef]
156. Li, F.; Sørensen, P.; Li, X.; Olesen, J.E. Carbon and nitrogen mineralization differ between incorporated shoots and roots of legume versus non-legume based cover crops. *Plant Soil* **2020**, *446*, 243–257. [CrossRef]
157. Sanchez, J.E.; Harwood, R.R.; Willson, T.C.; Kizilkaya, K.; Smeenk, J.; Parker, E.; Paul, E.A.; Knezek, B.D.; Robertson, G.P. Managing soil carbon and nitrogen for productivity and environmental quality. *Agron. J.* **2004**, *96*, 769–775. [CrossRef]
158. Dabney, S.M.; Delgado, J.A.; Reeves, D.W. Using winter cover crops to improve soil and water quality. *Commun. Soil Sci. Plant Anal.* **2001**, *32*, 1221–1250. [CrossRef]
159. Kuzyakov, Y. Priming effects: Interactions between living and dead organic matter. *Soil Biol. Biochem.* **2010**, *42*, 1363–1371. [CrossRef]
160. Clivot, H.; Mary, B.; Valé, M.; Cohan, J.-P.; Champolivier, L.; Piraux, F.; Laurent, F.; Justes, E. Quantifying in situ and modeling net nitrogen mineralization from soil organic matter in arable cropping systems. *Soil Biol. Biochem.* **2017**, *111*, 44–59. [CrossRef]
161. Nascente, A.S.; Stone, L.F. Cover crops as affecting soil chemical and physical properties and development of upland rice and soybean cultivated in rotation. *Rice Sci.* **2018**, *25*, 340–349. [CrossRef]
162. Damon, P.M.; Bowden, B.; Rose, T.; Rengel, Z. Crop residue contributions to phosphorus pools in agricultural soils: A review. *Soil Biol. Biochem.* **2014**, *74*, 127–137. [CrossRef]
163. Almeida, D.S.; Menezes-Blackburn, D.; Rocha, K.F.; de Souza, M.; Zhang, H.; Haygarth, P.M.; Rosolem, C.A. Can tropical grasses grown as cover crops improve soil phosphorus availability? *Soil Use Manag.* **2018**, *34*, 316–325. [CrossRef]
164. Almeida, D.S.; Menezes-Blackburn, D.; Zhang, H.; Haygarth, P.M.; Rosolema, C.A. Phosphorus availability and dynamics in soil affected by long-term ruzigrass cover crop. *Geoderma* **2019**, *337*, 434–443. [CrossRef]
165. de Queiroz Cunha, E.; Stone, L.F.; Moreira, J.A.A.; de Brito Ferreira, E.P.; Didonet, A.D.; Leandro, W.M. Soil tillage systems and cover crops in organic production of common bean and corn. I-soil physical properties. *Rev. Bras. Ciênc. Solo* **2011**, *35*, 589–602. [CrossRef]
166. Wanic, M.; Żuk-Gołaszewska, K.; Orzech, K. Catch crops and the soil environment—A review of the literature. *J. Elem.* **2019**, *24*, 31–45. [CrossRef]
167. Sharma, V.; Irmak, S.; Padhi, J. Effects of cover crops on soil quality: Part, I. Soil chemical properties-organic carbon, total nitrogen, pH, electrical conductivity, organic matter content, nitrate-nitrogen, and phosphorus. *J. Soil Water Conserv.* **2018**, *73*, 637–651. [CrossRef]
168. Moreti, D.; Alves, M.C.; Filho, W.V.V.; de Passos e Carvalho, M. Soil chemical attributes of a red latosol under different systems of preparation, management, and covering plants. *Rev. Bras. Ciênc. Solo* **2007**, *31*, 167–175. [CrossRef]
169. Breil, N.L.; Lamaze, T.; Bustillo, V.; Marcato-Romain, C.-E.; Coudert, B.; Queguiner, S.; Jarosz-Pellé, N. Combined impact of no-tillage and cover crops on soil carbon stocks and fluxes in maize crops. *Soil Tillage Res.* **2023**, *233*, 105782. [CrossRef]
170. Matichenkov, V.; Bocharnikova, E.; Campbell, J. Reduction in nutrient leaching from sandy soils by Si-rich materials: Laboratory, greenhouse and filed studies. *Soil Tillage Res.* **2020**, *196*, 104450. [CrossRef]
171. Bashagaluke, J.B.; Logah, V.; Opoku, A.; Sarkodie-Addo, J.; Quansah, C. Soil nutrient loss through erosion: Impact of different cropping systems and soil amendments in Ghana. *PLoS ONE* **2018**, *19*, e0208250. [CrossRef]
172. Tilman, D.; Cassman, K.G.; Matson, P.A.; Naylor, R.; Polasky, S. Agricultural sustainability and intensive production practices. *Nature* **2002**, *418*, 671–677. [CrossRef]
173. Böldt, M.; Taube, F.; Vogeler, I.; Reinsch, T.; Kluß, C.; Loges, R. Evaluating different catch crop strategies for closing the nitrogen cycle in cropping systems-field experiments and modelling. *Sustainability* **2021**, *13*, 394. [CrossRef]
174. Iqbal, R.; Raza, M.A.S.; Valipour, M.; Saleem, M.F.; Zaheer, M.S.; Ahmad, S.; Toleikiene, M.; Haider, I.; Aslam, M.U.; Nazar, M.A. Potential agricultural and environmental benefits of mulches—A review. *Bull. Natl. Res. Cent.* **2020**, *44*, 1–16. [CrossRef]
175. Mazarei, M.; Ahangar, A.G. The effects of tillage and geographic factors on soil erosion: A review. *Int. J. Agric. Crop Sci.* **2013**, *6*, 1024–1031.
176. Akplo, T.M.; Alladassi, F.K.; Zoundji, M.C.C.; Avakoudjo, J.; Houngnandan, P.; Dagbénonbakin, D.G.; Saïdou, A.; Benmansour, M.; Fulajtar, E.; Amadji, G.L.; et al. Impact of conservation tillage on runoff, soil loss, and soil properties on acrisols and ferralsols in central Benin. *Can. J. Soil Sci.* **2022**, *102*, 659–671. [CrossRef]
177. Korkanç, S.Y.; Şahin, H. The effects of mulching with organic materials on the soil nutrient and carbon transport by runoff under simulated rainfall conditions. *J. Afr. Earth Sci.* **2021**, *176*, 104152. [CrossRef]
178. Pittelkow, C.; Liang, X.; Linquist, B.; van Groenigen, K.; Lee, J.; Lundy, M.; van Gestel, N.; Six, J.; Venterea, R.; Kessel, C. Productivity limits and potentials of the principles of conservation agriculture. *Nature* **2015**, *517*, 365–368. [CrossRef] [PubMed]

179. Jug, D.; Jug, I.; Vukadinović, V.; Đurđević, B.; Stipešević, B.; Brozović, B. *Conservation Soil Tillage as a Measure for Climate Change Mitigation*; University Textbook; Croatian Soil Tillage Research Organization: Osijek, Croatia, 2017; p. 176, ISBN 978-953-7871-61-1. (In Croatian)

180. Jug, I.; Brozović, B.; Đurđević, B.; Wilczewski, E.; Vukadinović, V.; Stipešević, B.; Jug, D. Response of crops to conservation tillage and nitrogen fertilization under different agroecological conditions. *Agronomy* **2021**, *11*, 2156. Available online: https: //www.mdpi.com/2073-4395/11/11/2156 (accessed on 13 May 2024). [CrossRef]

181. Jug, I.; Đurđević, B.; Vukadinović, V.; Jug, D.; Brozović, B. Optimization of nitrogen fertilization in sustainable agriculture practices. *Glas. Zaštite Bilja* **2018**, *41*, 3. [CrossRef]

182. Van Den Bossche, A.; De Bolle, A.; De Neve, S.; Hofman, G. Effect of tillage intensity on N mineralization of different crop residues in a temperate climate. *Soil Tillage Res.* **2009**, *103*, 316–324. [CrossRef]

183. Paramesh, V.; Mohan Kumar, R.; Rajanna, G.A.; Gowda, S.; Nath, A.J.; Madival, Y.; Jinger, D.; Bhat, S.; Toraskar, S. Integrated nutrient management for improving crop yields, soil properties, and reducing greenhouse gas emissions. *Front. Sustain. Food Syst.* **2023**, *7*, 1173258. [CrossRef]

184. Hirel, B.; Tetu, T.; Lea, P.J.; Dubois, F. Improving nitrogen use efficiency in crops for sustainable agriculture. *Sustainability* **2011**, *3*, 1452–1485. [CrossRef]

185. Habbib, H.; Verzeaux, J.; Nivelle, E.; Roger, D.; Lacoux, J.; Catterou, M.; Hirel, B.; Dubois, F.; Tetu, T. Conversion to no-till improves maize nitrogen use efficiency in a continuous cover cropping system. *PLoS ONE* **2016**, *11*, e0164234. [CrossRef]

186. Habbib, H.; Hirel, B.; Verzeaux, J.; Roger, D.; Lacoux, J.; Lea, P.; Dubois, F.; Tétu, T. Investigating the combined effect of tillage, nitrogen fertilization and cover crops on nitrogen use efficiency in winter wheat. *Agronomy* **2017**, *7*, 66. [CrossRef]

187. Cameron, K.; Di, H.; Moir, J. Nitrogen losses from the soil/plant system: A review. *Ann. Appl. Biol.* **2013**, *162*, 145–173. [CrossRef]

188. Redin, M.; Recous, S.; Sita, C.; Dietrich, G.; Skolaude, A.C.; Ludke, W.H.; Schmatz, R.; Giacomini, S.J. How the chemical composition and heterogeneity of crop residue mixtures decomposing at the soil surface affects C and N mineralization. *Soil Biol. Biochem.* **2014**, *78*, 65–75. [CrossRef]

189. De Baets, S.; Poesen, J.; Meersmans, J.; Serlet, L. Cover crops and their erosion-reducing effects during concentrated flow erosion. *Catena* **2011**, *85*, 237–244. [CrossRef]

190. Thapa, R.; Mirsky, S.B.; Tully, K.L. Cover crops reduce nitrate leaching in agroecosystems: A global meta-analysis. *J. Environ. Qual.* **2018**, *47*, 1400–1411. [CrossRef] [PubMed]

191. Li, L.; Konkel, J.; Jin, V.L.; Schaeffer, S.M. Conservation management improves agroecosystem function and resilience of soil nitrogen cycling in response to seasonal changes in climate. *Sci. Total Environ.* **2021**, *779*, 146457. [CrossRef] [PubMed]

192. Quintarelli, V.; Radicetti, E.; Allevato, E.; Stazi, S.R.; Haider, G.; Abideen, Z.; Bibi, S.; Jamal, A.; Mancinelli, R. Cover Crops for Sustainable Cropping Systems: A Review. *Agriculture* **2022**, *12*, 2076. Available online: https://www.mdpi.com/2077-0472/12/1 2/2076 (accessed on 21 June 2024). [CrossRef]

193. Pereira, B.I.; Aparecida, C.F.L.; Guilherme de Araujo, A. Nitrate leaching and soil tillage practices: Global vs. Brazilian research trends for 2001–2011 and 2012–2022. *Pesqui. Agropecu. Trop.* **2023**, *53*, e76778. [CrossRef]

194. Issaka, F.; Zhang, Z.; Zhao, Z.Q.; Asenso, E.; Li, J.H.; Li, Y.T.; Wang, J.J. Sustainable Conservation Tillage Improves Soil Nutrients and Reduces Nitrogen and Phosphorous Losses in Maize Farmland in Southern China. *Sustainability* **2019**, *11*, 2397. [CrossRef]

195. Talgre, L.; Lauringson, E.; Makke, A.; Lauk, R. Biomass production and nutrient binding of catch crops. *Zemdirbyste* **2011**, *98*, 251–258.

196. Gollner, G.; Fohrafellner, J.; Friedel, J.K. Winter-hardy vs. freeze-killed cover crop mixtures before maize in an organic farming system with reduced soil cultivation. *Org. Agric.* **2020**, *10*, 5–11. [CrossRef]

197. Heuermann, D.; Gentsch, N.; Guggenberger, G.; Reinhold-Hurek, B.; Schweneker, D.; Feuerstein, U.; Heuermann, M.C.; Groß, J.; Kümmerer, R.; Bauer, B.; et al. Catch crop mixtures have higher potential for nutrient carry-over than pure stands under changing environments. *Eur. J. Agron.* **2022**, *136*, 126504. [CrossRef]

198. Hawkesford, M.; Horst, W.; Kichey, T.; Lambers, H.; Schjoerring, J.; Miller, I.S.; White, P. Functions of macronutrients. In *Marschner's Mineral Nutrition of Higher Plants*; Marschner, P., Ed.; Academic Press: London, UK, 2012; pp. 135–189.

199. Rekha, P.N.; Kanwar, R.S.; Nayak, A.K.; Hoang, C.K.; Pederson, C.H. Nitrate leaching to shallow groundwater systems from agricultural fields with different management practices. *J. Environ. Monit.* **2011**, *13*, 2550. [CrossRef]

200. Kraska, P.; Mielniczuk, E. The occurrence of fungi on the stem base and roots of spring wheat (*Triticum aestivum* L.) crown in monoculture depending on tillage systems and catch crops. *Acta Agrobot.* **2012**, *65*, 79–90. [CrossRef]

201. Mielniczuk, E.; Patkowska, E.; Jamiołkowska, A. The influence of catch crops on fungal diversity in the soil and health of oat. *Plant Soil Environ.* **2020**, *66*, 99–104. [CrossRef]

202. Wojciechowski, W.; Szałata, M.; Lehmann, A. Stubble catch crops cultivated in accordance with the principles of agri-environmental scheme "Soil and water protection" as a plant-health factor in spring wheat. *Progr. Plant Prot.* **2015**, *55*, 211–215. [CrossRef]

203. Kunellus, H.T.; Johnston, H.W.; Macleod, J.A. Effect of undersowing barley with Italian ryegrass or red clover on yield, crop composition and root biomass. *Agric. Ecosyst. Environ.* **1992**, *38*, 127–137. [CrossRef]

204. Jensen, B.; Munk, L. Nitrogen induced changes in colony density and spore production of Erysiphe *graminis* f.sp. hordei on seedlings of six spring barley cultivars. *Plant Pathol.* **1997**, *46*, 191–202. [CrossRef]

205. Lemańczyk, G.; Wilczewski, E. Effect of catch crop and cultivation intensity on the health of spring barley. *Acta Sci. Pol. Agric.* **2019**, *18*, 119–132. [CrossRef]
206. Wanic, M.; Majchrzak, B.; Nowicki, J.; Waleryś, Z.; Orzech, K. Role of undersown catch crops and crop rotation in state of health of spring barley. *Acta Sci. Pol. Agric.* **2012**, *11*, 113–124.
207. Kraska, P.; Andruszczak, S.; Kwiecińska-Poppe, E.; Pałys, E. Tillage systems and catch crops as factors determining weed infestation level in a spring wheat canopy (*Triticum aestivum* L.) sown in monoculture. *Acta Sci. Pol. Agric.* **2014**, *13*, 33–50.
208. Majchrzak, L.; Piechota, T. Wpływ technologii uprawy na zachwaszczenie pszenicy jarej. *Fragm. Agron.* **2014**, *31*, 94–101.
209. Carrera, L.M.; Abdul-Baki, A.A.; John, R.; Teasdale, J.R. Cover crop management and weed suppression in no-tillage sweet corn production. *Hort. Sci.* **2004**, *39*, 1262–1266. [CrossRef]
210. Darby, H.; Malone, R.; Krezinski, I. Impact of cover crops on no-till spring grain production. *Northwest Crops Soils Program* **2019**, *351*. Available online: https://scholarworks.uvm.edu/nwcsp/351 (accessed on 21 June 2024).
211. Majchrzak, L. *Influence of White Mustard Cover Crop and Method of Tillage on Soil Properties, Growth and Yield of Spring Wheat*; Monograph of Poznań University of Life Sciences: Poznan, Poland, 2015; Volume 480, p. 113. (In Polish)
212. Wittwer, R.A.; Dorn, B.; Marcel, W.J.; van der Heijden, M.G.A. Cover crops support ecological intensification of arable cropping systems. *Sci. Rep.* **2017**, *7*, 41911. [CrossRef] [PubMed]
213. Wittwer, R.A.; van der Heijden, M.G.A. Cover crops as a tool to reduce reliance on intensive tillage and nitrogen fertilization in conventional arable cropping systems. *Field Crops Res.* **2020**, *249*, 107736. [CrossRef]
214. Clark, K.; Boardman, D.; Staples, J.S.; Easterby, S.; Reinbott, T.M.; Kremer, R.J.; Kitchen, N.R.; Veum, K.S. Crop yield and soil organic carbon in conventional and no-till organic systems on a claypan soil. *Agron. J.* **2017**, *109*, 588–599. [CrossRef]
215. Lyon, D.J.; Blumenthal, J.M.; Burgener, P.A.; Harveson, R.M. Eliminating summer fallow reduces winter wheat yields, but not necessarily system profitability. *Crop Sci.* **2004**, *44*, 855–860. [CrossRef]
216. Nielsen, D.C.; Vigil, M.F. Legume green fallow effect on soil water content at wheat planting and wheat yield. *Agron. J.* **2005**, *97*, 684–689. [CrossRef]
217. Toom, M.; Tamm, S.; Talgre, L.; Tamm, I.; Tamm, Ü.; Narits, L.; Hiiesalu, I.; Mäe, A.; Lauringson, E. The Effect of Cover Crops on the Yield of Spring Barley in Estonia. *Agriculture* **2019**, *9*, 172. [CrossRef]
218. Yang, Y.; Ding, J.; Zhang, Y.; Wu, J.; Zhang, J.; Pan, X.; Gao, C.; Wang, Y.; He, F. Effects of tillage and mulching measures on soil moisture and temperature, photosynthetic characteristics and yield of winter wheat. *Agric. Water Manag.* **2018**, *201*, 299–308. [CrossRef]

Article

Effects of Different Straw Returning Periods and Nitrogen Fertilizer Combinations on Rice Roots and Yield in Saline–Sodic Soil

Yaoru Xie [1,†], Xiuli Zhang [1,†], Ya Gao [1], Jiaquan Li [1], Yanqiu Geng [2], Liying Guo [2], Xiwen Shao [2] and Cheng Ran [1,2,*]

[1] Heyuan Branch, Guangdong Laboratory for Lingnan Modern Agriculture, Heyuan 517000, China; yaoruxie@iclould.com (Y.X.); 18336338339@163.com (X.Z.); gy_elegance@163.com (Y.G.); dtsys_lijiaquan@163.com (J.L.)

[2] Agronomy College, Jilin Agricultural University, Changchun 130118, China; wangyibeiyongyx@163.com (Y.G.); tougaobeiyongyx@163.com (L.G.); shaoxiwen@126.com (X.S.)

* Correspondence: jlndrc@126.com

† These authors contributed equally to this work.

Abstract: Straw return is an effective management practice for improving physical and chemical properties of saline–sodic soil in Northeast China. Straw decomposition and nutrient release are deeply influenced by soil and climatic factors. In Northeast China, straw decomposes slowly due to the long winter with low temperatures. Therefore, the season of straw return may be a key issue affecting rice. However, the impact of returning straw in different seasons on rice is disregarded and not commonly researched. We conducted a 2-year field experiment, including two residue management treatments: spring straw return treatment (SR) and autumn straw return treatment (AR), each containing five different N rates (0, 90, 180, 270, and 360 kg ha^{-1}) as sub-treatments. The results reveal that, compared with the spring straw returning treatment, the autumn straw returning treatment significantly improved root morphology and root vigor and increased the number of spikes per unit area, which directly increased rice yield by 4.76% (2020) and 6.62% (2021). In addition, rice yield showed an increasing and then decreasing trend with the increase in N fertilizer application, and it was at its maximum when the N application rate was 270 kg ha^{-1}. Compared to the spring straw return treatment, the autumn straw return treatment was able to reduce 31.46% (2020) and 38.48% (2021) of N fertilizer application without decreasing rice yield. Our findings demonstrate that straw return combined with nitrogen fertilization may be a promising management practice for improving rice root systems and yield in saline–sodic soils, and under the conditions of the autumn straw returning treatment, the best nitrogen fertilizer application rate was 270 kg ha^{-1}.

Keywords: saline–sodic rice area; straw return period; nitrogen fertilizer; soil physicochemical properties; rice root system

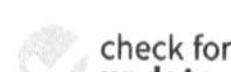

check for **updates**

Citation: Xie, Y.; Zhang, X.; Gao, Y.; Li, J.; Geng, Y.; Guo, L.; Shao, X.; Ran, C. Effects of Different Straw Returning Periods and Nitrogen Fertilizer Combinations on Rice Roots and Yield in Saline–Sodic Soil. *Agronomy* **2024**, *14*, 2463. https://doi.org/10.3390/agronomy14112463

Academic Editors: Mariola Staniak, Ewa Szpunar-Krok and Małgorzata Szostek

Received: 28 August 2024
Revised: 13 October 2024
Accepted: 17 October 2024
Published: 22 October 2024

1. Introduction

As an important land reserve resource for food production, saline–alkaline farmlands play an important role in ensuring national food security at a time when agricultural arable land is decreasing due to high urbanization [1]. China is one of the countries with the most serious soil salinization. The total area of saline–alkaline land in China is about 1.0×10^9 hectares, accounting for 10% of the world's saline–alkaline land area [2]. The salinized area on the west side of the Songnen Plain is about 3.73×10^6 hectares, which is one of the three largest concentrated distribution areas of soda saline–alkaline land in the world [3]. Saline–sodic soil exhibits high levels of soluble salts with a composition distinct from other types of saline–alkaline soils, predominantly comprising $NaHCO_3$ and Na_2CO_3, and typically has a soil pH above 8.5 [4]. Studies have shown that high pH can inhibit protein synthesis and produce cytotoxicity, leading to direct toxicity to plants [5]. In addition, high pH can reduce the effectiveness of phosphorus and impair nutrient uptake by rice roots [6]. Therefore, appropriate agronomic measures need to be taken to improve soil salinity, reduce damage to rice roots, and improve soil properties to increase the production potential of saline–sodic paddy fields.

Previous research has always focused on water, crops, chemical amendment, electric current, and tillage as prima amelioration tools for saline–alkaline soils [7]. Most of these methods are not suitable for large-scale promotion due to high costs, complex operations, and low acceptance among farmers. Recent studies have shown that straw return application to soil is more utilizable and sustainable [8,9]. Returning straw to the fields can not only solve the problem of resource waste and environmental pollution caused by agricultural wastes but can also significantly change the physical, chemical, and biological properties of the improved soil, thereby improving soil quality and crop yields [10,11], especially when applied together with nitrogen fertilizers [12]. Long-term straw return can significantly reduce soil bulk density, increase soil aggregate structure, increase soil total porosity and soil O^2 content, reduce N_2O emissions, improve soil ventilation and drainage capacity, and promote root growth in deep soil layers [13].

Nitrogen can serve both as a nutrient and for osmotic adjustment under saline–alkaline conditions, effectively mitigating the harm caused by salt and alkali stress to crops [14]. The effects of nitrogen application on root morphology, vigor, and distribution are significant and complex, and nitrogen use is well known for promoting root growth and downward rooting depth in the soil layer. Root length and number in mid-N treatment increased by 29.0% and 85.0%, compared with high-N treatment [15]. This also suggests that there is an optimal amount of nitrogen fertilizer application for root growth. However, in the saline–sodic rice area, farmers usually apply a large amount of nitrogen fertilizer in pursuit of high yields [3]. Chronic over-application of nitrogen fertilizer will cause a decline in soil quality (i.e., loss of soil organic matter, decreased soil fertility and inefficient nutrient use) and increase environmental pollution. [16,17].

Related studies have shown that under straw return in spring conditions, the rapid decomposition period of straw coincides with the rice rejuvenation and tillering stage, a process where significant amounts of nitrogen are absorbed and substantial toxic gases are produced, adversely impacting rice root growth and subsequently influencing the growth and development of rice [18,19]. However, the effects of seasonal differences in straw return on rice roots and yield in saline rice fields are unknown. Therefore, this study focuses on saline–sodic rice fields, examining the impacts of various straw return periods and nitrogen fertilization on rice root characteristics and yield under the conditions of full straw return. The aim is to identify the best straw return period and the most effective nitrogen application rate in these areas, thereby offering a theoretical foundation for enhancing the productivity of these fields and expanding the practice of rice straw return on a large scale in saline–sodic regions.

2. Materials and Methods

2.1. Study Site

The field experiment was carried out at Yixin Family Farm, Shili Town, Baicheng City, Jilin Province, China, from October 2019 to October 2021. Situated in the southwest of the Songnen Plain, Shili Town is a quintessential example of an area with moderate-to-severe saline–sodic soil. The average annual precipitation and evaporation are 413.7 and 1696.9 mm, respectively. The annual average sunshine duration, effective cumulative temperature, and frost-free period are 2996.2 h, 4.7 °C, and 144 days, respectively. Information on average precipitation and temperature during the experiment period is shown in Figure 1.

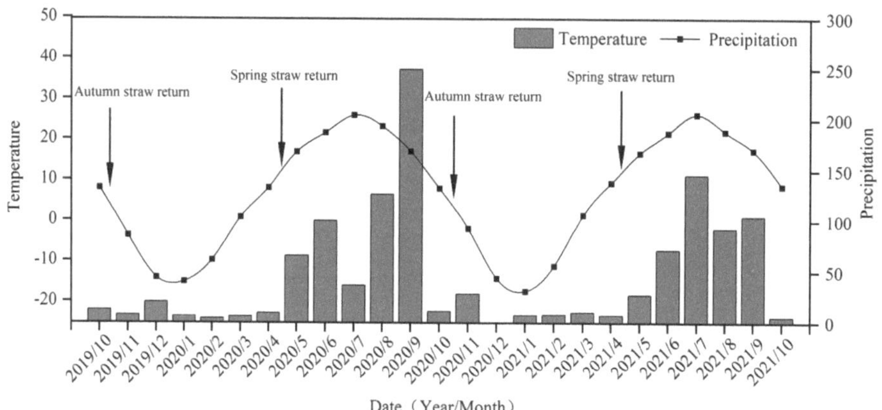

Figure 1. Monthly average temperature (°C) and monthly total precipitation (mm) from October 2019 to October 2021.

Prior to the experiment in October 2019, soil was collected from a depth of 0–20 cm to assess its physical and chemical properties. Table 1 shows the basic physical and chemical properties of the soil. At the experiment site, rice had been cultivated for five consecutive years. In non-experimental years, the rice straw produced was burned on-site before the spring ploughing in the second year. The tested variety was the local large-scale cultivar Baijing 1 (*Oryza saliva* subsp *keng*), which is characterized by its salt and alkali tolerance. The straw utilized in the experiment was harvested from the rice fields of the experimental site during the autumn, and its nutrient composition is detailed in Table 2.

Table 1. Basic physical and chemical properties of the tested soil.

Parameter	Mean	Parameter	Mean
BD (g cm^{-3})	1.52	Soil pH	8.91
EC$_e$ (dS m^{-1})	12.58	ENa$^+$ (cmol$_c$ kg^{-1})	5.34
ESP (%)	37.87	SOM (g kg^{-1})	6.08
Total N (g kg^{-1})	0.18	Available P (mg kg^{-1})	8.84
CEC (cmol$_c$ kg^{-1})	14.11	Available K (mg kg^{-1})	102.34

Note: BD: bulk density; EC$_e$: soil electrical conductivity of saturated paste extraction; ENa$^+$: exchangeable sodium; CEC: cation exchange capacity; ESP: exchangeable sodium percentage; SOM: soil organic matter.

Table 2. Nutrient content of tested straw.

Year	Total C (%)	Total N (mg g^{-1})	Total P (mg g^{-1})	Total K (mg g^{-1})	C/N Ratio	Cellulose (mg g^{-1})	Hemicellulose (mg g^{-1})	Lignin (mg g^{-1})
2019	38.09	4.56	1.72	8.09	83.40	366.07	265.57	54.95
2020	37.67	4.73	1.66	8.13	79.57	378.66	272.69	58.63

2.2. Experimental Design

The experiment was conducted using a split plot design, with the straw returning period as the main plot and nitrogen fertilizer as the secondary plot. The experiment was carried out continuously on the same plot for two years. The straw returning periods were spring straw returning (SR) and autumn straw returning (AR), and the five nitrogen fertilizer levels were 0 (N0), 90 (N90), 180 (N180), 270 (N270), and 360 (N360) kg ha^{-1}, for a total of 10 treatments with three replicates. The plot size was 30 m^2 (6 × 5 m). To prevent nutrient or water exchange between plots, the plots were separated by field ridges (0.6 m wide and 0.4 m high), and each plot had independent irrigation and drainage outlets. The nitrogen fertilizer for each treatment was applied according to a base fertilizer–tillering fertilizer–ear fertilizer ratio of 6:3:1. Phosphorus fertilizer (P$_2$O$_5$) and zinc fertilizer (ZnSO$_4$)

were applied as base fertilizer once, with application rates of 50 kg ha^{-1} and 20 kg ha^{-1}, respectively. Potassium fertilizer (K$_2$O) was applied at 45 and 30 kg ha^{-1} as a base fertilizer and ear fertilizer, respectively. The straw used was produced during the previous rice planting season. The straw was collected manually from the experimental field. After the straw was air-dried under natural conditions, it was cut into 5–7 cm long pieces with a straw chopper. The spring straw return treatment (SR) was carried out by rotary tillage in mid-April each year, and the straw was evenly spread on the soil surface together with basal fertilizer before being mixed into the soil with a reverse stubble rotary tiller. In the plants treated by autumn straw return (AR), all rice straw was applied after the previous year's harvest (mid to late October), and the same field was then tilled with a reverse stubble rotary tiller. All treatments were plowed again on May 9 of the following year, and base fertilizer was applied at the same time. The amount of straw returned to the field was converted to 8 t ha^{-1} based on the local rice yield and a rice-to-straw ratio of 1:1.1.

2.3. Plant Sampling Collection

At the tillering stage (MT), panicle initiation stage (PI), heading stage (HD), filling stage (FI), and physiological maturity stage (PM), nine rice samples were collected from each plot. Of these, three samples were designated for the assessment of root morphological traits and root-to-shoot ratio, while the remaining six were reserved for the measurement of root physiological traits. A sampler was employed to extract a cubic soil block (30.0 cm in length, 16.5 cm in width, and 30.0 cm in depth) encompassing each rice plant. This soil block contained about 95% of the total rice root biomass [20]. The roots in each soil block were then carefully rinsed with distilled water.

2.4. Morphological and Physiological Characteristics Analysis

The roots were evenly dispersed on a glass dish filled with shallow water and positioned optimally before being scanned using an Epson scanner (V850, Seiko Epson Corporation, Suwa, Nagano, Japan). The captured images were then processed and analyzed using WinRHIZO (2021a) software to derive metrics such as total root length (RL), total root surface area (RSA), and total root volume (RV). Subsequently, to ascertain the dry weight (RDW) of both the above-ground portion and the roots, the root samples were gathered following the scanning process. Both the above-ground and root samples were then blanched at 105 °C for half an hour and subsequently dried to a constant weight at 80 °C. Finally, the root-to-shoot ratio was computed based on the weights of the root and above-ground components.

The total root absorption surface area (RTA) and the root active absorption surface area (RAA) were measured using the methylene blue method. Following the protocol outlined by Yang et al., three rice plants were selected from each plot, based on the average number of tillers, for the collection of root bleeding sap [21]. This process was carried out as follows. At 18:00, during the tillering, panicle initiation, heading, filling, and physiological maturity stages, the rice plants were severed at an internode approximately 12 cm above the soil surface. A pre-weighed glass tube and absorbent cotton were then placed adjacent to the incision on the field stem, and the entire setup was wrapped in plastic film to prevent water infiltration. The absorbent cotton and glass tube were collected and re-weighed the following morning at 6:00. The difference in weight represented the root bleeding sap of each growth stage, expressed as the concentration per hour (mg h^{-1} plant^{-1}) per plant.

2.5. Nitrogen Use Efficiency and Rice Yield

During the mature stage, three rice plants with uniform growth were collected from each plot. The above-ground portion of the rice was subsequently divided into leaves, stem sheaths, and spikes. Samples were heated at 105 °C for half an hour to deactivate enzymes, followed by drying at 80 °C until a constant weight was achieved. Afterward, the samples were weighed and grounded. To ascertain the total nitrogen absorption, the above-ground samples from each part were sifted through a 0.5 mm sieve. The nitrogen content in various rice organs was determined using the Kjeldahl method through a Kjeldahl nitrogen analyzer (FOSS-8400) [22]. The nitrogen absorption of the plant was calculated based on the derived nitrogen concentration of rice and the weights of the different parts. The agronomic efficiency of nitrogen fertilizer (AEN), partial factor productivity of nitrogen fertilizer (PFP), and apparent utilization rate of nitrogen fertilizer (REN) were computed according to the methodology outlined by Wang et al. The corresponding calculation formulas are detailed as follows [23].

$$\text{AEN}\left(\text{kg kg}^{-1}\right) = \frac{Y - Y_0}{F} \tag{1}$$

$$\text{PFP}\left(\text{kg kg}^{-1}\right) = \frac{Y}{F} \tag{2}$$

$$\text{REN} = \frac{\text{TPN} - \text{TP0}}{F} \tag{3}$$

TPN: total nitrogen absorption of rice plants in the nitrogen application area; TP0: total nitrogen absorption of rice plants under nitrogen-free conditions; F: nitrogen application rate; Y: rice yield under nitrogen application conditions; Y_0: rice yield under nitrogen-free conditions.

2.6. Statistical Analysis

All the data were analyzed using SPSS 22.0 software (SPSS Inc., Chicago, IL, USA). The LSD method was employed to assess the significance of differences in the data across treatments, with a significance level set at $p < 0.05$. Additionally, Sigmaplot 14.0 software was utilized to generate graphical representations. All numerical values presented in the charts represent the mean $\pm$ standard error.

3. Results

3.1. Morphological Characteristics of the Rice Root System

Root length and root surface area initially increased and then decreased throughout the rice growth process, peaking in the heading phase and decreasing thereafter (Table 3). The application of nitrogen fertilizer significantly enhanced both root length and root surface area. However, no significant difference was observed between the N270 and N360 treatments during the same straw returning period (Table 3). It is indicated that excessive nitrogen application did not lead to a notable increase in total root length or root surface area. Furthermore, at equivalent nitrogen levels, the autumn straw returning treatments exhibited higher root length and root surface area than the spring straw returning treatments. This observation was indirectly corroborated by the field growth and root length of rice during the tillering stage (Figure 2). In comparison to spring straw returning, autumn straw returning resulted in significant increases in rice root length by 30.6%, 23.8%, 14.4%, 11.3%, and 13.5% (two-year average) during the tillering, panicle initiation, heading, filling, and physiological maturity stages, respectively. Similarly, the root surface area increased significantly by 14.5%, 20.7%, 28.5%, 25.8%, and 12.9% (two-year average) across the same stages.

As depicted in Table 4, the dry weight of rice roots exhibited a pattern of initial increase followed by a decline throughout the growth cycle, peaking at the heading stage. When comparing the same straw returning period, the application of nitrogen fertilizer resulted in significant enhancement in the dry weight of rice roots across all stages. Specifically, the dry weight of rice roots increased significantly with increasing nitrogen application throughout the life span of rice. However, there was no significant difference between the N270 and N360 treatments, indicating that excessive nitrogen application did not promote rice root growth under straw returning conditions. Furthermore, when comparing autumn straw returning with spring straw returning, the dry weight of rice roots increased significantly by 11.31%, 9.82%, 10.03%, and 10.26% on average over a two-year period during the tillering stage, panicle initiation stage, heading stage, and filling stage, respectively. It is noteworthy that the ratio of root to shoot in rice gradually decreased as the growth cycle progressed. The application of nitrogen fertilizer significantly reduced this ratio at each growth stage.

As the growth process of rice progressed (Table 4), the root volume reached a peak at the heading stage and then decreased. In the same straw returning period, the application of nitrogen fertilizer significantly increased the root volume of rice at various growth stages, but there was no significant difference between N270 and N360. Compared with SR, the root volume achieved with the autumn straw returning treatment at the tillering stage, panicle initiation stage, heading stage, and filling stage significantly increased by 9.4%, 13.9%, 13.5%, and 15.0% (two-year average), respectively.

Figure 2. Root length and field expression of rice in the spring straw return and autumn straw return rice fields at the tillering stage. Note: SR: straw returned to the field in spring; AR: straw returned to the field in autumn; N0: nitrogen application of 0 kg ha^{-1}; N180: nitrogen application of 180 kg ha^{-1}; N360: nitrogen application of 360 kg ha^{-1}.

Table 3. Effect of straw return period and nitrogen fertilizer application on the root length and root surface area of rice.

	Treatment		Root Length (km m^{-2})					Root Surface Area (m^2 m^{-2})				
			MT	PI	HD	FI	PM	MT	PI	HD	FI	PM
2020	SR	N0	0.46 c	1.05 c	2.16 c	2.00 d	1.40 c	0.96 c	1.89 d	3.15 d	2.83 d	2.14 d
		N90	0.51 c	1.18 bc	2.51 b	2.39 c	1.56 bc	1.05 bc	2.44 c	3.96 c	3.79 c	2.37 c
		N180	0.65 b	1.35 b	2.62 b	2.53 b	1.63 ab	1.15 b	2.85 b	4.19 b	3.91 b	2.84 b
		N270	0.78 a	1.68 a	2.98 a	2.89 a	1.72 a	1.27 a	3.51 a	4.60 a	4.48 a	3.15 a
		N360	0.80 a	1.76 a	3.05 a	2.91 a	1.79 a	1.33 a	3.66 a	4.72 a	4.57 a	3.26 a
	AR	N0	0.66 d	1.25 c	2.46 c	2.37 d	1.61 c	1.18 c	2.21 d	4.52 d	4.28 e	2.52 d
		N90	0.79 c	1.42 c	2.95 b	2.58 c	1.70 c	1.25 bc	2.71 c	5.37 c	4.97 d	3.12 c
		N180	0.86 b	1.79 b	3.06 b	2.85 b	1.77 bc	1.34 ab	3.30 b	5.93 b	5.48 c	3.25 bc
		N270	0.95 a	2.07 a	3.58 a	3.40 a	1.91 ab	1.45 a	4.41 a	6.49 a	6.30 a	3.44 ab
		N360	0.91 ab	2.05 a	3.49 a	3.36 a	1.97 a	1.40 a	4.33 a	6.31 a	5.95 b	3.64 a
2021	SR	N0	0.52 c	1.15 c	2.49 c	2.30 d	1.43 c	1.07 c	2.09 d	4.21 d	3.95 d	2.57 d
		N90	0.58 bc	1.21 c	2.86 b	2.65 c	1.63 b	1.13 c	2.61 c	4.68 c	4.55 c	2.72 c
		N180	0.68 b	1.49 b	2.98 b	2.88 b	1.73 ab	1.29 b	2.94 b	5.19 b	4.89 b	3.11 b
		N270	0.84 a	1.71 a	3.45 a	3.12 a	1.87 a	1.37 ab	3.62 a	5.48 a	5.25 a	3.67 a
		N360	0.88 a	1.80 a	3.53 a	3.27 a	1.92 a	1.46 a	3.68 a	5.57 a	5.39 a	3.73 a
	AR	N0	0.78 d	1.32 d	2.80 d	2.62 c	1.88 b	1.30 b	2.53 d	4.69 d	4.26 d	2.66 d
		N90	0.85 c	1.56 c	3.13 c	2.70 c	1.90 b	1.37 b	2.92 c	5.53 c	5.09 c	3.20 c
		N180	0.92 b	1.83 b	3.50 b	3.11 b	1.95 ab	1.39 b	3.81 b	6.39 b	5.95 b	3.68 b
		N270	0.99 a	2.41 a	3.91 a	3.50 a	2.09 a	1.57 a	4.60 a	6.82 a	6.43 a	3.87 ab
		N360	0.98 a	2.32 a	3.82 a	3.27 a	2.10 a	1.53 ab	4.51 a	6.73 ab	6.12 ab	3.93 a
ANOVA	Y		**	**	**	*	*	*	**	**	*	**
	N		**	**	**	**	**	**	**	**	**	**
	T		*	**	*	*	*	**	**	**	*	*
	N × T		*	*	*	*	ns	*	*	*	*	ns
	Y × N		*	*	*	*	ns	*	*	*	*	ns
	T × Y		ns	ns	ns	ns	ns	ns	ns	ns	ns	ns
	N × T × Y		ns	ns	ns	ns	ns	ns	ns	ns	ns	ns

Note: The data in the table are the mean values of three replications. Different lowercase letters in the column under the same straw treatment in the same year indicate significant differences between treatments ($p < 0.05$). SR: straw returned to the field in spring; AR: straw returned to the field in autumn; MT: mid-tillering stage; PI: panicle initiation stage; HD: heading stage; FI: filling stage; PM: physiological maturity stage; Y: year; N: nitrogen fertilizer; T: straw returning period; *, ** mean $p < 0.05$, $p < 0.01$; ns means non-significant.

Table 4. Effects of straw return period and nitrogen fertilizer combination on root dry weight and root–shoot ratio of rice.

Treatment			Root Dry Weight (g m^{-2})					Root–Shoot Ratio (%)				
			MT	PI	HD	FI	PM	MT	PI	HD	FI	PM
2020	SR	N0	19.53 c	40.17 d	89.48 d	88.40 d	73.73 d	26.5 a	25.7 a	21.2 a	16.2 a	9.1 a
		N90	20.87 c	67.82 c	128.44 c	124.07 c	99.47 c	25.8 ab	24.8 ab	20.4 a	16.0 a	8.8 ab
		N180	28.47 b	79.27 b	138.57 b	137.55 b	111.87 b	24.1 b	23.2 ab	20.1 a	15.8 ab	8.7 ab
		N270	30.41 ab	89.66 a	155.24 a	158.51 a	139.73 a	24.1 b	23.1 ab	20.0 ab	15.2 ab	8.5 ab
		N360	31.08 a	92.65 a	160.57 a	160.25 a	142.47 a	23.0 b	22.8 b	19.5 b	14.6 b	8.3 b
	AR	N0	19.92 d	46.72 c	96.80 d	92.15 d	74.13 e	27.9 a	28.2 a	22.8 a	16.5 a	9.0 a
		N90	23.81 c	79.00 b	146.81 c	145.14 c	102.20 d	26.1 b	26.1 b	21.8 a	16.5 a	8.3 ab
		N180	32.12 b	84.86 b	154.73 b	161.73 b	114.07 c	24.9 bc	24.7 c	21.7 a	15.9 a	7.9 b
		N270	36.60 a	99.80 a	167.10 a	182.30 a	141.40 b	24.2 cd	24.3 c	20.4 ab	15.1 ab	7.8 b
		N360	36.81 a	108.72 a	179.21 a	187.82 a	155.13 a	23.1 d	23.1 c	19.1 b	14.2 b	7.5 b
2021	SR	N0	18.60 d	50.81 d	90.47 d	95.35 c	77.35 d	28.5 a	27.7 a	21.0 a	16.0 a	9.5 a
		N90	21.73 c	67.55 c	140.59 c	144.22 b	100.67 c	25.0 ab	24.9 ab	20.9 a	15.8 a	8.7 ab
		N180	30.62 b	80.95 b	158.72 b	160.53 ab	114.33 b	24.8 b	23.1 b	19.9 ab	15.2 a	8.5 ab
		N270	33.48 a	94.58 a	166.77 a	163.21 a	139.85 a	23.1 b	22.1 bc	18.9 ab	14.3 ab	8.2 ab
		N360	34.80 a	99.72 a	170.21 a	165.37 a	146.73 a	22.1 b	20.9 c	18.7 b	13.6 b	7.9 b
	AR	N0	20.00 d	54.53 d	97.40 d	98.40 d	78.47 d	28.5 a	30.0 a	22.0 a	17.7 a	9.4 a
		N90	24.60 c	75.60 c	147.60 c	153.20 c	103.60 c	25.1 b	24.5 b	21.4 a	17.3 ab	8.6 ab
		N180	32.47 b	89.00 b	169.80 b	166.20 b	117.73 b	25.1 b	24.5 b	21.3 a	16.2 b	7.9 bc
		N270	36.67 a	98.80 ab	187.27 a	172.93 ab	147.87 a	23.5 c	23.9 bc	20.4 ab	14.9 c	7.8 bc
		N360	37.07 a	101.13 a	192.47 a	180.73 a	150.80 a	22.7 c	21.1 c	19.4 b	12.8 d	7.0 c
ANOVA	Y		ns	ns	ns	ns	ns	ns	ns	ns	ns	ns
	N		**	**	**	**	**	*	*	*	*	*
	T		*	*	*	*	ns	ns	ns	ns	ns	ns
	N × T		*	*	*	*	ns	ns	ns	ns	ns	ns
	Y × N		ns	ns	ns	ns	ns	ns	ns	ns	ns	ns
	T × Y		ns	ns	ns	ns	ns	ns	ns	ns	ns	ns
	N × T × Y		ns	ns	ns	ns	ns	ns	ns	ns	ns	ns

Note: The data in the table are the mean values of three replications. Different lowercase letters in the column under the same straw treatment in the same year indicate significant differences between treatments ($p < 0.05$). SR: straw returned to the field in spring; AR: straw returned to the field in autumn; MT: mid-tillering stage; PI: panicle initiation stage; HD: heading stage; FI: filling stage; PM: physiological maturity stage; Y: year; N: nitrogen fertilizer; T: straw returning period; *, ** mean $p < 0.05$, $p < 0.01$; ns means non-significant.

3.2. Physiological Characteristics of Rice Root System

As the growth process of rice progresses, the root bleeding sap of rice increased first and then decreased, reaching its maximum at the heading stage (Table 5). Compared with spring straw returning, autumn straw returning significantly increased the root bleeding sap of rice at various growth stages, with significant increases of 17.4%, 16.2%, 7.3%, 11.6%, and 16.7% (two-year average) in the tillering stage, panicle initiation stage, heading stage, grain filling stage, and physiological maturity stage, respectively. The application of nitrogen fertilizer significantly increased the root bleeding sap of rice at various growth stages. The amount of rice root bleeding sap increased significantly with increasing nitrogen application throughout the life span of rice. However, there was no significant difference in the root bleeding sap of rice between the N270 and N360 treatments at the same straw returning period.

As rice grows, the total root absorption area and active absorption area first increased and then decreased, reaching a peak at the heading stage, and then gradually decreased (Table 6). In each growth stage of rice, at the same nitrogen level, the total absorption area and active absorption area of rice roots in the autumn straw returning treatment were higher than those in the spring straw returning treatment and were significantly higher than those in the spring straw returning treatment in the tillering stage, panicle initiation stage, heading stage, and filling stage. Compared with the spring straw returning treatment, the total absorption area of the root system in the autumn straw returning treatment increased significantly by 18.3%, 30.8%, 24.8%, and 9.9% (average of two years) in the tillering stage, panicle initiation stage, heading stage, and filling stage, respectively. The active absorption area of the root system in the autumn straw returning treatment increased significantly by 10.6%, 6.7%, 6.9%, and 5.9% (average of two years) in the tillering stage, panicle initiation stage, heading stage, and filling stage, respectively. In the context of identical straw returning periods, the application of nitrogen fertilizer significantly enhanced the total root absorption surface area and the active absorption area across various growth stages of rice. Specifically, under spring straw returning conditions, both the total root absorption surface area and the active absorption area progressively increased with the augmentation of nitrogen application rates. Conversely, under autumn straw returning conditions, there was an initial increase followed by a decrease in both parameters with the rise in nitrogen application rates, peaking at the N270 level. However, regardless of whether it was spring or autumn straw returning, no significant differences were observed in the total root absorption surface area and the active absorption area between the N270 and N360 levels across the growth stages of rice. In summary, the integration of autumn straw returning with nitrogen fertilizer application demonstrated a superior promotional effect on the root exudates, the total root absorption surface area, and the active absorption area in soda saline-alkaline paddy fields. However, it is essential to note that the dosage of nitrogen fertilizer should not be excessively high.

Table 5. Effect of straw return period and nitrogen fertilizer allocation on rice root volume and root bleeding sap.

Treatment			Root Volume ($cm^3\ m^{-2}$)					Root Bleeding Sap ($mg\ h^{-1}\ plant^{-1}$)				
			MT	PI	HD	FI	PM	MT	PI	HD	FI	PM
2020	SR	N0	115.51 d	314.50 d	631.24 d	600.08 d	338.60 d	13.83 d	44.47 c	90.25 c	85.77 d	15.69 b
		N90	124.70 c	394.53 c	728.11 c	686.98 c	360.23 c	17.19 c	50.47 bc	100.21 b	92.55 c	16.37 b
		N180	136.85 b	468.57 b	806.40 b	752.47 b	442.07 b	20.58 b	58.00 b	108.87 ab	100.24 b	17.33 ab
		N270	158.83 a	485.67 ab	845.77 a	817.57 a	493.67 a	26.88 a	67.99 a	112.29 a	104.63 ab	18.19 a
		N360	160.25 a	509.72 a	869.52 a	822.12 a	517.22 a	27.85 a	70.25 a	114.58 a	111.19 a	18.83 a
	AR	N0	136.53 b	366.93 d	741.93 d	720.73 d	360.20 d	15.38 d	50.25 c	96.86 d	97.38 d	17.19 b
		N90	144.33 b	433.93 c	800.86 c	789.13 c	389.46 c	20.72 c	62.80 b	108.27 c	107.22 c	18.63 ab
		N180	157.66 a	532.80 b	933.80 b	892.13 b	461.86 b	26.91 b	69.05 b	118.77 b	114.66 b	19.94 a
		N270	168.13 a	631.93 a	995.46 a	931.00 a	513.46 a	31.19 b	85.80 a	128.50 a	127.27 a	23.08 a
		N360	163.60 a	569.13 b	952.53 b	906.53 ab	518.53 a	30.11 ab	82.19 a	124.19 a	122.19 a	23.52 a
2021	SR	N0	120.44 d	349.83 d	680.51 d	649.10 d	376.71 c	16.74 c	46.98 d	99.17 d	92.42 d	16.36 c
		N90	138.53 c	428.07 c	781.55 c	721.25 c	390.81 c	19.99 bc	55.73 c	118.36 c	105.58 c	17.72 bc
		N180	155.22 b	493.66 b	813.48 b	793.60 b	447.55 b	22.95 b	66.84 b	122.37 bc	115.63 b	18.28 ab
		N270	162.73 a	553.73 ab	853.82 a	810.40 ab	498.35 a	26.67 a	76.57 a	127.22 ab	123.49 a	19.67 a
		N360	167.53 a	571.22 a	890.75 a	836.98 a	518.23 a	27.71 a	78.84 a	130.65 a	125.57 a	20.25 a
	AR	N0	143.20 c	397.26 e	772.40 d	753.93 c	390.20 c	18.72 c	52.88 d	108.05 c	105.44 c	18.05 c
		N90	149.73 c	467.06 d	831.40 c	806.93 b	411.13 c	24.19 b	65.58 c	122.02 b	115.30 b	19.02 bc
		N180	164.73 b	547.33 c	951.40 b	921.80 a	471.20 b	27.08 b	77.36 b	130.16 ab	125.44 a	20.27 b
		N270	178.06 a	647.20 a	1020.06 a	969.53 a	514.93 a	33.44 a	85.55 a	134.97 a	133.52 a	23.72 a
		N360	170.40 ab	610.20 b	967.86 b	920.06 a	520.93 a	30.86 a	84.72 a	133.61 a	130.77 a	25.05 a
ANOVA	Y		*	*	*	*	*	*	*	*	*	*
	N		**	**	**	**	*	**	**	**	**	*
	T		*	*	*	*	ns	*	*	*	*	*
	N × T		*	*	*	*	ns	*	*	*	*	*
	Y × N		*	*	*	*	ns	*	*	*	*	ns
	T × Y		ns	ns	ns	ns	ns	ns	ns	ns	ns	ns
	N × T × Y		ns	ns	ns	ns	ns	ns	ns	ns	ns	ns

Note: The data in the table are the mean values of three replications. Different lowercase letters in the column under the same straw treatment in the same year indicate significant differences between treatments ($p < 0.05$). SR: straw returned to the field in spring; AR: straw returned to the field in autumn; MT: mid-tillering stage; PI: panicle initiation stage; HD: heading stage; FI: filling stage; PM: physiological maturity stage; Y: year; N: nitrogen fertilizer; T: straw returning period; *, ** mean $p < 0.05$, $p < 0.01$; ns means non-significant.

Table 6. Effect of straw return period combined with nitrogen fertilizer application on the root total absorbing surface area and root activity absorbing area of rice.

Treatment			RTA (m^2 m^{-2})					RAA (m^2 m^{-2})				
			MT	PI	HD	FI	PM	MT	PI	HD	FI	PM
2020	SR	N0	3.57 d	6.19 c	7.96 c	7.86 c	4.53 b	2.06 b	3.25 c	4.13 c	3.79 c	2.05 b
		N90	4.36 c	6.84 bc	8.01 c	8.03 c	5.04 ab	2.17 b	3.45 c	4.60 b	4.52 b	2.27 ab
		N180	5.08 b	7.35 b	9.97 b	9.69 b	5.15 a	2.45 ab	4.22 b	4.89 b	4.71 b	2.31 a
		N270	5.86 a	8.07 a	10.27 ab	10.08 ab	5.28 a	2.88 a	4.83 a	5.67 a	5.58 a	2.42 a
		N360	6.02 a	8.52 a	10.80 a	10.44 a	5.33 a	3.02 a	4.87 a	5.88 a	5.72 a	2.44 a
	AR	N0	4.02 c	7.49 c	9.69 d	8.36 d	4.85 c	2.27 c	3.51 d	4.35 d	4.26 d	2.15 c
		N90	5.10 b	8.22 c	10.92 c	9.35 c	5.43 b	2.52 b	3.70 d	4.91 c	4.77 c	2.36 b
		N180	6.64 a	9.67 b	12.21 b	10.19 b	5.53 b	2.95 ab	4.35 c	5.47 b	5.26 b	2.43 b
		N270	7.00 a	11.56 a	13.84 a	11.76 a	5.81 a	3.17 a	5.29 a	6.24 a	5.95 a	2.62 a
		N360	6.90 a	11.01 a	13.37 a	10.96 ab	5.93 a	2.95 ab	4.96 b	6.05 a	5.74 a	2.66 a
2021	SR	N0	3.78 d	6.50 d	8.21 d	8.05 c	5.09 b	2.21 c	3.53 c	4.48 d	4.21 d	2.24 b
		N90	4.69 c	7.18 c	9.29 c	9.03 b	5.23 b	2.41 c	3.86 bc	4.92 c	4.53 c	2.32 ab
		N180	5.36 b	7.79 b	10.02 b	9.89 b	5.82 a	2.80 b	4.34 ab	5.29 b	5.09 b	2.44 ab
		N270	6.19 a	8.72 a	11.35 a	11.03 a	6.22 a	3.01 ab	4.86 a	5.84 a	5.65 a	2.51 a
		N360	6.25 a	9.08 a	11.89 a	11.80 a	6.25 a	3.26 a	4.93 a	5.98 a	5.72 a	2.55 a
	AR	N0	4.43 d	8.12 d	10.03 d	9.72 b	5.40 c	2.52 c	3.61 d	4.65 d	4.37 d	2.30 c
		N90	5.53 c	8.92 c	11.55 c	10.22 b	5.54 c	2.73 c	4.01 c	5.11 c	4.75 c	2.37 c
		N180	6.73 b	10.44 b	12.80 b	11.41 a	6.11 b	3.15 b	4.68 b	5.66 b	5.37 b	2.53 b
		N270	7.27 a	12.28 a	13.94 a	12.30 a	6.35 ab	3.48 a	5.53 a	6.46 a	6.14 a	2.64 ab
		N360	6.92 ab	11.99 a	13.59 a	11.12 a	6.49 a	3.31 ab	5.31 a	6.32 a	5.85 a	2.67 a
ANOVA	Y		*	*	*	*	ns	*	*	*	*	ns
	N		**	**	**	**	**	**	**	**	**	*
	T		*	*	*	*	ns	*	*	*	*	ns
	N × T		*	*	*	*	ns	*	*	*	*	ns
	Y × N		*	*	*	*	ns	*	*	*	*	ns
	T × Y		ns	ns	ns	ns	ns	ns	ns	ns	ns	ns
	N × T × Y		ns	ns	ns	ns	ns	ns	ns	ns	ns	ns

Note: The data in the table are the mean values of three replications. Different lowercase letters in the column under the same straw treatment in the same year indicate significant differences between treatments ($p < 0.05$). SR: straw returned to the field in spring; AR: straw returned to the field in autumn; MT: mid-tillering stage; PI: panicle initiation stage; HD: heading stage; FI: filling stage; PM: physiological maturity stage; Y: year; N: nitrogen fertilizer; T: straw returning period; *, ** mean $p < 0.05$, $p < 0.01$; ns means non-significant.

3.3. Rice Yield

The year (Y), nitrogen fertilizer (N), straw returning period (T), the interaction between nitrogen fertilizer and straw returning periods (N × T), and the interaction between the year and different straw returning periods (Y × T) all exerted significant or extremely significant impacts on rice yield, as illustrated in Figure 3. Regardless of whether the straw was returned in spring or autumn, rice yield demonstrated a trend of initial increase followed by decrease with the augmentation of nitrogen application, peaking at the N270 level. Within the same straw returning period, no significant difference in rice yield was observed between the N180 and N360 treatments. Under the N270 condition, the two-year average yield for the autumn straw returning treatment was 8.01 t ha^{-1}, while the average yield for the spring straw returning treatment was 7.77 t ha^{-1}. On average across all nitrogen levels, the rice yield significantly increased by 4.8% (in 2020) and 6.6% (in 2021) with the autumn straw treatment compared to the spring straw returning treatment.

It can be seen from Table 7 that whether the straw was returned in spring or autumn, the theoretical yield of rice first increased and then decreased with the increase in nitrogen application rate and reached a maximum at N270, but there was no significant difference between N360 and N270 when using the same straw returning period. Compared with straw returning in spring, straw returning in autumn significantly increased the theoretical yield of rice by 6.6% in 2020 and 7.4% in 2021. Under the same straw returning period, the application of nitrogen fertilizer significantly increased the number of panicles and grains per panicle of rice and significantly reduced the seed setting rate, but there was no significant difference in the number of panicles, grains per panicle, and seed setting rate between N270 and N360. Compared with N0, the number of panicles in the N90, n180, N270 and N360 treatments increased significantly by 33.8%, 53.2%, 71.9% and 76.3% (two-year average), and the number of grains per panicle increased significantly by 18.1%, 33.7%, 41.9% and 42.9% (two-year average), respectively. Compared with spring straw returning, autumn straw returning increased the number of panicles, grain number per panicle, the seed setting rate, and the 1000-grain weight. However, straw returning in autumn only significantly increased the number of rice panicles, which increased by 6.9% and 4.4%, respectively, from 2020 to 2021. This shows that the theoretical yield of rice can be improved by increasing the number of rice panicles in the paddy field in autumn, but the amount of nitrogen fertilizer should not be too high when returning straw in autumn.

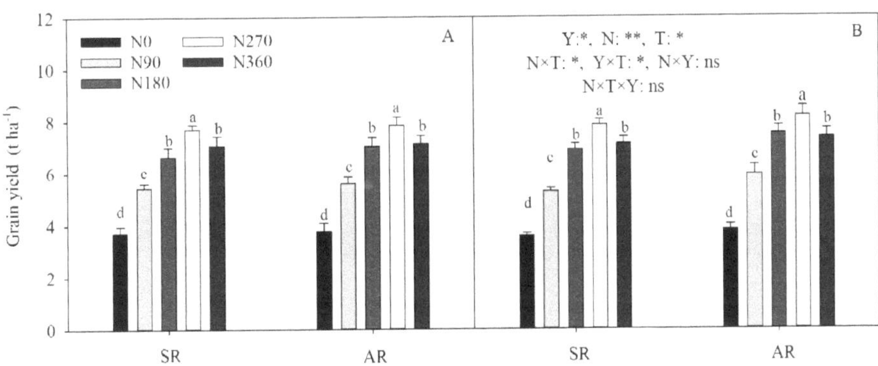

Figure 3. Effect of straw return period and combined application of nitrogen fertilizer on rice yield. Data in the figures are mean ± standard error. SR: straw returned to the field in spring; AR: straw returned to the field in autumn; Y: year; N: nitrogen fertilizer; T: straw returning period; N0: nitrogen application of 0 kg ha^{-1}; N90: nitrogen application of 90 kg ha^{-1}; N180: nitrogen application of 180 kg ha^{-1}; N270: nitrogen application of 270 kg ha^{-1}; N360: nitrogen application of 360 kg ha^{-1}; (**A**) denotes 2020, and (**B**) denotes 2021. Different lowercase letters in the graphs under the same straw management indicate that the values are significantly different at the 0.05 level; *, ** indicates significant differences at the 0.05 and 0.01 levels, respectively, and ns indicates no significant difference.

Table 7. Effect of straw return period and combined application of nitrogen fertilizer on the components of rice yield.

	Treatment		Spike Number (×10⁴ ha⁻¹)	Spikelets per Panicle	Seed Setting Rate (%)	1000-Grain Weight (g)	Theoretical Yield (t ha⁻¹)
2020	SR	N0	224.58 ± 20.38 d	66.05 ± 4.27 c	90.87 ± 3.25 a	25.87 ± 1.02 a	3.65 ± 0.23 d
		N90	313.66 ± 30.41 c	78.04 ± 3.58 b	89.58 ± 2.71 ab	25.43 ± 0.92 a	5.58 ± 0.51 c
		N180	353.53 ± 25.14 b	89.19 ± 2.92 a	88.42 ± 2.66 ab	24.88 ± 1.13 a	6.94 ± 0.32 b
		N270	396.57 ± 30.52 a	95.03 ± 4.22 a	86.45 ± 3.18 b	24.57 ± 0.89 a	8.01 ± 0.44 a
		N360	413.26 ± 29.58 a	95.57 ± 5.69 a	83.82 ± 1.50 c	23.85 ± 1.15 a	7.89 ± 0.35 a
	AR	N0	256.67 ± 30.18 d	66.55 ± 5.27 c	90.48 ± 2.42 a	25.07 ± 0.77 a	3.87 ± 0.29 d
		N90	327.67 ± 32.17 c	79.31 ± 6.35 b	89.29 ± 3.25 a	24.92 ± 0.92 a	5.78 ± 0.38 c
		N180	378.67 ± 40.22 b	90.91 ± 8.24 a	88.96 ± 2.71 ab	24.39 ± 1.15 ab	7.47 ± 0.55 b
		N270	423.33 ± 41.25 a	97.17 ± 9.11 a	88.25 ± 1.99 ab	23.93 ± 1.32 b	8.69 ± 0.23 a
		N360	433.33 ± 30.27 a	96.54 ± 7.23 a	86.13 ± 2.67 b	23.24 ± 0.89 b	8.37 ± 0.62 ab

Table 7. *Cont.*

	Treatment		Spike Number ($\times 10^4$ ha^{-1})	Spikelets per Panicle	Seed Setting Rate (%)	1000-Grain Weight (g)	Theoretical Yield (t ha^{-1})
2021	SR	N0	236.89 ± 24.37 d	67.18 ± 6.17 c	91.23 ± 1.88 a	25.89 ± 0.93 a	3.76 ± 0.38 d
		N90	321.22 ± 32.19 c	76.58 ± 6.88 b	89.89 ± 2.64 a	25.62 ± 1.25 a	5.67 ± 0.55 c
		N180	371.27 ± 22.19 b	84.59 ± 7.59 ab	88.79 ± 2.72 ab	25.58 ± 1.53 a	7.13 ± 0.29 b
		N270	408.22 ± 30.20 a	89.71 ± 8.69 a	86.45 ± 1.86 b	25.48 ± 1.62 a	8.07 ± 0.36 a
		N360	423.33 ± 25.27 a	93.25 ± 9.57 a	83.26 ± 1.77 c	24.33 ± 1.21 a	8.00 ± 0.38 a
	AR	N0	250.37 ± 20.08 d	63.99 ± 4.69 c	90.67 ± 2.68 a	26.89 ± 0.85 a	3.91 ± 0.24 e
		N90	333.58 ± 32.15 c	77.67 ± 7.66 b	90.38 ± 2.79 a	26.16 ± 0.93 a	6.13 ± 0.29 d
		N180	380.18 ± 33.59 b	87.99 ± 5.08 ab	89.76 ± 2.93 ab	25.68 ± 1.26 ab	7.71 ± 0.35 c
		N270	436.77 ± 26.42 a	92.24 ± 6.77 a	87.17 ± 1.55 b	25.18 ± 1.33 ab	8.84 ± 0.39 a
		N360	437.52 ± 30.59 a	91.53 ± 8.24 a	84.98 ± 1.72 c	24.79 ± 1.25 b	8.44 ± 0.42 b
ANOVA		Y	ns	ns	ns	ns	*
		N	**	**	*	ns	**
		T	*	ns	ns	ns	*
		N × T	*	ns	ns	ns	*
		Y × T	ns	ns	ns	ns	ns
		N × Y	ns	ns	ns	ns	ns
	N × T × Y		ns	ns	ns	ns	ns

Note: The data represent the mean ± standard error. Different lowercase letters in the column under the same straw treatment in the same year indicate significant differences between treatments ($p < 0.05$). SR: straw returned to the field in spring; AR: straw returned to the field in autumn; N0: nitrogen application of 0 kg $^{-2}$; N90: nitrogen application of 90 kg ha^{-1}; N180: nitrogen application of 180 kg ha^{-1}; N270: nitrogen application of 270 kg ha^{-1}; N360: nitrogen application of 360 kg ha^{-1}; Y: year; N: nitrogen fertilizer; T: straw returning period; *, ** mean $p < 0.05$, $p < 0.01$; ns means non-significant.

3.4. Nitrogen Fertilizer Utilization

The total nitrogen uptake and nitrogen use efficiency of rice were significantly or extremely significantly affected by nitrogen fertilizer (N), straw returning period (T), and their interaction ($p < 0.05$, $p < 0.01$) (Table 8). No matter when the straw was returned to the field, the total nitrogen absorption and apparent nitrogen use efficiency of rice gradually increased with the increase in nitrogen fertilizer application. Compared with N0, the total nitrogen accumulation of N90, N180, N270 and N360 increased by 47.7%, 113.9%, 191.6% and 240.7%, respectively (two-year average). Compared with the spring straw returning treatment, the total nitrogen accumulation of the autumn straw returning treatment increased significantly by 9.6% and 6.1% from 2020 to 2021, respectively. Although the application of nitrogen fertilizer significantly improved the apparent nitrogen use efficiency of rice, there was no significant difference in the apparent nitrogen use efficiency of rice among the n180, N270 and N360 treatments using the same straw returning period. Compared with spring straw returning treatment, the apparent nitrogen use efficiency of the autumn straw returning treatment increased by 10.3% (2020) and 12.2% (2021), respectively.

Whether straw was returned in spring or autumn, the agronomic efficiency and partial productivity of nitrogen fertilizer decreased with the increase in the nitrogen application rate. With the same nitrogen application rate, straw returning significantly improved the agronomic efficiency of nitrogen fertilizer and the partial productivity of nitrogen fertilizer. Compared with the spring straw returning treatment, the autumn straw returning treatment increased the agronomic efficiency of nitrogen fertilizer and the partial productivity of nitrogen fertilizer by 8.1% and 6.0%, respectively (two-year average). Compared with N90, the agronomic efficiency of nitrogen fertilizer in the N180, N270 and N360 treatments decreased significantly by 10.6%, 24.7% and 53.3% (two-year average), respectively. The partial productivity of nitrogen fertilizer in N180, N270 and N360 treatments decreased significantly by 36.9%, 52.5% and 67.8% (two-year average), respectively. To summarize, the total nitrogen accumulation and nitrogen use efficiency of the soda saline–alkaline rice field were higher when straw was returned to the field in autumn combined with nitrogen fertilizer, but under the condition of straw returning to the field in autumn, considering the yield and nitrogen use efficiency, the best nitrogen application rate was 270 kg ha^{-1}.

Regression analysis showed that there was a close quadratic relationship between rice yield and nitrogen application rate ($p < 0.01$) (Figure 4). To achieve the maximum theoretical rice yield (i.e., the maximum value of the two curves), the nitrogen fertilizer application rate in the SR treatment was 297.90 kg ha^{-1} (2020) and 284 kg ha^{-1} (2021), while the corresponding values of AR treatment were 281 kg ha^{-1} (2020) and 265.8 kg ha^{-1} (2021). Under the same straw returning period, the theoretical optimal nitrogen fertilizer application rate in 2021 was lower than that in 2020. This shows to a certain extent that the optimal nitrogen fertilizer application rate of continuous straw returning in soda saline–alkaline rice area tends to gradually decrease with the increase in years. The reduced amount of nitrogen fertilizer (N-reduced) is the difference between the amount of nitrogen fertilizer required to reach the maximum theoretical rice yield in the spring straw returning treatment and the corresponding amount of nitrogen fertilizer in the autumn straw returning treatment. According to the regression results, the corresponding N-reduced amounts from 2020 to 2021 were 93.73 (297.90–204.17) kg ha^{-1} and 109.28 (284–174.12) kg ha^{-1}, respectively. Concerning the figures in brackets, the former represent the amount of nitrogen fertilizer required for the theoretical maximum rice yield when returning straw in spring, and the latter represent the amount of nitrogen fertilizer required for the straw returning in autumn treatment to reach the "theoretical maximum of rice yield when returning straw in spring". Therefore, from 2020 to 2021, compared with spring straw returning, autumn straw returning was able to reduce the amount of nitrogen fertilizer application by 31.5% and 38.5% respectively, without reducing the rice yield in the saline–sodic paddy field.

Table 8. Effect of straw return period and combined application of nitrogen fertilizer on nitrogen fertilizer utilization in rice.

		Treatment	TPN kg ha^{-1}	REN %	AEN kg kg^{-1}	PFP kg kg^{-1}
2020	SR	N0	30.28 ± 2.38 e	—	—	—
		N90	44.69 ± 2.08 d	16.01 ± 1.72 b	19.33 ± 1.66 a	60.41 ± 1.77 a
		N180	63.97 ± 1.75 c	18.72 ± 0.54 a	16.33 ± 0.26 b	36.87 ± 0.78 b
		N270	82.31 ± 2.27 b	19.27 ± 1.32 a	14.80 ± 0.87 c	28.49 ± 0.49 c
		N360	99.55 ± 2.36 a	19.24 ± 0.83 a	9.27 ± 0.19 d	19.64 ± 0.44 d
	AR	N0	33.70 ± 1.65 e	—	—	—
		N90	51.34 ± 3.20 d	19.60 ± 1.69 b	20.30 ± 1.23 a	62.22 ± 1.11 a
		N180	70.17 ± 0.61 c	20.26 ± 1.25 ab	18.11 ± 1.79 b	39.07 ± 1.90 b
		N270	89.28 ± 2.31 b	20.59 ± 0.34 a	15.07 ± 0.62 b	29.62 ± 0.61 c
		N360	107.00 ± 1.52 a	20.36 ± 0.96 ab	9.32 ± 0.90 c	19.80 ± 0.86 d
2021	SR	N0	33.02 ± 4.52 e	—	—	—
		N90	48.83 ± 8.67 d	17.57 ± 1.25 c	18.97 ± 1.84 a	58.71 ± 1.29 a
		N180	75.18 ± 1.14 c	22.42 ± 2.79 b	18.47 ± 1.56 a	38.34 ± 1.06 b
		N270	107.85 ± 4.32 b	27.70 ± 3.23 a	15.81 ± 1.66 b	29.06 ± 0.20 c
		N360	125.39 ± 5.22 a	25.63 ± 1.47 a	9.81 ± 0.99 c	19.74 ± 0.93 d
	AR	N0	37.96 ± 2.14 e	—	—	—
		N90	54.41 ± 3.55 d	18.28 ± 1.22 c	23.52 ± 1.01 a	65.81 ± 1.97 a
		N180	79.29 ± 0.29 c	22.96 ± 2.67 b	20.49 ± 0.69 b	41.63 ± 0.53 b
		N270	114.10 ± 4.08 b	28.20 ± 2.67 a	16.17 ± 1.50 c	30.27 ± 1.36 c
		N360	128.04 ± 2.52 a	25.02 ± 1.36 a	9.85 ± 0.31 d	20.43 ± 0.54 d
ANOVA		Y	*	*	*	ns
		N	**	*	**	**
		T	*	*	*	*
		N × T	*	*	*	*
		Y × T	ns	ns	ns	ns
		N × Y	*	*	*	ns
		N × T × Y	ns	ns	ns	ns

Note: The data represent the mean $\pm$ standard error. Different lowercase letters in the column under the same straw treatment in the same year indicate significant differences between treatments ($p < 0.05$). SR: straw returned to the field in spring; AR: straw returned to the field in autumn; N0: nitrogen application of 0 kg ha^{-1}; N90: nitrogen application of 90 kg ha^{-1}; N180: nitrogen application of 180 kg ha^{-1}; N270: nitrogen application of 270 kg ha^{-1}; N360: nitrogen application of 360 kg ha^{-1}; Y: year; N: nitrogen fertilizer; T: straw returning period; *, ** mean $p < 0.05$, $p < 0.01$; ns means non-significant.

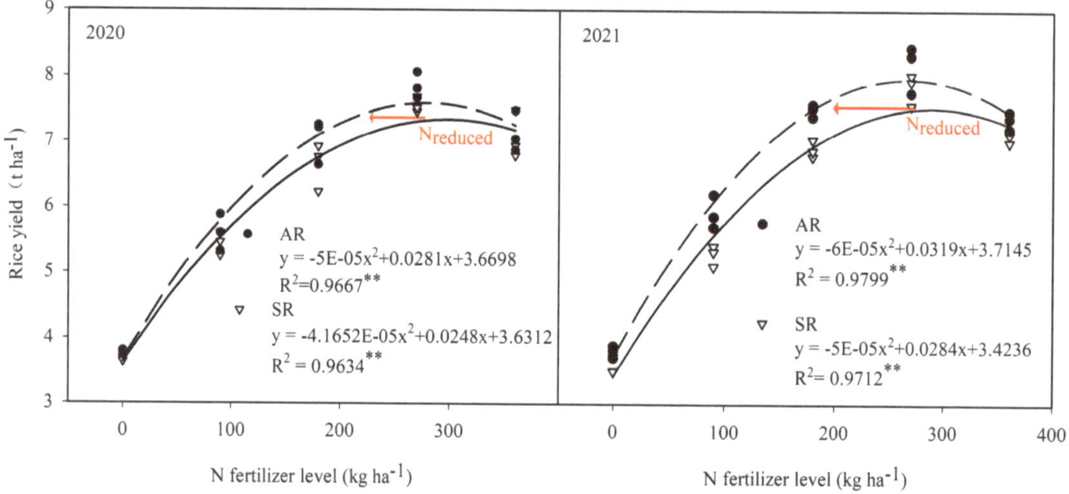

Figure 4. Relationship between nitrogen fertilizer and rice yield under different straw return periods. Note: SR: straw returned to the field in spring; AR: straw returned to the field in autumn; ** is signification correlation at $p < 0.01$.

4. Discussion

4.1. Effects of Combined Application of Straw Returning Period and Nitrogen Fertilizer on Root Characteristics in Saline–Sodic Paddy Field

In saline–sodic soil, crop growth inhibition is mainly affected by ion stress, osmotic stress, and high pH toxicity [24]. In response to salt stress, root growth regulates specific key features such as root length and branching, reorientation of root growth, and alteration of cell wall composition. Numerous studies have suggested that salinity stress significantly reduced plant root length, root volume, root surface area, and root dry weight, thus changing the root morphology [25,26], and root penetration obviously decreased in the seedling stage [27]. Under alkaline stress, it was observed that rice roots accumulated a large amount of Na^+, and the root cell membrane system was seriously damaged, resulting in a decrease in root activity. [28]. This seriously limits the absorption of nutrients and water by the root system, thus limiting the growth of above-ground plants and ultimately reducing the yield [29]. It is generally believed that moderate N application can increase root production, promote root penetration, improve N fertilizer utilization, and alleviate root growth inhibition induced by moderate soil salinity [30,31]. This is consistent with the data; N application was able to improve the root morphology and root growth, but there was no significant difference between the N270 and N360 treatments, indicating that beyond this range of N270, the effect of nitrogen fertilizer in alleviating saline stress and promoting root growth is inconspicuous (Tables 3–6). These changes may be attributed to the fact that moderate N application not only has a nutritive effect but also plays an important role in improving plant salt tolerance by increasing nutrient uptake and decreasing the accumulation of Na^+ in plant tissues [32]; it can also increase the accumulation of amino acids in plant tissues, which counteracts the increased osmotic potential of NaCl solution and protects membranes and metabolites by scavenging reactive oxygen species (ROS), which protects the cells from further damage [33].

Our previous research found that straw returning can have negative impacts on new germinating roots; on the contrary, in the later stage of rice growth, straw returning produces some positive impacts on rice roots, which enhances the yield of rice [34]. Methods of effectively alleviating the root growth inhibition caused by straw decay in the initial stages of rice growth are of great significance for straw returning in saline–sodic paddy areas. Therefore, this study compared straw returning in spring and autumn. This study found that compared with straw returning in spring, straw returning in autumn significantly improved the root morphology and physiological function of rice in saline–sodic paddy field (Tables 3–6). The effect of straw returning on rice root growth was better in autumn than in spring, which may be due to straw returned to the field in the autumn staying in the soil for 7 months longer than straw returned to the field in the spring. To prevent the return of salt in the paddy field, the paddy field is flooded for a long time in the rice growing season; thus, the aggregation of salts in saline–sodic rice field soils mainly occurs during the fallow period. Straw returning in autumn disturbs the continuity of soil capillary movement, hinders the upward movement of salt in groundwater or deep soil during the fallow period [35], improves the chemical properties of surface soil in sodic saline–sodic rice fields, and reduces the damage that saline–sodic stress causes to rice roots. In addition, the rapid decomposition of straw under flooding conditions led to a sharp decrease in soil oxygen content, releasing large amounts of reducing substances and harmful gases that worsen the growth conditions of rice roots [36,37]. Straw that is returned to the field in the autumn is partially decomposed during the fallow period; the amount of rice straw remaining in the soil at the rice growth stage the following year was lower than the amount of straw returned in spring. The damage to the root system was also less than that which occurred when straw was returned in spring. Kanal et al.'s research in Estonia also showed that the decomposition rate of wheat straw returned to the field in winter was significantly higher than that in spring, and 6–7% of the winter wheat straw returned to the field had decomposed from soil freezing (December) to thawing (April) in the second year [38].

Interestingly, this study found that the interaction between straw returning period and nitrogen fertilizer had a significant impact on the morphology and physiological function of rice roots from the tillering stage to the grain filling stage (Tables 3–6). This may be related to soil microorganisms, which take up and utilize large amounts of available nitrogen from the soil, compete with crops for nutrients, and are activated after straw returning [39,40]. The application of nitrogen fertilizer can effectively mitigate the phenomenon of "nitrogen competition" and reduce the inhibition caused by straw decomposition. At the same time, nitrogen application accelerates the straw decomposition process, allowing more nutrients enter the soil and promoting the growth of the rice root system [41,42]. In addition, the amount of nitrogen fertilizer application affects the rate of carbon and nitrogen exudation from the root system, limits microbial resources, and affects microbial carbon use efficiency [43]. Our study found that the amount of rice root bleeding sap increased significantly with increasing nitrogen application (Table 5), which made microbial diversity more complex, thus favoring soil N cycling and inorganic N accumulation [44]. This study also found that when using the same straw returning period, there was no significant difference between N360 and N270 treatments (Tables 3–6). This is probably because sufficient nitrogen fertilizer provides a sufficient nitrogen source for soil microorganisms, and nitrogen is no longer a factor limiting straw decomposition. In addition, the application of high amounts of nitrogen fertilizer significantly reduced the root–shoot ratio of rice [45]. A suitable root–shoot ratio favors the enhancement of rice yield. Therefore, it is necessary to find the optimum amount of nitrogen fertilizer to be applied when returning straw to the field in autumn in saline–sodic soil rice planting areas in order to obtain the maximum rice yield.

4.2. Effects of Straw Return Period and Nitrogen Fertilizer Application on Rice Yield and Nitrogen Use Efficiency in Saline–Sodic Paddy Fields

Many studies have shown that straw returning can improve the physical and chemical properties of saline–alkaline soil, root characteristics and photosynthetic characteristics, and rice yield in saline–sodic paddy fields [34,46,47]. However, there are few studies on rice yield in saline–alkaline paddy fields with different straw returning periods. This study found that compared with straw returning in spring, straw returning in autumn significantly increased rice yield and panicles per unit area (Figure 3, Table 7). The reason for this is that on the one hand, rice systems with autumn straw returning perform significantly better than those with spring straw returning (Tables 3–6), promote nutrient uptake, and increase yield. On the other hand, compared with straw returning in spring, autumn straw returning in alkaline soil in a saline–sodic paddy field undergoing a winter fallow period was equivalent to alkaline pretreatment and freezing–thawing treatment. The use of urea after alkaline pretreatment will accelerate the decomposition of straw [48]. Wang et al. suggested that the chemical bonds between lignin and carbohydrates in rice straw are further broken after winter freeze–thaw treatment, and organic nutrients are more quickly decomposed and released [49]. Furthermore, the rapid decomposition stage of straw overlaps with the tillering stage of rice, which can inhibit rice tillering, reduces the number of effective panicles of rice, and leads to yield loss [50]. The application of base fertilizer in spring soil preparation meets the above requirements, thus reducing the damage to rice in the tillering stage, which leads to a significantly higher number of ears per unit area when straw is returned in autumn rather than in spring [51,52]. This experiment also found that whenever the straw was returned to the field in spring or autumn, when the nitrogen fertilizer exceeded 270 kg ha^{-1}, the rice yield declined (Figure 3). This may be due to the fact that excessive application of nitrogen delayed the vegetative stage, reduced the efficiency of nitrogen fertilizer use and eventually led to lower rice yields [23]. The fact that rice yield first increases and then decreases with the increase in nitrogen fertilizer application also emphasizes the importance of scientific fertilizer application when returning straw (Figure 2).

Compared with straw returning in spring, straw returning in autumn significantly improved the total nitrogen content, apparent nitrogen use efficiency, agronomic nitrogen use efficiency and partial nitrogen productivity of rice plants (Table 8). This may be due to the improvement in soil physical and chemical properties [53,54], root morphology (Tables 3 and 4), and root physiological function (Tables 5 and 6) in the saline–sodic paddy field when returning straw in autumn compared to returning straw in spring; this promoted the absorption of nitrogen by roots and then improved the nitrogen use efficiency. This study also found that when using the same straw returning period, with the increase in the nitrogen application rate, the apparent utilization of nitrogen fertilizer increased first and then decreased and reached a maximum of 270 kg ha^{-1}. The agronomic efficiency and partial productivity of nitrogen fertilizer gradually decreased with the increase in the nitrogen fertilizer application rate (Table 8). This also emphasizes the importance of rational application of nitrogen fertilizer when returning straw. Compared with straw returning in spring, straw returning in autumn reduced the mineral nitrogen input by 31.5% (2020) and 38.5% (2021), which had no adverse effect on rice yield (Figure 3). Wang et al. also observed similar results in the study of other crops and found that continuous straw returning can increase the total yield of cotton while saving about 40% of nitrogen input [55]. In addition, according to Figure 4, when using the same straw returning period, the theoretical optimal nitrogen fertilizer application rate in 2021 was lower than that in 2020. This shows that the optimal nitrogen fertilizer application rate may tend to gradually decrease with the increase in years when there is continuous straw returning in saline–sodic rice areas. In addition, the theoretical optimal nitrogen fertilizer for straw returning in autumn was 281 kg ha^{-1} (2020) and 265.8 kg ha^{-1} (2021), and the theoretical optimal nitrogen fertilizer for straw returning in spring was 297.9 kg ha^{-1} (2020) and 284 kg ha^{-1} (2021); the rice yield of straw returning in autumn was higher than that of straw returning in spring (Figure 3). This shows that when returning straw to sodic saline–alkaline rice areas, the effect of straw returning in autumn is better, and less nitrogen fertilizer is required. However, as straw returning years increase, the optimal nitrogen application rate will require further study.

5. Conclusions

In the early stage of straw returning in saline–sodic rice area, the effect of straw returning in autumn was better than that in spring, and the effect of 270 kg ha^{-1} of nitrogen fertilizer combined with autumn straw returning in the rice growing season was better. Autumn straw returning combined with 270 kg ha^{-1} of nitrogen fertilizer significantly improved rice root morphology, root activity, nutrient uptake by rice roots, nitrogen use efficiency, and rice yield. The increase in effective panicles per unit area was the main factor of the increase in rice yield. The average yield of rice was 8.01 t ha^{-1} within two years when straw was returned to the field and 270 kg ha^{-1} of nitrogen fertilizer was applied in autumn. Compared with straw returning in spring, straw returning in autumn can reduce the amount of nitrogen fertilizer application by 31.5% (2020) and 38.5% (2021) without reducing rice yield. Moreover, the theoretical optimal nitrogen application rates of straw returned in autumn were 281 kg ha^{-1} (2020) and 265.8 kg ha^{-1} (2021), respectively. Therefore, for the sustainable development of saline–sodic rice planting areas, it is suggested that autumn straw returning measures are adopted after rice harvest and that 270 kg ha^{-1} of nitrogen fertilizer is applied in the following year. However, as the field experiment was only carried out for two years, a long-term study of straw return in autumn and the application of nitrogen fertilizer is needed to determine the optimum amount of nitrogen fertilizer that should be applied during the different stages of straw return.

Author Contributions: C.R. proposed the ideas, designed the experiments, and provided funding, Y.X. and X.Z. conducted the experiment and wrote the paper, Y.G. (Ya Gao) and J.L. analyzed thedata, X.S., Y.G. (Yanqiu Geng) and L.G. reviewed paper. All authors have read and agreed to the published version of the manuscript.

Funding: This study was supported by the Lingnan Modern Agriculture Guangdong Provincial Laboratory Heyuan Branch Autonomous Research Project (DT20240003), the Heyuan City Science and Technology Plan Project (Heke Platform 2022 Major Special Project 010), and Jilin Science and Technology Development Program Project (20230508001RC).

Data Availability Statement: Data will be made available on request.

Conflicts of Interest: The authors declare that this research was conducted in the absence of any commercial or financial relationships that could be construed as a potential conflict of interest.

References

1. Yang, J.; Zhang, S.; Li, Y.; Bu, K.; Zhang, Y.; Chang, L.; Zhang, Y. Dynamics of Saline-Alkali Land and Its Ecological Regionalization in Western Songnen Plain, China. *Chin. Geogr. Sci.* **2010**, *20*, 159–166. [CrossRef]
2. Montanarella, L.; Badraoui, M.; Chude, V.; Costa, I.; Mamo, T.; Yemefack, M.; Aulang, M.; Yagi, K.; Hong, S.Y.; Vijarnsorn, P. *Status of the World's Soil Resources: Main Report*; FAO: Rome, Italy, 2015.
3. Wang, L.; Seki, K.; Miyazaki, T.; Ishihama, Y. The Causes of Soil Alkalinization in the Songnen Plain of Northeast China. *Paddy Water Environ.* **2009**, *7*, 259–270. [CrossRef]
4. Chi, C.M.; Zhao, C.W.; Sun, X.J.; Wang, Z.C. Reclamation of Saline-Sodic Soil Properties and Improvement of Rice (*Oriza sativa* L.) Growth and Yield Using Desulfurized Gypsum in the West of Songnen Plain, Northeast China. *Geoderma* **2012**, *187–188*, 24–30. [CrossRef]
5. Jutras, P.V.; Goulet, M.; Lavoie, P.; D'Aoust, M.; Sainsbury, F.; Michaud, D. Recombinant Protein Susceptibility to Proteolysis in the Plant Cell Secretory Pathway Is pH-dependent. *Plant Biotechnol. J.* **2018**, *16*, 1928–1938. [CrossRef] [PubMed]
6. Liao, P.; Ros, M.B.H.; Van Gestel, N.; Sun, Y.; Zhang, J.; Huang, S.; Zeng, Y.; Wu, Z.; Van Groenigen, K.J. Liming Reduces Soil Phosphorus Availability but Promotes Yield and P Uptake in a Double Rice Cropping System. *J. Integr. Agric.* **2020**, *19*, 2807–2814. [CrossRef]
7. Qadir, M.; Schubert, S.; Ghafoor, A.; Murtaza, G. Amelioration Strategies for Sodic Soils: A Review. *Land Degrad. Dev.* **2001**, *12*, 357–386. [CrossRef]
8. Yin, H.; Zhao, W.; Li, T.; Cheng, X.; Liu, Q. Balancing Straw Returning and Chemical Fertilizers in China: Role of Straw Nutrient Resources. *Renew. Sustain. Energy Rev.* **2018**, *81*, 2695–2702. [CrossRef]
9. Ma, Y.; Shen, Y.; Liu, Y. State of the Art of Straw Treatment Technology: Challenges and Solutions Forward. *Bioresour. Technol.* **2020**, *313*, 123656. [CrossRef]
10. Jin, Z.; Shah, T.; Zhang, L.; Liu, H.; Peng, S.; Nie, L. Effect of Straw Returning on Soil Organic Carbon in Rice–Wheat Rotation System: A Review. *Food Energy Secur.* **2020**, *9*, e200. [CrossRef]
11. Hu, Y.; Sun, B.; Wu, S.; Feng, H.; Gao, M.; Zhang, B.; Liu, Y. After-Effects of Straw and Straw-Derived Biochar Application on Crop Growth, Yield, and Soil Properties in Wheat (*Triticum aestivum* L.)—Maize (*Zea mays* L.) Rotations: A Four-Year Field Experiment. *Sci. Total Environ.* **2021**, *780*, 146560. [CrossRef]
12. Chen, L.; Sun, S.; Yao, B.; Peng, Y.; Gao, C.; Qin, T.; Zhou, Y.; Sun, C.; Quan, W. Effects of Straw Return and Straw Biochar on Soil Properties and Crop Growth: A Review. *Front. Plant Sci.* **2022**, *13*, 986763. [CrossRef] [PubMed]

13. Wei, H.; Li, Y.; Zhu, K.; Ju, X.; Wu, D. The Divergent Role of Straw Return in Soil O2 Dynamics Elucidates Its Confounding Effect on Soil N2O Emission. *Soil Biol. Biochem.* **2024**, *199*, 109620. [CrossRef]
14. Song, X.; Zhou, G.; Ma, B.-L.; Wu, W.; Ahmad, I.; Zhu, G.; Yan, W.; Jiao, X. Nitrogen Application Improved Photosynthetic Productivity, Chlorophyll Fluorescence, Yield and Yield Components of Two Oat Genotypes under Saline Conditions. *Agronomy* **2019**, *9*, 115. [CrossRef]
15. Wu, B.; Yang, P.; Zuo, W.; Zhang, W. Optimizing Water and Nitrogen Management Can Enhance Nitrogen Heterogeneity and Stimulate Root Foraging. *Field Crops Res.* **2023**, *304*, 109183. [CrossRef]
16. Lal, R. Restoring Soil Quality to Mitigate Soil Degradation. *Sustainability* **2015**, *7*, 5875–5895. [CrossRef]
17. Virto, I.; Imaz, M.; Fernández-Ugalde, O.; Gartzia-Bengoetxea, N.; Enrique, A.; Bescansa, P. Soil Degradation and Soil Quality in Western Europe: Current Situation and Future Perspectives. *Sustainability* **2014**, *7*, 313–365. [CrossRef]
18. Devêvre, O.C.; Horwáth, W.R. Decomposition of Rice Straw and Microbial Carbon Use Efficiency under Different Soil Temperatures and Moistures. *Soil Biol. Biochem.* **2000**, *32*, 1773–1785. [CrossRef]
19. Hu, N.; Wang, B.; Gu, Z.; Tao, B.; Zhang, Z.; Hu, S.; Zhu, L.; Meng, Y. Effects of Different Straw Returning Modes on Greenhouse Gas Emissions and Crop Yields in a Rice–Wheat Rotation System. *Agric. Ecosyst. Environ.* **2016**, *223*, 115–122. [CrossRef]
20. Kukal, S. Puddling Depth and Intensity Effects in Rice–Wheat System on a Sandy Loam Soil I. Development of Subsurface Compaction. *Soil Tillage Res.* **2003**, *72*, 1–8. [CrossRef]
21. Yang, J. Correlation of Cytokinin Levels in the Endosperms and Roots with Cell Number and Cell Division Activity during Endosperm Development in Rice. *Ann. Bot.* **2002**, *90*, 369–377. [CrossRef]
22. Jia, Y.; Wang, J.; Qu, Z.; Zou, D.; Sha, H.; Liu, H.; Sun, J.; Zheng, H.; Wang, J.; Yang, L.; et al. Effects of Low Water Temperature during Reproductive Growth on Photosynthetic Production and Nitrogen Accumulation in Rice. *Field Crops Res.* **2019**, *242*, 107587. [CrossRef]
23. Wang, J.; Hussain, S.; Sun, X.; Zhang, P.; Javed, T.; Dessoky, E.S.; Ren, X.; Chen, X. Effects of Nitrogen Application Rate Under Straw Incorporation on Photosynthesis, Productivity and Nitrogen Use Efficiency in Winter Wheat. *Front. Plant Sci.* **2022**, *13*, 862088. [CrossRef] [PubMed]
24. Chen, W.; Feng, C.; Guo, W.; Shi, D.; Yang, C. Comparative Effects of Osmotic-, Salt- and Alkali Stress on Growth, Photosynthesis, and Osmotic Adjustment of Cotton Plants. *Photosynt.* **2011**, *49*, 417–425. [CrossRef]
25. Zou, Y.; Zhang, Y.; Testerink, C. Root Dynamic Growth Strategies in Response to Salinity. *Plant Cell Environ.* **2022**, *45*, 695–704. [CrossRef]
26. Van Zelm, E.; Zhang, Y.; Testerink, C. Salt Tolerance Mechanisms of Plants. *Annu. Rev. Plant Biol.* **2020**, *71*, 403–433. [CrossRef]
27. Ma, T.; Zeng, W.; Lei, G.; Wu, J.; Huang, J. Predicting the Rooting Depth, Dynamic Root Distribution and the Yield of Sunflower under Different Soil Salinity and Nitrogen Applications. *Ind. Crops Prod.* **2021**, *170*, 113749. [CrossRef]
28. Zhang, H.; Liu, X.-L.; Zhang, R.-X.; Yuan, H.-Y.; Wang, M.-M.; Yang, H.-Y.; Ma, H.-Y.; Liu, D.; Jiang, C.-J.; Liang, Z.-W. Root Damage under Alkaline Stress Is Associated with Reactive Oxygen Species Accumulation in Rice (*Oryza sativa* L.). *Front. Plant Sci.* **2017**, *8*, 1580. [CrossRef]
29. Torabian, S.; Farhangi-Abriz, S.; Rathjen, J. Biochar and Lignite Affect H+-ATPase and H+-PPase Activities in Root Tonoplast and Nutrient Contents of Mung Bean under Salt Stress. *Plant Physiol. Biochem.* **2018**, *129*, 141–149. [CrossRef]
30. Wang, C.; Liu, W.; Li, Q.; Ma, D.; Lu, H.; Feng, W.; Xie, Y.; Zhu, Y.; Guo, T. Effects of Different Irrigation and Nitrogen Regimes on Root Growth and Its Correlation with Above-Ground Plant Parts in High-Yielding Wheat under Field Conditions. *Field Crops Res.* **2014**, *165*, 138–149. [CrossRef]
31. Chen, W.; Hou, Z.; Wu, L.; Liang, Y.; Wei, C. Effects of Salinity and Nitrogen on Cotton Growth in Arid Environment. *Plant Soil* **2010**, *326*, 61–73. [CrossRef]
32. Zhang, D.; Li, W.; Xin, C.; Tang, W.; Eneji, A.E.; Dong, H. Lint Yield and Nitrogen Use Efficiency of Field-Grown Cotton Vary with Soil Salinity and Nitrogen Application Rate. *Field Crops Res.* **2012**, *138*, 63–70. [CrossRef]
33. Flores, P.; Carvajal, M.; Cerdá, A.; Martínez, V. Salinity and Ammonium/Nitrate Interactions on Tomato Plant Development, Nutrition, And Metabolites. *J. Plant Nutr.* **2001**, *24*, 1561–1573. [CrossRef]
34. Ran, C.; Gao, D.; Liu, W.; Guo, L.; Bai, T.; Shao, X.; Geng, Y. Straw and Nitrogen Amendments Improve Soil, Rice Yield, and Roots in a Saline Sodic Soil. *Rhizosphere* **2022**, *24*, 100606. [CrossRef]
35. Zhao, Y.; Wang, S.; Li, Y.; Zhuo, Y.; Liu, J. Effects of Straw Layer and Flue Gas Desulfurization Gypsum Treatments on Soil Salinity and Sodicity in Relation to Sunflower Yield. *Geoderma* **2019**, *352*, 13–21. [CrossRef]
36. Wang, X.; Samo, N.; Zhao, C.; Wang, H.; Yang, G.; Hu, Y.; Peng, Y.; Rasul, F. Negative and Positive Impacts of Rape Straw Returning on the Roots Growth of Hybrid Rice in the Sichuan Basin Area. *Agronomy* **2019**, *9*, 690. [CrossRef]
37. Zhang, H.; Liang, S.; Wang, Y.; Liu, S.; Sun, H. Greenhouse Gas Emissions of Rice Straw Return Varies with Return Depth and Soil Type in Paddy Systems of Northeast China. *Arch. Agron. Soil Sci.* **2021**, *67*, 1591–1602. [CrossRef]
38. Kanal, A. Effect of Incorporation Depth and Soil Climate on Straw Decomposition Rate in a Loamy Podzoluvisol. *Biol. Fertil. Soils* **1995**, *20*, 190–196. [CrossRef]
39. Cheshire, M.V.; Bedrock, C.N.; Williams, B.L.; Chapman, S.J.; Solntseva, I.; Thomsen, I. The Immobilization of Nitrogen by Straw Decomposing in Soil. *Eur. J. Soil Sci.* **1999**, *50*, 329–341. [CrossRef]
40. Lei, K.; Dai, W.; Wang, J.; Li, Z.; Cheng, Y.; Jiang, Y.; Yin, W.; Wang, X.; Song, X.; Tang, Q. Biochar and Straw Amendments over a Decade Divergently Alter Soil Organic Carbon Accumulation Pathways. *Agronomy* **2024**, *14*, 2176. [CrossRef]

41. Zhao, S.; Zhang, S. Linkages between Straw Decomposition Rate and the Change in Microbial Fractions and Extracellular Enzyme Activities in Soils under Different Long-Term Fertilization Treatments. *PLoS ONE* **2018**, *13*, e0202660. [CrossRef]
42. Xu, G.; Lu, D.-K.; Wang, H.-Z.; Li, Y. Morphological and Physiological Traits of Rice Roots and Their Relationships to Yield and Nitrogen Utilization as Influenced by Irrigation Regime and Nitrogen Rate. *Agric. Water Manag.* **2018**, *203*, 385–394. [CrossRef]
43. Lian, J.; Li, G.; Zhang, J.; Massart, S. Nitrogen Fertilization Affected Microbial Carbon Use Efficiency and Microbial Resource Limitations via Root Exudates. *Sci. Total Environ.* **2024**, *950*, 174933. [CrossRef] [PubMed]
44. Liu, D.; Xu, L.; Wang, H.; Xing, W.; Song, B.; Wang, Q. Root Exudates Promoted Microbial Diversity in the Sugar Beet Rhizosphere for Organic Nitrogen Mineralization. *Agriculture* **2024**, *14*, 1094. [CrossRef]
45. Xin, W.; Liu, H.; Zhao, H.; Wang, J.; Zheng, H.; Jia, Y.; Yang, L.; Wang, X.; Li, J.; Li, X.; et al. The Response of Grain Yield and Root Morphological and Physiological Traits to Nitrogen Levels in Paddy Rice. *Front. Plant Sci.* **2021**, *12*, 713814. [CrossRef] [PubMed]
46. Yuan, X.; Ran, C.; Gao, D.; Zhao, Z.; Meng, X.; Geng, Y.; Shao, X.; Chen, G. Changes in Soil Characteristics and Rice Yield under Straw Returning in Saline Sodic Soils. *Soil Sci. Plant Nutr.* **2022**, *68*, 563–573. [CrossRef]
47. Ran, C.; Gao, D.; Bai, T.; Geng, Y.; Shao, X.; Guo, L. Straw Return Alleviates the Negative Effects of Saline Sodic Stress on Rice by Improving Soil Chemistry and Reducing the Accumulation of Sodium Ions in Rice Leaves. *Agric. Ecosyst. Environ.* **2023**, *342*, 108253. [CrossRef]
48. Yan, C.; Yan, S.-S.; Jia, T.-Y.; Dong, S.-K.; Ma, C.-M.; Gong, Z.-P. Decomposition Characteristics of Rice Straw Returned to the Soil in Northeast China. *Nutr. Cycl. Agroecosystems* **2019**, *114*, 211–224. [CrossRef]
49. Wang, L.; Wang, Z.; Wang, Z.; Zheng, Y. Delignification Characteristics of Rice Straw Pretreatment Combining Biogas Slurry Immersion with Freeze–Thaw. *Biomass Convers. Biorefinery* **2024**, 1–13. [CrossRef]
50. Zhu, L.; Hu, N.; Zhang, Z.; Xu, J.; Tao, B.; Meng, Y. Short-Term Responses of Soil Organic Carbon and Carbon Pool Management Index to Different Annual Straw Return Rates in a Rice–Wheat Cropping System. *CATENA* **2015**, *135*, 283–289. [CrossRef]
51. Jing, X.; Chai, X.; Long, S.; Liu, T.; Si, M.; Zheng, X.; Cai, X. Urea/Sodium Hydroxide Pretreatments Enhance Decomposition of Maize Straw in Soils and Sorption of Straw Residues toward Herbicides. *J. Hazard. Mater.* **2022**, *431*, 128467. [CrossRef]
52. Kontogianni, N.; Barampouti, E.M.; Mai, S.; Malamis, D.; Loizidou, M. Effect of Alkaline Pretreatments on the Enzymatic Hydrolysis of Wheat Straw. *Environ. Sci. Pollut. Res.* **2019**, *26*, 35648–35656. [CrossRef] [PubMed]
53. Che, W.; Piao, J.; Gao, Q.; Li, X.; Li, X.; Jin, F. Response of Soil Physicochemical Properties, Soil Nutrients, Enzyme Activity and Rice Yield to Rice Straw Returning in Highly Saline-Alkali Paddy Soils. *J. Soil Sci. Plant Nutr.* **2023**, *23*, 4396–4411. [CrossRef]
54. Huang, Y.; Yan, Y.; Ma, Y.; Zhang, X.; Zhao, Q.; Men, M.; Huang, Y.; Peng, Z. The Effect of Low-Temperature Straw-Degrading Microbes on Winter Wheat Growth and Soil Improvement under Straw Return. *Front. Microbiol.* **2024**, *15*, 1391632. [CrossRef] [PubMed]
55. Wang, Z.; Wang, Z.; Ma, L.; Lv, X.; Meng, Y.; Zhou, Z. Straw Returning Coupled with Nitrogen Fertilization Increases Canopy Photosynthetic Capacity, Yield and Nitrogen Use Efficiency in Cotton. *Eur. J. Agron.* **2021**, *126*, 126267. [CrossRef]

MDPI AG

Grosspeteranlage 5

4052 Basel

Switzerland

Tel.: +41 61 683 77 34

Agronomy Editorial Office

E-mail: agronomy@mdpi.com

www.mdpi.com/journal/agronomy